Human Heredity

NINTH EDITION

Human Heredity

Principles & Issues

Michael R. Cummings

Illinois Institute of Technology

BROOKS/COLE
CENGAGE Learning

Australia • Brazil • Japan • Korea • Mexico • Singapore • Spain • United Kingdom • United States

BROOKS/COLE
CENGAGE Learning™

Human Heredity: Principles and Issues,
Ninth Edition
Michael R. Cummings

Senior Acquisitions Editor, Life Sciences:
  Peggy Williams

Development Editor: Suzannah Alexander

Assistant Editor: Elizabeth Momb

Editorial Assistants: Alexis Glubka, Shana Baldassari

Media Editor: Lauren Oliveira

Marketing Manager: Tom Ziolkowski

Marketing Assistant: Elizabeth Wong

Marketing Communications Manager: Linda Yip

Content Project Manager: Hal Humphrey

Design Director: Rob Hugel

Art Director: John Walker

Print Buyer: Karen Hunt

Rights Specialist, Text: Roberta Broyer

Rights Specialist, Images: Dean Dauphinais

Production Service: Elm Street Publishing Services

Text Designer: Riezbos Holzbaur Group (RHDG)

Photo Researcher: Chris Althof, Bill Smith Group

Illustrators: Precision Graphics, Integra Software
  Services Pvt. Ltd.

Cover Image: © Michele Constantini/PhotoAlto/
  Corbis

Compositor: Integra Software Services Pvt. Ltd.

For product information and technology assistance, contact us at
Cengage Learning Customer & Sales Support, 1-800-354-9706

For permission to use material from this text or product,
submit all requests online at **www.cengage.com/permissions**
Further permissions questions can be e-mailed to
permissionrequest@cengage.com

Library of Congress Control Number: 2010922222

International Edition:
ISBN-13: 978-0-8400-5318-3
ISBN-10: 0-8400-5318-5

Cengage Learning International Offices

Asia
www.cengageasia.com
tel: (65) 6410 1200

Australia/New Zealand
www.cengage.com.au
tel: (61) 3 9685 4111

Brazil
www.cengage.com.br
tel: (55) 11 3665 9900

India
www.cengage.co.in
tel: (91) 11 4364 1111

Latin America
www.cengage.com.mx
tel: (52) 55 1500 6000

UK/Europe/Middle East/Africa
www.cengage.co.uk
tel: (44) 0 1264 332 424

Represented in Canada by Nelson Education, Ltd.
www.nelson.com
tel: (416) 752 9100 / (800) 668 0671

Cengage Learning is a leading provider of customized learning solutions with office locations around the globe, including Singapore, the United Kingdom, Australia, Mexico, Brazil, and Japan. Locate your local office at: **www.cengage.com/global**

For product information: **www.cengage.com/international**
Visit your local office: **www.cengage.com/global**
Visit our corporate website: **www.cengage.com**

AVAILABILITY OF RESOURCES MAY DIFFER BY REGION. Check with your local Cengage Learning representative for details.

Printed in the United States of America
1 2 3 4 5 6 7 14 13 12 11 10

To my son Brendan, whose courage I admire and respect.

About the Author

MICHAEL R. CUMMINGS received his Ph.D. in Biological Sciences from Northwestern University. His doctoral work, conducted in the laboratory of Dr. R. C. King, centered on ovarian development in *Drosophila melanogaster*. After a year on the faculty at Northwestern, he moved to the University of Illinois at Chicago, where for many years he held teaching and research positions. In 2003, he joined the faculty in the Department of Biological, Chemical, and Physical Sciences at Illinois Institute of Technology, where he is currently a Research Professor.

At the undergraduate level, he focused on teaching genetics, human genetics for nonmajors, and general biology to majors and nonmajors. About fifteen years ago, Dr. Cummings developed a strong interest in scientific literacy. He is now working to integrate the use of the Internet and the World Wide Web into the undergraduate teaching of genetics and general biology and into textbooks. He has received awards given by the university faculty for outstanding teaching, has twice been voted by graduating seniors as the best teacher in their years on campus, and has received several teaching awards from student organizations.

His current research interests involve the organization of DNA sequences on the short-arm and centromere region of human chromosome 21. His laboratory is engaged in a collaborative effort to construct a physical map of this region of chromosome 21 for the purpose of exploring molecular mechanisms of chromosome interactions.

Dr. Cummings is the author and coauthor of a number of widely used college textbooks, including *Biology: Science and Life*; *Concepts of Genetics*; *Genetics: A Molecular Perspective*; *Essentials of Genetics*; and *Human Genetics and Society*. He has also written articles on aspects of genetics for the *McGraw-Hill Encyclopedia of Science and Technology* and has published a newsletter on advances in human genetics for instructors and students.

He and his wife, Lee Ann, are the parents of two adult children, Brendan and Kerry, and have two grandchildren, Colin and Maggie. He is an avid sailor, enjoys reading and collecting books (biography, history), appreciates music (baroque, opera, and urban electric blues), and is a long-suffering Cubs fan.

Contents

3 Transmission of Genes from Generation to Generation   44

4 Pedigree Analysis in Human Genetics   70

5 The Inheritance of Complex Traits   94

6 Cytogenetics: Karyotypes and Chromosome Aberrations 120

7 Development and Sex Determination 148

13 An Introduction to Genetic Technology 292

14 Biotechnology and Society 312

8 The Structure, Replication, and Chromosomal Organization of DNA 176

15 Genomes and Genomics 332

16 Reproductive Technology, Genetic Testing, and Gene Therapy 354

17 Genes and the Immune System 376

Preface

OVER THE YEARS, *Human Heredity* has developed in parallel with advances in human genetics. Although the text has changed a great deal from the first edition, its rationale and aims have remained constant. This book is written for a one-term human genetics course for students in the humanities, social sciences, business, engineering, and other fields. It assumes that the students in this course will have little or no background in biology, chemistry, or mathematics and will have personal, professional, or intellectual reasons for learning human genetics. The book is intended to serve those who will become *consumers* of genetic-based health care services and those who may become *providers* of health care services.

Genetic knowledge and technology is rapidly being transferred to many areas of our society. This transition makes it imperative that the public, elected officials, and policy makers outside the scientific community have a working knowledge of genetics to help shape applications of genetics in our society. *Human Heredity* is written to transmit the principles of genetics in a straightforward and accessible way, without unnecessary jargon, detail, or the use of anecdotal stories in place of content. Some descriptive chemistry is used after an appropriate introduction and definition of terms. In the same vein, no advanced math skills are required to calculate elementary probabilities or to calculate genotype and allele frequencies.

Goals of the Text

From the start, this book has held to a few simple goals for teaching students about human genetics. This edition continues that tradition, incorporating the following goals:

1. Present the concepts underlying human genetics in clear, concise language to give students a working knowledge of genetics. Each chapter presents a limited number of clearly stated, interconnected concepts to assist learning a complex topic.
2. Begin each chapter at a level that nonmajors can understand and provide relevant examples that students can apply to themselves, their families, and their work environments.
3. Examine the social, cultural, and ethical implications associated with the use of genetic technology.
4. Explain the origin, nature, and amount of genetic diversity present in the human population and how that diversity has been shaped by natural selection.

To achieve these goals, emphasis has been placed on clear writing and the use of accompanying photographs and artwork that teach rather than merely illustrate the ideas under discussion.

Organization

The text has four sections: Chapters 1 through 7 cover cell division, transmission of traits from generation to generation, and development. Chapters 8 through 12 emphasize molecular genetics, mutation, and cancer. Chapters 13 through 16 include recombinant DNA, genomics, and biotechnology. These chapters cover gene action, mutation, cloning, genomics, as well as genetic screening, genetic testing, and genetic counseling. Chapters 17 through 19 cover specialized topics: the immune system, the genetics of behavior, and population genetics and human evolution.

Instructors teaching genetics courses to nonmajors come from many different backgrounds and use a wide range of instructional formats, including active learning. To facilitate this array of approaches, the book is organized to allow both students and instructors to use the material no matter what order of topics is selected. After the first section, the chapters can be used in any order. Within each chapter, outlines and end-of-chapter activities let the instructor and students easily identify and explore central ideas.

What's New in the Ninth Edition

New and updated topics, sections, figures, and tables are the hallmarks of this edition. This edition includes more than 135 new and redrawn figures. Some of these changes are listed here:

Chapter 1: A Perspective on Human Genetics
- The text has been revised and updated to reflect the impact of genomics and genomic technology on human genetics, including the use of genetic markers.
- New Figures: 1.2 Example of a protein; 1.7 Transgenic corn.

Chapter 2: Cells and Cell Division
- New Section 2.1 (The Chemistry of Cells).
- New Table 2.1.
- New Figures: 2.4 Golgi complex; 2.15 Comparison of mitosis and meiosis; 2.16 Random assortment and crossing over; 2.17a Male gametogenesis; 2.17b Female gametogenesis.

Chapter 3: Transmission of Genes from Generation to Generation
- The text has been condensed to focus on Mendelian principles and their variations.
- New Figure: 3.13 Photo of an albino.

Chapter 4: Pedigree Analysis in Human Genetics
- The chapter sections have been extensively reorganized with a new order of presentation.
- New subsections in 4.1 (Pedigree Analysis Is a Basic Method in Human Genetics): There are five basic patterns of Mendelian inheritance; Analyzing a pedigree.
- New subsections in 4.4 (Sex-Linked Inheritance Involves Genes on the X and Y Chromosomes): X-linked dominant traits; X-linked recessive traits.
- New Section 4.8 (Many Factors Can Affect the Outcome of Pedigree Analysis). New subsection: Common recessive alleles can produce pedigrees that resemble dominant inheritance.
- New and Revised Figures: Chapter opener; Revised 4.2 Autosomal recessive pedigree; New 4.3 Organs affected in cystic fibrosis; Revised 4.6 Autosomal dominant pedigree; Revised 4.11 X-linked dominant pedigree; Revised 4.12 X-linked recessive pedigree; New 4.17 Dystrophin distribution; New 4.21 Woman with Huntington disease; New 4.23 Pedigree of common alleles.

Chapter 5: The Inheritance of Complex Traits (New Title)
- New subsection in 5.1 (Some Traits Are Controlled by Two or More Genes): What are complex traits?
- New subsections in 5.2 (Polygenic Inheritance and Variation in Phenotype): Defining the genetics behind continuous phenotypic variation; How many genes control a polygenic trait?
- Revised Section 5.3 (The Additive Model for Polygenic Inheritance).
- New Feature: The Genetic Revolution: Dissecting Genes and Environment in Spina Bifida.
- New subsection in 5.6 (Twin Studies and Multifactorial Traits): Scanning the genome for obesity-related genes.

- New Section 5.7 (Genetics of Height: A Closer Look). New subsections: Haplotypes and genome-wide association studies; Genes for human height: What have we learned so far?
- New Figures: 5.7 Inheritance of height; 5.8 Distribution of height phenotypes; 5.16 SNP haplotypes.

Chapter 6: Cytogenetics: Karyotypes and Chromosome Aberrations
- Text has been refined and updated.
- New Figures: 6.10 Photo of amniocentesis; 6.11 Photo of chorionic villus biopsy; 6.14 Nondisjunction; 6.24 Structural aberrations.

Chapter 7: Development and Sex Determination
- New discussion of embryonic stem cells in Section 7.2 (A Survey of Human Development from Fertilization to Birth).
- New subsection in 7.4 (How Is Sex Determined?): Environmental interactions can help determine sex.
- New Figure: 7.10 Sex determination in animals.

Chapter 8: The Structure, Replication, and Chromosomal Organization of DNA (New Title)
- Section 8.3 (The Watson-Crick Model of DNA Structure) has been renamed and reorganized, with several new figures.
- New Feature: The Genetics Revolution: What Happens When Your Genes Are Patented?
- New Section 8.6 (The Organization of DNA in Chromosomes). New subsections: Chromosomes have a complex structure; Centromeres and telomeres are specialized chromosomal regions.
- New Figures: 8.3 Hershey-Chase experiment; 8.7 Polynucleotide chains; 8.9 Antiparallel chains in DNA; 8.10 DNA model; 8.11 Structure of RNA; 8.13 DNA replication; 8.15 Scanning electron micrograph of chromosome; 8.16 Chromosome structure; 8.17a Centromere; 8.17b Telomere.

Chapter 9: Gene Expression and Gene Regulation (New Title)
- New subsections in 9.1 (The Link Between Genes and Proteins): Genetic information is stored in DNA; The relationship between genes and proteins.
- New subsections in 9.5 (Translation Requires the Interaction of Several Components): Amino acids are subunits of proteins; Messenger RNA, ribosomal RNA, and transfer RNA interact during translation.
- New subsections in 9.6 (Polypeptides Are Processed and Folded to Form Proteins): How many proteins can human cells make?; Proteins are sorted and distributed to their cellular locations.
- New Section 9.8 (Several Mechanisms Regulate the Expression of Genes). New subsections: Chromatin remodeling and access to promoters; DNA methylation can silence genes; RNA interference is one mechanism of post-transcriptional regulation; Translational and post-translational mechanisms regulate the production of proteins.
- New Figures: 9.3 Transcription; 9.5 Alternative splicing; 9.9 Translation; 9.10 Polysome; 9.14 Regulation of gene expression; 9.15 Chromatin remodeling; 9.16 RNAi regulation of gene expression.

Chapter 10: From Proteins to Phenotypes
- New subsections in 10.3 (Phenylketonuria: A Mutation That Affects an Enzyme): How does the buildup of phenylalanine produce mental retardation?; How effective is testing for PKU in newborns?
- Section 10.7 (Pharmacogenetics and Pharmacogenomics) renamed and updated. New material on genetics of taste and smell. Expanded coverage of allele variations and breast cancer therapy.
- New Feature: The Genetic Revolution: PKU.
- New Figures: 10.15 Mutations and sickle cell anemia; 10.16 Population distribution of taste abilities; 10.17 Taste receptor; 10.20 Tamoxifen metabolism; 10.21 Interaction of tamoxifen with other drugs; 10.22 Photo of parathion in use.

Chapter 11: Mutation: The Source of Genetic Variation
- Topics have been condensed and reworked.
- New and Revised Figures: Revised 11.1; 11.10 New photo of fragile-X chromosome; 11.12 New pedigree of anticipation in myotonic dystrophy; Revised 11.18 Genomic imprinting and normal development; Revised 11.19 Genomic imprinting is reversible.

Chapter 12: Genes and Cancer
- Expanded Section 12.5 (Mutant Cancer Genes Affect DNA Repair Systems and Genome Stability). New subsections: Mutant DNA repair genes cause a predisposition to breast cancer; Breast cancer risks depend on genotype.
- Expanded Section 12.7 (Hybrid Genes, Epigenetics, and Cancer).
- New Section 12.8 (Genomics and Cancer). New subsections: Sequencing cancer genomes identifies cancer-associated genes; Epigenetics and cancer; Targeted therapy offers a new approach to treating cancer.
- New Feature: The Genetic Revolution: Cancer Stem Cells.
- New Feature: Exploring Genetics: The Cancer Genome Atlas (TCGA).
- New Figures: 12.1 Cancer deaths by age; 12.8 Action of retinoblastoma protein; 12.10 Photo of breast tumor; 12.12 Colon polyps; Redrawn 12.13 Colon-cancer model; 12.17 Cancer-genome sequencing; 12.18 Epigenetic mechanisms.

Chapter 13: An Introduction to Genetic Technology
- New Section 13.7 (DNA Microarrays Are Used to Analyze Gene Expression).
- New Table 13.2 Uses of Microarrays.
- New Figures: 13.1 Cloning Dolly; 13.4 Restriction-enzyme cutting and ligation; 13.7 Summary of cloning; 13.10 Library screening; 13.15 DNA sequencing; 13.16 Nucleotide and dideoxynucleotide; 13.18 Microarray experiment.

Chapter 14: Biotechnology and Society
- Revised Section 14.1 (Biopharming: Making Human Proteins in Animals). New subsection on enzyme replacement therapy for Pompe disease. New subsection on transgenic plants as sources of human proteins.
- New Section 14.2 (Using Stem Cells to Treat Disease). New subsections: Stem cells provide insight into basic biological processes; Stem-cell-based therapies may treat many diseases.
- Expanded Section 14.3 (Genetically Modified Foods). New subsection: Functional foods and health.
- New Figures: 14.2 Photo of insulin produced by recombinant DNA technology; 14.3 Photo of bioreactor for producing human therapeutic proteins; 14.4 Blastocyst with inner cell mass and stem cell-culture; 14.5 iPS cells; 14.6 Photo of mice generated from iPS cells; 14.7 Stem-cell therapy; 14.8 Photo of skin stem cells and burn therapy; 14.13 DNA profile and family identification; 14.14 Forensic DNA profile.

Chapter 15: Genomes and Genomics
- New opening vignette: Exome sequencing to diagnose a genetic disorder.
- Revised and updated Section 15.4 (Genomics: Sequencing, Identifying, and Mapping Genes). New subsection: As genes are discovered, the function of their encoded proteins are studied.
- Revised and expanded Section 15.5 (What Have We Learned So Far About the Human Genome?). New subsections: New disease-related types of mutations have been discovered; Nucleotide variation in genomes is common (SNPs and copy number variations).
- Updated Section 15.8 (Ethical Concerns About Human Genomics) to include GINA.
- New Figure: 15.1a Photo of nail-patella syndrome.

Chapter 16: Reproductive Technology, Genetic Testing, and Gene Therapy
- Revised Section 16.1 (Infertility Is a Common Problem). New subsections: Infertility is a complex problem; Infertility in women has many causes; Infertility in men involves sperm defects; Other causes of infertility.

- Updated Section 16.2 (Assisted Reproductive Technologies (ART) Expand Childbearing Options). New subsections: Intrauterine insemination uses donor sperm; Egg retrieval or donation is an option; *In vitro* fertilization (IVF) is a widely used form of ART; GIFT and ZIFT are based on IVF; Surrogacy is a controversial form of ART.
- Section 16.4 (Genetic Testing and Screening) moved here from Chapter 14. New subsection: The use of PGD raises ethical issues.
- New Feature: The Genetic Revolution: Should I Save Cord Blood?
- New Figures: 16.1 Age and female infertility; 16.2 Causes of female infertility; 16.3 Male infertility disorders; 16.5 Intrauterine sperm transfer; 16.9 GIFT and ZIFT; 16.10 Photo of surrogate mother and parents; 16.11 IVF risks; 16.15 Reasons for PGD; 16.18 EPO.

Chapter 17: Genes and the Immune System
- The chapter has been reorganized, a new section has been added, and other sections have been reordered to provide a more comprehensive overview. The art program has been reworked to simplify the concepts.
- New Section 17.1 (The Body Has Three Levels of Defense Against Infection). New subsections: The skin is not part of the immune system but is a physical barrier; There are two parts to the immune system that protect against infection.
- Revised Section 17.6 (Organ Transplants Must Be Immunologically Matched). New subsection: Copy number variation (CNV) and transplant success.
- New Figures: 17.1 Inflammatory response; 17.2 Complement system; 17.5 Antibody-mediated immune response; 17.6 T-cell activation; 17.7 B-cell activation; 17.10 Cell-mediated immune response; 17.12 Immunological memory; 17.16 Photo of transgenic pig for xenotransplants.

Chapter 18: Genetics of Behavior
- The chapter has been reorganized and streamlined to emphasize genomic approaches to the study of behavior. New information has been added about the role of the *FOXP2* gene in language.
- Section 18.5 (The Genetics of Schizophrenia and Bipolar Disorder) has been reorganized. New subsections: Genetic models for schizophrenia and bipolar disorders; Genomic approaches to schizophrenia and bipolar disorder.
- New Figures: 18.1 Pedigree of Lesch-Nyhan syndrome; 18.13 Alcohol metabolism.

Chapter 19: Population Genetics and Human Evolution
- The chapter has been reorganized to more closely integrate population genetics and evolutionary changes.
- Reorganized Section 19.1 (How Can We Measure Allele Frequencies in Populations?). New subsection: We can use the Hardy-Weinberg Law to calculate allele and genotype frequencies.
- Reorganized Section 19.4 (Natural Selection Affects the Frequency of Genetic Disorders). New subsection: Selection can rapidly change allele frequencies.
- New Table 19.5 Human Genome Variations.
- Reorganized Section 19.6 (The Evolutionary History and Spread of Our Species (*Homo sapiens*). New subsections: Our evolutionary heritage begins with hominoids; Early humans emerged almost 5 million years ago; Our species, *Homo sapiens*, originated in Africa; Ancient migrations dispersed humans across the globe.
- New Section 19.7 (Genomics and Human Evolution). New subsections: The human and chimpanzee genomes are similar in many ways; Neanderthals are not closely related to us; Chimpanzees, modern humans, and Neanderthals share a gene important in language development.
- New Feature: The Genetic Revolution: Tracing Ancient Migrations.
- New Figures: 19.7 G6PD deficiency in Jewish populations: 19.8 Gradient of B blood type allele in Europe; 19.10 Molecular phylogeny of hominoids; 19.11 Timeline of human evolution; 19.12 Genetic relationships among human populations; 19.14 Neanderthal phylogeny; 19.15 Neanderthal and human divergence; 19.16 Language genes in humans and chimps.

Features of the Book

Numbered Chapter Outlines

At the beginning of each chapter, an outline of the primary chapter headings provides an overview of the main concepts, secondary ideas, and examples. To help students grasp the central points, many of the headings are written as narratives or summaries of the ideas that follow. These outlines also serve as convenient starting points for students to review the material in each chapter. To make the outlines more useful, they have been numbered and used to organize the summary, the questions, and the problems at the end of each chapter. In this way, students can relate examples and questions to specific topics in the chapter more easily and clearly.

Opening Case Study

Each chapter begins with a short prologue directly related to the main ideas of the chapter, often drawn from real life. Topics include the use of DNA fingerprinting in court cases, the cloning of milk cows, the use of exome sequencing to diagnose a genetic disorder, and the development of *in vitro* fertilization (IVF) and the birth of Louise Brown—the first IVF baby. These vignettes are designed to promote student interest in the topics covered in the chapter and to demonstrate that laboratory research often has a direct impact on everyday life. In this edition, many of the opening stories are new or rewritten, and all are tied to the *How Would You Vote?* feature.

How Would You Vote?

To stimulate thought and discussion, each chapter has a *How Would You Vote?* section that presents an issue directly related to the opening story. It asks students to think about the topic and then visit the book's website, where they can explore related links and cast a pro or con vote on the question that has been posed. At the end of the chapter, the question is posed again against the information presented in the chapter, after students have had a chance to learn more about the concepts related to the issue. These questions—which are applicable in a variety of ways both in and out of the classroom—are intended to encourage students to think seriously about the genetic issues and concerns presented, provoking individual reflection and group discussion.

Keep in Mind as You Read Points

To keep students focused on the basic concepts in the chapter, a *Keep in Mind as You Read* box in the margin of the chapter opener contains a bulleted list of the main topics and key concepts presented in the chapter. Each item on the list is repeated in a highlighted box at the conclusion of the section related to the concept, reinforcing the importance of the concept and providing students with an aid for focusing their studies on fundamental points.

Active Figures

Active Figures link art in the text to animations of important concepts, processes, and technologies discussed throughout the book. These animations convey an immediate appreciation of how a process works in a way that cannot be shown effectively in a static series of illustrations. These *Active Figure* animations can be found on the password-protected Biology CourseMate website or the CengageNOW Homework site.

The Genetic Revolution

NEW! *The Genetic Revolution* is a new feature that emphasizes the past, present, and future impact of genetic technology on our daily lives, from genetic testing at birth to the future of cancer therapy.

Exploring Genetics

Exploring Genetics feature boxes present ideas and applications that are related to and extend the central concepts in a chapter. Some of these examine controversies that arise as genetic knowledge is transferred into technology and services.

Spotlight on...

Located in margins throughout the book, *Spotlight on...* sidebars highlight applications of concepts, present the latest findings, and point out controversial ideas without interrupting the flow of the text.

Marginal Glossary

A glossary in the page margins gives students immediate access to definitions of terms as they are introduced in the text. This format also allows definitions to be identified when students are studying or preparing for examinations. The definitions have been gathered into an alphabetical glossary at the back of the book. Because an understanding of the concepts of genetics depends on understanding the relevant terms, more than 350 terms are included in the glossary. These glossary terms are also available on the website as flashcards.

End-of-Chapter Features

Genetics in Practice: Relevant Case Studies

Case studies are included at the end of each chapter, illustrating the impact of genetics in our society. These contain scenarios and examples of genetic issues related to health, reproduction, personal decision making, public health, and ethics. Many of the case studies and the accompanying questions can be used for classroom discussions, student papers and presentations, and role playing. The cases and their questions are also located on the book's website along with links to resources for further research and exploration.

Summary

Each chapter ends with a summary that restates the major ideas covered in the chapter. The beginning outline and ending summary for each chapter use the same content and order to emphasize major concepts and their applications. Each point of the summary outline is followed by a brief restatement of the chapter material covered under the same heading. This helps students recall the concepts, topics, and examples presented in the chapter. It is hoped that this organization will minimize the chance that they will attempt to learn by rote memorization.

Questions and Problems

The summary's focus on the chapter's main points is continued in the *Questions and Problems* at the end of each chapter. The questions and problems are presented under the headings from the chapter outline. This allows students to relate the problems and questions to specific topics presented in the chapter, focus on concepts they find difficult, and work the problems that illustrate those topics. The questions and problems are designed to test students' knowledge of the facts and their ability to reason from the facts to conclusions. To this end, they use an objective question format and a problem-solving format. Because some quantitative skills are necessary in human genetics, almost all chapters include some problems that require students to organize the concepts in the chapter and use those concepts in reasoning to a conclusion. Answers to selected problems are provided in an appendix. Answers to all questions and problems are available in the Instructor's Manual and on the password-protected Biology CourseMate website.

Internet Activities

Internet Activities at the end of each chapter use websites to engage the student in activities related to the concepts discussed in the text. Internet resources are now an essential part of teaching genetics, and this section introduces students to the many databases, instructional sites, and support groups available to them. The activities are repeated and expanded on in the book's website.

Pedagogical Features

Personalized Learning Resources and Learning Assessment

To help students solve genetics problems, the end-of-chapter questions and problems are supplemented by CengageNOW, a password-protected website integrated with each chapter. All the *Active Figures* from the text are located on this site, along with dozens more animations, interactive media, and tutorials. On CengageNOW, students can take diagnostic pre-tests, which guide them to text, art, and animations that help them learn what they haven't yet mastered. After finishing this personalized course of study, students can take post-learning quizzes to assess their grasp of the new knowledge.

The results of both pre-tests and post-tests can be sent to instructors, who can keep track of students' progress through their own access to the site.

Genomic Databases as Resources

To make students aware of the array of genomic resources available to them, genetic disorders mentioned in the book are referenced by their indexing numbers from the comprehensive catalog available online as *Online Mendelian Inheritance in Man*™ (OMIM). OMIM™ (updated daily) contains text, pictures, and videos, along with literature references. Through Entrez, OMIM is cross-linked to databases containing DNA sequences, protein sequences, chromosome maps, and other resources. Students and an informed public need to be aware of the existence and relevance of such databases, and to be up to date, textbooks must incorporate these resources.

Students can use OMIM to obtain detailed information about a genetic disorder, its mode of inheritance, its phenotype and clinical symptoms, mapping information, biochemical properties, the molecular nature of the disorder, and a bibliography of relevant papers. In the classroom, OMIM and its links are valuable resources for student projects and presentations.

For further reading about genetics, students can log on to InfoTrac® College Edition, an online library of articles from nearly 5,000 periodicals, which is offered as a part of CengageNOW. This resource can be used in conjunction with electronic databases as material for papers, class discussions, and presentations.

Expanded Internet Activities

The Internet is an important and valuable resource in teaching human genetics, and both the *Human Heredity* website and CengageNOW host quizzes, a glossary, activities, and links that can be used to expand on concepts and topics covered in the text. The website content can also be used to introduce the social, legal, and ethical aspects of human genetics into the classroom and to serve as a point of contact with support groups and testing services. All the website features, exercises, and activities described below can easily be completed online and e-mailed to instructors, making them ideal for assignments.

This edition continues a popular feature, the *How Would You Vote?* questions that follow every chapter's opening vignette. These are targeted questions about an issue related to the story and the chapter content. On the website, each *How Would You Vote?* question is accompanied by background information, links to helpful sites and materials, questions for thought, and a chance to cast an online vote on the topic and view the resulting tallies.

Another online feature is the *Genetics in Practice* case studies. The cases, which are found at the end of each chapter, are repeated in their entirety on the website and accompanied by helpful links to resources for further exploration. This extra information makes them ideal as starting points for research projects and presentations.

In addition to these features, the ninth edition of *Human Heredity* contains end-of-chapter *Internet Activities* for students. These activities use resources to enhance the topics covered in the chapter and are designed to develop critical-thinking skills and generate interaction and thought rather than passive observation. As with the other features, these activities are repeated on the book's website, along with links to other websites and resources.

Ancillary Materials

The ancillary materials that accompany this edition are designed to aid student learning as well as to assist the instructor in preparing lectures and examinations and in keeping abreast of the latest developments in the field. Instructor materials are available to qualified adopters. Please consult your local Cengage Learning sales representative for details. You may also visit the Brooks/Cole biology site at **www.cengage.com/biology** to see samples of these materials, request a desk copy, locate your sales representative, or purchase a copy online.

Electronic Test Bank
Over 1,100 test items consisting of multiple-choice, true/false, fill-in-the-blank, and short-answer questions. Included in Microsoft® Word format on the PowerLecture DVD. Prepared by Carl Frankel of Pennsylvania State University, Hazleton Campus.

Online Instructor's Manual
Includes chapter outlines, chapter summary, teaching/learning objectives, definitions of in-text terms, teaching hints, and answers to in-text questions. Included in Microsoft® Word format on the PowerLecture DVD. Prepared by Carl Frankel of Pennsylvania State University, Hazleton Campus.

Study Guide
Chapter summaries, learning objectives, and key terms, along with multiple-choice, fill-in-the-blank, true/false, discussion, and case-study questions to help students with retention and better test results. Prepared by Nancy Shontz of Grand Valley State University.

PowerLecture
This convenient tool makes it easy for you to create customized lectures. Each chapter includes the following features, all organized by chapter: lecture slides, all chapter art and photos, animations, videos, Instructor's Manual, Test Bank, Examview® testing software, and JoinIn polling and quizzing slides. This single disc places all the media resources at your fingertips.

The Brooks/Cole Biology Video Library, featuring BBC Motion Gallery
Looking for an engaging way to launch your lectures? The Brooks/Cole series features short high-interest segments: Bone Marrow as a New Source for the Creation of Sperm, Repairing Damaged Hearts with Patients' Own Stem Cells, Genetically Modified Virus Used to Fight Cancer, and much more.

CengageNOW
Students: Save time, learn more, and succeed in the course with CengageNOW, an online set of resources (including Personalized Study Plans) that gives you the choices and tools you need to study smarter and get the grade. You will have access to hundreds of animations that clarify the illustrations in the text, videos, and quizzing to test your knowledge. You can also access live online tutoring from an experienced biology instructor. New to this edition are pop-up tutors that help clarify key topics by providing short video explanations. Get started today!

WebTutors for WebCT and BlackBoard

Jumpstart your course with customizable, rich, text-specific content. Whether you want to Web-enable your class or put an entire course online, WebTutor delivers. WebTutor offers a wide array of resources, including media assets, quizzing, Web links, exercises, flashcards, and more. Visit *webtutor.cengage.com* to learn more. New to this edition are pop-up tutors that help clarify key topics by providing short video explanations.

Biology CourseMate

Cengage Learning's Biology CourseMate brings course concepts to life with interactive learning, study, and exam preparation tools that support the printed textbook or the included eBook. With CourseMate, professors can use the included Engagement Tracker to assess student preparation and engagement. Use the tracking tools to see progress for the class as a whole or for individual students.

Premium eBook

This complete online version of the text is integrated with multimedia resources and special study features, providing the motivation that so many students need for studying and the interactivity they need for learning. New to this edition are pop-up tutors that help clarify key topics by providing short video explanations.

A Problem-Based Guide to Basic Genetics

Provides students with a thorough and systematic approach to solving transmission genetics problems, along with numerous solved problems and practice problems. Written and illustrated by Donald Cronkite of Hope College.

Virtual Biology Laboratories: Genetics and Genetics 2 (Pedigree Analysis) Modules

These "virtual" online experiments expose students to the tools used in modern biology, support and illustrate lecture material, and allow students to "do" science by performing experiments, acquiring data, and using the data to explain biological phenomena.

Gene Discovery Lab

This is a CD-ROM lab manual that provides a virtual laboratory experience for the student in doing experiments in molecular biology. It includes experiments that use nine of the most common molecular techniques in biology, an overview of scientific method and experimental techniques, and Web links to provide access to data and other resources.

Acknowledgments

Over the course of nine editions, many reviewers, including those who helped with this edition have given their time to improve the pedagogy, presentation of concepts, and ways of inspiring students. In past editions, I have been fortunate enough to have three reviewers who went to extraordinary lengths to keep my ideas and writing on the straight and narrow path. In addition, with their guidance, I was able to learn and re-learn many of the nuances involved in writing about genetics. These individuals generously gave me access to their collective wisdom: George Hudock of Indiana University, H. Eldon Sutton of the University of Texas, and Werner Heim of Colorado College. I am most grateful for their efforts.

For this edition, I was privileged to have two more individuals whose extensive and detailed reviews have greatly improved the book. Daniel Friderici of Michigan State University examined the text, figures, and problems from a student's point of view, and helped me present each chapter's important concepts in a straightforward and engaging way. In addition, I greatly appreciate his many suggestions on how to improve the end-of-chapter questions, problems, and how to frame the answers so that the questions become effective teaching tools. Some of his ideas will be implemented in future editions, but others have been incorporated here. I am also very grateful to Patricia Matthews of Grand Valley State University who spent many hours scrutinizing the text, helping me clarify and streamline my writing, pointing out inconsistencies in word use, and improving the flow of ideas throughout the text.

To all the reviewers who helped in the preparation of this edition, I offer my thanks and gratitude for their efforts.

Ted W. Fleming, *Bradley University*
Daniel Friderici, *Michigan State University*
Pamela L. Hanratty, *Indiana University*
Bradley Isler, *Ferris State University*
Mary King Kananen, *Pennsylvania State University, Altoona*
Brenda Knotts, *Eastern Illinois University*
Clint Magill, *Texas A&M University*
Robert L. Snyder, *State University of New York, Potsdam*
Jan Trybula, *State University of New York, Potsdam*
Jo Ann Wilson, *Florida Gulf Coast University*
Elizabeth T. Wood, *University of Arizona*
Denise Woodward, *Pennsylvania State University*

At Brooks/Cole, I am grateful for the direction and encouragement offered by my editors, Yolanda Cossio and Peggy Williams, who helped me reexamine and reinforce the strengths of the book. Hal Humphrey was the project manager who coordinated and directed the diverse array of individuals who put this edition together. I would also like to thank others at Brooks/Cole, including Brandy Radoias, who analyzed the reviews; Alexis Glubka, editorial assistant; and Elizabeth Momb, assistant editor, for their many contributions. Suzannah Alexander oversaw the preparation of this edition and supported and helped shape the significant revisions that became part of the book.

Lauren Oliveira coordinated the Web-based features of the book. The layout was designed by Riezebos Holzbaur Design Group, and the cover design was done by Irene Morris. Photo research was handled by Chris Althof of Bill Smith Group. His persistence in finding the right photo is evident throughout the text.

Amanda Zagnoli at Elm Street Publishing Services was the project manager for production. She made the difficult task of bringing together all the elements of the book look easy.

Contacting the Author

I welcome questions and comments from faculty and students about the book or about human genetics. Please contact me at: cummings.chicago@gmail.com.

Michael R. Cummings

Human Heredity

1

A Perspective on Human Genetics

Just over 10 years ago, the Icelandic Parliament (Althingi) passed a bill allowing deCODE, a biotech company, to establish and operate a database called HSD (Health Sector Database) containing the medical records of all residents of Iceland. The unusual step of allowing medical records to be examined by a public corporation was part of a project to identify genes that predispose to complex disorders, such as diabetes and heart disease. To accompany this database, deCODE also compiled the genealogies of the approximately 800,000 Icelanders who have lived there since the colonization of the island in the ninth and tenth centuries. deCODE also set up a bank of blood and tissue samples (for DNA extraction) provided by patients. These resources are powerful tools in the hunt for disease-causing genes. The law grants the company the right to sell this information (and the DNA samples) to third parties—including the research labs of pharmaceutical companies—with the hope that once disease genes are identified, diagnostic tests and therapies will follow quickly.

Why establish such a database in Iceland? Among the nations of the world, Iceland has one of the smallest and most genetically isolated populations, as well as very little genetic variation among members of the population. Small founding populations came to Iceland in the ninth and tenth centuries, and until about 50 years ago, Iceland was almost completely isolated from outside immigration. Plague (in the 1400s) and volcanoes (in the 1700s) decimated the population, further reducing genetic variation. The present-day population of 290,000 inhabitants has a remarkably similar set of genes, providing fertile ground for gene hunters seeking to identify disease genes.

Why the controversy? Opponents point out that the privacy provisions of the law are inadequate and may violate the ethical principle that health records must be kept confidential. Abuses and misunderstandings may affect employment, insurance, and even marriage. In addition, critics question whether a single company should have exclusive rights to medical information and whether the Icelandic population will derive health benefits from this arrangement.

Since deCODE began the project, they have analyzed the medical records and DNA from over 100,000 individuals (more than half the country's adult population). Coupling these data with the

© Bartek Wrzesniowski/Alamy

A crowd in Iceland reflects its narrow range of genetic diversity.

genealogical information, deCODE scientists have identified genetic risk factors for dozens of complex diseases, including cardiovascular disease, asthma, stroke, osteoporosis, and cancer. The company has developed and marketed DNA-based tests and genome scans for diseases including type II diabetes, cardiovascular disease, glaucoma, and several forms of cancer. In spite of their success in identifying mutations that cause complex diseases, deCode filed for bankruptcy in November, 2009. The problem is that many of these diseases are too complex; each disease is caused by mutations in a large number of different genes, each of which has only a very small effect. Because each mutation is rare, there is little reason to develop diagnostic tests or drugs to treat such a small number of cases. In January 2010, deCode was sold to Saga Investments, which plans to redirect the company's research program and drop efforts to develop drugs based on its discoveries.

deCODE's efforts have spurred development of similar projects elsewhere. In Great Britain, a government-sponsored project—the UK Biobank—is screening 1.2 million volunteers to establish a database of medical records and DNA samples from 500,000 Britons, ages 40–69, whose health will be followed for 25 years. The Biobank will use information gathered in this study to investigate the role of genetic and environmental factors in the development of disease—especially complex disorders such as hypertension and heart disease. After a series of pilot studies, the study began in 2007, although volunteers will continue to be recruited for several more years. Similar screening programs are also being developed in other countries, including Estonia, Latvia, Singapore, and the Kingdom of Tonga. In the United States, programs using medical records and DNA samples from tens of thousands of individuals are under way at the Marshfield Clinic in Marshfield, Wisconsin; Northwestern University in Chicago; and Howard University in Washington, D.C.

Underlying all these programs are serious bioethical issues centered on privacy, informed consent, and commercialization and corporate profit—profit derived from information gained through the medical records of and DNA from individuals. These important issues are at the heart of discussions and disagreements arising from the application of genetic technology. Scientists, physicians, politicians, and others are debating the control and use of genetic information as well as the role of policy, law, and society in decisions about how and when genetic technology should be used. Addressing the legal, ethical, and social questions surrounding an emerging technology is now as important as the information gained from that technology.

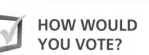

HOW WOULD YOU VOTE?

Several different countries, organizations, and corporations are compiling genetic databases using medical records and DNA samples from individuals within a population. Generally, these databases are intended as resources for medical research; however, the extremely private nature of the information being gathered causes many people to be concerned about its misuse. If a major medical center asked you to provide a DNA sample and give researchers access to your medical records, how would you respond? What if they explained that the information would be used in a project to develop diagnostic tests and treatments for diseases such as Alzheimer disease, hypertension, cardiovascular disease, and mental illness? Visit the *Human Heredity* companion website for this edition at *www.cengage.com/biology/cummings* to find out more on the issue; then cast your vote online.

1.1 Genetics Is the Key to Biology

With gene-based programs like deCODE's becoming common, as we begin this book, we might pause and remember that genetics is more than a laboratory science; unlike some other areas of science, genetics and biotechnology have a direct impact on society.

Perhaps as the first step in studying human genetics, we should ask, what *is* genetics? As a working definition, we can say that **genetics** is the scientific study of heredity. Like all definitions, this leaves a lot unsaid. To be more specific, what geneticists do is study how **traits** (such as eye color and hair color) and diseases (such as cystic fibrosis and sickle cell anemia) are passed from generation to generation. They also study the molecules that make up genes and gene products, as well as the way in which genes are turned on and off. Some geneticists study why variants of some genes occur more frequently in one population than in others. Other geneticists work in industry to develop products for agricultural and pharmaceutical firms. This work is part of the biotechnology industry, which is now a multi-billion-dollar component of the U.S. economy.

In a sense, genetics is the key to all of biology; genes control what cells look like and what they do, as well as how babies develop and how we reproduce. An understanding of what genes are, how they are passed from generation to generation, and how they work is essential to our understanding of all life on Earth, including our species, *Homo sapiens*.

In the chapters that follow, we will ask and answer many questions about genetics: How are genes passed from parents to their children? What are genes made of? Where are they located? How do they encode products called proteins, and how do proteins create the differences among individuals that we can see and study? Because this book is about human genetics, we will use human genetic disorders as examples of inherited traits (see Exploring Genetics: Genetic Disorders in Culture and Art). We will also examine how genetic knowledge and genetic technology interact with and shape many of our social, political, legal, and ethical institutions and policies.

Almost every day, the media carry a story about human genetics. These stories may report the discovery of a gene responsible for a genetic disorder, a controversy about genetic testing or a debate on the wisdom of genetically modifying our children. In many cases, as we will see, technology is far ahead of public policy and laws. To make informed decisions about genetics and biotechnology in your personal and professional life, you will need to have a foundation based on a knowledge of genetics. In the rest of this chapter, we will preview some of the concepts of human genetics that will be covered in more detail later in the book and introduce some of the social issues and controversies generated by genetic research. Many of these concepts and issues will be explored in more detail in the chapters that follow.

Genetics The scientific study of heredity.

Trait Any observable property of an organism.

1.2 What Are Genes and How Do They Work?

Simply put, a **gene** is the basic structural and functional unit of genetics. In molecular terms, a gene is a string of chemical subunits (nucleotides) in a **DNA** molecule (Figure 1.1). (DNA is shorthand for deoxyribonucleic acid.) There are four different nucleotides in DNA, each composed of a sugar, a base, and a phosphate group. The nucleotides are abbreviated as single letters:

- A for adenine
- T for thymine
- G for guanine
- C for cytosine

Combinations of these four nucleotides in the form of genes store all the genetic information carried by an individual. The nucleotide sequence encoded in a gene defines the chemical subunits (amino acids) that make up gene products (proteins). When a gene is activated, its

Gene The fundamental unit of heredity and the basic structural and functional unit of genetics.

DNA A helical molecule consisting of two strands of nucleotides that is the primary carrier of genetic information.

Genetic Disorders in Culture and Art

It is difficult to pinpoint when the inheritance of specific traits in humans was first recognized. Descriptions of people with heritable disorders appear in myths and legends of many cultures. In some of these cultures, assigned social roles—from prophets and priests to kings and queens—were hereditary. The belief that certain traits were heritable helped shape the development of many social customs.

In some societies, the birth of a deformed child was regarded as a sign of impending war or famine. Clay tablets excavated from Babylonian ruins record more than 60 types of birth defects, along with the dire consequences thought to accompany such births. Later societies, from Roman to those of eighteenth-century Europe, regarded malformed individuals (such as dwarfs) as curiosities rather than figures of impending doom; they were highly prized by royalty as courtiers and entertainers.

Over the millennia, artists have portrayed both famous and anonymous individuals with genetic disorders in paintings, sculptures, and other forms of the visual arts. These portrayals are detailed, highly accurate, and easily recognizable today. In fact, across time, culture, and artistic medium, affected individuals in these portraits often resemble each other more closely than they do their siblings, peers, or relatives. In some cases, the representations allow a disorder to be diagnosed at a distance of several thousand years.

Throughout the book, you will find fine-art representations of individuals with genetic disorders. These portraits represent a long-standing link between science and the arts in many cultures. They are not intended as a gallery of freaks or monsters but as a reminder that being human encompasses a wide range of conditions. A more thorough discussion of genetic disorders in art is in *Genetics and Malformations in Art* by J. Kunze and I. Nippert, published by Grosse Verläg, Berlin, 1986.

Rubberball Productions/Getty Images

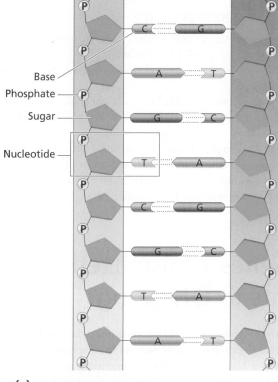

FIGURE 1.1 (a) Genes are composed of a sequence of nucleotides in a DNA molecule. (b) The double helix structure of DNA.

Base
Phosphate
Sugar
Nucleotide

(a) Genes are sequences of nucleotides in DNA

(b) DNA molecule showing arrangement of polynucleotide strands

KEY

A Adenine	G Guanine	Sugar–phosphate backbone
T Thymine	C Cytosine	

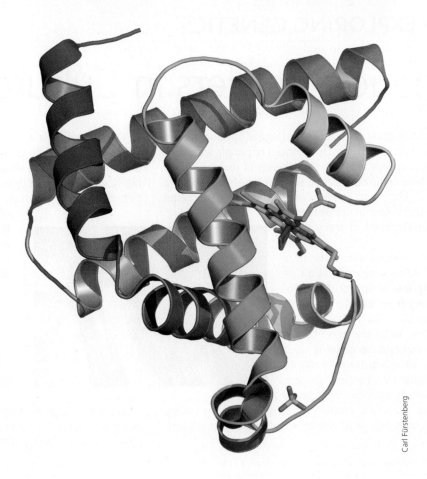

FIGURE 1.2 The three-dimensional structure formed by a protein.

Carl Fürstenberg

stored information is decoded and used to make a polypeptide, which folds into a three-dimensional shape and becomes a functional protein (Figure 1.2). The action of proteins produces characteristics we can see (such as eye color or hair color) or measure (blood proteins or height). Understanding how different proteins are produced and how they work in the cell are important parts of genetics. We will cover these topics in Chapters 9 and 10.

KEEP IN MIND

Genes control cellular function and link generations together.

We can also define genes by their properties. Genes are copied (replicated), they undergo change (mutate), they are expressed (they can be switched on or off), and they can move from one chromosome to another (recombine). In later chapters, we will explore these properties and see how they are involved in genetic diseases.

National Library of Medicine

FIGURE 1.3 Gregor Mendel, the Augustinian monk whose work on pea plants provided the foundation for genetics as a scientific discipline.

1.3 How Are Genes Transmitted from Parents to Offspring?

Thanks to the work of Gregor Mendel (Figure 1.3), a European monk who lived in the nineteenth century, we know how genes are passed from parents to offspring in plants and animals, including humans. When Mendel began his experiments, many people thought that traits carried by parents were blended together in their offspring. According to this idea, crossing a plant with red flowers and one with white flowers should produce plants with pink flowers (the pink color is a blend of red and white). Mendel's experiments on pea plants showed that genes are passed intact from generation to generation and that traits are not blended. As we will see, however, things are not always simple. There are cases in which crossing plants with red flowers and plants with white flowers *does* produce plants with pink flowers. We will discuss these cases in Chapter 3 and show that crosses between plants with red flowers and plants with white flowers that produce plants with pink flowers do not contradict the principles of inheritance discovered by Mendel.

Working at a monastery in what is now the Czech Republic, Mendel conducted research on the inheritance of traits in pea plants for more than a decade. He chose parental plants that each had a different distinguishing characteristic, called a trait. For example, Mendel bred tall pea plants with short pea plants. Plant height is the trait in this case and has two variations: tall and short. He also bred plants carrying green seeds with plants having yellow seeds. In this work, seed color is the trait; green and yellow are the variations of the trait he studied. In these breeding experiments, he wanted to see how seed color was passed from generation to generation.

Mendel kept careful records of the number and type of traits present in each generation. He also recorded the number of individual plants that carried each trait. He discovered patterns in the way traits were passed from parent to offspring through several generations. On the basis of those patterns, Mendel developed clear ideas about how traits are inherited. He concluded that traits such as plant height and flower color are passed from generation to generation by "factors" that are passed from parent to offspring. What he called "factors" we now call genes. Mendel reasoned that each parent carries two genes (a gene pair) for a specific trait (flower color, plant height, etc.) but that each parent contributes only one of those genes to its offspring; otherwise, the number of genes for a trait would double in each generation and soon reach astronomical numbers.

Mendel proposed that the two copies of a gene separate from each other during the formation of egg and sperm. As a result, only one copy of each gene is present in a sperm or egg. When an egg and sperm fuse at fertilization, the genes from the mother and father become members of a new gene pair in the offspring. In the mid-twentieth century, researchers discovered that genes are made of DNA and that this molecule is part of cellular structures known as chromosomes. Chromosomes (Figure 1.4) are found in the nucleus of human cells and other higher organisms. As we will see in Chapter 2, the

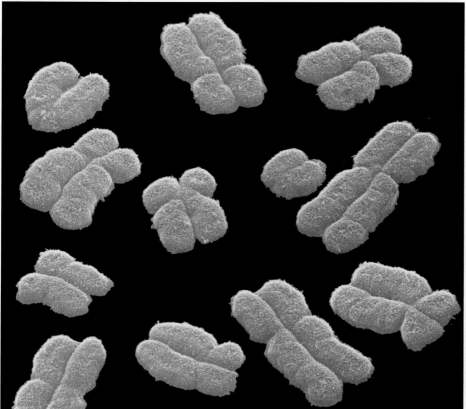

FIGURE 1.4 Replicated human chromosomes as seen by scanning electron microscopy.

Andrew Syred/Photo Researchers, Inc.

separation of genes during the formation of the sperm and egg and the reunion of genes at fertilization is explained by the behavior of chromosomes in a form of cell division called meiosis.

When Mendel published his work on the inheritance of traits in pea plants (discussed in Chapter 3), there was no well-accepted idea of how traits were transmitted from parents to offspring; his evidence changed that situation. To many, Mendel was the first geneticist and the founder of genetics, a field that has expanded in numerous directions in the last 125 years. If you want to read more about the beginnings of genetics, the story of Mendel's work is told in an engaging book entitled *The Monk in the Garden: The Lost and Found Genius of Gregor Mendel, the Father of Genetics* by Robin M. Henig.

Transmission genetics The branch of genetics concerned with the mechanisms by which genes are transferred from parent to offspring.

Pedigree analysis The construction of family trees and their use to follow the transmission of genetic traits in families. It is the basic method of studying the inheritance of traits in humans.

Cytogenetics The branch of genetics that studies the organization and arrangement of genes and chromosomes by using the techniques of microscopy.

Karyotype A complete set of chromosomes from a cell that has been photographed during cell division and arranged in a standard sequence.

Molecular genetics The study of genetic events at the biochemical level.

Recombinant DNA technology A series of techniques in which DNA fragments from an organism are linked to self-replicating vectors to create recombinant DNA molecules, which are replicated or cloned in a host cell.

Clones Genetically identical molecules, cells, or organisms, all derived from a single ancestor.

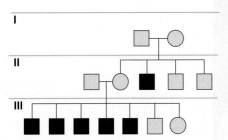

FIGURE 1.5 A pedigree represents the inheritance of a trait through several generations of a family. In this pedigree, males are symbolized by squares, females by circles. Darker symbols indicate those expressing the trait being studied; lighter symbols indicate unaffected individuals.

1.4 How Do Scientists Study Genes?

Ideas that form the foundation of genetics were discovered by studying many different organisms, including bacteria, yeast, insects, and plants, as well as humans. Because genetic mechanisms (and often genes) are the same across species, discoveries made in one organism (such as yeast) can be applied to other species, including humans. This close genetic relationship allows researchers to study human genetic disorders using experimental organisms, including insects, yeast, and mice. Although geneticists study many different species, they use a relatively small set of investigative methods, some of which are outlined in the following section.

Some basic methods in genetics.

The most basic approach studies the pattern of inheritance when traits are passed from generation to generation; this is called **transmission genetics** (Chapters 3 and 4). Using experimental organisms, geneticists study how traits such as height, eye color, flower color, and so on, are passed from parents to offspring. These experimental results are analyzed to establish how a trait is inherited. As we discussed in an earlier section, Gregor Mendel did the first significant work in transmission genetics, using pea plants as his experimental organism. His methods form the foundation of transmission genetics—methods that are still used today.

To study the inheritance of traits in humans, a more indirect method called **pedigree analysis** is used. Pedigree analysis begins by examining records to reconstruct the pattern followed by a trait as it passes through several generations. These results are used to determine how a trait is inherited and to establish the risk of having affected children (Figure 1.5). Pedigrees are constructed from information obtained from interviews, medical files, letters, diaries, photographs, and family records.

Cytogenetics is a branch of genetics that studies chromosome number and structure (discussed in Chapter 6). At the beginning of the twentieth century, observations on chromosome behavior were used to propose (correctly) that genes are located on chromosomes. Cytogenetics is one of the most important investigative approaches in human genetics and is used, among other things, to map genes and study chromosome structure and abnormalities. In clinical settings, cytogeneticists prepare **karyotypes** (Figure 1.6), standardized arrangements of chromosomes that are used to diagnose or rule out certain genetic disorders. In a karyotype, chromosomes are arranged by size, shape, and other characteristics that we will describe in Chapter 6.

A third approach, **molecular genetics**, has had the greatest impact on human genetics over the last several decades. Molecular genetics uses **recombinant DNA technology** to identify, isolate, **clone** (produce multiple copies), and analyze genes.

Genetics in Practice

Genetics in Practice case studies are critical-thinking exercises that allow you to apply your new knowledge of human genetics to real-life problems. You can find these case studies and links to relevant websites at *www.cengage.com/biology/cummings*

Summary

1.1 Genetics Is the Key to Biology

- Genetics is the scientific study of heredity. In a sense, genetics is the key to all of biology, because genes control what cells look like and what they do. Understanding how genes work is essential to our understanding of how life works.

1.2 What Are Genes and How Do They Work?

- The gene is the basic structural and functional unit of genetics. It is a string of chemical building blocks (nucleotides) in a DNA molecule. When a gene is turned on, the information stored in the gene is decoded and used to make a molecule that folds into a three-dimensional shape. This molecule is known as a protein (Figure 1.2). The actions of proteins produce the traits we see (such as eye color and hair color).

1.3 How Are Genes Transmitted from Parents to Offspring?

- From his experiments on pea plants, Mendel concluded that pairs of genes separate from each other during the formation of egg and sperm. When the egg and sperm fuse during fertilization to form a zygote, the genes from the mother and the father become members of a new gene pair in the offspring. The separation of genes during formation of the sperm and egg and the reunion of genes at fertilization are explained by the behavior of chromosomes in a form of cell division called meiosis.

1.4 How Do Scientists Study Genes?

- Genes are studied using several different methods. Transmission genetics studies how traits are passed from generation to generation. Cytogenetics studies chromosome structure and the location of genes on chromosomes. Molecular geneticists study the molecular makeup of genes, gene products, and the function of genes. Population genetics focuses on the dynamics of populations and their interaction with the environment that results in changing gene frequencies over several generations.

1.5 Has Genetics Affected Social Policy and Law?

- Eugenics was an attempt to improve the human race by using the principles of genetics. In the early years of the twentieth century, eugenics was a powerful force in shaping laws and public policy in the United States. This use of genetics was based on the mistaken assumption that genes alone determined human behavior and disorders, and it neglected the role of the environment. Eugenics fell into disfavor when it became part of the social programs of the Nazis in Germany.

1.6 What Impact Is Genomics Having?

- The development of recombinant DNA technology is the foundation for DNA cloning, genome projects, and biotechnology. These developments are causing large-scale changes in many aspects of life and are affecting medicine, agriculture, and the legal system.

1.7 What Choices Do We Make in the Era of Genomics and Biotechnology?

- With the completion of the Human Genome Project, the ability to manipulate human reproduction, and the ability to transfer genes, we are faced with many personal and social decisions. The ethical use of genetic information and biotechnology will require participation by a broad cross section of society.

Questions and Problems

CENGAGENOW Preparing for an exam? Assess your understanding of this chapter's topics with a pre-test, a personalized learning plan, and a post-test by logging on to *login.cengage.com/sso* and visiting CengageNOW's Study Tools.

1. Summarize Mendel's conclusions about traits and how they are passed from generation to generation.
2. What is population genetics?
3. What is hereditarianism, and what is the invalid assumption it makes?
4. What impact has recombinant DNA technology had on genetics and society?
5. What are genomes?
6. What is genomics?
7. In what way has biotechnology had an impact on agriculture in the United States?
8. We each carry 20,000–25,000 genes in our genome. Genes can be patented, and over 6,000 human genes have been patented. Do you think that companies or individuals should be able to patent human genes? Why or why not?
9. If your father were diagnosed with an inherited disease that develops around the age of 50, would you want to be tested to know if you would develop this disease? If so, when would you want to be tested? As a teenager or sometime in your 40s? If not, would you have children?

Internet Activities

Internet Activities are critical-thinking exercises using the resources of the World Wide Web to enhance the principles and issues covered in this chapter. For a full set of links and questions investigating the topics described below, visit *www.cengage.com/biology/cummings*

1. *Learning Styles.* You can learn more from your studies in any subject if you know something about your personal learning preferences. At the *Active Learning Site*, you may take a simple, informal assessment of your learning style. After completing the VARK learning-style inventory, explore the tips for using your preferred style(s) to enhance learning.
2. *How to Study Biology.* The University of Texas maintains a web-site that provides suggestions on how to approach the study of biology, including genetics. Check out the general study suggestions for biology courses. Try developing a concept map, as outlined on this website, for some of the topics being covered in your genetics course.
3. *Genetics as a Contemporary Field of Research.* Genetics is one of the most active research fields in biology today. Go to the website for the Genetics Society of America and browse the information on the journal, meetings, and awards. Using the link to the "Careers Brochure," read what a number of prominent geneticists have to say about their careers.
4. *The Ongoing Eugenics Debate.* For a history of the eugenics movement in the United States, take a look at the "Eugenics Slide Show." Although the eugenics movement in the United States declined by the mid-1930s, there are those who argue that eugenicists are alive and active among us. Check out the "Eugenics" page for links to several points of view on this issue.

HOW WOULD YOU VOTE NOW?

Our understanding of genetics, as well as the application of this understanding and its impact on society, is growing rapidly. Not all applications of genetic knowledge are for the good, and individuals in our society need to be aware of the principles and issues involved so that they can make informed decisions about their own genetic issues. At the beginning of this chapter, you were asked how you would respond if a major medical center asked you to donate a sample of DNA and allow access to your medical records for a project searching for genes that control complex traits such as hypertension, cardiovascular disease, and mental retardation. Now that you know more about how genetics and genetic information have been used and abused, what do you think? Visit the *Human Heredity* companion website for this edition at *www.cengage.com/biology/cummings* to find out more on the issue; then cast your vote online.

2 Cells and Cell Division

Leah G., 22, went to the hospital emergency room with severe leg pain after a fall while rollerblading. Despite the fact she had only a minor fall, X-rays showed a broken bone in her lower leg. There was no family history of brittle bones, but when asked about her general health, she reported that over the past few months she had tired easily and her abdomen was tender and sometimes painful. The emergency room physician ordered an abdominal MRI, which showed enlargement of Leah's spleen and liver. She left the hospital with a cast on her leg and an appointment with a genetic counselor.

The genetic counselor told Leah that because of her age, her Eastern European Jewish heritage, and her symptoms, she might have a genetic disorder called Gaucher (pronounced go-SHAY) disease. The counselor explained that affected individuals lack an enzyme and cannot break down a particular type of fat, which then accumulates in the liver, spleen, and bone marrow, forming distinctive cells called Gaucher cells in these tissues. Her symptoms of fatigue, enlargement of the liver and spleen, as well as her easily fractured leg and age of onset in early adulthood, are all symptoms of this disease. The counselor also explained that a liver biopsy and a blood test could confirm whether or not she had this disorder. While the disorder is rare in the general population, as many as 1 in 450 individuals of Eastern European Jewish descent have Gaucher disease. Leah arranged for the biopsy and blood test and for a follow-up visit with the counselor.

During her second visit, the counselor informed Leah that the biopsy and blood test confirmed that she had Gaucher disease. Leah was told that treatment for the disease was available, involving a recombinant DNA–produced form of the missing enzyme given intravenously. Each treatment was done on an outpatient basis, took about 1–2 hours, and was usually done every 2 weeks at a cost of $125,000–$150,000 a year. After discussing the situation with her parents, Leah began treatments; after 6 months, most of her symptoms had disappeared.

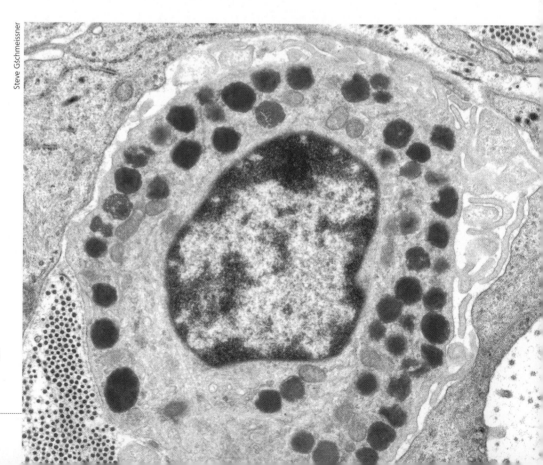

Steve Gschmeissner

A white blood cell containing enlarged lysosomes (stained brown) associated with a lysosomal storage disease.

2.1 The Chemistry of Cells

Cells are the basic structural and functional unit of living systems. But cells themselves are partly constructed from four classes of large molecules, often called **macromolecules** (Table 2.1). These cell components are carbohydrates, lipids, proteins, and nucleic acids. To understand how cell structure and function are related, we will briefly examine some of the structural and functional properties of cellular macromolecules. In later chapters, we will discuss how mutations disrupt the synthesis or function of these molecules, resulting in genetic disorders.

Carbohydrates include small, water-soluble sugars and large polymers made of sugars. In the cell, carbohydrates have three important functions: They are structural components of cells; they act as energy sources for the cell; and, in combination with proteins on the surface, they give cells a molecular identity.

Lipids are a structurally and functionally diverse class of biological molecules partially defined by their insolubility in water. Lipids have many functions: They are structural components of membranes, some serving as energy reserves, while others act as hormones and vitamins. Lipids are classified into three major groups: fats and oils, phospholipids, and steroids. The phospholipids play important roles in the structure and function of the cell membrane.

Proteins are the most functionally diverse class of macromolecules. Proteins are polymers, made up of one or more chains of subunits, called amino acids. The varied structures of proteins are reflected in their diversity of functions. Some of these are listed in Table 2.1.

Nucleic acids are polymers made from nucleotide subunits. Nucleotides themselves have important functions in energy transfer, but nucleic acids are the storehouses of genetic information in the cell. The information is encoded in the nucleotide sequence.

The combinations of various types of these four macromolecules are the foundation for the structural and functional diversity seen in the more than 200 cell types in the human body. In the next section, we will describe some of the fundamental structural and functional aspects of cells and their contents.

Macromolecules Large cellular polymers assembled by chemically linking monomers together.

Carbohydrates Macromolecules including sugars, glycogen, and starches composed of sugar monomers linked and cross-linked together.

Lipids A class of cellular macromolecules including fats and oils that are insoluble in water.

Proteins A class of cellular macromolecules composed of amino acid monomers linked together and folded into a three-dimensional shape.

Nucleic acids A class of cellular macromolecules composed of nucleotide monomers linked together. There are two types of nucleic acids, deoxyribonucleic acid (DNA) and ribonucleic acid (RNA), which differ in the structure of the monomers.

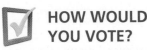

HOW WOULD YOU VOTE?

Bone marrow transplantation is an alternative treatment for Gaucher disease and offers a permanent cure in place of costly twice-monthly enzyme infusions. Some have argued that bone marrow donors are in short supply and that because Gaucher disease is not life threatening and can be treated by other means, these patients should have a lower priority as candidates for transplantation than those with high-risk diseases such as leukemia. Do you think candidates for transplants should be prioritized according to their illness? Visit the *Human Heredity* companion website at *www.cengage.com/biology/cummings* to find out more about the issue; then cast your vote online.

Molecules Structures composed of two or more atoms held together by chemical bonds.

Organelles Cytoplasmic structures that have a specialized function.

Table 2.1 The Main Biomolecules in Cells

Class	Subclasses	Examples	Functions
CARBOHYDRATES	**Monosaccharides** (simple sugars)	Glucose	Energy source
	Oligosaccharides (short-chain carbohydrates)	Sucrose	A common sugar
	Polysaccharides (complex carbohydrates)	Starch, glycogen	Energy storage
LIPIDS	**Glycerides** Glycerol plus fatty acids	Fats	Energy storage
	Phospholipids Glycerol, fatty acids, phosphate group	Lecithin	Structure of cell membranes
	Sterols Carbon-ring structures	Cholesterol	Membrane structure, precursor to steroid hormones
PROTEINS	**Mostly fibrous** (sheets of polypeptide chains; mostly water insoluble)	Keratin Collagen	Structure of hair Structure of bones
	Mostly globular (protein chains folded into globular shapes; mostly water soluble)	Enzymes Hemoglobin Insulin Antibodies	Catalysts Oxygen transport Hormone Immune system
NUCLEIC ACIDS	**Adenosine phosphates** **Nucleic Acids** (polymers of nucleotides)	ATP DNA, RNA	Energy carrier Storage, transmission of genetic information

2.2 Cell Structure Reflects Function

We will review some of the basic aspects of human cell structure and discuss the functions of cell components and how these functions are disrupted in genetic disorders. Although cells differ widely in their size, shape, function, and life cycle, at a structural level they are fundamentally similar—they all have a plasma membrane, cytoplasm, membranous organelles, and a membrane-bound nucleus. An idealized human cell is shown in Figure 2.1. A cell's shape, internal organization, and function are under genetic control, and many genetic disorders cause changes in cellular structure and/or function.

There are two cellular domains: the plasma membrane and the cytoplasm.

A double-layered plasma membrane separates the cell from the external environment. Lipids in the membrane provide a structural component, and a patchwork of different proteins gives the membrane many of its functional characteristics. The plasma membrane controls the exchange of materials with the environment outside the cell (Figure 2.2). Gases, water, and some small **molecules** pass through the membrane easily, but others are transported by energy-requiring systems. Proteins with attached carbohydrates in and on the plasma membrane provide cells with a form of molecular identity. The type and number of these molecules are genetically controlled and are responsible for many important properties of cells, including blood type and compatibility for organ transplants. Several genetic disorders, including cystic fibrosis (OMIM 219700; see Spotlight on A Fatal Membrane Flaw), are caused by defects in the plasma membrane (see Chapter 4 for an explanation of OMIM numbers and the catalog of human genetic disorders). The plasma membrane encloses the cytoplasm, which is a complex mixture of molecules and membrane-enclosed structures known collectively as **organelles**.

KEEP IN MIND

Many genetic disorders alter cellular structure or function.

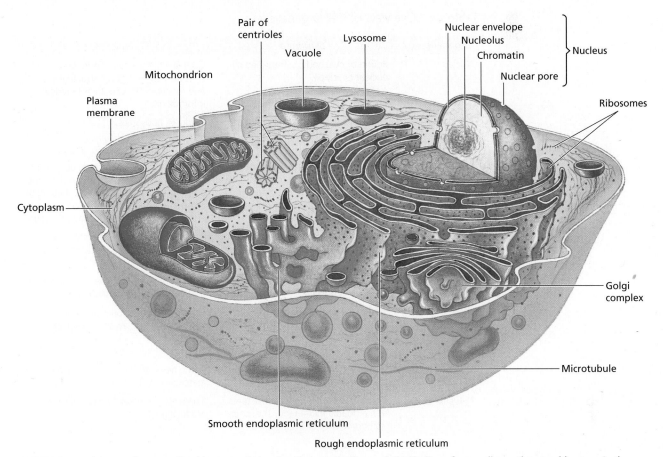

FIGURE 2.1 A diagram of a generalized human cell showing the organization and distribution of organelles as they would appear in the transmission electron microscope. The type, number, and location of organelles are related to cell function.

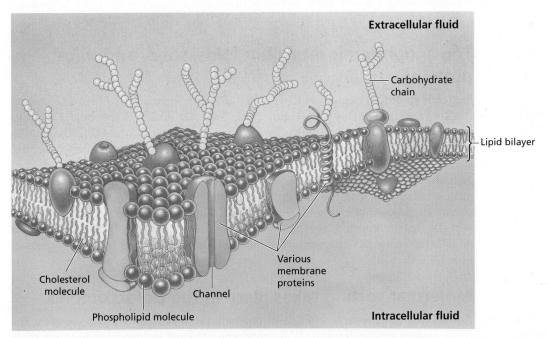

FIGURE 2.2 The plasma membrane. Proteins are embedded in a double layer of lipids. Short carbohydrate polymers are attached to some proteins on the outer surface of the membrane.

Table 2.2 Overview of Cell Organelles

Organelle	Structure	Function
Nucleus	Round or oval body; surrounded by nuclear envelope.	Contains the genetic information necessary to control cell structure and function. DNA contains heredity information.
Nucleolus	Round or oval body in the nucleus containing DNA and RNA.	Produces ribosomes.
Endoplasmic reticulum	Network of membranous tubules in the cytoplasm of the cell. Smooth endoplasmic reticulum contains no ribosomes. Rough endoplasmic reticulum is studded with ribosomes.	Smooth endoplasmic reticulum (SER) is involved in producing phospholipids and has many different functions in different cells. Rough endoplasmic reticulum (RER) is the site of the synthesis of lysosomal enzymes and proteins for extracellular use.
Ribosomes	Small particles found in the cytoplasm; made of RNA and protein.	Aid in the production of proteins on the RER and in ribosome complexes (polysomes).
Golgi complex	Series of flattened sacs and associated vesicles.	Sorts, chemically modifies, and packages proteins produced on the RER.
Secretory vesicles	Membrane-bound vesicles containing proteins produced by the RER and repackaged by the Golgi complex; contain protein hormones or enzymes.	Store protein hormones or enzymes in the cytoplasm, awaiting a signal for release.
Lysosome	Membrane-bound structure containing digestive enzymes.	Combines with food vacuoles and digests materials engulfed by cells.
Mitochondria	Round, oval, or elongated structures with a double membrane. The inner membrane is extensively folded.	Complete the breakdown of glucose, producing ATP.

Organelles are specialized structures in the cytoplasm.

The cytoplasm in a human cell has an organization that is related to its function, which is reflected in the number and type of organelles it contains. In eukaryotes, cytoplasmic organelles divide the cell into a number of functional compartments. Table 2.2 summarizes the major organelles and their functions. We will review some of them here.

The endoplasmic reticulum folds, sorts, and ships proteins.

Endoplasmic reticulum (ER) A system of cytoplasmic membranes arranged into sheets and channels whose function it is to synthesize and transport gene products.

Ribosomes Cytoplasmic particles that aid in the production of proteins.

The **endoplasmic reticulum (ER)** is a network of membrane channels and pockets (vesicles) within the cytoplasm (Figure 2.3a). The outer surface of the rough ER (RER) is covered with **ribosomes**, another cytoplasmic component (Figure 2.3b). The smooth ER (SER) has no ribosomes on its surface; the RER and SER, although inter-connected, have different functions. Ribosomes are the most numerous cellular structures and can be found in the cytoplasm or attached to the outer surface of the RER. Ribosomes are involved in protein synthesis (discussed in Chapter 9). The space inside the ER is called the lumen. Ribosomes on the ER surface synthesize amino acid chains known as polypeptides that are inserted into the lumen where they are folded and modified to form proteins and prepared for transport to other locations in the cell, or tagged for export from the cell.

Molecular sorting takes place in the Golgi complex.

Golgi complex Membranous organelles composed of a series of flattened sacs. They sort, modify, and package proteins synthesized in the ER.

Animal cells contain clusters of flattened membrane sacs called the **Golgi complex**. The Golgi receives vesicles from the RER containing proteins, and modifies and sorts them into other vesicles (Figure 2.4), which then deliver their contents to destinations inside and outside the cell. Functional abnormalities of the Golgi are responsible for a number of genetic disorders, including Menkes disease (OMIM 309400). The Golgi complex is also a source of membranes for other organelles, including lysosomes.

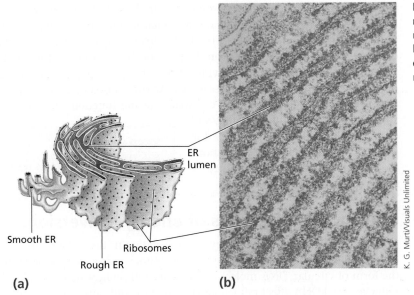

FIGURE 2.3 (a) Three-dimensional representation of the endoplasmic reticulum (ER) showing the relationship between the smooth and rough ER. (b) An electron micrograph of ribosome-studded rough ER.

K. G. Murti/Visuals Unlimited

ER lumen

Smooth ER

Ribosomes

Rough ER

(a)

(b)

Lysosomes are cytoplasmic disposal sites.

The **lysosomes** are membrane-enclosed vesicles containing digestive enzymes made in the RER. In the RER, these enzymes are packaged into vesicles and transported to the Golgi, where they are modified and repackaged into vesicles that bud off the Golgi to form lysosomes (Figure 2.4). Lysosomes are the processing and recycling centers of the cell. Proteins, fats, carbohydrates, and worn-out organelles in the cell that are marked for

Lysosomes Membrane-enclosed organelles in eukaryotic cells that contain digestive enzymes.

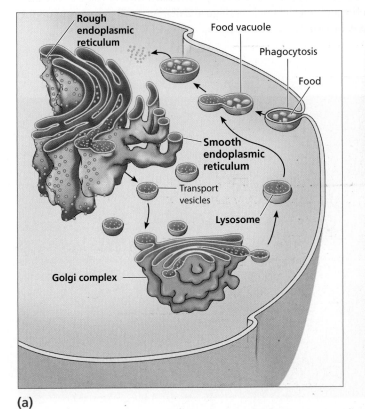

Rough endoplasmic reticulum

Food vacuole

Phagocytosis

Food

Smooth endoplasmic reticulum

Transport vesicles

Lysosome

Golgi complex

(a)

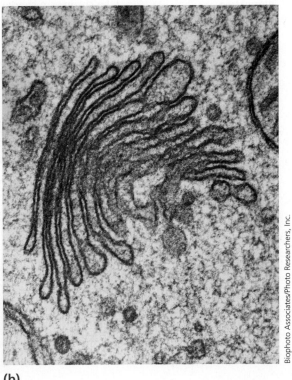

Biophoto Associates/Photo Researchers, Inc.

(b)

FIGURE 2.4 (a) The relationship between the Golgi complex and lysosomes. Digestive enzymes are synthesized on ribosomes attached to the ER, internalized, and moved to the Golgi in transport vesicles. In the Golgi, the enzymes are modified and packaged. Lysosomes pinch off the ends of the Golgi membrane. In the cytoplasm, lysosomes fuse with and digest the contents of vesicles that are internalized at the plasma membrane. (b) A transmission electron micrograph of the Golgi complex.

destruction end up in lysosomes, where they are broken down and recycled or exported for disposal. Lysosomes are important in cellular maintenance, and about 40 genetic disorders, including Gaucher disease (OMIM 230800)—described at the beginning of this chapter—disrupt lysosome function. In most of these disorders, the mutation disrupts production or function of an enzyme. When this happens, specific molecules are not digested and accumulate in the lysosomes. As the lysosomes enlarge, they become distorted, eventually altering normal cell structure and function. Disorders that affect the structure or function of lysosomes and other cellular organelles reinforce the point made earlier that the functioning of the organism can be explained by events that occur within its cells.

Mitochondria (singular: mitochondrion) Membrane-bound organelles, present in the cytoplasm of all eukaryotic cells, that are the sites of energy production.

Nucleus The membrane-bound organelle in eukaryotic cells that contains the chromosomes.

Nucleolus (plural: nucleoli) A nuclear region that functions in the synthesis of ribosomes.

Chromatin The DNA and protein components of chromosomes, visible as clumps or threads in nuclei.

Chromosomes The threadlike structures in the nucleus that carry genetic information.

> **KEEP IN MIND**
> Gaucher disease affects lysosomal function.

Mitochondria are sites of energy conversion.

Mitochondria are centers of energy transformation in the cell and are composed of an outer and an inner membrane (Figure 2.5). Mitochondria carry genetic information in the form of circular DNA molecules; they are self-replicating organelles. Mutations in mitochondrial DNA affect mitochondrial function and cause a number of genetic disorders, including Kearns-Sayre syndrome (OMIM 530000) and MELAS syndrome (OMIM 535000). These and other genetic disorders affecting mitochondria are discussed in Chapter 4.

The nucleus contains chromosomes.

The largest organelle is the **nucleus** (Figure 2.6a). It is enclosed by a double membrane called the nuclear envelope, which is studded with pores that allow communication between the nucleus and cytoplasm (Figure 2.6b). Within the nucleus, dense regions known as **nucleoli** (singular: **nucleolus**; Figure 2.6a) synthesize ribosomes. Dark strands of **chromatin** are seen throughout the nucleus (Figure 2.6c). As a cell prepares to divide, the chromatin condenses to form the **chromosomes**.

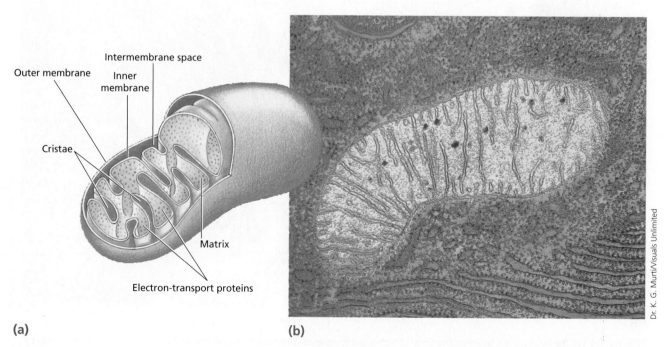

(a)

(b)

Dr. K. G. Murti/Visuals Unlimited

FIGURE 2.5 The mitochondrion is the center of energy transformation in the cell. (a) The infolded inner membrane forms two compartments where chemical reactions transfer energy from one form to another, allowing the cell to power many of its biochemical reactions. (b) A colorized transmission electron micrograph of a mitochondrion.

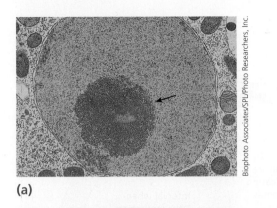

(a)

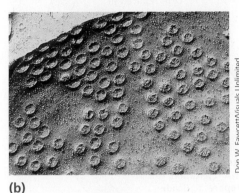

(b)

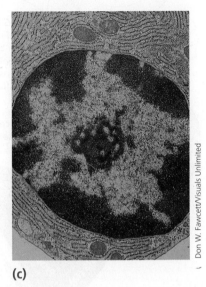

(c)

FIGURE 2.6 (a) The nucleus is bounded by a double membrane called the nuclear envelope. The nucleolus (arrow) is a prominent structure in the nucleus and is the site of ribosome synthesis. (b) The nuclear envelope is studded with pores, which allow exchange of materials between the nucleus and the cytoplasm. (c) When the cell is not dividing, the chromosomes are uncoiled and dispersed throughout the nucleus as clumps of chromatin, clustered just inside the nuclear envelope.

In humans, chromosomes exist in pairs. Most human cells, called somatic cells, carry 23 pairs, or 46 chromosomes, but certain cells, such as sperm and eggs, carry only one copy of each chromosome and have 23 unpaired chromosomes. Human males have one pair of chromosomes that are not completely matched. Members of this pair are known as **sex chromosomes** and are involved in sex determination (see Chapter 7 for a discussion of this topic). There are two types of sex chromosomes: X and Y. Males carry an X chromosome and a Y chromosome, and females carry two X chromosomes. All other chromosomes are known as **autosomes**.

Chromosomes carry genetic information that ultimately determines the structure, shape, and functions of the cell. This genetic information is contained in the sequence of nucleotide subunits in DNA and organized into genes.

Sex chromosomes In humans, the X and Y chromosomes that are involved in sex determination.

Autosomes Chromosomes other than the sex chromosomes. In humans, chromosomes 1–22 are autosomes.

Cell cycle The sequence of events that takes place between successive mitotic divisions.

Interphase The period of time in the cell cycle between mitotic divisions.

Mitosis Form of cell division that produces two cells, each of which has the same complement of chromosomes as the parent cell.

Cytokinesis The process of cytoplasmic division that accompanies cell division.

2.3 The Cell Cycle Describes the Life History of a Cell

Cells in the body alternate between two states: division and non-division. The sequence of events from division to division is called the **cell cycle**. The time between divisions varies from minutes to months or even years. A cycle consists of three parts: **interphase**, **mitosis**, and **cytokinesis** (Active Figure 2.7). The first part of the cycle, interphase, is the time between divisions. The other two parts—mitosis (division of the chromosomes) and cytokinesis (division of the cytoplasm)—define cell division.

ACTIVE FIGURE 2.7 The cell cycle has three stages: interphase, mitosis, and cytokinesis. Interphase has three parts: G1, S, and G2. Some cells can opt out of the cycle and enter a resting stage called G0. Times shown for the stages are representative for cells grown in the laboratory.

CENGAGENOW Learn more about the cell cycle by viewing the animation by logging on to *login.cengage.com/sso* and visiting CengageNOW's Study Tools.

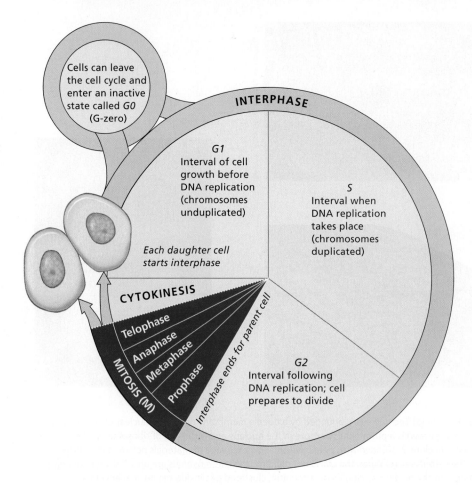

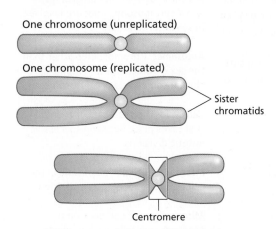

FIGURE 2.8 Chromosomes replicate during the S phase. While attached to a common the centromere, the replicated chromosomes have two sister chromatids.

Interphase has three stages.

Let's begin with a cell that has just finished division, when the two daughter cells are about one-half the size of the parental cell (Active Figure 2.7). Before they can divide again, these cells must grow to the size of the parental cell. Growth takes place in the G1 stage, which begins immediately after division; during this stage, many cytoplasmic components, including organelles, membranes, and ribosomes, are made. This synthetic activity doubles the cell's size and replaces organelles distributed to the other daughter cell during separation of the cytoplasm. G1 is followed by the S (synthesis) phase, during which a copy of each chromosome is made (Figure 2.8). A period known as G2 takes place before the cell is ready to begin a new round of division. By the end of G2, the cell is ready to divide.

In human cells grown in the laboratory, the time spent in interphase (stages G1, S, and G2) varies from 18–24 hours. Mitosis (the M phase) usually takes less than 1 hour, and so cells spend most of their time in interphase. Table 2.3 summarizes the phases of the cell cycle.

The life history and cell cycles vary for different cell types. Some cells, such as those in bone marrow, pass through the cell cycle continuously and divide regularly to form blood cells. At the other extreme, in response to internal and external signals, some cell types permanently leave the cell cycle and enter an inactive state called G0 and never divide. In between these extremes are cell types, such as white blood cells, that enter G0 but can reenter G1 and divide.

When cells escape from the controls that are part of the cell cycle, they can become cancerous.

Table 2.3 Phases of the Cell Cycle

Phase	Characteristics
Interphase	
G1 (Gap 1)	Stage begins immediately after mitosis. RNA, proteins, and organelles are synthesized.
S (Synthesis)	DNA is replicated, and chromosomes form sister chromatids.
G2 (Gap 2)	Mitochondria divide. Precursors of spindle fibers are synthesized.
Mitosis	
Prophase	Chromosomes condense. Nuclear envelope disappears. Centrioles divide and migrate to opposite poles of the dividing cell. Spindle fibers form and attach to chromosomes.
Metaphase	Chromosomes line up on the midline of the dividing cell.
Anaphase	Chromosomes begin to separate.
Telophase	Chromosomes reach opposite poles. New nuclear envelope forms. Chromosomes decondense.
Cytokinesis	Cleavage furrow forms and deepens. Cytoplasm divides.

Cell division by mitosis occurs in four stages.

When a cell reaches the end of G2, it enters the second part of the cell cycle. During division, two important processes take place. A complete set of chromosomes is distributed to each daughter cell (mitosis), and the cytoplasm is distributed more or less equally to the two daughter cells (cytokinesis). Although cytoplasmic division can be somewhat imprecise and still be operational, the division and distribution of the chromosomes must be accurate and unfailing for the cell to function properly.

The result of division is two daughter cells. In humans, each daughter cell receives a set of 46 chromosomes derived from a single parental cell with 46 replicated chromosomes. Although the distribution of chromosomes in cell division is usually precise, errors in this process occur. Those mistakes often have serious genetic consequences and are discussed in detail in Chapter 6.

Although mitosis is a continuous process, for the sake of discussion, it is divided into four stages: prophase, metaphase, anaphase, and telophase (Active Figure 2.9). These stages are accompanied by changes in chromosome organization, as described in the following sections.

Prophase Prophase marks the beginning of mitosis. In the preceding interphase (Active Figure 2.9a), the cell has replicated its chromosomes. Chromosomes are not usually visible in the nuclei of nondividing cells because they are uncoiled and dispersed throughout the nucleus. At the beginning of **prophase**, the chromosomes condense and become recognizable (Active Figure 2.9b) as long, thin, intertwined threads. As prophase continues, the chromosomes become shorter and thicker (Active Figure 2.9c). In human cells, 46 chromosomes are present. Near the end of prophase, each replicated chromosome consists of two strands called **chromatids**, held together by a structure called the **centromere**. Chromatids joined together by a centromere are known as **sister chromatids**. Near the end of prophase, the nuclear membrane breaks down and a network of spindle fibers forms in the cytoplasm. When fully formed, the spindle fibers stretch across the cell (Active Figure 2.9d).

Prophase A stage in mitosis during which the chromosomes become visible and contain sister chromatids joined at the centromere.

Chromatid One of the strands of a duplicated chromosome, joined by a single centromere to its sister chromatid.

Centromere A region of a chromosome to which spindle fibers attach during cell division. The location of a centromere gives a chromosome its characteristic shape.

Sister chromatids Two chromatids joined by a common centromere. Each chromatid carries identical genetic information.

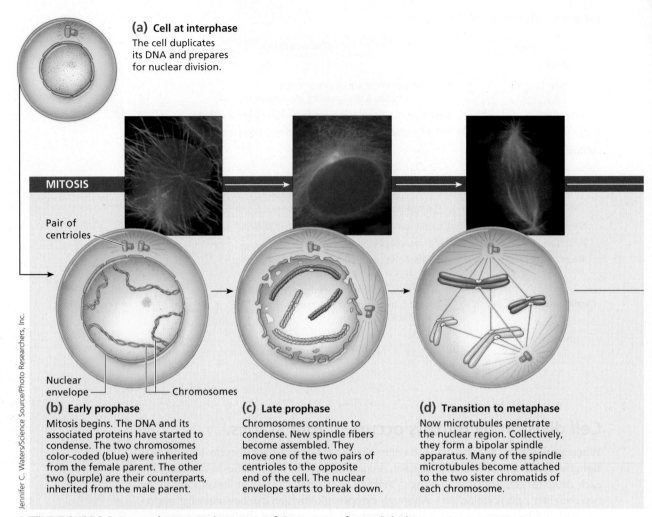

(a) Cell at interphase

The cell duplicates its DNA and prepares for nuclear division.

MITOSIS

Pair of centrioles

Nuclear envelope — Chromosomes

(b) Early prophase

Mitosis begins. The DNA and its associated proteins have started to condense. The two chromosomes color-coded (blue) were inherited from the female parent. The other two (purple) are their counterparts, inherited from the male parent.

(c) Late prophase

Chromosomes continue to condense. New spindle fibers become assembled. They move one of the two pairs of centrioles to the opposite end of the cell. The nuclear envelope starts to break down.

(d) Transition to metaphase

Now microtubules penetrate the nuclear region. Collectively, they form a bipolar spindle apparatus. Many of the spindle microtubules become attached to the two sister chromatids of each chromosome.

Jennifer C. Waters/Science Source/Photo Researchers, Inc.

ACTIVE FIGURE 2.9 Stages of mitosis. Only two pairs of chromosomes from a diploid (*2n*) cell are shown here. The photographs show mitosis in a mouse cell; the DNA is stained blue and the microtubules of the spindle fibers are stained green.

CENGAGE**NOW** Learn more about mitosis by viewing the animation by logging on to *login.cengage.com/sso* and visiting CengageNOW's Study Tools.

FIGURE 2.10 Roberts syndrome is a genetic disorder caused by the malfunction of centromeres during mitosis. In this painting by Goya (1746–1828), the child on the woman's lap has greatly shortened arms and legs—one of the characteristics of this disorder.

Metaphase Metaphase begins when the chromosomes, with spindle fibers attached, have moved to the middle, or equator, of the cell (Active Figure 2.9d and e). In human cells, at this stage there are 46 centromeres, each attached to two sister chromatids.

Anaphase In **anaphase**, the centromeres divide, converting each sister chromatid into a chromosome (Active Figure 2.9f). A genetic disorder called Roberts syndrome is caused by a malfunction in centromere splitting during development (OMIM 268300; Figure 2.10). Late in anaphase, the chromosomes migrate toward opposite ends of the cell. By the end of anaphase, there is a complete set of chromosomes at each end of the cell. Although anaphase is the briefest stage of mitosis, it is essential for ensuring that each daughter cell receives a complete and identical set of 46 chromosomes.

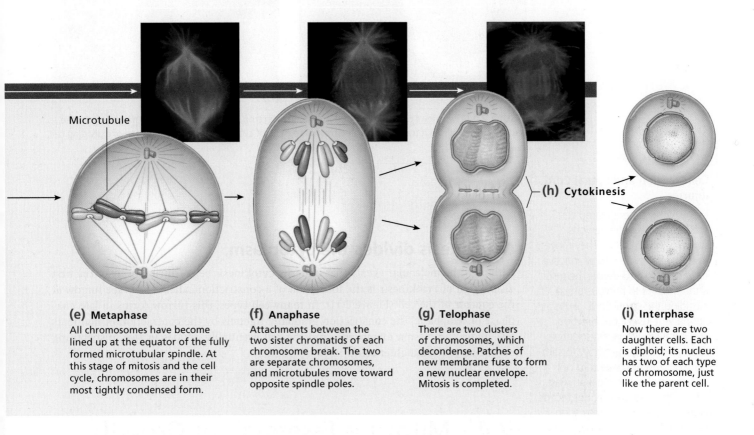

Microtubule

(e) Metaphase
All chromosomes have become lined up at the equator of the fully formed microtubular spindle. At this stage of mitosis and the cell cycle, chromosomes are in their most tightly condensed form.

(f) Anaphase
Attachments between the two sister chromatids of each chromosome break. The two are separate chromosomes, and microtubules move toward opposite spindle poles.

(g) Telophase
There are two clusters of chromosomes, which decondense. Patches of new membrane fuse to form a new nuclear envelope. Mitosis is completed.

(h) Cytokinesis

(i) Interphase
Now there are two daughter cells. Each is diploid; its nucleus has two of each type of chromosome, just like the parent cell.

Telophase At **telophase**, the chromosomes have reached opposite ends of the cell, they begin to uncoil, the spindle fibers break down, and membranes from the ER begin to form a new nuclear envelope (Active Figure 2.9g). At this point, mitosis is completed (Active Figure 2.9h). The major features of mitosis are summarized in Table 2.4.

Table 2.4 Summary of Mitosis

Stage	Characteristics
Prophase	Chromosomes become visible as threadlike structures. As they continue to condense, they are seen as double structures, with sister chromatids joined at a single centromere.
Metaphase	Chromosomes become aligned at equator of cell.
Anaphase	Centromeres divide, and chromosomes move toward opposite poles.
Telophase	Chromosomes decondense; nuclear membrane forms.

Metaphase A stage in mitosis during which the chromosomes become arranged near the middle of the cell.

Anaphase A stage in mitosis during which the centromeres split and the daughter chromosomes begin to separate.

Telophase The last stage of mitosis, during which the chromosomes of the daughter cells decondense and the nucleus re-forms.

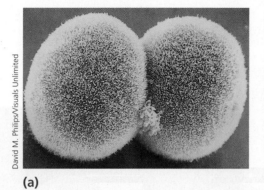 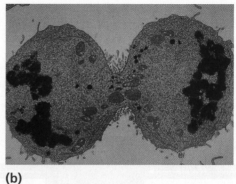

David M. Philips/Visuals Unlimited

(a) **(b)**

FIGURE 2.11 Two views of cytokinesis. (a) A scanning electron microscope view of cytokinesis from the outside of the cell. (b) A transmission electron micrograph of cytokinesis in a cross section of a dividing cell.

Cytokinesis divides the cytoplasm.

Although the molecular events that underlie cytokinesis begin during mitosis, the first visible sign of cytokinesis is the formation of a constriction called a cleavage furrow at the equator of the cell (Figure 2.11). In many cell types, this furrow forms in late anaphase or telophase. The constriction gradually tightens by the contraction of filaments just under the plasma membrane, and the cell eventually divides in two, distributing organelles to the daughter cells.

2.4 Mitosis Is Essential for Growth and Cell Replacement

Mitosis is an essential process in humans and all multicellular organisms. Some cells retain the capacity to divide throughout their life cycle, whereas others do not divide in adulthood. For example, cells in bone marrow continually move through the cell cycle, producing about 2 million red blood cells each second. Skin cells constantly divide to replace dead cells that are sloughed off the surface of the body. By contrast, other cells, including many cells in the nervous system, leave the cell cycle, enter G0, and do not divide in adulthood (see Spotlight on Cell Division and Spinal Cord Injuries).

Occasionally, cells escape from cell cycle regulation and grow uncontrollably, forming cancerous tumors. The major mechanisms that regulate the cell cycle operate in G1. Much is known about how these regulatory systems work, and they will be described in Chapter 12, Genes and Cancer.

Cells grown in the laboratory undergo a specific number of divisions (known as the Hayflick limit) and then stop dividing. Cells from human embryos have a limit of about 50 divisions—enough to produce an adult and for cell replacement during a lifetime. Cells from adults can divide only about 10–30 times. However, embryonic stem cells have unlimited proliferative capacity.

KEEP IN MIND

Cancer is a disease of the cell cycle.

In human cells, the maximum number of divisions is under genetic control; several genetic disorders that affect control of cell division are associated with accelerated aging. One of these is progeria (OMIM 176670), in which 7- or 8-year-old affected children look like they are 70 or 80 years old (Figure 2.12). Affected individuals usually die of coronary artery disease in their teens. Werner syndrome (OMIM 277700) is another genetic disorder associated with premature aging. In this case, the disease process begins between the ages of 15 and 20 years, and affected individuals die of age-related problems by 45–50 years. Both disorders are associated with defects in DNA repair and switch cells from a growth to a maintenance mode, halting divisions far short of the Hayflick limit.

FIGURE 2.12 John Tackett in spring of 2003 at age 15. He died in 2004 as the oldest person with progeria.

2.5 Cell Division by Meiosis: The Basis of Sex

The genetic information we inherit comes from two cells: a sperm and an egg. These cells are produced by a form of cell division called **meiosis** (Active Figure 2.13). Cells in the testis and ovary called germ cells undergo meiosis and produce gametes. Recall that in mitosis, each daughter cell receives two copies of each chromosome. Cells with two copies of each chromosome are **diploid** (**2n**) and have 46 chromosomes. In meiosis, members of a chromosome pair separate from each other, and each daughter cell receives a **haploid** (**n**) set of 23 chromosomes. These haploid cells form gametes (sperm and egg). Fusion of two haploid gametes in fertilization restores the chromosome number to the diploid number of 46, providing a full set of genetic information to the fertilized egg.

The distribution of chromosomes in meiosis is an exact process. Each gamete contains one member of each chromosome pair—not a random selection of 23 of the 46 chromosomes. The two rounds of division (meiosis I and meiosis II) accomplish this precise reduction in the chromosome number.

Meiosis The process of cell division during which one cycle of chromosomal replication is followed by two successive cell divisions to produce four haploid cells.

Diploid (2n) The condition in which each chromosome is represented twice as a member of a homologous pair.

Haploid (n) The condition in which each chromosome is represented once in an unpaired condition.

KEEP IN MIND

Meiosis maintains a constant chromosome number from generation to generation.

Meiosis I reduces the chromosome number.

Before cells enter meiosis, the chromosomes replicate during interphase. In prophase I, the chromosomes condense and become visible in the microscope (Active Figure 2.13a). As the chromosomes condense, the nuclear envelope disappears and the spindle becomes organized. Each chromosome physically associates with the other member of its pair. Members of a chromosome pair are **homologous chromosomes**. Once paired, the sister chromatids of each chromosome are visible, showing that each consists of two sister chromatids joined by a single centromere.

In metaphase I (Active Figure 2.13b), paired homologous chromosomes line up at the equator of the cell, with each chromosome attached to spindle fibers from opposite poles of the cell. In anaphase I, members of each homologous pair separate from each other and move toward opposite sides of the cell (Active Figure 2.13c). Cytokinesis (division of the cytoplasm) occurs after telophase I, producing two haploid cells (Active Figure 2.13d).

Homologous chromosomes Chromosomes that physically associate (pair) during meiosis. Homologous chromosomes have identical gene loci.

Meiosis II begins with haploid cells.

In prophase II, the unpaired chromosomes (Active Figure 2.13e) consist of two sister chromatids joined by a centromere. At metaphase II (Active Figure 2.13f), the 23 unpaired chromosomes are at the equator of the cell, with spindle fibers from opposite poles of the cell attached to their centromeres. Anaphase II (Active Figure 2.13g) begins when the centromeres of each chromosome divide for the first time. The 46 chromatids become chromosomes and move to opposite poles of the cell.

In telophase II, the chromosomes uncoil, the nuclear envelope re-forms (Active Figure 2.13h), and the process of meiosis is complete. Cytokinesis then divides the cytoplasm, producing four haploid cells. In meiosis, one diploid cell with 46 chromosomes

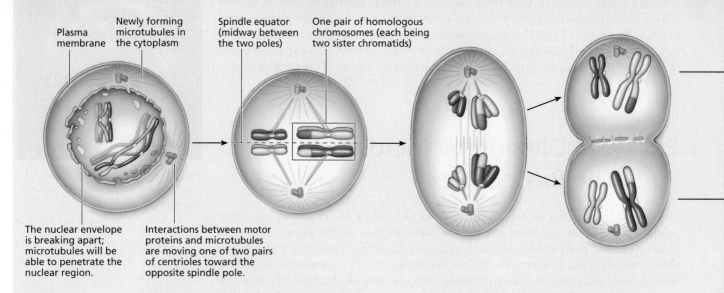

Plasma membrane

Newly forming microtubules in the cytoplasm

Spindle equator (midway between the two poles)

One pair of homologous chromosomes (each being two sister chromatids)

The nuclear envelope is breaking apart; microtubules will be able to penetrate the nuclear region.

Interactions between motor proteins and microtubules are moving one of two pairs of centrioles toward the opposite spindle pole.

(a) Prophase I
At the end of interphase, chromosomes are duplicated and in threadlike form. Now they start to condense. Each pairs with its homologue, and the two usually swap segments. The swapping, called crossing over, is indicated by the break in color on the pair of larger chromosomes. Newly forming spindle microtubules become attached to each chromosome.

(b) Metaphase I
Motor proteins projecting from the microtubules move the chromosomes and spindle poles apart. Chromosomes are tugged into position midway between the spindle poles. The spindle becomes fully formed by the dynamic interactions among motor proteins, microtubules, and chromosomes.

(c) Anaphase I
Some microtubules extend from the spindle poles and overlap at the equator. These lengthen and push the poles apart. Other microtubules extending from the poles shorten and pull each chromosome away from its homologous partner. These motions move the homologous partners to opposite poles.

(d) Telophase I
Cytokinesis divides the cytoplasm of the cell after telophase. There are now two haploid (*n*) cells with one of each type of chromosome that was present in the parent (*2n*) cell. *All chromosomes are still in the duplicated state.*

ACTIVE FIGURE 2.13 The stages of meiosis. In this form of cell division, replicated homologous chromosomes physically associate to form a chromosome pair. Members of each chromosome pair separate from each other at the first meiotic division (meiosis I). In meiosis II, the centromeres of unpaired chromosomes divide, resulting in four cells, each with the haploid (*n*) number of chromosomes.

CENGAGENOW Learn more about meiosis by viewing the animation by logging on to *login.cengage.com/sso* and visiting CengageNOW's Study Tools.

undergoes one round of chromosome replication and two rounds of division to produce four haploid cells, each of which contains one copy of each chromosome (Active Figure 2.13h).

The characteristics of each stage of meiosis are presented in Table 2.5, and the movement of chromosomes during meiosis is summarized in Figure 2.14. Figure 2.15 compares the events of mitosis and meiosis.

Table 2.5 Summary of Meiosis

Stage	Characteristics
Prophase I	Chromosomes become visible, homologous chromosomes pair, and sister chromatids become apparent. Recombination takes place.
Metaphase I	Paired chromosomes align at equator of cell.
Anaphase I	Paired homologous chromosomes separate. Members of each chromosome pair move to opposite poles.
Telophase I	Chromosomes uncoil, become dispersed.
Cytokinesis	Cytoplasm divides, forming two cells.
Prophase II	Chromosomes re-coil, shorten.
Metaphase II	Unpaired chromosomes become aligned at equator of cell.
Anaphase II	Centromeres separate. Daughter chromosomes, which were sister chromatids, pull apart.
Telophase II	Chromosomes uncoil, nuclear envelope re-forms. Meiosis ends.
Cytokinesis	The cytoplasm divides, forming daughter cells.

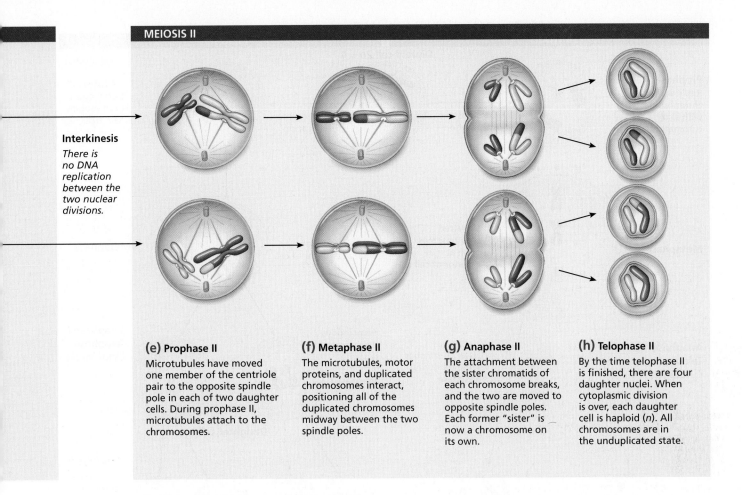

Interkinesis

There is no DNA replication between the two nuclear divisions.

(e) Prophase II
Microtubules have moved one member of the centriole pair to the opposite spindle pole in each of two daughter cells. During prophase II, microtubules attach to the chromosomes.

(f) Metaphase II
The microtubules, motor proteins, and duplicated chromosomes interact, positioning all of the duplicated chromosomes midway between the two spindle poles.

(g) Anaphase II
The attachment between the sister chromatids of each chromosome breaks, and the two are moved to opposite spindle poles. Each former "sister" is now a chromosome on its own.

(h) Telophase II
By the time telophase II is finished, there are four daughter nuclei. When cytoplasmic division is over, each daughter cell is haploid (*n*). All chromosomes are in the unduplicated state.

Meiosis produces new combinations of genes in two ways.

Meiosis produces new combinations of parental genes in two ways: by random **assortment** of maternal and paternal chromosomes; and by **crossing over**, the exchange of chromosome segments between homologues. Remember: In each pair of chromosomes, one copy was inherited from each parent.

When pairs of homologous chromosomes line up in metaphase I, the maternal and paternal members of each pair line up at random with respect to all other pairs

Assortment The result of meiosis I that puts random combinations of maternal and paternal chromosomes into gametes.

Crossing over A process in which chromosomes physically exchange parts.

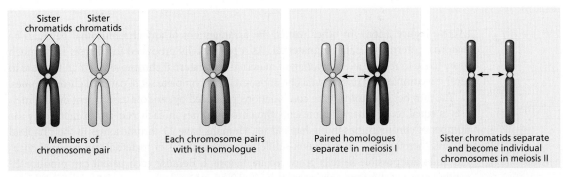

Sister chromatids | Sister chromatids

Members of chromosome pair

Each chromosome pairs with its homologue

Paired homologues separate in meiosis I

Sister chromatids separate and become individual chromosomes in meiosis II

FIGURE 2.14 Summary of chromosome movements in meiosis. Replicated homologous chromosomes become visible in prophase I. At metaphase I, homologous pairs of chromosomes line up at the equator of the cell and separate from each other in anaphase I. In meiosis II, the centromeres split, and sister chromatids form individual chromosomes. Each of the resulting haploid cells has one set of chromosomes.

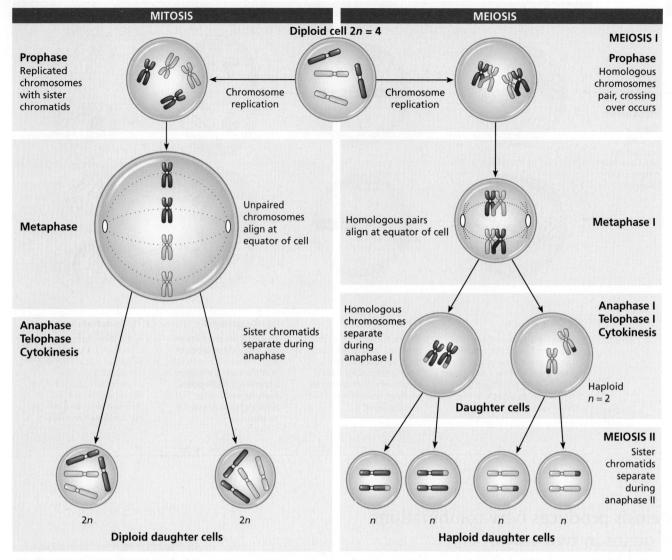

FIGURE 2.15 A comparison of the events in mitosis (*left*) and meiosis (*right*). In mitosis, a diploid parental cell undergoes chromosome replication. When the cell enters prophase, the chromosomes become visible as replicated structures with sister chromatids held together by a common centromere. Unpaired chromosomes line up at the equator of the cell in metaphase. In anaphase, the centromeres separate, converting the sister chromatids into chromosomes. The result is two daughter cells, each of which is genetically identical to the parental cell. In meiosis, the parental diploid cell undergoes chromosome replication. When the cell enters prophase I, the chromosomes become visible as replicated structures with sister chromatids held together by a common centromere. Members of a chromosome pair physically associate with each other and line up at the equator of the cell at metaphase I. Members of a chromosome pair separate in anaphase I and move to opposite poles of the cell. In meiosis II, the unpaired chromosomes in each cell line up on the equator of the cell. During anaphase II, the centromeres split, converting the sister chromatids into chromosomes, which are distributed to daughter cells. The result is four haploid daughter cells, each with one copy of the chromosomes.

(Active Figure 2.16a). In other words, the arrangement of any chromosomal pair can be maternal:paternal or paternal:maternal. As a result, cells produced in meiosis I are much more likely to receive a *combination* of maternal and paternal chromosomes than they are to receive a complete set of maternal chromosomes or a complete set of paternal chromosomes.

The number of chromosome combinations produced by random assortment during meiosis is equal to 2^n, where 2 represents the chromosomes in each pair and n represents the number of chromosomes in the haploid set. Humans have 23 chromosomes in their haploid set, and therefore 2^{23}—or 8,388,608—different combinations of maternal and paternal chromosomes are possible in cells produced in meiosis I. Because each parent can produce 2^{23} combinations of chromosomes, more than 7×10^{13} combinations are possible in their children, each of whom would carry a different combination of maternal and paternal chromosomes.

Crossing over involves the physical exchange of parts between non-sister chromatids (Active Figure 2.16b). This process adds to the genetic variation produced by random

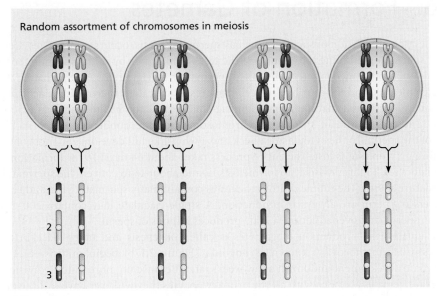

Random assortment of chromosomes in meiosis

(a)

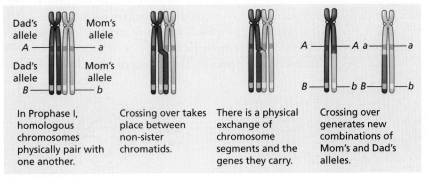

Dad's allele A — Mom's allele a

Dad's allele B — Mom's allele b

In Prophase I, homologous chromosomes physically pair with one another.

Crossing over takes place between non-sister chromatids.

There is a physical exchange of chromosome segments and the genes they carry.

A — A a — a

B — b B — b

Crossing over generates new combinations of Mom's and Dad's alleles.

(b)

ACTIVE FIGURE 2.16 Two ways of generating genetic variation in meiosis. (a) Random assortment of maternal (blue) and paternal (purple) chromosomes at metaphase I. Here, three chromosome pairs (1, 2, and 3) have four possible orientations at metaphase I. In the haploid cells produced, there are eight possible combinations of maternal and paternal chromosomes. (b) Crossing over in prophase I increases genetic variation by generating new combinations of maternal and paternal alleles. The paternal chromosome (purple) carries the *A* and *B* alleles, while the maternal chromosome (blue) carries the *a* and *b* alleles. Crossing over between non-sister chromatids produces chromosomes carrying *A* and *b*, and chromosomes carrying *a* and *B*.

CENGAGENOW Learn more about the assortment of chromosomes and crossing over by viewing the animation by logging on to *login.cengage.com/sso* and visiting CengageNOW's Study Tools.

assortment. Members of a chromosome pair carry identical genes but may carry different versions of the gene. These different versions are called **alleles**. For example, a chromosome may carry a gene for eye color. One copy of the chromosome may carry an allele for blue eyes, while the other carries an allele for brown eye color. The exchange of chromosome parts during crossing over makes new combinations of alleles inherited from each parent.

When the variability generated by crossing over is added to that produced by the random combination of maternal and paternal chromosomes, the number of different chromosome combinations that a couple can produce in their offspring has been estimated at 8×10^{23}. Obviously, the offspring of a couple represent only a very small fraction of these possible gamete combinations. For this reason, it is almost impossible for any two children (aside from identical twins) to be genetically identical.

Allele One of the possible alternative forms of a gene, usually distinguished from other alleles by its phenotypic effects.

2.6 Formation of Gametes

Spermatogonia Mitotically active cells in the gonads of males that give rise to primary spermatocytes.

Spermatids The four haploid cells produced by meiotic division of a primary spermatocyte.

Oogonia Mitotically active cells that produce primary oocytes.

Secondary oocyte The large cell produced by the first meiotic division.

In males, the production of sperm, known as spermatogenesis, occurs in the testes. Cells called **spermatogonia** line the tubules of the testes and divide by mitosis from puberty until death, producing daughter cells called spermatocytes (Figure 2.17a). Spermatocytes undergo meiosis, and the four haploid cells that result are called **spermatids**. Each spermatid develops into a mature sperm. During this period, the haploid nucleus (sperm carry 22 autosomes and an X or a Y sex chromosome) becomes condensed and forms the head of the sperm. In the cytoplasm, a neck and a whip-like tail develop, and most of the remaining cytoplasm is lost. The entire process takes about 64 days: 16 for formation of spermatocytes, 16 for meiosis I, 16 for meiosis II, and about 16 to convert the spermatids into mature sperm. The tubules within the testes contain many spermatocytes, and large numbers of sperm are always in production. A single ejaculate may contain 200–300 million sperm, and over a lifetime a male produces billions of sperm.

In females, the production of gametes is called oogenesis and takes place in the ovaries. Cells in the ovary known as **oogonia** (Figure 2.17b) begin mitosis early in embryonic development and finish a few weeks later. Because no more mitotic divisions take place, females are born with all the primary oocytes they will ever have. All the primary oocytes begin meiosis during embryonic development and then stop. They remain in meiosis I until a female undergoes puberty. After puberty, usually one primary oocyte per month completes the first meiotic division, and the secondary oocyte is released from the ovary and moves into the oviduct. Cytokinesis following meiosis I does not produce cells of equal size. One cell, destined to become the female gamete, receives about 95% of the cytoplasm and is called a **secondary oocyte**. The larger cell becomes the functional

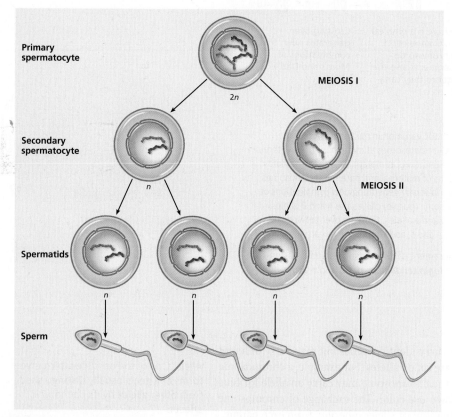

(a)

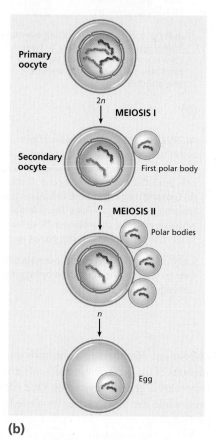

(b)

FIGURE 2.17 A comparison of sperm production and egg production in humans. (a) Beginning at puberty, some germ cells enter meiosis as primary spermatocytes. After meiosis I, the secondary spermatocytes contain 23 chromosomes composed of sister chromatids joined by a common centromere. After the second division (meiosis II), the haploid spermatids undergo developmental changes and become mature sperm. (b) In oogenesis, cells enter meiosis as primary oocytes during embryonic development and arrest in meiosis I. After puberty, usually one oocyte per menstrual cycle completes meiosis I just before ovulation. Formation of the haploid secondary oocyte is accompanied by unequal cytokinesis to produce a polar body, which is nonfunctional. Meiosis is completed only if the secondary oocyte is fertilized, when penetration of the sperm stimulates meiosis II and the second division.

Table 2.6 A Comparison of the Duration of Meiosis in Males and Females

Spermatogenesis		Oogenesis	
Begins at Puberty		**Begins During Embryogenesis**	
Spermatogonium ↓	} 16 days	Oogonium ↓	} Forms at 2–3 months after conception
Primary spermatocyte ↓	} 16 days	Primary oocyte ↓	} Forms at 2–3 months of gestation. Remains in meiosis I until ovulation, 12–50 years after formation.
Secondary spermatocyte ↓	} 16 days	Secondary oocyte ↓	
Spermatid ↓	} 16 days	Ootid	} Less than 1 day, when fertilization occurs
Mature sperm Total time	64 days	Mature egg-zygote Total time	12–50 years

gamete (the **ovum**) and the nonfunctional, smaller cells are known as **polar bodies**. If the secondary oocyte is fertilized, meiosis II is completed quickly and the haploid nuclei of the ovum and sperm fuse to produce a diploid zygote.

The timing of meiosis and gamete formation in human females is different from what it is in males (Table 2.6).

Unfertilized oocytes are lost during menstruation, along with uterine tissue. Each month until menopause, one or more primary oocytes complete meiosis I and are released from the ovary. Altogether, a female produces about 450 secondary oocytes during the reproductive phase of her life.

In females, then, meiosis takes years to complete or may never be completed. Meiosis begins with prophase I, while she is still an embryo. Meiosis I is completed at ovulation, and meiosis II stops at metaphase. If the egg is fertilized, meiosis is completed—a process that can take from 12–50 years.

Oocyte A cell from which an ovum develops by meiosis.

Ovum The haploid cell produced by meiosis that becomes the functional gamete.

Polar bodies Cells produced in the first and second meiotic division in female meiosis that contain little cytoplasm and will not function as gametes.

Genetics in Practice

Genetics in Practice case studies are critical-thinking exercises that allow you to apply your new knowledge of human genetics to real-life problems. You can find these case studies and links to relevant websites at **www.cengage.com/biology/cummings**

CASE 1

It is May 1989, and the scene is a research laboratory, with beakers, flasks, and pipettes covering the lab bench. People and equipment take up every possible space. One researcher, Joe, passes a friend staring into a microscope. Another student wears gloves while she puts precisely measured portions of various liquids into tiny test tubes. Joe glances at the DNA sequence results he is carrying. Something is wrong. There it is: a unique type of genetic mutation in a DNA sequence. The genetic information required to make a complete protein is missing, as if one bead had fallen from a precious necklace. Instead of returning to his station, Joe rushes to tell his supervisor, Dr. Tsui (pronounced "Choy"), that he has found a specific mutation in a person with cystic fibrosis (CF) and he does not

see that mutation in a normal person's gene. CF is a fatal disease that kills about 1 out of every 2,000 Caucasians (mostly children). Dr. Tsui examines the findings and is impressed but wants more evidence to prove that the result is real. He has had false hopes before, so he is not going to celebrate until they check this out carefully. Maybe the difference between the two gene sequences is just a normal variation among individuals. Five months later, Dr. Tsui and his team identify a "signature" pattern of DNA on either side of the mutation, and using that as a marker, they compare the DNA sequence of 100 unaffected people with the DNA sequence from 100 CF patients.

By September 1989, they are sure they have identified the CF gene. After several more years, Tsui and his team discover that the DNA sequence with the mutation encodes the information for a protein called CFTR (cystic fibrosis transmembrane conductance regulator), a part of the plasma membrane in cells that make mucus. This protein regulates a channel for chloride ions. Proteins are made of long chains of amino acids. The CFTR protein has 1,480 amino acids. Most children with CF are missing a single amino acid in their CFTR. Because of this, their mucus becomes too thick, causing all the other symptoms of CF. Thanks to Tsui's research, scientists now have a much better idea of how the disease works. We can easily predict when a couple is at risk for having a child with CF. With increasing understanding, scientists also may be able to devise improved treatments for children born with this disease.

CF is the most common genetic disease among persons of European ancestry. Children who have CF are born with it. Half of them will die before they are 25, and few make it past age 30. It affects all parts of the body that secrete mucus: the lungs, stomach, nose, and mouth. The mucus of children with CF is so thick that sometimes they cannot breathe. Why do 1 in 25 Caucasians carry the mutation for CF? Tsui and others think that people who carry it may also have resistance to diarrhea-like diseases.

1. Dr. Tsui's research team discovered the gene for cystic fibrosis. What medical advances can be made after a gene is cloned?

2. How can a change in one amino acid in the CF gene cause such severe effects in CF patients? Relate your answer to the CFTR protein function and the cell membrane.

CASE 2

Jim, a 37-year-old construction worker, and Sally, a 42-year-old business executive, were eagerly preparing for the birth of their first child. They, like more and more couples, chose to wait to have children until they were older and more financially stable. Sally had an uneventful pregnancy, with prenatal blood tests and an ultrasound indicating that the baby looked great and everything seemed "normal." Then, a few hours after Ashley was born, they were told she had been born with Down syndrome. In shock and disbelief, the couple questioned how that could have happened to them. It has long been recognized that the risk of having a child with Down syndrome increases with maternal age. For example, the risk of having a child with Down syndrome when the mother is 30 years old is 1 in 1,000; at maternal age 40, it is 9 in 1,000.

Well-defined and distinctive physical features characterize Down syndrome, which is the most common form of mental retardation caused by a chromosomal aberration. Most individuals (95%) with Down syndrome, or trisomy 21, have three copies of chromosome 21. Errors in meiosis that lead to trisomy 21 are almost always of maternal origin; only about 5% occur during spermatogenesis. It has been estimated that meiosis I errors account for 76% to 80% of maternal meiotic errors. In about 5% of patients, one copy is translocated to another chromosome—most often chromosome 14 or 21. No one is at fault when a child is born with Down syndrome, but the chances of it occurring increase with advanced maternal age.

Children with Down syndrome often have specific major congenital malformations, such as those of the heart (30% to 40% in some studies), and have an increased incidence (10–20 times higher) of leukemia compared with the normal population. Ninety percent of all Down syndrome patients have significant hearing loss. The frequency of trisomy 21 in the population is 1 in 650–1,000 live births.

1. What prenatal tests could have been done to detect Down syndrome before birth? Should they have been done?

2. Down syndrome is characterized by mental retardation. Can individuals with Down syndrome go to school or hold a job?

3. Should people with mental disabilities be integrated into the community? Why or why not?

Summary

2.1 The Chemistry of Cells

- Cells contain four classes of macromolecules: carbohydrates, lipids, proteins, and nucleic acids. These molecules provide the structural and functional framework for all cells. Mutations in genes that affect the structure or function of these macromolecules create genetic disorders.

2.2 Cell Structure Reflects Function

- The cell is the basic unit of structure and function in all organisms, including humans. Because genes control the number, size, shape, and function of cells, the study of cell structure helps us understand how genetic disorders disrupt cellular processes.

  In humans, 46 chromosomes—the $2n$, or diploid, number—are present in most cells, whereas specialized cells known as gametes contain half that number—the n, or haploid, number—of chromosomes.

2.3 The Cell Cycle Describes the Life History of a Cell

- At some point in their life, cells pass through the cell cycle, a period of non-division (interphase) that alternates with division of the nucleus (mitosis) and division of the cytoplasm (cytokinesis). Cells must contain a complete set of genetic information. This is ensured by replication of each chromosome and by the distribution of a complete chromosomal set in the process of mitosis.

  Mitosis (division) is one part of the cell cycle. During interphase (non-division), a duplicate copy of each chromosome is made. The process of mitosis is divided into four stages: prophase, metaphase, anaphase, and telophase. In mitosis, one diploid cell divides to form two diploid cells. Each cell has an exact copy of the genetic information contained in the parental cell.

2.4 Mitosis Is Essential for Growth and Cell Replacement

- Human cells are genetically programmed to divide about 50 times. This limit allows growth to adulthood and repairs such as wound healing. Alterations in this program can lead to genetic disorders of premature aging or to cancer.

2.5 Cell Division by Meiosis: The Basis of Sex

- Meiosis is a form of cell division that produces haploid cells containing only one copy of each chromosome. In an early stage of meiosis, members of a chromosome pair physically associate. At this time, each chromosome consists of two sister chromatids joined by a common centromere. In metaphase I, pairs of homologous chromosomes line up at the equator of the cell. In anaphase I, members of a chromosome pair separate from each other. Meiosis I produces cells that contain one member of each chromosome pair. In meiosis II, the unpaired chromosomes line up at the middle of the cell. In anaphase II, the centromeres divide, and the daughter chromosomes move to opposite poles. The four cells produced in meiosis contain the haploid number (23 in humans) of chromosomes.

2.6 Formation of Gametes

- In males, cells in the testes (spermatagonia) divide by mitosis to produce spermatocytes, which undergo meiosis to form spermatids. Spermatids undergo structural changes to convert them into functional sperm.

    In females, ovarian cells (oogonia) divide by mitosis to form primary oocytes. The primary oocytes undergo meiosis. In female meiosis, division of the cytoplasm is unequal, leading to the formation of one functional gamete and three smaller cells known as polar bodies.

Questions and Problems

CENGAGENOW Preparing for an exam? Assess your understanding of this chapter's topics with a pre-test, a personalized learning plan, and a post-test by logging on to *login.cengage.com/sso* and visiting CengageNOW's Study Tools.

Cell Structure Reflects Function

1. What advantages are there in having the interior of the cell divided into a number of compartments such as the nucleus, the ER, lysosomes, and so forth?
2. Assign a function(s) to the following cellular structures:
    a. plasma membrane
    b. mitochondrion
    c. nucleus
    d. ribosome
3. How many autosomes are present in a body cell of a human being? In a gamete?
4. Define the following terms:
    a. chromosome
    b. chromatin
5. Human haploid gametes (sperm and eggs) contain:
    a. 46 chromosomes, 46 chromatids
    b. 46 chromosomes, 23 chromatids
    c. 23 chromosomes, 46 chromatids
    d. 23 chromosomes, 23 chromatids

The Cell Cycle Describes the Life History of a Cell

6. What are sister chromatids?
7. Draw the cell cycle. What is meant by the term *cycle* in the cell cycle? What is happening at the S phase and the M phase?
8. In the cell cycle, at which stages do *two* chromatids make up *one* chromosome?
    a. beginning of mitosis
    b. end of G1
    c. beginning of S
    d. end of mitosis
    e. beginning of G2
9. Does the cell cycle refer to mitosis as well as meiosis?
10. It is possible that an alternative mechanism for generating germ cells could have evolved. Consider meiosis in a germ cell precursor (a cell that is diploid but will go on to make gametes). If the S phase were skipped, which meiotic division (meiosis I or meiosis II) would no longer be required?
11. Identify the stages of mitosis, and describe the important events that occur during each stage.
12. Why is cell furrowing important in cell division? If cytokinesis did not occur, what would be the end result?
13. A cell from a human female has just undergone mitosis. For unknown reasons, the centromere of chromosome 7 failed to divide. Describe the chromosomal contents of the daughter cells.
14. During which phases of the mitotic cycle would the terms *chromosome* and *chromatid* refer to identical structures?
15. Describe the critical events of mitosis that are responsible for ensuring that each daughter cell receives a full set of chromosomes from the parent cell.

Mitosis Is Essential for Growth and Cell Replacement

16. Mitosis occurs daily in a human being. What type of cells do humans need to produce in large quantities on a daily basis?
17. Speculate on how the Hayflick limit may lead to genetic disorders such as progeria and Werner syndrome. How is this related to cell division?
18. How can errors in the cell cycle lead to cancer in humans?

Cell Division by Meiosis: The Basis of Sex

19. List the differences between mitosis and meiosis in the following chart:

Attribute	Mitosis	Meiosis
Number of daughter cells produced		
Number of chromosomes per daughter cell		
Do chromosomes pair? (Y/N)		
Does crossing over occur? (Y/N)		
Can the daughter cells divide again? (Y/N)		
Do the chromosomes replicate before division? (Y/N)		
Type of cell produced		

20. In the following diagram, designate each daughter cell as diploid (2n) or haploid (n).

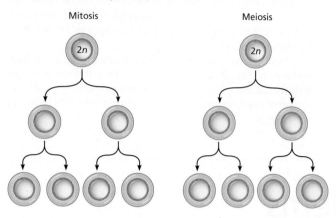

Mitosis   Meiosis

21. Which of the following statements is *not* true in comparing mitosis and meiosis?
  a. Twice the number of cells are produced in meiosis as in mitosis.
  b. Meiosis is involved in the production of gametes, unlike mitosis.
  c. Crossing over occurs in meiosis I but not in meiosis II or mitosis.
  d. Meiosis and mitosis both produce cells that are genetically identical.
  e. In both mitosis and meiosis, the parental cell is diploid.

22. Match the phase of cell division with the following diagrams. In these cells, 2n = 4.

  a. anaphase of meiosis I
  b. interphase of mitosis
  c. metaphase of mitosis
  d. metaphase of meiosis I
  e. metaphase of meiosis II

23. A cell has a diploid number of 6 (2n = 6).
  a. Draw the cell in metaphase of meiosis I.
  b. Draw the cell in metaphase of mitosis.
  c. How many chromosomes are present in a daughter cell after meiosis I?
  d. How many chromatids are present in a daughter cell after meiosis II?
  e. How many chromosomes are present in a daughter cell after mitosis?
  f. How many pairs of homologous chromosomes are visible in the cell in metaphase of meiosis I?

24. A cell (2n = 4) has undergone cell division. Daughter cells have the following chromosome content. Has this cell undergone mitosis, meiosis I, or meiosis II?

a.

b.

c.

25. We are following the progress of human chromosome 1 during meiosis. At the end of prophase I, how many chromosomes, chromatids, and centromeres are present to ensure that chromosome 1 faithfully traverses meiosis?

26. What is physically exchanged during crossing over?

27. Compare meiotic anaphase I with meiotic anaphase II. Which meiotic anaphase is most similar to the mitotic anaphase?

28. Provide two reasons why meiosis leads to genetic variation in diploid organisms.

Internet Activities

Internet Activities are critical-thinking exercises using the resources of the World Wide Web to enhance the principles and issues covered in this chapter. For a full set of links and questions investigating the topics described below, visit *www.cengage.com/biology/cummings*

1. *Structure and Function of the Nucleus.* The *Cell Biology Topics* website maintained by the University of Texas presents basic information about cell biology arranged by organelle system.
   a. Choose the "Nucleus" link and explore the numerous structures within the nucleus.
   b. Within the "Nucleus" topic, choose the "chromosome" link to compare heterochromatin and euchromatin, the two different forms of DNA in the nucleus.
2. *Diversity of Cell Types.* The cellular world is almost unimaginably diverse, and modern technology has not only permitted new ways of viewing this diversity, but it has also made it possible to share this information worldwide. At the *Molecular Expressions Photo Gallery,* check out any of the "Galleries" on the contents page to view a variety of cells, organisms, cellular structures, and (occasionally) everyday objects photographed using a variety of different photomicrographic techniques. For an overview of different types of cellular structure (with colorful line drawings but no photomicrographs), follow the "Cell and Virus Structure" Link.
3. *Mitosis Overview.* The "Mitosis" link at the *Molecular Expressions Photo Gallery* has both photomicrographs and an interactive tutorial for reviewing the phases of mitosis.
4. *Cell Size—and More Mitosis.* At the *Cells Alive!* website, follow the "Cell Biology" link and compare the sizes of different cells at the "How Big Is a…?" page.

HOW WOULD YOU VOTE NOW?

It is possible to treat Gaucher disease, a genetic disorder resulting from a missing enzyme, with bone marrow transplantation. Transplanted bone marrow allows a Gaucher patient to produce the missing enzyme and inhibits the formation of the abnormal Gaucher cells. But bone marrow donors are in short supply, and there are other life-threatening diseases that can be treated only with such a transplant. Now that you know more about cells, what do you think? Should candidates for transplants be prioritized according to their illnesses? Visit the *Human Heredity* companion website at *www.cengage.com/biology/cummings* to find out more on the issue; then cast your vote online.

3 Transmission of Genes from Generation to Generation

One Friday evening in July, Patricia Stallings fed her 3-month-old son Ryan his bottle and put him to bed. Ryan soon became ill and threw up, but the next day he seemed better. By Sunday, however, Ryan was vomiting and having trouble breathing. Patricia drove him to a hospital in St. Louis. Tests there uncovered high levels of ethylene glycol, a component of antifreeze, in his blood. A pediatrician at the hospital believed that Ryan had been poisoned and had the infant placed in foster care. Patricia and David, her husband, could see him only during supervised visits. On one of those visits early in September, Patricia was left alone with Ryan briefly, gave him a bottle, and went home. After she left, Ryan became ill and died. The next day, Patricia Stallings was arrested and charged with murder. Authorities found large quantities of ethylene glycol in Ryan's blood and traces of it on a bottle Patricia used to feed Ryan during her visit. At trial, Patricia was found guilty of first-degree murder and sentenced to life in prison.

While in jail, she gave birth to another son, David Jr., who immediately was placed in foster care. Within 2 weeks, the baby developed similar symptoms, but because Patricia had had no contact with her baby, she could not have poisoned him. Hearing of the case, two scientists at St. Louis University performed additional tests on blood samples taken from Ryan when he was hospitalized. They found no ethylene glycol in his blood and consulted with a human geneticist from Yale University, who conducted additional tests on Ryan's blood. His results also showed no traces of ethylene glycol, but he did find other compounds present, which helped solve the mystery. Based on those and further tests done on David Jr., the scientists presented evidence that previous testing had been done improperly and that both Ryan and his brother suffered from a rare genetic disorder called methylmalonic acidemia (MMA). Biochemical evidence from blood samples supported their conclusion. Symptoms of MMA are similar to those seen in ethylene glycol poisoning, but the cause is an inability to break down proteins in food. In light of that evidence, Patricia Stallings's conviction was overturned, and she was released from jail after serving 14 months for a crime she did not commit.

Transmission genetics began with the study of pea plants.

Image copyright Valentyn Volkov, 2010. Used under license from Shutterstock.com/

3.1 Heredity: How Are Traits Inherited?

Before we get to a discussion of how traits in humans such as eye color and hair color are passed from generation to generation, let's ask the obvious question: Why are we starting

with Gregor Mendel and pea plants if we are going to discuss human genetics? The answer won't be fully evident for a chapter or two, but there are two main reasons for starting with pea plants. First, Mendel used experimental genetics to uncover the fundamental principles of genetics—principles that apply to pea plants as well as to humans—and, for ethical reasons, humans can't be used in experimental genetics. Second, following the inheritance of traits in humans is difficult; we have very few offspring compared with pea plants, and there is a big difference in generation time (20 or so years in humans compared with about 100 days in peas). As you will see in Chapters 4 and 5, how traits are inherited in humans can be somewhat ambiguous. Thus, we begin with an experimental system in which the mechanisms of inheritance are clearly defined.

At a young age, Johann Mendel entered the Augustinian monastery at Brno for the purpose of continuing his studies in natural history (see Spotlight on Mendel and Test Anxiety). After completing his monastic studies, Mendel enrolled at the University of Vienna in the fall of 1851. There he encountered the new idea that cells are the fundamental unit of all living things. The cell theory raised several questions about inheritance. Does each parent contribute equally to the traits of the offspring? In plants and most animals, the female gametes are much larger than those of the male, so this was a logical and widely

Mendel and Test Anxiety

Mendel entered the Augustinian monastery in 1843 and took the name Gregor. While studying at the monastery, he served as a teacher at the local technical high school. In the summer of 1850, he decided to take the examinations that would allow him to have a permanent appointment as a teacher. The exam was in three parts. Mendel passed the first two parts but failed one of the sections in the third part. In the fall of 1851, he enrolled at the University of Vienna to study natural science (the section of the exam he flunked). He finished his studies in the fall of 1853, returned to the monastery, and again taught at a local high school.

In 1855, he applied to take the teacher's examination again. The test was held in May 1856, and Mendel became ill while answering the first question on the first essay examination. He left and never took another examination. As a schoolboy and again as a student at the monastery, Mendel had experienced bouts of illness, all associated with times of stress.

In an analysis of Mendel's illnesses made in the early 1960s, a physician concluded that Mendel had a psychological condition that today would probably be called "test anxiety." If you are feeling stressed at exam time, take some small measure of comfort in knowing that it was probably worse for Mendel.

debated question. Related to it was the question of whether the traits in the offspring result from blending of parental traits or if they are inherited as discrete units. In 1854, Mendel returned to Brno to teach physics and began a series of experiments that resolved those questions.

3.2 Mendel's Experimental Design Resolved Many Unanswered Questions

Mendel's success in discovering the fundamental principles of inheritance was the result of carefully planned experiments. His first step was selecting an organism for his experiments. Near the beginning of his landmark paper on inheritance, Mendel wrote:

> The value and validity of any experiment are determined by the suitability of the means as well as by the way they are applied. In the present case as well, it cannot be unimportant which plant species were chosen for the experiments and how these were carried out. Selection of the plant group for experiments of this kind must be made with the greatest possible care if one does not want to jeopardize all possibility of success from the very outset.

He then listed the properties that an experimental organism should have:
- It should have a number of different traits that can be studied.
- The plant should be self-fertilizing and have a flower structure that minimizes accidental pollination.
- Offspring of self-fertilized plants should be fully fertile so that further crosses can be made.

After evaluating several plant species, he selected pea plants for his work (Figure 3.1). Peas had many of the properties Mendel was seeking in an experimental organism:
- More than 30 varieties with different traits were available from seed dealers.
- The plants can be self-fertilized or artificially fertilized by hand, and the offspring are fully fertile.
- Peas have a relatively short life cycle and can be grown outside or in the greenhouse.

Mendel then collected and tested all available varieties of peas for two years to ensure that the traits they carried were true-breeding; that is, self-fertilization produced the same

FIGURE 3.1 The study of how traits such as flower color are passed from generation to generation provided the material for Mendel's studies.

(a)

(b)

James W. Richardson/Visuals Unlimited

R.Caleutine/Visuals Unlimited

Table 3.1 Traits Selected for Study by Mendel

Trait Studied	Dominant	Recessive
PEAS		
Shape	Smooth	Wrinkled
Color	Yellow	Green
Pea coat color	Gray	White
PODS		
Shape	Full	Constricted
Color	Green	Yellow
FLOWERS		
Position	Axial (along stems)	Terminal (top of stems)
STEMS		
Length	Tall	Short

traits in all the offspring, generation after generation. From those varieties he tested, he selected 22 to use in his experiments (Figure 3.2). Mendel studied seven traits that affected the peas, pods, flowers, and stems of the plant (Table 3.1). Each trait was represented by two variations: For example, plant height is the trait, tall and short are the variations; pea shape is the trait, and the variations are wrinkled and smooth peas; and so forth. In his experiments, he wanted to see how traits such as height or pea shape were passed from generation to generation.

To avoid errors caused by small sample sizes, he planned experiments on a large scale, using more than 28,000 pea plants in his experiments. He began by studying one pair of traits at a time and repeated his experiments for each trait, to confirm the results. Using his training in physics and mathematics, Mendel analyzed his data according to the principles of probability. His methodical and thorough approach to his work and his lack of preconceived notions were the secrets of his success.

3.3 Crossing Pea Plants: Mendel's Study of Single Traits

To show how Mendel developed his ideas about how traits are inherited, we will first describe his experiments and his results. Then we will follow his reasoning in drawing conclusions and outline some of the further experiments that confirmed his ideas.

In his first set of experiments, Mendel studied the inheritance of shape. Plants with smooth peas were crossed to plants with wrinkled peas. In making the cross, flowers from one variety were fertilized using pollen from the other variety. The peas that formed as a result of those crosses were all smooth. This was true whether the pollen came from a plant with smooth peas or a plant with wrinkled peas. Mendel planted the smooth peas from this cross; when the plants matured, the flowers were self-fertilized, and he collected 7,324 peas for analysis. The results showed that 5,474 peas were smooth and 1,850 were wrinkled. Using the terminology Mendel established, this experiment can be diagrammed as follows:

P1: Smooth × wrinkled
F1: All Smooth
F2: 5,474 Smooth and 1,850 wrinkled

Mendel called the parental generation P1; the offspring were called the F1 (first filial) generation. The second generation, produced by self-fertilizing the F1 plants, was called the F2 (or second filial) generation. His experiments with pea shape are diagrammed in Figure 3.3.

Malcolm Gutter/Visuals Unlimited

FIGURE 3.2 The monastery garden where Mendel carried out his experiments on plant genetics.

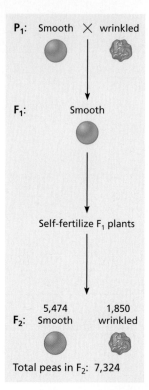

P₁: Smooth ✕ wrinkled

F₁: Smooth

Self-fertilize F₁ plants

F₂: 5,474 Smooth   1,850 wrinkled

Total peas in F₂: 7,324

FIGURE 3.3 A diagram showing one of Mendel's crosses. True-breeding strains of pea plants with smooth and wrinkled peas were used as the P₁ generation. All the offspring in the F₁ generation had smooth peas. Self-fertilization of F₁ plants gave rise to plants with smooth and wrinkled peas in the F₂ generation. About three-fourths of the peas were smooth, and about one-fourth were wrinkled.

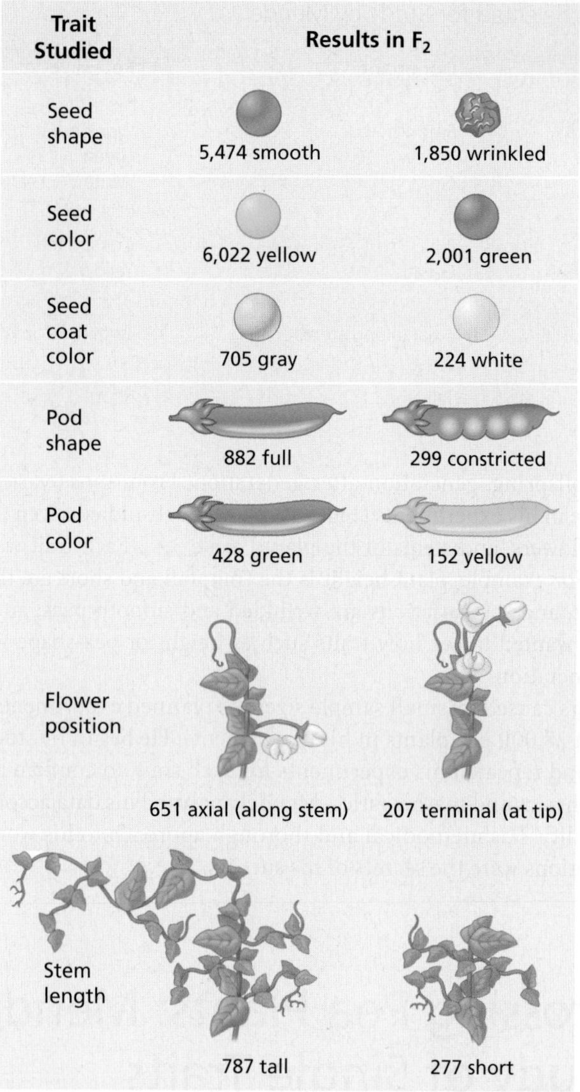

Trait Studied	Results in F₂	
Seed shape	5,474 smooth	1,850 wrinkled
Seed color	6,022 yellow	2,001 green
Seed coat color	705 gray	224 white
Pod shape	882 full	299 constricted
Pod color	428 green	152 yellow
Flower position	651 axial (along stem)	207 terminal (at tip)
Stem length	787 tall	277 short

FIGURE 3.4 Results of Mendel's crosses using pea plants. The numbers represent the F₂ plants showing a given trait. On average, three-fourths of the offspring showed one trait, and one-fourth showed the other trait (a 3:1 ratio). Mendel called crosses that involve a single trait monohybrid crosses.

What were the results and conclusions from Mendel's first series of crosses?

The results from experiments with all seven traits were the same as those seen with smooth and wrinkled peas (Figure 3.4). In all crosses, the following results were obtained:

- Only one of the parental traits was present in the F1 plants.
- The trait not present in the F1 plants reappeared in about 25% of the F2 plants.
- In all crosses, it did not matter which trait was present in the plant that contributed the pollen; the results were always the same.

The results of these experiments were Mendel's first discoveries. He concluded that traits were not blended as they passed from parent to offspring; they remained unchanged, even though they might not be expressed in a specific generation. This convinced him that inheritance did not work by blending the traits of the parents in the offspring. Instead, he concluded that traits were inherited as if they were separate units that did not blend together. In all his experiments, it did not matter whether the male or female plant in the P1 generation had smooth or wrinkled peas; the results were the same. From these results he concluded that each parent made an equal contribution to the genetic makeup of the offspring.

Based on the results of his crosses with each of the seven characteristics, Mendel came to several conclusions:

- **Genes** (Mendel called them factors) determine traits and can be present but not expressed. For example, if you cross plants with smooth peas to plants with wrinkled peas, all the F1 peas will be smooth. When these peas are planted and the mature plants are self-fertilized, the next generation of plants (the F2) will have some wrinkled peas. This means that the peas from the F1 plants carry a gene for wrinkled that was present but not expressed. Mendel called the trait expressed in F1 plants a **dominant trait**. He called the trait not expressed in the F1 but expressed in some F2 plants a **recessive trait**.

- Mendel concluded that despite their identical appearances, the P1 and F1 plants were genetically different. When P1 plants with smooth peas are self-fertilized, all the plants in the next generation have only smooth peas. But when F1 plants with smooth peas are self-fertilized, the F2 plants have both smooth and wrinkled peas. Mendel realized that it was important to make a distinction between the appearance of an organism and its genetic constitution. We now use the term **phenotype** to describe the appearance of an organism and **genotype** to describe the genetic makeup of an organism. In our example, the P1 and F1 plants with smooth peas have identical phenotypes but different genotypes.

- The results of self-fertilization experiments show that the F1 plants must carry genes for smooth and wrinkled traits because both types of peas are present in the F2 generation. The question is: How many genes for shape are carried in the F1 plants? Mendel had already reasoned that the male parent and female parent contributed equally to the traits of the offspring. The simplest explanation is that each F1 plant carried two genes for pea shape: one for smooth that was expressed and one for wrinkled that was unexpressed (see Exploring Genetics: Ockham's Razor).

- If each F1 plant carries two genes for shape, then each P1 and F2 plant must also contain two genes for shape. To symbolize genes, uppercase letters are used to represent forms of a gene with a dominant pattern of inheritance, and lowercase letters are used to represent those with a recessive pattern of inheritance (S = smooth, s = wrinkled). Using this shorthand, we can reconstruct the genotypes and phenotypes of the P1 and F1, as shown in Figure 3.5.

Gene The fundamental unit of heredity and the basic structural and functional unit of genetics.

Dominant trait The trait expressed in the F1 (or heterozygous) condition.

Recessive trait The trait unexpressed in the F1 but re-expressed in some members of the F2 generation.

Phenotype The observable properties of an organism.

Genotype The specific genetic constitution of an organism.

Segregation The separation of members of a gene pair from each other during gamete formation.

The principle of segregation describes how a single trait is inherited.

If genes exist in pairs, there must be some way to prevent gene number from doubling in each succeeding generation. (If each parent has two genes for a given trait, why doesn't the offspring have four?) Mendel reasoned that members of a gene pair must separate or segregate from each other during gamete formation. As a result, each gamete receives only one of the two genes that control a particular trait. The separation of members of a gene pair during gamete formation is called the principle of **segregation**, or Mendel's First Law.

Active Figure 3.6 diagrams the separation of a gene pair, demonstrating that only one member of that pair is included in each gamete. In our example, each member of the F1 generation can make two kinds of gametes in equal proportions (S gametes and s gametes). At fertilization, the random combination of these gametes produces the genotypic combinations shown in the Punnett square

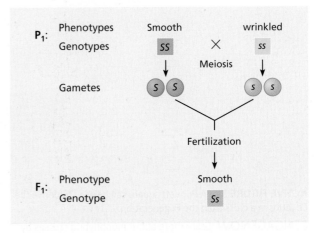

FIGURE 3.5 The phenotypes and genotypes of the parents (P1) and the offspring (F1) in a cross involving seed shape.

Ockham's Razor

When Mendel proposed the simplest explanation for the number of factors contained in the F1 plants in his monohybrid crosses, he was using a principle of scientific reasoning known as parsimony, or Ockham's razor.

William of Ockham (also spelled Occam) was a Franciscan monk and scholastic philosopher who lived from about 1300 to 1349. He had a strong interest in the study of thought processes and logical methods. He is the author of the maxim known as Ockham's razor: *Entia non sunt multiplicanda praeter necessitatem* which translates from the Latin as "Entities must not be multiplied without necessity." This was taken to mean that when constructing an argument, you should use the smallest number of steps possible. In other words, never go beyond the simplest argument. Although Ockham was not the first to use this approach, he employed this tool of logic so well and so often to dissect the arguments of his opponents that it became known as Ockham's razor.

The principle was transferred from philosophy to science in the fifteenth century. Galileo used the principle of parsimony to argue that because his model of the solar system was the simplest, it was probably correct (he was right). In modern terms, the phrase is used as a rule of thumb to mean that in proposing a mechanism or hypothesis, we should use the smallest number of steps possible. The simplest mechanism is not necessarily correct, but it is usually the easiest to disprove by doing experiments and the most likely to produce scientific progress.

For a given trait, Mendel concluded that both parents contribute an equal number of genes to the offspring. In this case, the simplest assumption is that each parent contributed one gene and that the F1 offspring contained two copies of that gene. Further experiments proved this conclusion correct.

(a method for analyzing genetic crosses devised by R. C. Punnett). The F2 has a genotypic ratio of 1 *SS*:2 *Ss*:1 *ss* and a phenotypic ratio of 3 smooth : 1 wrinkled (dominant to recessive).

Mendel's reasoning allows us to predict the genotypes of the F2 generation. One-fourth of the F2 plants should carry only genes for smooth peas (*SS*), and, when self-fertilized, all the offspring will have smooth peas. Half (two-fourths) of the F2 plants should carry genes for both smooth and wrinkled (*Ss*) and give rise to plants with smooth and wrinkled peas in a 3:1 ratio when self-fertilized (Figure 3.7). Finally, one-fourth of the F2 plants should carry only genes for wrinkled (*ss*) and have all wrinkled progeny if self-fertilized. In fact, Mendel self-fertilized plants from the F2 generation and five succeeding generations to confirm these predictions.

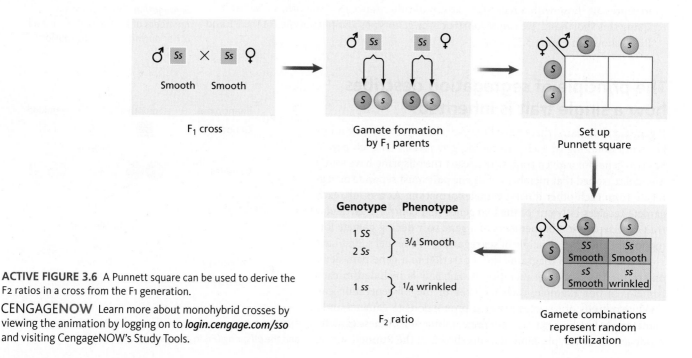

ACTIVE FIGURE 3.6 A Punnett square can be used to derive the F2 ratios in a cross from the F1 generation.

CENGAGENOW Learn more about monohybrid crosses by viewing the animation by logging on to *login.cengage.com/sso* and visiting CengageNOW's Study Tools.

Mendel carried out his experiments before the discovery of mitosis and meiosis and before the discovery of chromosomes. As we discuss in a later section, his conclusions about how traits are inherited are, in fact, indirect descriptions of the way in which chromosomes behave in meiosis. Seen in this light, his discoveries are all the more remarkable.

Today, we call Mendel's factors *genes* and refer to the alternative forms of genes as **alleles**. In the example we have been discussing, the gene for pea shape (S) has two alleles: smooth (S) and wrinkled (s). Individuals that carry identical alleles of a given gene (SS or ss) are **homozygous** for the gene in question. Similarly, when two different alleles are present in a gene pair (Ss), the individual has a **heterozygous** genotype. The SS homozygotes and the Ss heterozygotes show the dominant smooth phenotype (because S is dominant to s), and ss homozygotes show the recessive wrinkled phenotype.

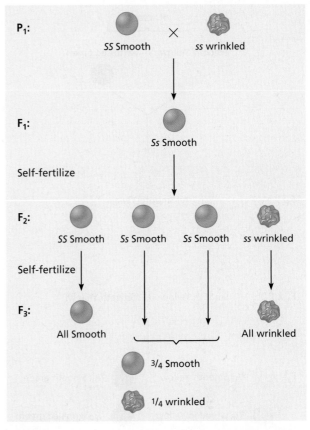

FIGURE 3.7 Self-crossing F2 plants to produce an F3 generation shows that there are two different genotypes among the plants in the F2 generation.

3.4

3.4 More Crosses with Pea Plants: The Principle of Independent Assortment

Mendel realized the need to extend his studies on inheritance from crosses involving one trait to more complex situations. Following his work on the inheritance of single traits, he wrote:

> In the experiments discussed above, plants were used which differed in only one essential trait. The next task consisted of investigating whether the law of development thus found would also apply to a pair of differing traits.

For these experiments, he selected pea shape and color as traits to be studied, because, as he put it, "Experiments with seed traits lead most easily and assuredly to success."

Mendel performed crosses involving two traits.

As before, we will analyze the actual experiments of Mendel, outline his results, and summarize the conclusions he drew from them. From previous crosses, Mendel knew that for peas, smooth is dominant to wrinkled and yellow is dominant to green. In our reconstruction of these experiments, we will use the following symbols: S for smooth, s for wrinkled, Y for yellow, and y for green. Mendel selected true-breeding plants with smooth yellow peas and crossed them with true-breeding plants with wrinkled green peas (Figure 3.8). The F1 plants had all smooth yellow peas.

Analyzing the results and drawing conclusions.

The F1 plants were crossed, producing an F2 generation with four phenotypic combinations. Mendel counted a total of 556 peas with these phenotypes:

315 smooth and yellow
108 smooth and green
101 wrinkled and yellow
 32 wrinkled and green

The F2 included the parental phenotypes (smooth yellow, wrinkled green) and two new phenotypes (smooth green and wrinkled yellow). These four phenotypic classes occurred in a 9:3:3:1 ratio.

Allele One of the possible alternative forms of a gene, usually distinguished from other alleles by its phenotypic effects.

Homozygous Having identical alleles for one or more genes.

Heterozygous Carrying two different alleles for one or more genes.

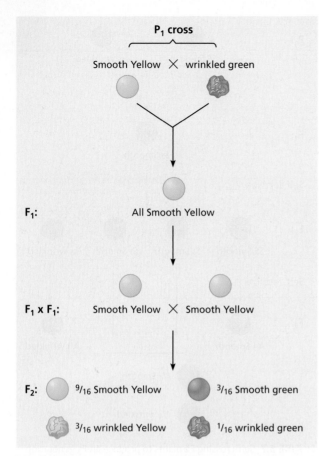

P₁ cross

Smooth Yellow ✕ wrinkled green

F₁: All Smooth Yellow

F₁ x F₁: Smooth Yellow ✕ Smooth Yellow

F₂: ⁹/₁₆ Smooth Yellow   ³/₁₆ Smooth green

³/₁₆ wrinkled Yellow   ¹/₁₆ wrinkled green

FIGURE 3.8 The phenotypic distribution in a cross with two traits: seed color and seed shape. Plants in the F2 generation show the parental phenotypes *and* two new phenotype combinations. Mendel called crosses involving two traits dihybrid crosses.

To determine how the two genes in this cross were inherited, Mendel first analyzed the F2 results for each trait separately, as if the other trait were not present (Figure 3.9). If we look only at shape (smooth or wrinkled) and ignore color, we expect three-fourths smooth and one-fourth wrinkled peas in the F2. Analyzing the results, we find that the total number of smooth peas is 315 + 108 = 423. The total number of wrinkled peas is 101 + 32 = 133. The proportion of smooth to wrinkled peas (423:133) is very close to a 3:1 ratio. Similarly, if we consider only color (yellow or green), there are 416 yellow peas (315 + 101) and 140 green peas (108 + 32) in the F2 generation. These results are also close to a 3:1 ratio.

Once he established a 3:1 ratio for each trait separately (consistent with the principle of segregation), then, using probability, Mendel considered the inheritance of both traits simultaneously. By combining the individual probabilities (¾ of the peas are smooth, and ¾ of the peas are yellow), then ⁹/₁₆ of the peas should be smooth *and* yellow—which turns out to be true. By doing this for all combinations of traits, the phenotypic ratio in the F2 generation is 9:3:3:1 (Figure 3.9).

The principle of independent assortment explains the inheritance of two traits.

Before we discuss what is meant by independent assortment, let's see how the phenotypes and genotypes of the F1 and F2 were generated. The F1 plants with smooth yellow peas were heterozygous for both shape and color. The genotype of the F1 plants was *SsYy*, with *S* and *Y* alleles dominant to *s* and *y*. Mendel already had concluded that members of a gene pair separate or segregate from each other during gamete formation.

During meiosis in the F1 plants, the *S* and *s* alleles went into gametes independently of the *Y* and *y* alleles (Active Figure 3.10). Because each gene pair segregated independently, the F1 plants produced gametes with all combinations of those alleles in equal proportions: *SY*, *Sy*, *sY*, and *sy*. If fertilizations occur at random (as expected), 16 combinations result (Active Figure 3.10).

The Punnett square shows the following:

- Nine combinations have at least one copy of *S* and *Y*.
- Three combinations have at least one copy of *S* and are homozygous for *yy*.
- Three combinations have at least one copy of *Y* and are homozygous for *ss*.
- One combination is homozygous for *ss* and *yy*.

FIGURE 3.9 Analysis of a dihybrid cross involving two traits. Each trait is analyzed separately, then the frequencies of each are combined to yield the observed phenotypic ratios, confirming that the genes for these traits assort independently.

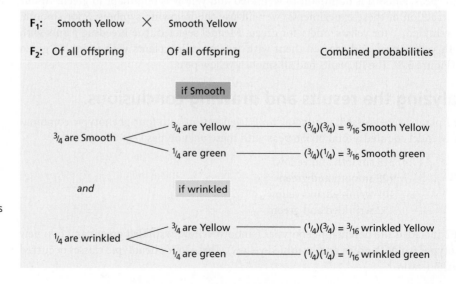

F₁: Smooth Yellow   ✕   Smooth Yellow

F₂: Of all offspring   Of all offspring   Combined probabilities

if Smooth

¾ are Smooth

³/₄ are Yellow ——— (³/₄)(³/₄) = ⁹/₁₆ Smooth Yellow

¹/₄ are green ——— (³/₄)(¹/₄) = ³/₁₆ Smooth green

and

if wrinkled

¼ are wrinkled

³/₄ are Yellow ——— (¹/₄)(³/₄) = ³/₁₆ wrinkled Yellow

¹/₄ are green ——— (¹/₄)(¹/₄) = ¹/₁₆ wrinkled green

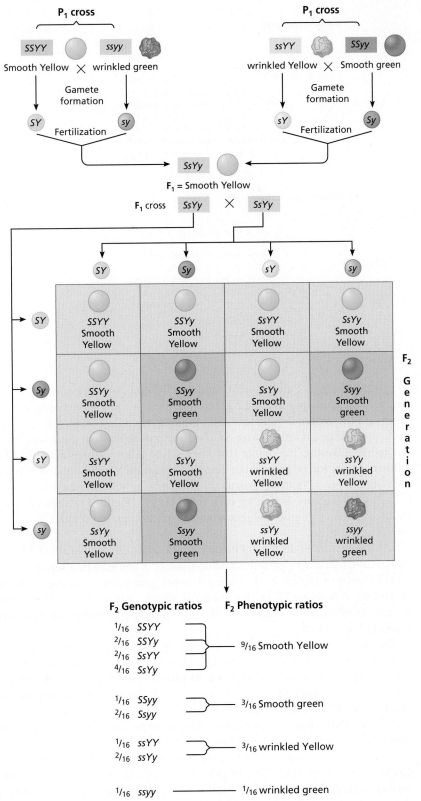

ACTIVE FIGURE 3.10 Punnett square of the dihybrid cross shown in Figure 3.8. There are two combinations of dominant and recessive traits that can result in double heterozygotes in the F1 plants. One (*left*) involves plants with smooth yellow peas crossed to plants with wrinkled green peas. The other (*right*) is a cross between plants with wrinkled yellow peas and plants with smooth green peas.

CENGAGENOW Learn more about dihybrid crosses by viewing the animation by logging on to *login.cengage.com/sso* and visiting CengageNOW's Study Tools.

P₁ cross

SSYY ○ Smooth Yellow × ssyy ● wrinkled green

Gamete formation

SY sy

Fertilization

P₁ cross

ssYY ● wrinkled Yellow × SSyy ● Smooth green

Gamete formation

sY Sy

Fertilization

SsYy ○
F₁ = Smooth Yellow

F₁ cross SsYy × SsYy

SY Sy sY sy

F₂ Generation

	SY	Sy	sY	sy
SY	SSYY Smooth Yellow	SSYy Smooth Yellow	SsYY Smooth Yellow	SsYy Smooth Yellow
Sy	SSYy Smooth Yellow	SSyy Smooth green	SsYy Smooth Yellow	Ssyy Smooth green
sY	SsYY Smooth Yellow	SsYy Smooth Yellow	ssYY wrinkled Yellow	ssYy wrinkled Yellow
sy	SsYy Smooth Yellow	Ssyy Smooth green	ssYy wrinkled Yellow	ssyy wrinkled green

F₂ Genotypic ratios

1/16 SSYY
2/16 SSYy
2/16 SsYY
4/16 SsYy

F₂ Phenotypic ratios

9/16 Smooth Yellow

1/16 SSyy
2/16 Ssyy

3/16 Smooth green

1/16 ssYY
2/16 ssYy

3/16 wrinkled Yellow

1/16 ssyy —— 1/16 wrinkled green

In other words, the 16 combinations of genotypes fall into 4 phenotypic classes in a 9:3:3:1 ratio:

9 smooth and yellow (*S-Y-*)
3 smooth and green (*S-yy*)
3 wrinkled and yellow (*ssY-*)
1 wrinkled and green (*ssyy*)

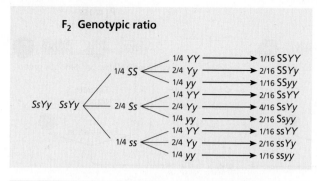

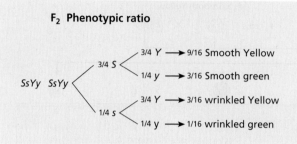

FIGURE 3.11 The phenotypic and genotypic ratios of a dihybrid cross can be derived using a branched-line method instead of a Punnett square.

Instead of using a Punnett square to determine the distribution and frequency of phenotypes and genotypes in the F2 generation, we can use a branch diagram (also called the forked-line method) that is based on probability. In the F2 generation, the probability that a pea will be smooth is three-fourths. The probability that a pea will be wrinkled is one-fourth. Likewise, the chance that a pea will be yellow is three-fourths and the probability that a pea will be green is one-fourth. Because each trait is inherited independently, each smooth pea has a three-quarters chance of being yellow and a one-fourth chance of being green. The same is true for each wrinkled pea. Figure 3.11 shows how these probabilities combine to give the genotypic and phenotypic ratio characteristic of a cross involving two traits.

The results of Mendel's cross involving two traits can be explained by assuming (as Mendel did) that during gamete formation, alleles of one gene pair segregate into gametes independently of the alleles belonging to other gene pairs, resulting in the production of gametes containing all combinations of alleles. This second fundamental principle of genetics outlined by Mendel is called the principle of **independent assortment**, or Mendel's Second Law.

The results of this cross raise an interesting question: How can we be sure that the number of offspring in each phenotypic class is close enough to what we expect? For example, if we do a cross and expect a 3:1 phenotypic ratio in the offspring, finding 75 plants with the dominant phenotype and 25 with the recessive phenotype in every 100 offspring would be ideal. What happens if 80 offspring have the dominant phenotype and 20 have the recessive phenotype, or what if the results are 65 dominant and 35 recessive? Is this close enough to a 3:1 ratio, or is our expectation wrong?

To determine whether the observed results of an experiment meet expectations, geneticists use a statistical test; in this case, a test called the chi-square test would be used to evaluate how closely the results of the cross fit our expectations (see Exploring Genetics: Evaluating Results: The Chi-Square Test on page 56).

After 10 years of work, Mendel presented his results in 1865 at the meeting of the local Natural Science Society and published his paper the following year in the *Proceedings* of the society. Although copies of the journal were widely circulated, the significance of Mendel's findings was not appreciated. Finally, in 1900, three scientists independently confirmed Mendel's work and brought his paper to widespread attention. This stimulated great interest in the branch of biology called **genetics**. Unfortunately, Mendel died in 1884—unaware that he had founded an entire scientific discipline.

Independent assortment The random distribution of alleles into gametes during meiosis.

Genetics The scientific study of heredity.

3.5 Meiosis Explains Mendel's Results: Genes Are on Chromosomes

When Mendel was working with pea plants, the behavior of chromosomes in mitosis and meiosis was unknown. By 1900, however, the details of mitosis and meiosis had been described. As scientists confirmed that Mendelian inheritance operated in many organisms, it became obvious that genes and chromosomes had much in common (Table 3.2). Both chromosomes and genes occur in pairs. In meiosis, members of a chromosome pair separate from each other, and members of a gene pair separate from each other during gamete formation (Active Figure 3.12). Finally, the fusion of gametes during fertilization

Table 3.2 Genes, Chromosomes, and Meiosis

Genes	Chromosomes
Occur in pairs (alleles)	Occur in pairs (homologues)
Members of a gene pair separate from each other during meiosis	Members of a homologous pair separate from each other during meiosis
Members of one gene pair independently assort from other gene pairs during meiosis	Members of one chromosome pair independently assort from other chromosome pairs during meiosis

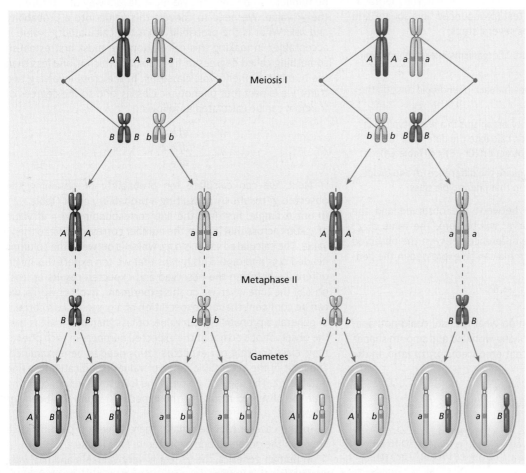

ACTIVE FIGURE 3.12 Mendel's observations about segregation and independent assortment are explained by the behavior of chromosomes during meiosis. The random arrangement of maternal and paternal chromosome pairs at metaphase I produces four combinations of the two genes in gametes.

CENGAGENOW Learn more about independent assortment by viewing the animation by logging on to *login.cengage.com/sso* and visiting CengageNOW's Study Tools.

Evaluating Results: The Chi-Square Test

One of Mendel's innovations was the application of mathematics and combinatorial theory to biological research. That allowed him to predict the genotypic and phenotypic ratios in his crosses and follow the inheritance of several traits simultaneously. If the cross involved two alleles of a gene (e.g., A and a), the expected outcome was an F2 phenotypic ratio of 3 dominant: 1 recessive and a genotypic ratio of 1 AA:2 Aa:1 aa.

What Mendel was unable to analyze mathematically was how well the observed outcome of the cross fulfilled his predictions. He apparently recognized this problem and compensated for it by conducting his experiments on a large scale, counting substantial numbers of offspring in each experiment to reduce the chance for error. Shortly after the turn of the twentieth century, an English scientist named Karl Pearson developed a statistical test to determine whether the observed distribution of individuals in phenotypic categories is as predicted or occurs by chance. This simple test, regarded as one of the fundamental advances in statistics, is a valuable tool in genetic research. The method is known as the chi-square (χ^2) test (pronounced "kye square"). In practical use, this test requires several steps:

1. Record the observed numbers of organisms in each phenotypic class.

2. Calculate the expected values for each phenotypic class on the basis of the predicted ratios.

3. If O is the observed number of organisms in a phenotypic class and E is the expected number, calculate the difference (d) in each phenotypic class by subtraction ($O - E$) = d (Table 3.3).

4. For each phenotypic class, square the difference (d) and divide by the expected number (E) in that phenotypic class.

If there is no difference between the observed and the expected ratios, the value for χ^2 will be zero. The value of χ^2 increases with the size of the difference between the observed and expected classes. The formula can be expressed in the general form:

$$\chi^2 = \Sigma \frac{d^2}{E}$$

Using this formula, we can do what Mendel could not: analyze his data for the cross involving wrinkled and smooth shapes and yellow and green colors that produced a 9:3:3:1 ratio. In the

F2, Mendel counted a total of 556 peas. The number in each phenotypic class is the observed number (Table 3.3). Using the total of 556 peas, we can calculate that the expected number in each class for a 9:3:3:1 ratio would be 313:104:104:35 (9/16 of 556 is 313, 3/16 of 556 is 104, and so on). Substituting these numbers into the formula, we arrive at:

$$\chi^2 = \frac{2^2}{313} + \frac{4^2}{104} + \frac{3^2}{104} + \frac{3^2}{35} = 0.371$$

This χ^2 value is very low, confirming that there is very little difference between the number of peas observed and the number expected in each class. In other words, the results are close enough to the expectation that we need not reject them as occurring by chance alone.

The question remains, however: How much deviation from the expected numbers is permitted before we decide that the observations do not fit our expectation that a 9:3:3:1 ratio will be fulfilled? To decide this, we must have a way of interpreting the χ^2 value. We need to convert this value into a probability and ask: What is the probability that the calculated χ^2 value is acceptable? In making this calculation, we must first establish something called degrees of freedom, df, which is one less than the number of phenotypic classes, n. In the cross involving two traits, we expect four phenotypic classes, and so the degrees of freedom can be calculated as follows:

$$df = n-1$$
$$df = 4-1$$
$$df = 3$$

Next, we can calculate the probability of obtaining the observed χ^2 results by consulting a probability chart (Table 3.4). In our example, first find the line corresponding to a df value of 3. Look across that line for the number corresponding to the χ^2 value. The calculated value is 0.37, which is between the columns headed 0.95 and 0.90. This means that we can expect this much difference between the observed and expected results at least 90% of the time when we do this experiment. In other words, we can be confident that our expectation of a 9:3:3:1 ratio is correct. In general, a probability, or p value, of less than 0.05 means that the observations do not fit the expected numbers in each phenotypic class and that our expected ratios need to be reexamined. The limit of the acceptable range of values is indicated by a line in Table 3.4. The use of $p = 0.05$ as the lowest value for accepting that the observed results fit the expected results has been set arbitrarily.

In the case of Mendel's data, there is very little difference between the observed and expected results (Table 3.3).

In human genetics, the χ^2 test is very valuable and has wide application. It is used to determine how a trait is inherited (autosomal or sex-linked), whether the pattern of inheritance shown by two genes indicates that they are on the same chromosome, and whether marriage patterns have produced genetically divergent groups in a population.

Table 3.3 Chi-Square Analysis of Mendel's Data

Speed Shape	Seed Color	Observed Numbers	Expected Numbers (based on a 9:3:3:1 ratio)	Difference (d) (O–E)
Smooth	Yellow	315	313	+2
Smooth	Green	108	104	+4
Wrinkled	Yellow	101	104	−3
Wrinkled	Green	32	35	−3

Table 3.4 Probability Values for Chi-Square Analysis

df	0.95	0.90	0.70	0.50	0.30	0.20	0.10	0.05	0.01
1	0.004	0.016	0.15	0.46	1.07	1.64	2.71	3.84	6.64
2	0.10	0.21	0.71	1.39	2.41	3.22	4.61	5.99	9.21
3	0.35	0.58	1.42	2.37	3.67	4.64	6.25	7.82	11.35
4	0.71	1.06	2.20	3.36	4.88	5.99	7.78	9.49	13.28
5	1.15	1.61	3.00	4.35	6.06	7.29	9.24	11.07	15.09
6	1.64	2.20	3.83	5.35	7.23	8.56	10.65	12.59	16.81
7	2.17	2.83	4.67	6.35	8.38	9.80	12.02	14.07	18.48
8	2.73	3.49	5.53	7.34	9.52	11.03	13.36	15.51	20.09
9	3.33	4.17	6.39	8.34	10.66	12.24	14.68	16.92	21.67
10	3.94	4.87	7.27	9.34	11.78	13.44	15.99	18.31	23.21

← ———————————————— Acceptable ————————————————→ Unacceptable

Note: From *Statistical Tables for Biological, Agricultural and Medical Research* (6th ed.), Table IV, by R. Fisher and F. Yates, Edinburgh: Longman Essex, 1963.

restores the diploid number of chromosomes and two copies of each gene to the zygote, producing the genotypes of the next generation.

In 1903, Walter Sutton and Theodore Boveri independently proposed the idea that because genes and chromosomes behave in similar ways, genes must be located on chromosomes. This idea, the chromosome theory of inheritance, has been confirmed in many different ways and is one of the foundations of modern genetics. Each gene is located at a specific site—called a **locus**—on a chromosome, and each chromosome carries many genes. In humans, it is estimated that about 20,000 to 25,000 genes are carried on the 24 different chromosomes (22 autosomes, the X, and the Y).

Locus The position occupied by a gene on a chromosome.

3.6 Mendelian Inheritance in Humans

Now that we know how segregation and independent assortment work in pea plants, let's turn our attention to humans. After Mendel's work was rediscovered, some scientists believed that inheritance of traits in humans might not work the same way as it did in plants and other animals. However, the first trait (a hand deformity called brachydactyly; OMIM 112500) analyzed in humans (in 1905) was found to follow the rules of Mendelian inheritance, and so have all the 5,000-plus traits described since then.

Segregation and independent assortment occur with human traits.

To illustrate that segregation and independent assortment apply to human traits, let's follow the inheritance of a recessive trait called albinism (OMIM 203100). Homozygotes (*aa*) cannot make a skin pigment called melanin. Melanin is the principal pigment in skin, hair, and eye color. Albinos cannot make melanin and as a result have very pale white skin, white hair, and colorless eyes (Figure 3.13). Anyone carrying at least one dominant allele (*A*) can make enough pigment to have colored skin, hair, and eye color.

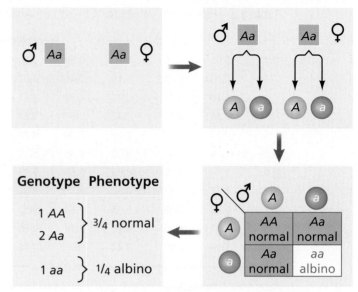

FIGURE 3.14 The segregation of albinism, a recessively inherited trait in humans. As in pea plants, alleles of a human gene pair separate from each other during gamete formation.

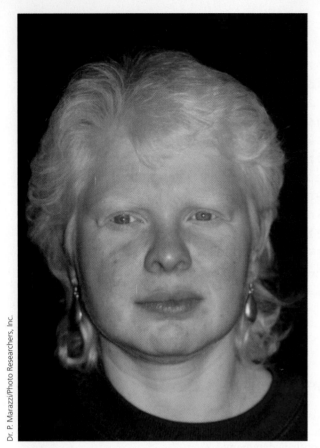

FIGURE 3.13 Albinos lack pigment in the skin, hair, and eyes.

To apply Mendelian inheritance to humans, let's diagram the genotypes and phenotypes for heterozygous (*Aa*) parents with normal pigmentation (Figure 3.14). During meiosis, the dominant and recessive alleles for this trait separate from each other and end up in different gametes. Because each parent can produce two different types of gametes (one with *A* and another with *a*), there are four possible combinations of gametes at fertilization. If they have enough children (say, 20 or 30), we will see something close to the predicted phenotypic ratio of 3 pigmented:1 albino offspring and a genotypic ratio of 1*AA*:2*Aa*:1*aa* (Figure 3.14). In other words, segregation of alleles during gamete formation produces the same outcome in both pea plants and humans. Important: This does not mean that there will be one albino child and three normally pigmented children in every such family with four children. It does mean that if the parents are heterozygotes, each child has a 25% chance of being albino and a 75% chance of having normal pigmentation.

The inheritance of two traits in humans also follows the Mendelian principle of independent assortment (Figure 3.15). To illustrate, let's examine a family in which the parents are each heterozygous for albinism (*Aa*) and heterozygous for hereditary deafness (OMIM 220290), another recessive trait. Homozygous dominant (*DD*) or heterozygous individuals (*Dd*) can hear, but homozygous recessive (*dd*) individuals are deaf. During meiosis, alleles for skin color and alleles for hearing assort into gametes independently. As a result, each parent produces equal proportions of 4 different gametes (*AD*, *Ad*, *aD*, and *ad*). There are 16 possible combinations of gametes at fertilization (4 types of gametes in all possible combinations), resulting in 4 different phenotypic classes. An examination of the possible genotypes shows that there is a 1 in 16 chance that a child will be both deaf and an albino.

In pea plants and other experimental organisms, genetic analysis is done using crosses that start with predetermined genotypes. In humans, experimental crosses are not possible, and geneticists must often infer genotypes from the pattern of inheritance observed in a family. In human genetics, the study of a trait begins with a family history, as outlined in the following section.

KEEP IN MIND

We can identify genetic traits because they have a predictable pattern of inheritance worked out by Gregor Mendel.

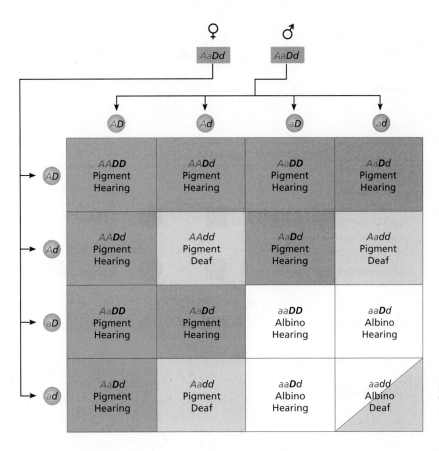

FIGURE 3.15 Independent assortment for two traits in humans follows the same pattern of inheritance as traits in pea plants.

Pedigree construction is an important tool in human genetics.

The fundamental method of genetic analysis in humans begins with the collection of a family history to follow the inheritance of a trait. This method is called **pedigree construction**. A **pedigree** results in the presentation of family information in the form of an easily readable chart. With a pedigree, the inheritance of a trait can be followed through several generations. Analysis of the pedigree using the principles of Mendelian inheritance can determine whether a trait has a dominant or recessive pattern of inheritance.

Pedigree construction Use of family history to determine how a trait is inherited and to estimate risk factors for family members.

Pedigree A diagram listing the members and ancestral relationships in a family; used in the study of human heredity.

Pedigrees use a standardized set of symbols, some of which are shown in Figure 3.16. In pedigrees, squares represent males and circles represent females. Someone with the phenotype in question is represented by a filled-in (darker) symbol. Heterozygotes, when identifiable, are indicated by a shaded dot inside a symbol or a half-filled symbol. Relationships between individuals in a pedigree are shown as a series of lines. Parents are connected by a horizontal line, and a vertical line leads to their offspring. If the parents are closely related (such as first cousins), they are connected by a double line. The offspring are connected by a horizontal sibship line and listed in birth order from left to right along the sibship line:

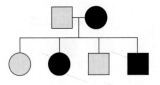

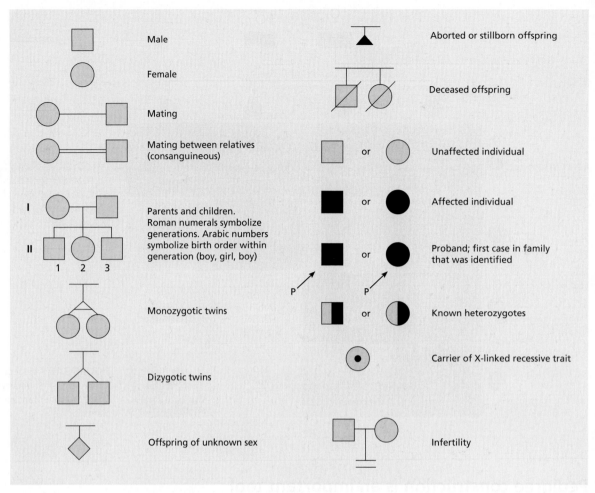

FIGURE 3.16 Some of the symbols used in pedigree construction.

A numbering system is used in pedigree construction. Each generation is identified by a Roman numeral (I, II, III, and so on), and each individual within a generation is identified by an Arabic number (1, 2, 3, and so on):

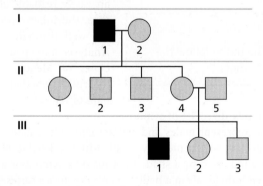

Proband First affected family member who seeks medical attention for a genetic disorder.

Pedigrees are often constructed after a family member afflicted with a genetic disorder has been identified. This individual, known as the **proband**, is indicated on the pedigree by an arrow and the letter P:

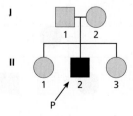

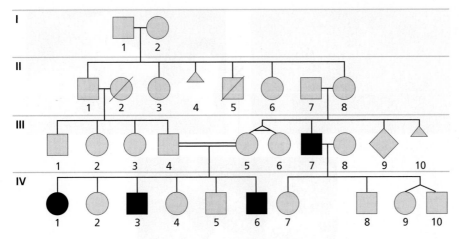

ACTIVE FIGURE 3.17 A pedigree showing the inheritance of a trait through several generations of a family.
CENGAGENOW Learn more about pedigree analysis by viewing the animation by logging on to *login.cengage .com/sso* and visiting CengageNOW's Study Tools.

Because pedigree construction is a family history, details about earlier generations may be uncertain as memories fade. If the sex of a person is unknown, a diamond is used. If there is doubt that a family member had the trait in question, that is indicated by a question mark above the symbol.

A pedigree is a form of symbolic communication used by clinicians and researchers in human genetics (Active Figure 3.17). It contains information that can help establish how a trait is inherited and identify those at risk of developing or transmitting the trait; it is also a resource for establishing biological relationships within a family. In Chapter 4, we will see how pedigree analysis is used to establish the genotypes of individuals and predict the chances of having children affected with a genetic disorder.

3.7 Variations on a Theme by Mendel

After Mendel's work became widely known, geneticists turned up cases in which the F1 phenotypes were not identical to one of the parents. In some cases, the offspring had a phenotype intermediate to that of the parents or a phenotype in which the traits of both parents were expressed. This led to a debate about whether such cases could be explained by Mendelian inheritance or whether there might be another, separate mechanism of inheritance that did not follow the laws of segregation and independent assortment.

Eventually, research showed that although phenotypes can be somewhat complex, these cases were not exceptions to Mendelian inheritance at the level of genotypes. In this section, we will discuss some of these cases and show that although phenotypes may not follow predicted ratios, genotypes do obey the principles of Mendelian inheritance.

Incomplete dominance has a distinctive phenotype in heterozygotes.

One case in which phenotypes do not follow the predicted ratios for a Mendelian trait is the inheritance of flower color in snapdragons. If snapdragons with red flowers (Figure 3.18) are crossed with plants carrying white flowers, the F1 will have pink flowers. In this case, the F1 phenotype is intermediate to the parental phenotypes, and neither the red nor the white color is dominant. This condition is called **incomplete dominance**. In snapdragons, flower color is controlled by a single gene, with two alleles. Because neither allele is recessive, we will call the alleles R^1 (red) and R^2 (white). The cross between red and white flowers is as follows:

Incomplete dominance Expression of a phenotype that is intermediate to those of the parents.

$$P1: \quad R^1R^1 \text{ (red)} \times R^2R^2 \text{ (white)}$$
$$\downarrow$$
$$F1: \quad R^1R^2 \text{ (pink)}$$

FIGURE 3.18 Incomplete dominance in snapdragon flower color. Red-flowered snapdragons crossed with white-flowered snapdragons produce offspring with pink flowers in the F1 generation. In heterozygotes, the allele for red flowers is incompletely dominant over the allele for white flowers.

Given the genotype of the F1, we can predict the outcome of an F1 × F1 cross:

F1 × F1: R^1R^2 (pink) × R^1R^2 (pink)
↓

F2: ¼ R^1R^1 (red) : ½ R^1R^2 (pink) : ¼ R^2R^2 (white)

Each genotype in this cross has a distinct phenotype (R^1R^1 is red, R^1R^2 is pink, and R^2R^2 is white), and the phenotypic ratio of 1 red:2 pink:1 white is the same as the expected genotypic ratio of $1R^1R^1:2R^1R^2:1R^2R^2$. To explain this outcome, let's assume that the R^1 allele encodes a gene product that synthesizes red pigment and that the gene product encoded by the R^2 allele cannot make red pigment. Let's also assume that each copy of the R^1 allele makes one unit of red pigment. In homozygotes (R^1R^1), two units of pigment are produced and the flower is red. Heterozygotes (R^1R^2) produce one unit of red pigment, and the result is pink flowers. The R^2 allele produces no pigment, so homozygous R^2R^2 plants have white flowers.

Easily observable examples of incomplete dominance in humans are rare, but a close examination of the phenotype, often at the cellular or molecular level, can reveal an incompletely dominant situation. Sickle cell anemia, a disorder we will discuss in Chapter 4, is one such condition.

Codominant alleles are fully expressed in heterozygotes.

In some cases, both alleles in a heterozygote are fully expressed. This situation is called **codominance**. In humans, the MN blood group is an example of this phenomenon. The MN blood group is controlled by a single gene, L, which directs the synthesis of a glycoprotein, found on the surface of red blood cells and other cells of the body. This gene has two alleles: L^M and L^N. Each allele directs the synthesis of a different form of glycoprotein. Depending on genotype, an individual may carry the M glycoprotein, the N glycoprotein, or both the M and N glycoproteins:

Genotype	Blood Type (Phenotype)
L^ML^M	M
L^ML^N	MN
L^NL^N	N

Codominance Full phenotypic expression of both members of a gene pair in the heterozygous condition.

This means that heterozygous parents may produce children with all three blood types:

$$L^M L^N \times L^M L^N$$
$$\downarrow$$
$$\tfrac{1}{4} L^M L^M : \tfrac{1}{2} L^M L^N : \tfrac{1}{4} L^N L^N$$

In this case, the expected Mendelian genotypic ratio of 1:2:1 is observed, showing that codominance does not violate the expectations of Mendel's laws.

In codominance, full expression of both alleles is seen in heterozygotes. This distinguishes codominance from incomplete dominance, in which the phenotype of heterozygotes is an intermediate phenotype to those of the parents.

Many genes have more than two alleles.

For the sake of simplicity, we have been discussing only genes with two alleles. However, because alleles are different forms of a gene, there is no reason why a gene has to have only two alleles. In fact, many genes have more than two alleles. Any *individual* can carry only two alleles of a gene, but members of a population can carry many different alleles of a gene. In humans, the gene for ABO blood types is a gene with more than two alleles; in this case, the gene has three alleles. Such genes are said to have **multiple alleles**. Your ABO blood type is determined by genetically encoded molecules (called antigens) present on the surface of your red blood cells (Figure 3.19). These molecules are an identity tag recognized by the body's immune system.

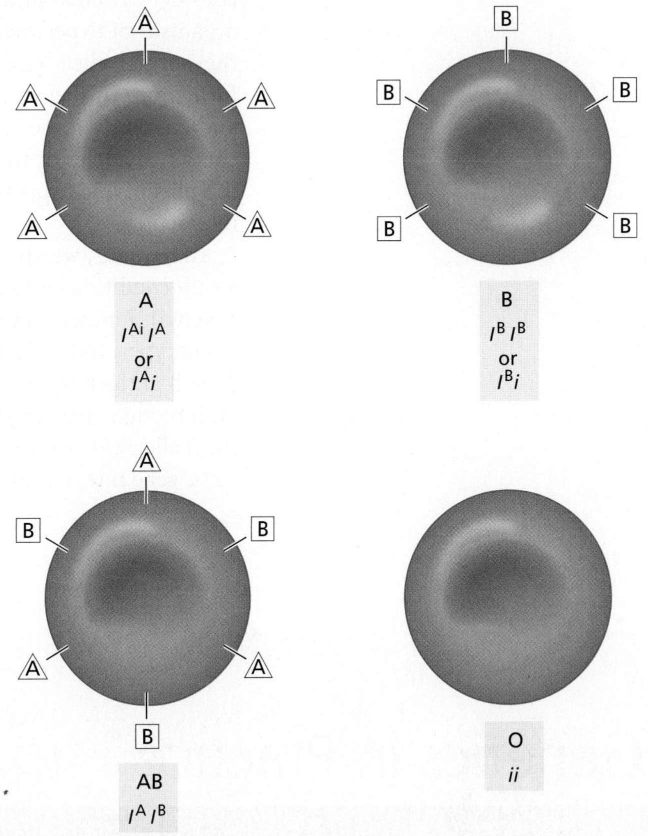

There is one gene (*I*) for the ABO blood types, and it has three alleles, I^A, I^B, and *i*. The I^A and I^B alleles control the formation of slightly different forms of the antigen. If you are homozygous for the *A* allele ($I^A I^A$), you carry the A antigen on cells and have blood type A. If you are homozygous for the *B* allele ($I^B I^B$), you carry the B antigen and are type B. The third allele does not make any antigen, and individuals homozygous for this allele carry no encoded antigen on their cells. The allele for the O blood type is recessive to both the *A* and *B* alleles. Because there are three alleles, there are six possible genotypes, including homozygotes and heterozygotes (Table 3.5). In Chapter 17, we will see how the multiple-allele system in the ABO blood type is used in blood transfusions. Through an understanding of the genetics of ABO blood types, people with a certain genotype can safely receive blood from any other genotype (these individuals are called universal recipients), whereas others with a different genotype are able to donate blood to anyone (and are called universal donors).

Genes can interact to produce phenotypes.

Soon after Mendel's work was rediscovered, it became clear that some traits are controlled by the interaction of two or more genes. This interaction is not necessarily direct; rather, the cellular function of several gene products may contribute to the development of a common phenotype.

One of the best examples of gene interaction is a phenomenon called **epistasis**, a term derived from the Greek word for *stoppage*. In epistasis, the action of one gene masks or prevents the expression of another gene. As an example of epistasis, let's

FIGURE 3.19 Each allele of codominant genes is fully expressed in heterozygotes. Type A blood has A antigens on the cell surface, and type B has B antigens on the surface. In type AB blood, both the A and the B antigen are present on the cell surface. Thus, the *A* and the *B* alleles of the *I* gene are codominant. In type O blood, no antigen is present. The *i* allele is recessive to both the *A* and the *B* alleles.

Multiple alleles Genes that have more than two alleles.

Epistasis The interaction of two or more non-allelic genes to control a single phenotype.

Table 3.5 ABO Blood Types

Genotypes	Phenotypes
$I^A I^A$, $I^A i$	Type A
$I^B I^B$, $I^B i$	Type B
$I^A I^B$	Type AB
ii	Type O

consider eye color and eye formation in the fruit fly, *Drosophila melanogaster*, a favorite organism of experimental geneticists. Eye color in adults is genetically controlled, and the normal allele produces a brick-red color. An unrelated gene, *eyeless*, controls eye formation. In flies homozygous for the mutant allele of *eyeless*, flies do not have eyes, and obviously there is no expression of the gene for eye color even though the fly may carry two copies of the gene for normal eye color. Thus, the *eyeless* gene interferes with the phenotypic expression of the gene for eye color and is an example of epistatic gene interaction.

In humans, we already have discussed the genetic basis for the ABO blood types. In a rare condition called the Bombay phenotype (named for the city in which it was discovered), a mutation in an unrelated gene prevents phenotypic expression of the A and B phenotypes. Individuals homozygous for a recessive allele *h* are blocked from adding the A or B antigen to the surface of their cells, making them phenotypically blood type O, even though genotypically they carry I^A or I^B alleles. In this case, being homozygous for the *h* allele (*hh*) prevents phenotypic expression of the I^A or I^B alleles and is a case of epistatic gene interaction.

Genetics in Practice

Genetics in Practice case studies are critical-thinking exercises that allow you to apply your new knowledge of human genetics to real-life problems. You can find these case studies and links to relevant websites at *www.cengage.com/biology/cummings*

CASE 1

Pedigree analysis is a fundamental tool for investigating whether or not a trait is following a Mendelian pattern of inheritance. It can also be used to help identify individuals within a family who may be at risk for the trait.

Adam and Sarah, a young couple of Eastern European Jewish ancestry, went to a genetic counselor because they were planning a family and wanted to know what their chances were for having a child with a genetic condition. The genetic counselor took a detailed family history from both of them and discovered several traits in their respective families.

Sarah's maternal family history is suggestive of an inherited form of breast and ovarian cancer with an autosomal dominant pattern of cancer predisposition from her grandmother to mother because of the young ages at which they were diagnosed with their cancers. If a mutant allele that predisposed to breast and ovarian cancer was inherited in Sarah's family, she, her sister, and any of her own future children could be at risk for inheriting this mutation. The counselor told her that genetic testing is available that may help determine if this mutant allele is present in her family members.

Adam's paternal family history has a very strong pattern of early-onset heart disease. An autosomal dominant condition known as

familial hypercholesterolemia may be responsible for the large number of deaths from heart disease. As with hereditary breast and ovarian cancer, genetic testing is available to see if Adam carries the mutant allele. Testing will give the couple more information about the chances that their children could inherit this mutation. Adam had a first cousin who died from Tay-Sachs disease (TSD), a fatal autosomal recessive condition most commonly found in people of Eastern European Jewish descent. For his cousin to have TSD, both parents must have been heterozygous for the disease-causing allele. If that is the case, Adam's father could be a carrier as well. If Adam's father carries the mutant TSD allele, it is possible that Adam inherited this mutation. Because Sarah is also of Eastern European Jewish ancestry, she could also be a carrier of the gene, even though no one in her family has been affected with TSD. If Adam and Sarah are both carriers, each of their children would have a 25% chance of being afflicted with TSD.

A simple blood test performed on both Sarah and Adam could determine whether they are carriers of this mutation.

1. If Sarah carries the mutant cancer allele and Adam carries the mutant heart disease allele, what is the chance that they would have a child who is free of both diseases? Are these good odds?

2. Would you want to know the results of the cancer, heart disease, and TSD tests if you were Sarah and Adam? Is it their responsibility as potential parents to gather this type of information before they decide to have a child?

3. Would you decide to have a child if the test results said that you carry the mutation for breast and ovarian cancer? The heart disease mutation? The TSD mutation? The heart disease *and* the TSD mutant alleles?

Summary

3.1 Heredity: How Are Traits Inherited?

- In the centuries before Gregor Mendel experimented with the inheritance of traits using the garden pea, blending of traits was widely accepted as an explanation of how traits were passed from generation to generation, but this and other ideas were not successful in explaining heredity.

3.2 Mendel's Experimental Design Resolved Many Unanswered Questions

- Mendel carefully selected an organism to study, kept careful records, and studied the inheritance of traits over several generations. In his decade-long series of experiments, Mendel established the foundation for the science of genetics.

3.3 Crossing Pea Plants: Mendel's Study of Single Traits

- Mendel studied crosses in the garden pea that involved one pair of alleles and demonstrated that the phenotypes associated with those traits are controlled by pairs of factors, now known as genes. Those factors separate or segregate from each other during gamete formation and exhibit dominant/recessive relationships. This is known as the principle of segregation.

3.4 More Crosses with Pea Plants: The Principle of Independent Assortment

- In later experiments, Mendel discovered that the members of one gene pair separate or segregate independently of other gene pairs. This principle of independent assortment leads to the formation of all possible combinations of gametes, with equal probability in a cross between two individuals.

3.5 Meiosis Explains Mendel's Results: Genes Are on Chromosomes

- Segregation and independent assortment of genes result from the behavior of chromosomes in meiosis. At the turn of the twentieth century, it became apparent that genes are located on chromosomes.

3.6 Mendelian Inheritance in Humans

- Because genes for human genetic disorders exhibit segregation and independent assortment, the inheritance of certain human traits is predictable, making it possible to provide genetic counseling to those at risk of having children afflicted with genetic disorders.

  Instead of direct experimental crosses, traits in humans are traced by constructing pedigrees that follow a trait through several generations.

3.7 Variations on a Theme by Mendel

- Codominant alleles are both expressed in the phenotype, whereas in incomplete dominance, the heterozygote has a phenotype intermediate to that of the parents. Although any individual can carry only two alleles of a gene, many genes have multiple alleles, carried by members of a population. Gene interaction can affect the phenotypic expression of some genes.

Questions and Problems

CENGAGENOW Preparing for an exam? Assess your understanding of this chapter's topics with a pre-test, a personalized learning plan, and a post-test by logging on to **login.cengage.com/sso** and visiting CengageNOW's Study Tools.

Crossing Pea Plants: Mendel's Study of Single Traits

1. Explain the differences between the following terms:
   a. Gene versus allele versus locus
   b. Genotype versus phenotype
   c. Dominant versus recessive
   d. Complete dominance versus incomplete dominance versus codominance
2. Of the following, which are phenotypes and which are genotypes?
   a. *Aa*
   b. Tall plants
   c. *BB*
   d. Abnormal cell shape
   e. *AaBb*
3. Define Mendel's Law of Segregation.
4. Define Mendel's Law of Independent Assortment.
5. Suppose that organisms have the following genotypes. What types of gametes will these organisms produce, and in what proportions?
   a. *Aa*
   b. *AA*
   c. *aa*

6. Given the following matings, what are the predicted genotypic ratios of the offspring?
    a. *Aa* × *aa*
    b. *Aa* × *Aa*
    c. *AA* × *Aa*
7. Wet ear wax (*W*) is dominant over dry ear wax (*w*).
    a. A 3:1 phenotypic ratio of F1 progeny indicates that the parents are of what genotype?
    b. A 1:1 phenotypic ratio of F1 progeny indicates that the parents are of what genotype?
8. An unspecified characteristic controlled by a single gene is examined in pea plants. Only two phenotypic states exist for this trait. One phenotypic state is completely dominant to the other. A heterozygous plant is self-crossed. What proportion of the progeny of plants exhibiting the dominant phenotype is homozygous?
9. Sickle cell anemia (SCA) is a human genetic disorder caused by a recessive allele. A couple plan to marry and want to know the probability that they will have an affected child. With your knowledge of Mendelian inheritance, what can you tell them if (1) each has one affected parent and a parent with no family history of SCA or (2) the man is affected by the disorder but the woman has no family history of SCA?
10. If you are informed that tune deafness is a heritable trait, and that a tune deaf couple is expecting a child, can you conclude that the child will be tune deaf?
11. Stem length in pea plants is controlled by a single gene. Consider the cross of a true-breeding long-stemmed variety to a true-breeding short-stemmed variety in which long stems are completely dominant.
    a. If 120 F1 plants are examined, how many plants are expected to be long stemmed? Short stemmed?
    b. Assign genotypes to both P1 varieties and to all phenotypes listed in (a).
    c. A long-stemmed F1 plant is self-crossed. Of 300 F2 plants, how many should be long stemmed? Short stemmed?
    d. For the F2 plants mentioned in (c), what is the expected genotypic ratio?

More Crosses with Pea Plants: The Principle of Independent Assortment

12. Organisms have the following genotypes. What types of gametes will these organisms produce, and in what proportions?
    a. *Aabb*
    b. *AABb*
    c. *AaBb*
13. Given the following matings, what are the predicted phenotypic ratios of the offspring?
    a. *AABb* × *Aabb*
    b. *AaBb* × *aabb*
    c. *AaBb* × *AaBb*
14. A woman is heterozygous for two genes. How many different types of gametes can she produce, and in what proportions?
15. Two traits are examined simultaneously in a cross of two pure-breeding pea-plant varieties. Pod shape can be either

swollen or pinched. Pea color can be either green or yellow. A plant with the traits swollen and green is crossed with a plant with the traits pinched and yellow, and a resulting F1 plant is self-crossed. A total of 640 F2 progeny are phenotypically categorized as follows:

360 swollen yellow
120 swollen green
120 pinched yellow
40 pinched green

    a. What is the phenotypic ratio observed for pod shape? Pea color?
    b. What is the phenotypic ratio observed for both traits considered together?
    c. What is the dominance relationship for pod shape? Pea color?
    d. Deduce the genotypes of the P1 and F1 generations.
16. Consider the following cross in pea plants, in which smooth pea shape is dominant to wrinkled, and yellow pea color is dominant to green. A plant with smooth yellow peas is crossed to a plant with wrinkled green peas. The offspring produced peas that were all smooth and yellow. What are the genotypes of the parents? What are the genotypes of the offspring?
17. Consider another cross in pea plants involving the genes for pea color and shape. As before, yellow is dominant to green and smooth is dominant to wrinkled. A plant with smooth yellow peas is crossed to a plant with wrinkled green peas. The peas produced by the offspring are as follows: one-fourth are smooth, yellow; one-fourth are smooth, green; one-fourth are wrinkled yellow; and one-fourth are wrinkled green.
    a. What is the genotype of the smooth yellow parent?
    b. What are the genotypes of the four classes of offspring?
18. Determine the possible genotypes of the following parents by analyzing the phenotypes of their children. In this case, we will assume that brown eyes (*B*) is dominant to blue (*b*) and that right-handedness (*R*) is dominant to left-handedness (*r*).
    a. Parents: brown eyes, right-handed × brown eyes, right-handed
       Offspring: 3/4 brown eyes, right-handed
       1/4 blue eyes, right-handed
    b. Parents: brown eyes, right-handed × blue eyes, right-handed
       Offspring: 6/16 blue eyes, right-handed
       2/16 blue eyes, left-handed
       6/16 brown eyes, right-handed
       2/16 brown eyes, left-handed
    c. Parents: brown eyes, right-handed × blue eyes, left-handed
       Offspring: 1/4 brown eyes, right-handed
       1/4 brown eyes, left-handed
       1/4 blue eyes, right-handed
       1/4 blue eyes, left-handed

19. Think about this one carefully. Albinism and hair color are governed by different genes. A recessively inherited form of albinism causes affected individuals to lack pigment in their skin, hair, and eyes. In hair color, red hair is inherited as a recessive trait and brown hair is inherited as a dominant trait relative to red hair. An albino woman whose parents both have red hair has two children with a man who is normally pigmented and has brown hair. The brown-haired partner has one parent who has red hair. The first child is normally pigmented and has brown hair. The second child is albino.

   a. What is the hair color (phenotype) of the albino parent?

   b. What is the genotype of the albino parent for hair color?

   c. What is the genotype of the brown-haired parent with respect to hair color? Skin pigmentation?

   d. What is the genotype of the first child with respect to hair color and skin pigmentation?

   e. What are the possible genotypes of the second child for hair color? What is the phenotype of the second child for hair color? Can you explain this?

20. Consider the following cross: P1: *AABBCCDDEE* × *aabbccddee*
F1: *AaBbCcDdEe* (self-cross to get F2)
What is the chance of getting an *AaBBccDdee* individual in the F2 generation?

21. In the following trihybrid cross, determine the chance that an individual could be phenotypically *A*, *b*, *C* in the F1 generation.
P1: *AaBbCc* × *AabbCC*

22. In pea plants, long stems are dominant to short stems, purple flowers are dominant to white, and round peas are dominant to wrinkled. Each trait is determined by a single, different gene. A plant that is heterozygous at all three loci is self-crossed, and 2,048 progeny are examined. How many of these plants would you expect to be long stemmed with purple flowers, producing wrinkled peas?

Meiosis Explains Mendel's Results: Genes Are on Chromosomes

23. Discuss the pertinent features of meiosis that provide a physical correlate to Mendel's abstract genetic laws of random segregation and independent assortment.

24. The following diagram shows a hypothetical diploid cell. The recessive allele for albinism is represented by *a*, and *d* represents the recessive allele for deafness. The normal alleles for these conditions are represented by *A* and *D*, respectively.

   a. According to the principle of segregation, what is segregating in this cell?

   b. According to Mendel's principle of independent assortment, what is independently assorting in this cell?

   c. How many chromatids are in this cell?

   d. Write the genotype of the individual from whom this cell was taken.

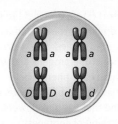

   e. What is the phenotype of this individual?

   f. What stage of cell division is represented by this cell (prophase, metaphase, anaphase, or telophase of meiosis I, meiosis II, or mitosis)?

   g. After meiosis is complete, how many chromatids and chromosomes will be present in one of the four progeny cells?

Mendelian Inheritance in Humans

25. Define the following pedigree symbols:

   a.

   b.

   c.

   d.

   e.

26. Draw the following simple pedigree. A man and a woman have three children: a daughter, then two sons. The daughter marries and has monozygotic (identical) twin girls. The youngest son in generation II is affected with albinism.

27. Construct a pedigree, given the following information. Mary is 16 weeks pregnant and was referred for genetic counseling because of advanced maternal age. Mary has one daughter, Sarah, who is 5 years old. Mary has three older sisters and four younger brothers. The two oldest sisters are married, and each has one son. All her brothers are married, but none has any children. Mary's parents are both alive, and she has two maternal uncles and three paternal aunts. Mary's husband, John, has two brothers, one older and one younger, neither of whom is married. John's mother is alive, but his father is deceased.

Variations on a Theme by Mendel

28. A characteristic of snapdragons amenable to genetic analysis is flower color. Imagine that a true-breeding red-flowered variety is crossed to a pure line having white flowers. The progeny are exclusively pink-flowered. Diagram this cross, including genotypes for all P1 and F1 phenotypes. What is the mode of inheritance? Let *F* = red and *f* = white.

29. In peas, straight stems (*S*) are dominant to gnarled (*s*), and round peas (*R*) are dominant to wrinkled (*r*). The following cross (a test cross) is performed: *SsRr × ssrr*. Determine the expected phenotypes of the progeny and what fraction of the progeny should exhibit each phenotype.

30. Pea plants usually have white or red flowers. A strange pea-plant variant is found that has pink flowers. A self-cross of this plant yields the following phenotypes:

30 red flowers

62 pink flowers

33 white flowers

What are the genotypes of the parents? What is the genotype of the progeny with red flowers?

31. A plant geneticist is examining the mode of inheritance of flower color in two closely related species of exotic plants. One species may have two pure-breeding lines—one produces a distinct red flower, and the other produces flowers with no color at all, or very pale yellow flowers—however, she cannot be sure. A cross of these varieties produces all pink-flowered progeny. The second species exhibits similar pure-breeding varieties; that is, one variety produces red flowers, and the other produces an albino or very pale yellow flower. A cross of these two varieties, however, produces orange-flowered progeny exclusively. Analyze the mode of inheritance of flower color in these two plant species.

32. What are the possible genotypes for the following blood types?

a. type A

b. type B

c. type O

d. type AB

33. A man with blood type A and a woman with blood type B have three children: a daughter with type AB and two sons, one with type B and one with type O blood. What are the genotypes of the parents?

34. What is the chance that a man with type AB blood and a woman with type A blood whose mother is type O can produce a child that is:

a. type A

b. type AB

c. type O

d. type B

35. A hypothetical human trait is controlled by a single gene. Four alleles of this gene have been identified: *a*, *b*, *c*, and *d*. Alleles *a*, *b*, and *c* are all codominant; allele *d* is recessive to all other alleles.

a. How many phenotypes are possible?

b. How many genotypes are possible?

36. In homozygotes, the recessive allele *h* prevents the A and B antigens from being placed on the surface of cells in individuals carrying either the I^A or I^B allele (or both alleles). The normal *H* allele allows these antigens to be placed on cell surfaces.

a. Predict all possible blood-type phenotypes and their ratios in a cross between *HhAB × HhAB* individuals.

b. Among those individuals with type O blood, what genotypes are present, and in what ratios?

Internet Activities

Internet Activities are critical-thinking exercises using the resources of the World Wide Web to enhance the principles and issues covered in this chapter. For a full set of links and questions investigating the topics described below, visit *www.cengage.com/biology/cummings*

1. *Mendelian Genetics and Plant Genetics.* Gregor Mendel crossed pea plants to investigate the results of hybridization experiments. Now you give it a try!

At the CUNY Brooklyn Mendelian Genetics site, read the Introduction carefully (some of the steps are a little tricky) and then click on the "Plant Hybridization" link at the site to choose and perform some crosses of your own.

2. *Mendel's Discoveries in His Own Words.* You can read Mendel's original paper in English and German at the MendelWeb website. In addition to Mendel's original text, this site has links to essays and commentary on his works and writings as well as on the state of knowledge about heredity before Mendel.

3. *Meet Gregor Mendel?* Check out Professor John Blamire's fictionalized account of Mendel's life.

HOW WOULD YOU VOTE NOW?

Using the principles Mendel discovered and modern pedigree analysis, it is possible for couples planning to have families to determine the approximate risk their children have of inheriting certain genetic disorders. To know for certain whether a child has inherited a genetic disorder, genetic testing can be performed. However, in the United States, some genetic testing is required by law and is performed on all newborns, regardless of their individual risk. Some states test for only a few genetic disorders, but others test for nearly three dozen diseases. Not everyone is comfortable with mandatory testing, feeling that it is an invasion of privacy and fearing that the results could be misused to restrict reproductive rights. Now that you know more about inheritance, what do you think? Should all states be required to test for as many genetic conditions as possible, or should this be left up to the parents? If genetic testing is mandatory, who should have access to the results? Visit the *Human Heredity* companion website at *www.cengage.com/biology/cummings* to find out more on the issue; then cast your vote online.

4 Pedigree Analysis in Human Genetics

W as Abraham Lincoln, the 16th president of the United States, affected with a genetic disease? Several authors have suggested that Lincoln may have had an inherited disorder called Marfan syndrome. This genetic condition affects the connective tissue of the body, causing visual problems, blood vessel defects, and loose joints. Evidence in support of this idea is based on two observations: Lincoln's physical appearance and the report of an inherited disorder in a distant relative. Photographs, written descriptions, and medical reports give us detailed information about Lincoln's physical appearance. He was 6 feet 4 inches tall and thin, weighing between 160 and 180 pounds for most of his adult life. He had long arms and legs, with large, narrow hands and feet. Contemporary descriptions of his appearance indicate that he was stoop-shouldered and loose-jointed and walked with a shuffling gait. In addition, he wore eyeglasses to correct a visual problem.

In addition to the physical evidence, pedigree analysis and genealogy research discovered that a child diagnosed with Marfan syndrome in the 1960s had ancestors in common with Lincoln (the common ancestor was Lincoln's great-great-grandfather). In the mid-1960s, these two sets of observations led to widespread speculation that Lincoln had Marfan syndrome.

Other experts disagree with that idea, arguing that Lincoln's long arms and legs and body proportions were well within the normal limits for tall, thin individuals. In addition, although Lincoln wore eyeglasses, he was farsighted, whereas those with the usual form of Marfan syndrome are nearsighted. Lastly, Lincoln showed no outward signs of problems with major blood vessels such as the aorta. Lincoln had only one son, Robert, who lived to adulthood; he showed no signs of Marfan syndrome.

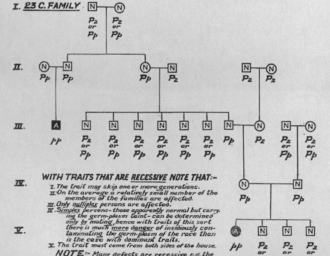

A pedigree for albinism collected by Charles Davenport, one of the leaders of the eugenics movement in the United States.

The gene for Marfan syndrome was identified and cloned in 1991. The gene, which maps to chromosome 15, encodes a protein that anchors and strengthens connective tissue. Using DNA testing, it is possible to determine whether Lincoln or anyone else carries the gene for Marfan syndrome. Soon after the gene was isolated, a group of scientists proposed extracting DNA from fragments of Lincoln's skull (preserved in the National Museum of Health and Medicine in Washington, D.C.) for DNA analysis to see if he had Marfan syndrome. In response, the U.S. Congress asked a panel of experts to review the request to determine whether such a request was ethical and scientifically possible. As described later in this chapter, this test has not yet been done, but the proposal raises several important questions related to the emerging field of biohistory. Is there an overriding public interest in knowing if Lincoln had a genetic disorder that had no bearing on his performance in office? Is there any justifiable scientific or societal gain from such knowledge? Does genetic testing violate Lincoln's right to privacy or that of his family from the disclosure of medical information?

HOW WOULD YOU VOTE?

In 1991, the committee of scientists, historians, and Lincoln scholars convened by the U.S. Congress recommended testing tissue samples from Abraham Lincoln to determine if he had Marfan syndrome. One bioethicist called the proposal a form of voyeurism, but others pointed out that public officials do not have the same expectation of privacy as the rest of us and supported the idea of testing. Do you think there is a compelling reason to determine whether Lincoln, who died in 1865, had Marfan syndrome? Is there a scientific or social benefit to having such information, or is it simply an invasion of privacy? Visit the *Human Heredity* companion website at **www.cengage.com/biology/cummings** to find out more on the issue; then cast your vote online.

4.1 Pedigree Analysis Is a Basic Method in Human Genetics

As outlined in Chapter 3, a pedigree is a diagram showing genetic information from a family, using standardized symbols. Analysis of pedigrees using knowledge of Mendelian principles has two initial goals:

- to determine whether the trait has a dominant or a recessive pattern of inheritance
- to discover whether the gene in question is located on an X or a Y chromosome or on an autosome (chromosomes 1 to 22)

For several reasons, it is important to establish how a trait is inherited. If the pattern of inheritance can be established, it can be used to predict genetic risk in several situations, including

- pregnancy outcomes
- adult-onset disorders
- recurrence risks in future offspring

These applications will be discussed in later chapters.

The collection, storage, and analysis of pedigree information can be done manually or by using software such as Cyrillic (Figure 4.1), Kindred, Peddraw, and Progeny. These programs give on-screen displays of pedigrees and genetic information that can be used to analyze patterns of inheritance.

There are five basic patterns of Mendelian inheritance.

Once a pedigree has been constructed, the principles of Mendelian inheritance are used to follow the trait through a family to determine whether it is inherited as a dominant or recessive trait and whether it is located on an autosome or a sex chromosome. The five basic Mendelian patterns of inheritance for traits controlled by single genes are:

- autosomal recessive inheritance
- autosomal dominant inheritance
- X-linked dominant inheritance
- X-linked recessive inheritance
- Y-linked inheritance

In addition, there is a distinctive non-Mendelian pattern of inheritance observed in traits controlled by single genes encoded by mitochondrial genes, which will be discussed later in the chapter.

Analyzing a pedigree.

Pedigree analysis proceeds in several steps. If you are analyzing a pedigree, first try to rule out all patterns of inheritance that are inconsistent with the pedigree. For example, only males carry a Y chromosome, and for traits controlled by a gene on the Y, only males will be affected. If the pedigree shows affected females, Y-linked inheritance can be ruled out. Second, if only one pattern of inheritance is supported by the information in the pedigree, it is accepted as the pattern of inheritance for the trait being examined.

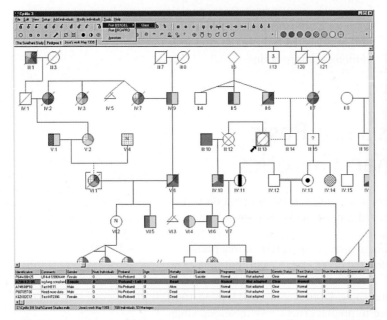

FIGURE 4.1 Software programs such as Cyrillic 3 can be used to prepare pedigrees, store information, and analyze pedigrees.

However, because of the relatively small size of human families, the pedigree may not provide enough information to rule out all but one possible pattern of inheritance. For example, the information in the pedigree may indicate that both autosomal dominant and X-linked dominant inheritance are possible explanations. If this is the case, the next step is to determine whether one pattern of transmission is more likely than the other. If so, the most likely type of inheritance is used as a working hypothesis for further work with this and other families with the trait.

If one pattern is as likely as the other, the only conclusion from the information available is that the trait can be explained by autosomal dominant or X-linked dominant inheritance and that more work is necessary to identify the pattern of inheritance. This may require adding more family members to the pedigree or analyzing pedigrees from other families with the same trait. There are other problems that often complicate pedigree analysis, including variations in gene expression, the degree of phenotypic expression (called penetrance), and traits controlled by alleles that are common in the population. We will discuss some of these in a later section of the chapter. For the sake of simplicity, in this chapter we will limit the discussion to traits controlled by single genes. In Chapter 5, we will discuss traits that are controlled by two or more genes.

4.2 Autosomal Recessive Traits

Although human families are relatively small, analysis of affected and unaffected members over several generations usually provides enough information to determine whether a trait has a recessive pattern of inheritance and is carried on an autosome or a sex chromosome. Recessive traits carried on autosomes have several distinguishing characteristics:

- For rare or relatively rare traits, affected individuals have unaffected parents.
- All the children of two affected (homozygous) individuals are affected.
- The risk of an affected child from a mating of two heterozygotes is 25%.
- Because the trait is autosomal, it is expressed in both males and females, who are affected in roughly equal numbers. Both the male and the female parent will transmit the trait.
- In pedigrees involving rare traits, the unaffected (heterozygous) parents of an affected (homozygous) individual may be related to each other.

A number of autosomal recessive genetic disorders are listed in Table 4.1. A pedigree illustrating a pattern of inheritance typical of autosomal recessive genes is shown in Active Figure 4.2. Characteristic for a rare recessive trait, the trait appears in individuals (II-2 and III-9) who have unaffected parents, and all children of affected parents.

Some autosomal recessive traits represent minor variations in phenotype, such as hair color and eye color (see Exploring Genetics: Was Noah an Albino? on page 76). Others, such as cystic fibrosis, can be life threatening or even fatal.

Autosomal recessive inheritance

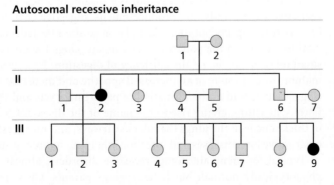

ACTIVE FIGURE 4.2 A pedigree for a rare autosomal recessive trait. In these pedigrees, most affected individuals have normal parents, there is a 25% chance that a child of heterozygotes will be affected, and both sexes are affected in roughly equal numbers.

CENGAGENOW Learn more about autosomal recessive inheritance by viewing the animation by logging on to *login.cengage.com/sso* and visiting CengageNOW's Study Tools.

Table 4.1 Some Human Traits Controlled by Single Genes

AUTOSOMAL RECESSIVE TRAITS		
Trait	**Phenotype**	**OMIM Number**
Albinism	Absence of pigment in skin, eyes, hair	203100
Ataxia telangiectasia	Progressive degeneration of nervous system	208900
Bloom syndrome	Dwarfism; skin rash; increased cancer rate	210900
Cystic fibrosis	Mucous production that blocks ducts of certain glands, lung passages; often fatal by early adulthood	219700
Fanconi anemia	Slow growth; heart defects; high rate of leukemia	227650
Galactosemia	Accumulation of galactose in liver; mental retardation	230400
Phenylketonuria	Excess accumulation of phenylalanine in blood; mental retardation	261600
Sickle cell anemia	Abnormal hemoglobin, blood vessel blockage; early death	141900
Thalassemia	Improper hemoglobin production; symptoms range from mild to fatal	141900/ 141800
Xeroderma pigmentosum	Lack of DNA repair enzymes, sensitivity to UV light; skin cancer; early death	278700
Tay-Sachs disease	Improper metabolism of gangliosides in nerve cells; early death	272800

AUTOSOMAL DOMINANT TRAITS		
Trait	**Phenotype**	**OMIM Number**
Achondroplasia	Dwarfism associated with defects in growth regions of long bones	100800
Brachydactyly	Malformed hands with shortened fingers	112500
Camptodactyly	Stiff, permanently bent little fingers	114200
Crouzon syndrome	Defective development of midfacial region, protruding eyes, hook nose	123500
Ehlers-Danlos syndrome	Connective tissue disorder, elastic skin, loose joints	130000
Familial hypercholesterolemia	Elevated levels of cholesterol; predisposes to plaque formation, cardiac disease; may be most prevalent genetic disease	144010
Adult polycystic kidney disease	Formation of cysts in kidneys; leads to hypertension, kidney failure	173900
Huntington disease	Progressive degeneration of nervous system; dementia; early death	143100
Marfan syndrome	Connective tissue defect; death by aortic rupture	154700
Nail-patella syndrome	Absence of nails, kneecaps	161200

Cystic fibrosis is an autosomal recessive trait.

Cystic fibrosis An often fatal recessive genetic disorder associated with abnormal secretions of the exocrine glands.

Cystic fibrosis (CF; OMIM 219700) is a disabling autosomal recessive genetic disorder that affects the glands that produce mucus, digestive enzymes, and sweat. This disease has far-reaching phenotypic effects because the affected glands perform a number of vital functions. In the pancreas, thick mucus clogs the ducts that carry enzymes to the small intestine, reducing the efficiency of digestion. As a result, affected children can be malnourished in spite of an increased appetite and increased food intake. Eventually, the clogged ducts lead to the formation of pancreatic cysts and the organ degenerates into a fibrous structure, giving rise to the name of the disease. CF also causes the production of thick mucus in the lungs that blocks airways, and most cystic fibrosis patients develop obstructive lung diseases and infections that lead to premature death (Figure 4.3).

Typical for many autosomal recessive disorders, almost all children with CF have phenotypically normal, but heterozygous, parents. CF is relatively common in some

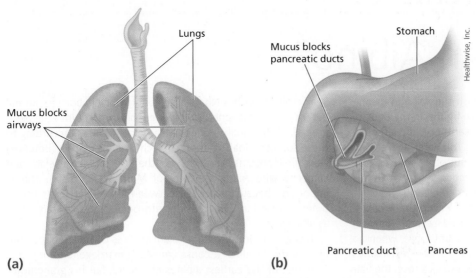

Lungs

Mucus blocks
pancreatic ducts

Stomach

Mucus blocks
airways

Pancreatic duct

Pancreas

Healthwise, Inc.

(a)

(b)

FIGURE 4.3 Organ systems affected by cystic fibrosis. (a) Cystic fibrosis affects both the upper respiratory tract (the nose and sinuses) and the lungs. Thick, sticky mucus clogs the bronchial tubes and the lungs, making breathing difficult. It also slows the removal of viruses and bacteria from the respiratory system, resulting in lung infections. (b) Thick mucus blocks the transport of digestive enzymes in the pancreas. The lack of digestive enzymes results in poor nutrition and slow growth. The trapped digestive enzymes gradually break down the pancreas into a fibrous structure.

populations but rare in others (Figure 4.4). Among the U.S. population of European origins, CF has a frequency of 1 in 2,000 births, and 1 in 22 members of this group are heterozygous carriers. U.S. populations with origins in west or central Africa have a lower frequency of CF (about 1 in 18,000). Among U.S. citizens with origins in Asia, CF is a rare disease whose frequency is about 1 in 90,000. Heterozygous carriers are extremely rare in this population.

Jeff Greenberg/Visuals Unlimited

FIGURE 4.4 About 1 in 25 Americans of European descent, 1 in 46 Hispanics, 1 in 60 to 65 African Americans, and 1 in 150 Asian Americans are carriers for cystic fibrosis. A crowd such as this may contain a carrier.

Was Noah an Albino?

The biblical character Noah, along with the Ark and its animals, is among the most recognizable figures in the Book of Genesis. His birth is recorded in a single sentence, and although the story of the Ark and a great flood is told later, there is no mention of Noah's physical appearance. But other sources contain references to Noah consistent with the idea that Noah was one of the first albinos mentioned in recorded history.

Noah's birth is recorded in the Book of Enoch the Prophet, written about 200 BCE. This book, quoted several times in the New Testament, was regarded as lost until 1773, when an Ethiopian version of the text was discovered. The text relates that Noah's "flesh was white as snow, and red as a rose; the hair of whose head was white like wool, and long, and whose eyes were beautiful." A reconstructed fragment of one of the Dead Sea Scrolls describes Noah as an abnormal child born to normal parents. This fragment also provides some insight into the pedigree of Noah's family, as does the Book of Jubilees. According to these sources, Noah's father (Lamech) and his mother (Betenos) were first cousins. Lamech was the son of Methuselah, and Lamech's wife was a daughter of Methuselah's sister. This is important, because marriage between close relatives is sometimes involved in pedigrees of autosomal recessive traits such as albinism.

If this interpretation of ancient texts is correct, Noah's albinism is the result of a consanguineous marriage, and not only is he one of the earliest albinos on record, but his grandfather Methuselah and Methuselah's sister are the first recorded heterozygous carriers of a recessive genetic trait.

The CF gene was identified in 1989. Using recombinant DNA techniques, a research team first located the gene in a region on the long arm of chromosome 7. Exploring that region using genomic sequencing techniques, they eventually identified the CF gene by comparing the DNA sequence of CF genes in normal and affected individuals.

The CF gene encodes a protein called the cystic fibrosis transmembrane conductance regulator (CFTR), which is normally present in the plasma membrane of secretory gland cells (Figure 4.5). In CF, the protein is either absent or only partially functional. An absent or defective CFTR protein changes the transport of chloride ions, which reduces the amount of fluid added to glandular secretions, making them thicker. This results in blocked ducts and obstructed airflow in the lungs.

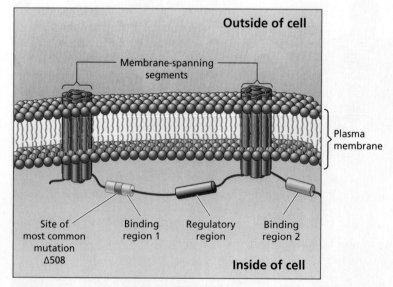

FIGURE 4.5 The cystic fibrosis protein is a membrane protein. The CFTR protein is located in the plasma membrane of the cell and regulates the movement of chloride ions and water across the cell membrane. The regulatory region controls the activity of the CFTR molecule in response to signals from inside the cell. In most cases (about 70%), the protein is defective in binding region 1.

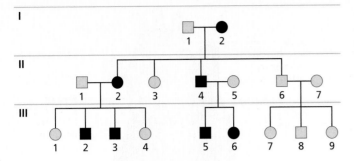

ACTIVE FIGURE 4.6 A pedigree for an autosomal dominant trait. This pedigree shows many of the characteristics of autosomal dominant inheritance. Affected individuals have at least one affected parent, about one-half of the children who have one affected parent are affected, both sexes are affected with roughly equal frequency, and affected parents can have unaffected children.

CENGAGENOW Learn more about autosomal dominant inheritance by viewing the animation by logging on to *login.cengage.com/sso* and visiting CengageNOW's Study Tools.

4.3 Autosomal Dominant Traits

In autosomal dominant disorders, heterozygotes have an abnormal phenotype. Unaffected individuals carry two recessive alleles and have a normal phenotype. Dominant traits have a distinctive pattern of inheritance and usually have affected family members in each generation:

- Every affected individual has at least one affected parent. Exceptions occur when the gene has a high mutation rate. (Mutation is a heritable change in a gene.)
- Most affected individuals are heterozygotes with a homozygous recessive (unaffected) spouse, so each child has a 50% chance of being affected.
- Because the trait is autosomal, the numbers of affected males and females are roughly equal.
- Two affected individuals may have unaffected children (because most affected individuals are heterozygous).
- The phenotype in homozygous dominant individuals is often more severe than the heterozygous phenotype.

The pedigree in Active Figure 4.6 is typical of the pattern found in autosomal dominant conditions. A number of autosomal dominant traits are listed in Table 4.1.

Marfan syndrome is inherited as an autosomal dominant trait.

Marfan syndrome (OMIM 154700) is an autosomal dominant disorder affecting the skeletal system, the eyes, and the cardiovascular system. Like Abraham Lincoln, those with Marfan syndrome are tall and thin, with long arms and legs and long, thin fingers. Because of their height and long limbs, these individuals often excel in basketball and volleyball (Figure 4.7).

The disorder affects males and females with equal frequency and is found in all ethnic groups, with a frequency of about 1 in 10,000 individuals. About 25% of affected individuals appear in families with no previous history of Marfan syndrome, indicating that this gene has a high mutation rate.

The gene responsible for Marfan syndrome, called *FBN1*, is located on chromosome 15 and encodes a protein, fibrillin, which is a component of connective tissue. The normal fibrillin protein also binds to a protein called TGF-β that regulates growth and development of muscle fibers. In Marfan syndrome, the mutant fibrillin produces defective connective tissue and excess TGF-β accumulates, further weakening connective tissue.

The most dangerous effects of Marfan syndrome are on the aorta, the main blood-carrying vessel in the body. As it leaves the heart, the aorta arches back and downward, feeding blood to all the major organ systems. Marfan syndrome weakens the connective tissue around the base of the aorta, causing it to enlarge and eventually split open (Figure 4.8). The enlargement can be repaired by surgery if it is detected in time.

As outlined at the beginning of the chapter, some experts suggested that Abraham Lincoln had Marfan syndrome. A group of research scientists met in 1991 to formulate

Marfan syndrome An autosomal dominant genetic disorder that affects the skeletal system, the cardiovascular system, and the eyes.

FIGURE 4.7 Flo Hyman was a 6 foot 5 inch star on the U.S. women's volleyball team that won a silver medal in the 1984 Olympics. Two years later, at the age of 31, she died in a volleyball game from a ruptured aorta caused by Marfan syndrome.

©Steven E. Sutton/Duomo/PCN Photography

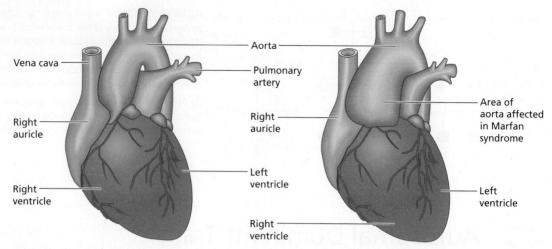

FIGURE 4.8 (*Left*) The heart and its major blood vessels. Oxygen-rich blood is pumped from the lungs to the left side of the heart. From there, blood is pumped through the aorta to all parts of the body. (*Right*) In Marfan syndrome, defective connective tissue causes the base of the aorta to enlarge, and potentially rupture, leading to death.

a proposal to use bone fragments from Lincoln's body as a source of DNA to determine whether Lincoln did, in fact, have Marfan syndrome. The next year, it was decided that testing should be delayed until more was known about the fibrillin gene. In 2001, scientists met again and concluded that enough was known about the gene and that testing should go forward; but as of this writing, no testing has been done.

4.4 Sex-Linked Inheritance Involves Genes on the X and Y Chromosomes

X-linked The pattern of inheritance that results from genes located on the X chromosome.

Y-linked The pattern of inheritance that results from genes located only on the Y chromosome.

The X and Y chromosomes (Figure 4.9) are called sex chromosomes because they play major roles in determining the sex of an individual.

Genes on the X chromosome are called **X-linked,** and genes on the Y chromosome are called **Y-linked.** Female humans have two X chromosomes and, therefore, two

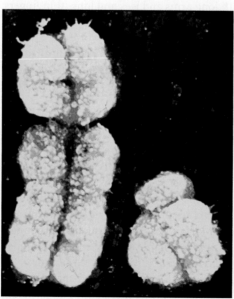

FIGURE 4.9 The human X chromosome (*left*) and the Y chromosome (*right*). This false-color scanning electron micrograph shows the differences between these chromosomes. Most genes on the X chromosome are not found on the Y chromosome. This gives rise to unique patterns of inheritance for genes on the X and Y chromosomes.

Biophoto/Photo Researchers, Inc.

copies of all X-linked genes and can be heterozygous or homozygous for any of them. Males, in contrast, are XY and carry only one copy of the X chromosome. Most genes on the X chromosome are not found on the Y chromosome. This means that males carrying a gene for a recessive disorder such as hemophilia or color blindness cannot carry a dominant allele to mask expression of the recessive allele. This explains why males are affected by X-linked recessive genetic disorders far more often than are females.

Because a male cannot be homozygous or heterozygous for genes on the X chromosome, males are said to be **hemizygous** for all genes on the X chromosome. Traits controlled by genes on the X chromosome are defined as dominant or recessive by their phenotype in females.

Males give an X chromosome to all daughters and a Y chromosome to all sons. Females give an X chromosome to all daughters and all sons (Figure 4.10). As a result, the X and Y chromosomes and the genes on these chromosomes have a distinctive pattern of inheritance. Males pass X-linked traits only to their daughters (who may be heterozygous or homozygous for the condition). Most females are heterozygous for X-linked traits, and her sons have a 50% chance of receiving the recessive allele from their mother. In the following sections, we consider examples of sex-linked inheritance and explore the characteristic pedigrees in detail.

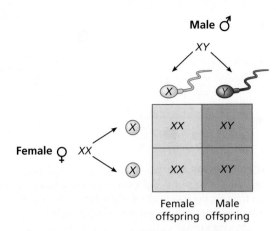

FIGURE 4.10 Distribution of sex chromosomes from generation to generation. All children receive an X chromosome from their mothers. Fathers pass their X chromosome to all their daughters and a Y chromosome to all their sons. The sex-chromosome content of the sperm determines the sex of the child.

X-Linked dominant traits.

Only a small number of dominant traits are carried on the X chromosome. Dominant X-linked traits have a distinctive pattern of inheritance:

- Affected males transmit the trait to all their daughters but none of their sons.
- A heterozygous affected female will transmit the trait to half of her children, with sons and daughters affected equally.
- On average, twice as many females are affected as males (females can be heterozygous or homozygous).

A pedigree for an X-linked dominant trait is shown in Figure 4.11. To determine whether a trait is X-linked dominant or autosomal dominant, the children of affected males must be analyzed carefully. Because males pass their X chromosome only to daughters, affected males transmit the trait *only* to daughters, never to sons. In contrast, if the condition is inherited as an autosomal dominant trait, heterozygous affected males pass the trait to daughters *and* sons, so that about half of all daughters and about half of all sons are affected. As seen in the pedigree (Figure 4.11), males affected with X-linked dominant traits transmit the trait to all their daughters, but affected females have affected sons *and* affected daughters.

Hemizygous A gene present on the X chromosome that is expressed in males in both the recessive and the dominant conditions.

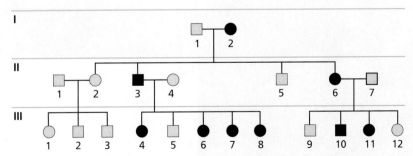

FIGURE 4.11 A pedigree for an X-linked dominant trait. Affected males produce all affected daughters and no affected sons; affected females transmit the trait to roughly half their children, with males and females equally affected; and overall, twice as many females as males are affected with the trait.

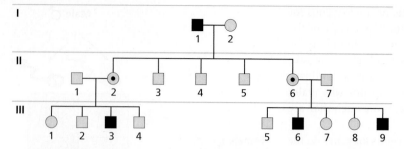

ACTIVE FIGURE 4.12 A pedigree for an X-linked recessive trait. This pedigree shows the characteristics of X-linked recessive traits: Hemizygous males are affected and transmit the trait to all their daughters, who become heterozygous carriers, and phenotypic expression is much more common in males than in females.

CENGAGENOW Learn more about X-linked recessive inheritance by viewing the animation by logging on to *login.cengage.com/sso* and visiting Cengage-NOW's Study Tools.

Color blindness Defective color vision caused by reduction or absence of visual pigments. There are three forms: red, green, and blue blindness.

X-Linked recessive traits.

Recall that there are two important characteristics associated with the inheritance of the X chromosome and the Y chromosome:

1. Males give an X chromosome to all their daughters but do not give an X chromosome to their sons.
2. Females give an X chromosome to each of their children. In addition, males are hemizygous for all genes on the X chromosome and show phenotypes for all X-linked genes.

These two factors produce a distinctive pattern of inheritance for X-linked recessive traits. This pattern can be summarized as follows:

- Hemizygous males and females homozygous for the recessive allele are affected.
- Phenotypic expression is much more common in males than in females. In the case of rare alleles, males are almost exclusively affected.
- Affected males receive the mutant allele from their mothers and transmit it to all their daughters but not to any of their sons.
- Daughters of affected males are usually heterozygous and therefore unaffected, but sons of heterozygous females have a 50% chance of receiving the recessive gene.

A pedigree for an X-linked recessive trait is shown in Active Figure 4.12.

Color blindness is an X-linked recessive trait.

The most common form of **color blindness**, known as red-green blindness, affects about 8% of the male population in the United States. Those with red blindness (OMIM 303900) do not see red as a distinct color (Figure 4.13), whereas those with green blindness

FIGURE 4.13 People who are color-blind see colors differently. (a) Those with normal vision see the red leaves. (b) Someone who is red-green color-blind sees the leaves as gray.

(OMIM 303800) cannot distinguish green or other colors in the middle of the visual spectrum (Figure 4.14). Both forms of color blindness are inherited as X-linked recessive traits. A rare form of blue color blindness (OMIM 190900) is inherited as an autosomal dominant condition that maps to chromosome 7.

The three genes controlling color vision encode three different but related proteins found in retinal cells (Active Figure 4.15). These proteins are normally found in cells sensitive to red, green, or blue wavelengths of light. If, for example, the protein for red color vision is defective or absent, cells that respond to red light are nonfunctional, resulting in red color blindness. Similarly, defects in the green or blue color vision proteins produce green and blue blindness.

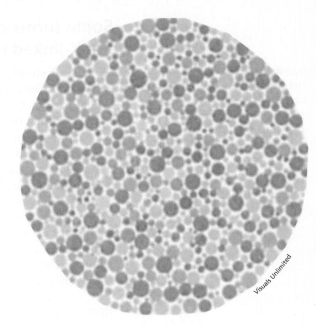

FIGURE 4.14 People with normal color vision see the number 29 in the chart; however, those who are color-blind cannot see any number.

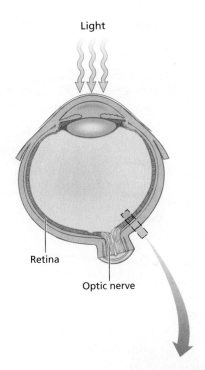

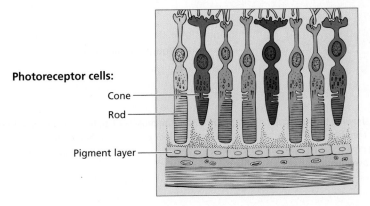

ACTIVE FIGURE 4.15 In the retina, there are two types of light receptor cells: Rods are sensitive to differences in light intensity, and cones are sensitive to differences in color. There are three types of cones: red sensitive, green sensitive, and blue sensitive. Defects in the cones cause color blindness.

CENGAGENOW Learn more about eye structure and function by viewing the animation by logging on to *login.cengage.com/sso* and visiting CengageNOW's Study Tools.

Some forms of muscular dystrophy are X-linked recessive traits.

Muscular dystrophy is a group of inherited diseases characterized by progressive weakness and loss of muscle tissue. The most common form of muscular dystrophy is an X-linked disorder, Duchenne muscular dystrophy (DMD; OMIM 310200), which affects 1 in 3,500 males in the United States. DMD males appear healthy at birth but develop symptoms between 1 and 6 years of age. Progressive muscle weakness is one of the first signs of DMD, and affected individuals use a distinctive set of maneuvers to get up from a prone position. The disease progresses rapidly, and affected individuals are usually confined to wheelchairs by 12 years of age because of muscle degeneration. Death usually occurs by age 20 as a result of respiratory infection or cardiac failure.

The DMD gene encodes a protein called dystrophin. Normal forms of dystrophin stabilize the membrane of muscle cells during the mechanical stresses of muscle contraction (Figure 4.16). In DMD, dystrophin is not present (Figure 4.17), and the muscle cell membranes are torn apart by the forces generated during muscle contraction, eventually causing the death of muscle tissue.

In another form of the disease, called Becker muscular dystrophy (BMD; OMIM 310200), a shortened and partially functional form of dystrophin is made, producing a less severe form of the disease. These two diseases are caused by different mutations in the dystrophin gene.

There are over 850 X-linked recessive traits, including color blindness, muscular dystrophy, and hemophilia (see Spotlight on Hemophilia, HIV, and AIDS; see also Exploring Genetics: Hemophilia and History on page 85), among many others (Table 4.2).

Muscular dystrophy A group of genetic diseases associated with progressive degeneration of muscles. Two of these, Duchenne and Becker muscular dystrophy, are inherited as X-linked allelic recessive traits.

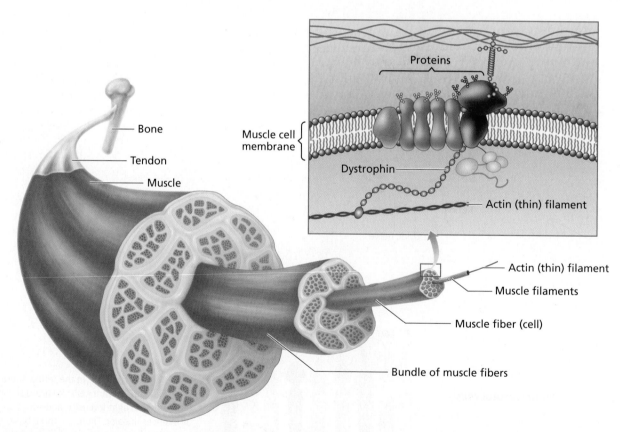

FIGURE 4.16 A cross section of muscle showing the molecular organization within the muscle fiber. In normal muscle (*inset*), dystrophin provides a flexible and elastic connection between actin and the muscle fiber plasma membrane that helps dissipate the force of muscle contraction. In Duchenne muscular dystrophy, dystrophin is absent, resulting in tearing of the plasma membrane during contraction and the subsequent death of muscle fibers.

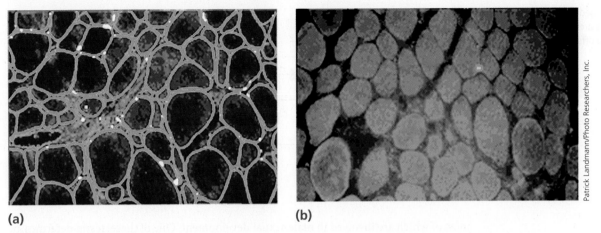

(a)   **(b)**

Patrick Landmann/Photo Researchers, Inc.

FIGURE 4.17 Distribution of dystrophin in muscle cells. (a) In normal muscle cells, all the dystrophin is located in the plasma membrane, and the cytoplasm looks dark. (b) In muscle cells of a boy with DMD, there is no dystrophin in the plasma membranes (so the membranes are invisible), and defective copies of the dystrophin molecules accumulate in the cytoplasm, which stains a light blue. When dystrophin is not present in the cell membrane, muscle contractions eventually tear the membrane and the cell dies.

Table 4.2   Some X-Linked Recessive Traits

Trait	Phenotype	OMIM Number
Adrenoleukodystrophy	Atrophy of adrenal glands; mental deterioration; death 1 to 5 years after onset	300100
Color blindness		
Green blindness	Insensitivity to green light; 60 to 75% of color-blindness cases	303800
Red blindness	Insensitivity to red light; 25 to 40% of color-blindness cases	303900
Fabry disease	Metabolic defect caused by lack of enzyme alpha-galactosidase A; progressive cardiac and renal problems; early death	301500
Glucose-6-phosphate dehydrogenase deficiency	Benign condition that can produce severe, even fatal, anemia in presence of certain foods, drugs	305900
Hemophilia A	Inability to form blood clots; caused by lack of clotting factor VIII	306700
Hemophilia B	"Christmas disease"; clotting defect caused by lack of factor IX	306900
Ichthyosis	Skin disorder causing large, dark scales on extremities, trunk	308100
Lesch-Nyhan syndrome	Metabolic defect caused by lack of enzyme hypoxanthine-guanine phosphoribosyl transferase (HGPRT); causes mental retardation, self-mutilation, early death	308000
Muscular dystrophy	Duchenne-type, progressive; fatal condition accompanied by muscle wasting	310200

4.5 | Paternal Inheritance: Genes on the Y Chromosome

Because only males have Y chromosomes, traits encoded by genes on the Y are passed directly from father to son and have a unique pattern of inheritance. In addition, all Y-linked traits should be expressed because males are hemizygous for all genes on the Y chromosome. To date, only about three dozen Y-linked traits have been discovered,

Spotlight on...

Hemophilia, HIV, and AIDS

In the 1980s, males with hemophilia who received blood and blood components to control bleeding episodes were exposed to HIV, the virus that causes AIDS. Some blood donors unknowingly had an HIV infection, which contaminated the blood supply. The result was that many people, including more than half the hemophilia patients in the United States, developed an HIV infection. Most of the blood contamination took place before the cause of AIDS was discovered and before a test to identify HIV-infected blood was developed.

Fortunately, blood-donor screening and new clotting products made by biotechnology have virtually eliminated the risk of HIV transmission through blood products. As of January 1991, there have been no reports that anyone who received donor-screened blood products has been infected with HIV.

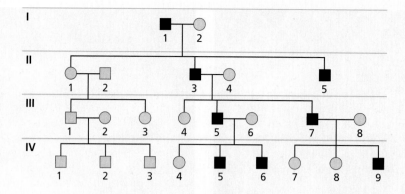

FIGURE 4.18 A pedigree for a Y-linked trait. These traits are transmitted from males to all male offspring in each generation.

most of which are involved in male sexual development. One of these, testis-determining factor (*TDF/SRY*; OMIM 480000), is involved in determining maleness in developing embryos. Its role in early male development is discussed in Chapter 7. Figure 4.18 shows a pedigree for Y-linked inheritance.

4.6 Non-Mendelian Inheritance: Maternal Mitochondrial Genes

Mitochondria are cytoplasmic organelles that convert energy from food molecules into ATP, a molecule that powers many cellular functions (review the structure and function of mitochondria in Chapter 2). Billions of years ago, ancestors of mitochondria were free-living bacteria that adapted to live inside the cells of primitive eukaryotes. Over time, most of the genes carried on the bacterial chromosome have been lost, but as an evolutionary relic of their free-living ancestry, mitochondria carry DNA molecules that encode information for 37 mitochondrial genes. These genes encode proteins that function in energy production.

Mitochondria are cellular organelles transmitted from mothers to all their children through the cytoplasm of the egg (sperm do not contribute mitochondria at fertilization). As a result, genetic disorders caused by mutations in mitochondrial genes have the following properties:

- They are maternally inherited and produce a distinctive pattern of inheritance.
- All the children of affected females are affected. Affected females will transmit the disorder to all their offspring, but affected males cannot transmit the mutations to any of their children.

A pedigree illustrating inheritance of a mitochondrial trait is shown in Figure 4.19.

Because mitochondria are energy producers, mutations in mitochondrial genes reduce the amount of energy available for cellular functions. As a result, the phenotypic effects of mitochondrial disorders can be highly variable. In general, tissues with the highest energy requirements are affected most often. These include muscles and the nervous system. Disorders that mainly affect the muscles are grouped together and called mitochondrial myopathies (*myo* = muscle, *pathy* = disease). Those that affect both muscles and the nervous system are called mitochondrial encephalomyopathy (*encephalo* = brain). Some genetic disorders associated with mitochondria are listed in Table 4.3.

Symptoms of mitochondrial myopathy include muscle weakness and death of muscle tissue, often affecting the movement of the eyes and causing droopy eyelids. These myopathies can also cause problems with swallowing and speech difficulties.

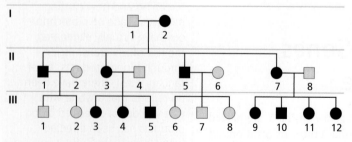

FIGURE 4.19 A pedigree showing the pattern of inheritance associated with mitochondrial genes. Both males and females can be affected by mitochondrial disorders, but only females can transmit the traits to their children.

Hemophilia and History

Hemophilia, an X-linked recessive disorder, is characterized by defects in the mechanism of blood clotting. This form of hemophilia, called hemophilia A, occurs with a frequency of 1 in 10,000 males. Because only homozygous recessive females can have hemophilia, the frequency in females is much lower—on the order of 1 in 100 million.

Pedigree analysis indicates that Queen Victoria of England carried this gene. Because she passed the mutant allele on to several of her children, it is likely that the mutation occurred in an X chromosome she received from one of her parents. Although this mutation spread through the royal houses of Europe, the present royal family of England is free of hemophilia because it is descended from Edward VII, an unaffected son of Victoria.

Perhaps the most important case of hemophilia among Victoria's offspring involved the royal family of Russia.

Victoria's granddaughter Alix, a carrier, married Czar Nicholas II of Russia. She gave birth to four daughters and then a son, Alexis, who had hemophilia. Frustrated by the failure of the medical community to cure Alexis, the royal couple turned to a series of spiritualists, including the monk Rasputin. While under Rasputin's care, Alexis recovered from several episodes of bleeding, and Rasputin became a powerful adviser to the royal family. Some historians have argued that the czar's preoccupation with Alexis's health and the resulting insidious influence of Rasputin contributed to the revolution that overthrew the throne. Other historians point out that Nicholas II was a weak czar and that revolution was inevitable, but it is nonetheless interesting to note that a mutation carried by an English queen had a considerable influence on twentieth-century Russian history.

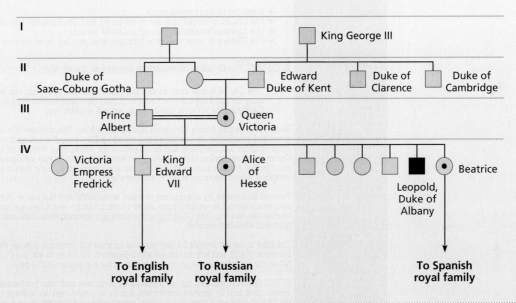

When someone is affected by encephalomyopathy, problems with the nervous system are added to the clinical symptoms that affect muscles. For example, in addition to effects on the muscles of the eyes, the disorder may affect the eye itself and the regions of the brain associated with vision.

Table 4.3 Some Mitochondrial Disorders

Trait	Phenotype	OMIM Number
Kearns-Sayre syndrome	Short stature; retinal degeneration	530000
Leber optic atrophy (LHON)	Loss of vision in center of visual field; adult onset	535000
Leigh syndrome	Degradation of motor skills	256000
MELAS syndrome	Episodes of vomiting, seizures, and stroke-like episodes	540000
MERRF syndrome	Deficiencies in the enzyme complexes associated with energy transfer	545000
Progressive external ophthalmoplegia (PEO)	Paralysis of the eye muscles	157640

4.7 An Online Catalog of Human Genetic Traits Is Available

A catalog of human genetic traits, developed and maintained by researchers at Johns Hopkins University, is available on the Internet (Figure 4.20) as OMIM (Online Mendelian Inheritance in Man). Each trait listed in OMIM is assigned a number (the OMIM number). In this chapter and throughout the book, the OMIM number for each trait discussed is listed. You can obtain more information about any of more than 10,000 inherited traits by accessing the OMIM page and using this number. OMIM is part of a larger series of integrated databases called Entrez, that provides information about genes, chromosome location, DNA sequence, and protein sequence. Access to Entrez and OMIM is available through the book's home page or through search engines.

FIGURE 4.20 Online Mendelian Inheritance in Man (OMIM) is an online database that contains information about human genetic disorders.

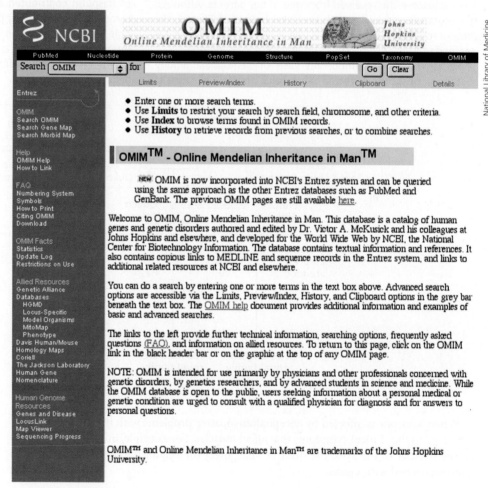

National Library of Medicine

4.8 Many Factors Can Affect the Outcome of Pedigree Analysis

Making medical and reproductive decisions derived from pedigree analysis is based on the assumption that the pattern of inheritance and assignment of phenotypes in the pedigree are correct. While in most cases the assumptions are valid, there are several factors that can influence the outcome of pedigree analysis, skewing the outcome. Many mutant alleles produce regular and consistent phenotypes, but others can produce a wide range of seemingly unrelated phenotypes, have only partial expression, or have expression delayed until middle age. Any of these variations can cause problems in pedigree analysis. In a few cases, a mutant genotype may not be expressed at all, resulting in a deceptively normal phenotype but an incorrect genotype assigned to one or more individuals in the pedigree.

In Chapter 3, we briefly discussed genotype-phenotype interactions and examined how incomplete dominance, codominance, and gene interaction affect the expression of a genotype. Variation in phenotypic expression can be caused by a number of factors, including age, interactions with other genes in the genotype, interactions between genes and the environment, and variations in the environment alone.

Phenotypes are often age-related.

Although many genes are expressed early in development or shortly after birth, the phenotype of some disorders does not develop until adulthood. One of the best-known examples is **Huntington disease** (HD; OMIM 143100), an autosomal dominant trait. The symptoms of HD first appear sometime between the ages of 30 and 50 years. Affected individuals initially develop uncontrolled jerky movements of the head and limbs. Additional neurodegenerative symptoms appear over time (Figure 4.21); the disease progresses slowly, with death occurring some 5 to 15 years after symptoms first appear. By the time the phenotype becomes apparent, the affected individual (who is heterozygous) usually has had children, each of whom has a 50% chance of developing the disease.

The gene for HD has been identified and cloned using recombinant DNA techniques, making it possible to test family members of any age to identify those who carry the mutant allele and will develop this untreatable and fatal disorder later in life.

FIGURE 4.21 People with Huntington disease develop problems with neuromuscular control and gradually lose the use of their limbs. The disease is progressive and fatal, with death occurring 5 to 15 years after symptoms begin.

Penetrance and expressivity cause variations in phenotype.

The terms *penetrance* and *expressivity* define two different aspects of phenotypic variation. **Penetrance** is the probability that a disease phenotype will be present when the disease genotype is present. For example, if all individuals carrying the allele for a dominant disorder have the mutant phenotype, the gene has 100% penetrance. If only 25% of those with the mutant allele show the mutant phenotype, penetrance is 25%. If the phenotype of a trait is not present in 100% of those with the related genotype, the trait is said to show incomplete penetrance.

Expressivity refers to the degree of a gene's phenotypic expression. The following example shows the relationship between penetrance and expressivity, using a single human trait.

An autosomal dominant trait called **camptodactyly** (OMIM 114200) causes an unmovable, bent little finger. Because the trait is dominant, all heterozygotes and all homozygotes should have a bent little finger on both hands. However, the pedigree in Figure 4.22 shows that only one family member (IV-8) has both little fingers bent; others (II-3, III-2, IV-5, IV-6, IV-7, and IV-9) have only one bent finger. One family member (III-4) has a normal phenotype, but must carry the mutant allele because he passed the trait to his children, all of whom have some level of expression.

We can see that at least nine people in the pedigree in Figure 4.22 carry the dominant mutant allele for camptodactyly, but phenotypic expression is seen only in eight, giving a preliminary estimate of 88% penetrance (8/9 individuals). We can only estimate the degree of penetrance in this pedigree, because individuals II-1, II-2, and III-1

Huntington disease (HD) An autosomal dominant disorder associated with progressive neural degeneration and dementia. Adult onset is followed by death 10 to 15 years after symptoms appear.

Penetrance The probability that a disease phenotype will appear when a disease-related genotype is present.

Expressivity The range of phenotypes resulting from a given genotype.

Camptodactyly A dominant human genetic trait that is expressed as immobile, bent, little fingers.

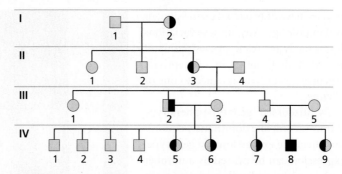

FIGURE 4.22 Penetrance and expressivity. This pedigree shows the transmission of camptodactyly in a family with both variable penetrance and variable expressivity. Fully shaded symbols indicate members with two affected hands. Those with affected left hands are indicated by shading in the left half of the symbol, and those affected only in the right hand have the right half of the symbol shaded. Symbols with light shading indicate unaffected family members. There is no penetrance in individual III-4 even though he passed the gene for camptodactyly to all of his children. Variable expressivity includes several phenotypes, including no phenotypic expression, expression in one hand, and one individual (IV-8) with both hands affected.

have normal phenotypes but no children; for these, we cannot be sure whether or not they carry the mutant allele for camptodactyly. As you can see from this example, incomplete penetrance can be a problem in interpreting the information from pedigrees and the assignment of genotypes to family members. For example, in this case, it is not clear whether II-1, II-2, and III-1 carry the mutant allele, and without genomic testing, it is not possible to say whether they are at risk of having affected children.

Expressivity defines the *degree* of expression for a particular trait. If a trait does not have a uniform level of expression, it is said to have variable expressivity. Because camptodactyly is a dominant trait, we would expect that everyone carrying the mutant allele would have both little fingers affected. However, there is clearly variation in phenotypic expression in this family. Some members are affected only on the left hand and others only on the right hand; in one case, both hands are affected; in another, neither hand is affected. This variable expression of the phenotype results from interactions with alleles of other genes and with nongenetic environmental factors.

Common recessive alleles can produce pedigrees that resemble dominant inheritance.

Pedigrees for rare autosomal recessive traits are usually clear and unmistakable. However, if an allele is common in the population, there is a chance that it will enter the pedigree from outside the family and may do so more than once. In some cases, this can result in a pedigree that looks like autosomal dominant inheritance (Figure 4.23). The allele for O blood type (*i*) is a common allele, often found in more than 50% of the members of some populations. In this pedigree, there are three copies of the *i* allele in the P1 generation (one homozygote and one heterozygote). In the second generation, outsiders bring two more copies of the allele for type O blood into the family. There are *ii* homozygotes in each generation, making it appear that the *i* allele is dominant, when in fact it is recessive.

From these examples, it should be clear that analysis of pedigrees is subject to many factors that can influence interpretation. For many traits, establishing the pattern of inheritance from one or a small number of pedigrees is often only a working hypothesis that must be confirmed by examination of additional pedigrees and by direct testing of genotypes, using genomic techniques.

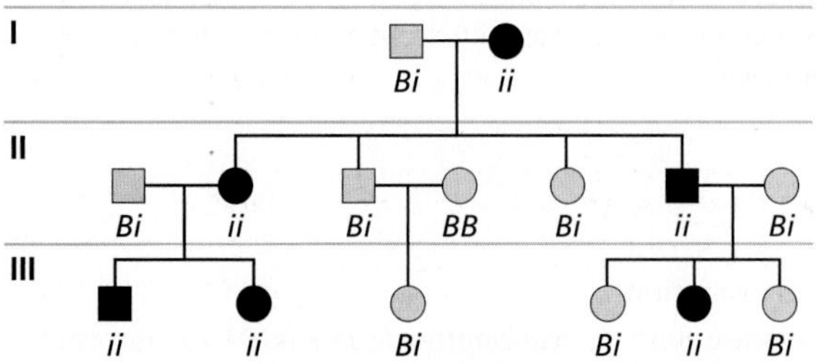

FIGURE 4.23 A common autosomal recessive allele can produce a pedigree that looks like an autosomal dominant trait. In generation I, one parent is homozygous for the type O recessive allele (*ii*) and the other is heterozygous. In generation II, two new copies of the *i* allele are introduced into the family (II-1 and II-7), producing three *ii* homozygotes in generation III. As a result, inheritance of this common autosomal recessive allele produces a pseudo-dominant pedigree, with members of each generation showing the trait and about one-half of the offspring in each generation exhibiting the trait.

KEEP IN MIND

Patterns of gene expression are influenced by many different environmental factors.

Genetics in Practice

Genetics in Practice case studies are critical-thinking exercises that allow you to apply your new knowledge of human genetics to real-life problems. You can find these case studies and links to relevant websites at *www.cengage.com/biology/cummings*

CASE 1

Florence is an active 44-year-old elementary school teacher who began experiencing severe headaches and nausea. She told her physician that her energy level had been reduced dramatically in the last few months, and her arms and legs felt like they "weighed 100 pounds each," particularly after she worked out in the gym. The doctor performed a complete physical and noticed that she did have reduced strength in her arms and legs and that her left eyelid was droopier than her right. He referred her to an ophthalmologist, who discovered that she had an unusual pigment accumulation on her retina that had not yet affected her vision. She then visited a clinical geneticist, who examined the mitochondria in her muscles. She was diagnosed with a mitochondrial genetic disorder known as Kearns-Sayre syndrome.

Mitochondria are responsible for the conversion of food molecules into energy to meet the cell's energy needs. In mitochondrial disorders, these biochemical processes are abnormal, and energy production is reduced. Muscle tends to be affected particularly because it requires a lot of energy, but other tissues, such as the brain, may also be involved. In the the microscope, the mitochondria in muscle from people with mitochondrial disorders look abnormal, and often accumulate around the edges of muscle fibers. This produces a particular staining pattern known as a "ragged red" appearance, and this is usually how mitochondrial disorders are diagnosed.

Mitochondrial disorders affect people in many ways. The most common problem is a combination of mild muscle weakness in the arms and legs, together with droopy eyelids and difficulty in moving the eyes. Some people do not have problems with their eye muscles but have arm and leg weakness that gets worse after exertion. This weakness may be associated with nausea and headaches. Sometimes muscle weakness is obvious in small babies if the illness is severe, and those babies may have difficulty feeding and swallowing. Other parts of the body may be involved, including the electrical conduction system of the heart. Most mitochondrial disorders are mildly disabling, particularly in people who have eye-muscle and limb weakness. The age at which the first symptoms develop is variable, ranging from early childhood to late adult life.

About 20% of those with mitochondrial disorders have similarly affected relatives. Because only mothers transmit this disorder, it was suspected that some of these conditions are caused by a mutation in the genetic information carried by mitochondria. Mitochondria have their own genes, separate from the genes in the chromosomes of the nucleus. Only mothers pass mitochondria and their genes to children, whereas the nuclear genes come from both parents. In about one-third of people with mitochondrial disorders, substantial chunks of the mitochondrial genes are deleted. Most of these individuals do not have affected relatives, and it seems likely that the deletions arise either during development of the egg or during very early development of the embryo.

Deletions are particularly common in people with eye muscle weakness and the Kearns-Sayre syndrome.

1. Why would mitochondria have their own genomes?

2. How would mitochondria be passed from mother to offspring during egg formation? Why doesn't the father pass mitochondria to offspring?

CASE 2

The Smiths had just given birth to their second child and were eagerly waiting to take the newborn home. At that moment, their obstetrician walked into the hospital room with some news about their daughter's newborn screening tests. The physician told them that the state's mandatory newborn screening test had detected an abnormally high level of phenylalanine in their daughter's blood. The Smiths asked if this was just a fleeting effect, like newborn jaundice, that would "go away" in a few days. When they were told that that was unlikely, they were even more confused. The pregnancy had progressed without any complications, and their daughter was born looking perfectly "normal." Mrs. Smith even had a normal amniocentesis early in the pregnancy. The physician asked a genetic counselor to come to their room to explain their daughter's newly diagnosed condition.

The counselor began her discussion with the Smiths by taking a family history from each of them. She explained that phenylketonuria (PKU) is a genetic condition that results when an individual inherits an altered gene from each parent. The counselor wanted to make this point early in the session in case either parent were to try to cast blame for their daughter's condition. She explained that PKU is characterized by an increased concentration of phenylalanine in blood and urine and that mental retardation can be part of this condition if it is not treated at an early age.

To prevent the development of mental retardation, after early diagnosis, dietary therapy must begin before the child is 30 days old. The newborn needs to follow a special diet in which the bulk of protein in the infant's formula is replaced by an artificial amino acid mixture low in phenylalanine. The child must stay on this diet indefinitely for it to be maximally effective.

PKU is one of several diseases known as the hyperphenylalaninemias, which occur with a frequency of 1 in 10,000 births. Classic PKU accounts for two-thirds of these cases. PKU is an autosomal recessive disorder that is distributed widely among whites and Asians but is rare in blacks. Heterozygous carriers do not show symptoms but may have slightly increased phenylalanine concentrations. If untreated, children with classic PKU can experience progressive mental retardation, seizures, and hyperactivity. EEG abnormalities; mousy odor of the skin, hair, and urine; a tendency to have light-colored skin; and eczema complete the clinical picture.

1. Why did amniocentesis fail to detect PKU? What disorders can amniocentesis detect?

2. Assume that you are the genetic counselor. How would you counsel the parents to help them cope with their situation if one or both were blaming themselves for the child's condition?

3. What foods contain phenylalanine? How disruptive do you think the diet therapy will be to everyday life?

Summary

4.1 Pedigree Analysis Is a Basic Method in Human Genetics

- Instead of direct experimental crosses, human traits are traced by constructing pedigrees that follow a trait through several generations of a family.

- Information in the pedigree is used to determine how a trait is inherited and to assign genotypes to members of the family. These patterns include autosomal dominant, autosomal recessive, X-linked dominant, X-linked recessive, Y-linked, and mitochondrial.

4.2 Autosomal Recessive Traits

- Autosomal recessive traits have several characteristics: For rare traits, most affected individuals have unaffected parents; all children of affected parents are affected; the risk of an affected child with heterozygous parents is 25%.

4.3 Autosomal Dominant Traits

- Dominant traits have several characteristics: Except in traits with high mutation rates, every affected individual has at least one affected parent; because most affected individuals are heterozygous and have unaffected mates, each child has a 50% risk of being affected; two affected individuals can have unaffected children.

4.4 Sex-Linked Inheritance Involves Genes on the X and Y Chromosomes

- Males give an X chromosome to all their daughters but not to their sons, and females pass an X chromosome to all their children; genes on the sex chromosomes have a distinct pattern of inheritance.

- In X-linked dominant traits, affected males produce all affected daughters and no affected sons. Heterozygous affected females transmit the trait to half of their children, with sons and daughters equally affected.

- In X-linked recessive inheritance, affected males receive the mutant allele from their mother and transmit it to all their daughters but not to their sons; daughters of affected males are usually heterozygous; sons of heterozygous females have a 50% chance of being affected.

4.5 Paternal Inheritance: Genes on the Y Chromosome

- Because only males have Y chromosomes, genes on the Y chromosome are passed directly from father to son. All Y-linked genes are expressed because males are hemizygous for genes on the Y chromosome.

4.6 Non-Mendelian Inheritance: Maternal Mitochondrial Genes

- Mitochondria are transmitted from mothers to all their offspring through the cytoplasm of the egg. As a result, mitochondria and genetic disorders caused by mutations in mitochondrial genes are maternally inherited. Genetic disorders in mitochondrial DNA are associated with defects in energy conversion.

4.7 An Online Catalog of Human Genetic Traits Is Available

- Genetic traits are described, cataloged, and numbered in a database called "Online Mendelian Inheritance in Man" (OMIM). This online resource is updated on a daily basis and contains information about all known human genetic traits.

4.8 Many Factors Can Affect the Outcome of Pedigree Analysis

- Several factors can affect phenotypic expression, including interactions with other genes in the genotype and interactions between genes and the environment. As a result, some phenotypes develop only in adulthood. Penetrance is the probability that a disease phenotype will appear when the disease-producing genotype is present; expressivity is the range of phenotypic variation associated with a given genotype. Common alleles can make autosomal recessive traits appear to be dominantly inherited. These and other variations in phenotypic expression can affect pedigree analysis and the assignment of genotypes to members of the pedigree.

Questions and Problems

CENGAGE**NOW** Preparing for an exam? Assess your understanding of this chapter's topics with a pre-test, a personalized learning plan, and a post-test by logging on to *login.cengage.com/sso* and visiting CengageNOW's Study Tools.

Pedigree Analysis Is a Basic Method in Human Genetics

1. What are the reasons that pedigree charts are used?
2. Pedigree analysis permits all of the following except:
   a. an orderly presentation of family information
   b. the determination of whether a trait is genetic
   c. the determination of whether a trait is dominant or recessive
   d. an understanding of which gene is involved in a heritable disorder
   e. the determination of whether a trait is sex-linked or autosomal
3. Using the pedigree provided, answer the following questions.

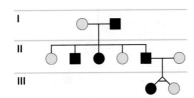

   a. Is the proband male or female?
   b. Is the grandfather of the proband affected?
   c. How many siblings does the proband have, and where is he or she in the birth order?
4. What does OMIM stand for? What kinds of information are in this database?

Analysis of Autosomal Recessive and Dominant Traits

5. a. What pattern of inheritance is suggested by the following pedigree?

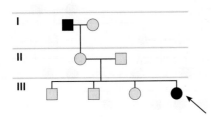

   b. For genotype assignment, assume that the pedigree is for an autosomal dominant trait and that the affected male in the first generation is heterozygous. Assign genotypes to all other individuals in the pedigree.

6. Does the indicated individual (III-5) show the trait in question?

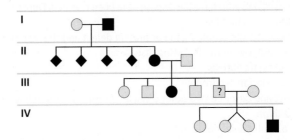

7. Use the following information to respond to the three question posed below: (1) The proband (affected individual who led to the construction of the pedigree) exhibits the trait. (2) Neither her husband nor her only sibling, an older brother, exhibits the trait. (3) The proband has five children by her current husband. The oldest is a boy, followed by a girl, then another boy, and then identical twin girls. Only the second oldest fails to exhibit the trait. (4) Both parents of the proband show the trait.
   a. Construct a pedigree of the trait in this family.
   b. Determine how the trait is inherited (go step by step to examine each possible pattern of inheritance).
   c. Can you deduce the genotype of the proband's husband for this trait?
8. In the following pedigree, assume that the father of the proband is homozygous for a rare trait. What pattern of inheritance is consistent with this pedigree? In particular, explain the phenotype of the proband.

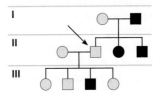

9. Using the following pedigree, deduce a compatible pattern of inheritance. Identify the genotype of the individual in question.

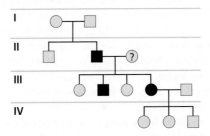

10. A proband female with an unidentified disease seeks the advice of a genetic counselor before starting a family. On the basis of the following data, the counselor constructs a pedigree encompassing three generations: (1) The maternal grandfather of the proband has the disease. (2) The mother of the proband is unaffected and is the youngest of five children, the three oldest being male. (3) The proband has an affected older sister, but the youngest siblings are unaffected twins (boy and girl). (4) All the individuals who have the disease have been revealed. Duplicate the counselor's feat.

11. Describe the primary gene or protein defect and the resulting phenotype for the following diseases:
    a. cystic fibrosis
    b. Marfan syndrome

12. List and describe two other diseases inherited in the following fashion:
    a. autosomal dominant
    b. autosomal recessive

13. The father of 12 children begins to show symptoms of Huntington disease.
    a. What is the probability that Sam, the man's second-oldest son (II-2), will suffer from the disease if he lives a normal life span? (Sam's mother and her ancestors do not have the disease.)
    b. Can you infer anything about the presence of the disease in Sam's paternal grandparents?

14. Huntington disease is a rare, fatal disease that usually develops in the fourth or fifth decade of life. It is caused by a single autosomal dominant allele. A phenotypically normal man in his twenties who has a 2-year-old son of his own learns that his father has developed Huntington disease. What is the probability that he himself will develop the disease? What is the chance that his young son will eventually develop the disease?

Analysis of X-Linked Dominant and Recessive Traits

15. The X and Y chromosomes are structurally and genetically distinct. However, they do pair during meiosis at a small region near the tips of their short arms, indicating that the chromosomes are homologous in this region. If a gene lies in this region, will its pattern of transmission be more like that of a sex-linked gene or an autosomal gene? Why?

16. What is the chance that a color-blind male and a carrier female will produce:
    a. a color-blind son?
    b. a color-blind daughter?

17. A young boy is color-blind. His one brother and five sisters are not. The boy has three maternal uncles and four maternal aunts. None of his uncles' children or grandchildren is color-blind. One of the maternal aunts married a color-blind man, and half of her children, both male and female, are color-blind. The other aunts married men who have normal color vision. All their daughters have normal vision, but half of their sons are color-blind.
    a. Which of the boy's four grandparents transmitted the gene for color blindness?
    b. Are any of the boy's aunts or uncles color-blind?
    c. Is either of the boy's parents color-blind?

18. Describe the phenotype and primary gene or protein defect of the X-linked recessive disease muscular dystrophy.

19. Suppose a couple, both phenotypically normal, have two children: one unaffected daughter and one son affected with a genetic disorder. The phenotype ratio is 1:1, making it difficult to determine whether the trait is autosomal or X-linked. With your knowledge of genetics, what are the genotypes of the parents and children in the autosomal case? In the X-linked case?

20. The following is a pedigree for a common genetic trait. Analyze the pedigree to determine whether the trait is inherited as:
    a. autosomal dominant
    b. autosomal recessive
    c. X-linked dominant
    d. X-linked recessive
    e. Y-linked

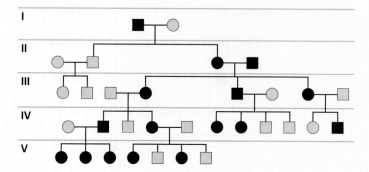

21. As a genetic counselor investigating a genetic disorder in a family, you are able to collect a four-generation pedigree that details the inheritance of the disorder in question. Analyze the information in the pedigree to determine whether the trait is inherited as:
    a. autosomal dominant
    b. autosomal recessive
    c. X-linked dominant
    d. X-linked recessive
    e. Y-linked

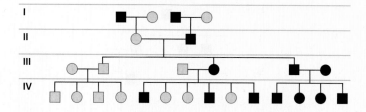

22. In the eighteenth century, a young boy with a skin condition known as ichthyosis hystrix gravior was identified. The phenotype of this disorder includes thickening of skin and the formation of loose spines that are sloughed off periodically. This man married and had six sons, all of whom had the same condition. He also had several daughters, all of whom were unaffected. In all succeeding generations, the condition was passed on from father to son. What can you theorize about the location of the gene that causes ichthyosis hystrix gravior?

Maternal Inheritance: Mitochondrial Genes

23. What are the unique features of mitochondria that are not present in other cellular organelles in human cells?

24. How is mitochondrial DNA transmitted?

Variations in Phenotype Expression

25. Define penetrance and expressivity.

26. Suppose space explorers discover an alien species governed by the same genetic principles that apply to humans.

Although all 19 aliens analyzed to date carry a gene for a third eye, only 15 display this phenotype. What is the penetrance of the third-eye gene in this population?

27. A genetic disorder characterized by falling asleep in genetics lectures is known to be 20% penetrant. All 90 students in a genetics class are homozygous for this gene. Theoretically, how many of the 90 students will fall asleep during the next lecture?

28. Explain how camptodactyly is an example of expressivity.

Internet Activities

Internet Activities are critical-thinking exercises using the resources of the World Wide Web to enhance the principles and issues covered in this chapter. For a full set of links and questions investigating the topics described below, visit *www.cengage.com/biology/cummings*

1. *A Database of Human Genetic Disorders and Traits.* The Internet site *Online Mendelian Inheritance in Man*, or OMIM, is an online database of human genetic disorders and genetically controlled traits that is updated daily. For any specific disorder or trait, information on symptoms, mode of inheritance, molecular genetics, diagnosis, therapies, and more is given.

2. *Genetic Disorders and Support.* Information about many genetic disorders and support groups and organizations for persons with genetic disorders is available on the World Wide Web. You can find information about a particular disorder, its treatments, or parent groups through Web search engines. In addition, this text's home page has a link to a list of genetic support groups

from which you can obtain more information about a specific genetic disorder.

3. *Would You Want to Know If You Carried the Gene for a Disorder?* Not all dominant genetic disorders are obvious in early life, and, of course, an individual may be a carrier for a recessive disorder without displaying the characteristics of the trait. *Do You Really Want to Know If You Have a Disease Gene?* Journalist and author Robin Henig explores this question, which we will return to in Chapter 16, Reproductive Technology, Gene Therapy, and Genetic Counseling. *Further Exploration.* For a simple version of the genetics of left-handedness, in addition to a look at what life is like for a southpaw, check out *Lorin's Left-handness Site.*

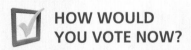 **HOW WOULD YOU VOTE NOW?**

In the emerging field of biohistory, researchers use genetic testing to investigate the lives and deaths of historical figures. On the basis of a pedigree analysis and some contemporary accounts, some scientists and historians believe that Abraham Lincoln had the genetic disorder Marfan syndrome. Genetic testing could provide the final answer, but there has been debate about the value of such information and the ethics of researching it. Now that you know more about pedigree analysis and inheritance in humans, what do you think? Should scientists perform tests to determine whether Lincoln had Marfan syndrome? Visit the *Human Heredity* companion website at *www.cengage.com/biology/cummings* to find out more on the issue; then cast your vote online.

5 The Inheritance of Complex Traits

I n 1713, a new king was crowned in Prussia, and immediately began one of the largest military buildups of the eighteenth century. In the space of 20 years, King Frederick William I, ruler of fewer than 2 million citizens, enlarged his army from around 38,000 men to slightly less than 100,000 troops. Compare Frederick's army with that of the neighboring kingdom of Austria, with 20 million citizens and an army about the same size as Prussia's, and you will understand why Frederick William was regarded as a military monomaniac. The crowning glory of this military machine was his personal troops, the Potsdam Grenadier Guards. This unit was composed of the tallest men that could be found. Frederick William was obsessed with having giants in his guard unit, and his recruiters used bribery, kidnapping, and smuggling to fill the ranks. It is said that while marching, members of the guard could lock arms across the top of the king's carriage. The minimum height requirement was 5 feet 11 inches, but some of the soldiers were close to 7 feet tall (the king was about 4 feet 11 inches). Although someone 7 feet tall is not much of a novelty in today's NBA, any man taller than about 5 feet 4 inches in eighteenth-century Prussia was above average height.

King Frederick William was also rather miserly, and his recruiting campaign was very expensive. To save money, he decided that it would be more economical to breed giants to serve in his elite unit. So he ordered that every tall man in the kingdom marry a tall, robust woman, expecting that the offspring would all be tall and that some would be giants. Unfortunately, his expectations fell far short of reality. Not only was his program slow (with humans, it takes 18 to 20 years to produce mature adults), but most of the children were actually shorter than their parents. While continuing this breeding program, the king reverted to kidnapping and bounties. He also let it be known that the best way for foreign governments to gain his favor was to send giants to be members of his guard. This human breeding experiment continued until shortly after the king's death in 1740, when his son, Frederick the Great, disbanded the Potsdam Guards.

Image copyright Monkey Business Images, 2010. Under license from Shutterstock.com.

The phenotypic differences seen among brothers and sisters is due to both genetic and environmental factors.

5.1 Some Traits Are Controlled by Two or More Genes

What exactly went wrong with King Frederick William's experiment in human genetics? After all, when Mendel intercrossed true-breeding tall pea plants, the offspring were all tall. Even when heterozygous tall pea plants are crossed, three-fourths of the offspring are tall. The situation in humans is more complex than King Frederick imagined, and, as we will see, the chances that his breeding program would have worked were very small.

 HOW WOULD YOU VOTE?

King Frederick William's program of selective breeding of tall humans was a failure, and today such programs would be condemned as unethical. In our time, it is possible to fertilize eggs outside the womb and test the resulting embryos for their genetic characteristics before implanting them in a woman's uterus (discussed in Chapter 14). One possible application of this technology would be to test for genetic markers associated with traits such as high IQ levels or body muscle patterns associated with athletes. Would you consider having such tests done and implanting only those embryos carrying such markers? Visit the *Human Heredity* companion website at *www.cengage.com/biology/cummings* to find out more on the issue; then cast your vote online.

KEEP IN MIND AS YOU READ

- Many human diseases are controlled by the actions of several genes.
- Environmental factors interact with genes to produce variations in phenotype.
- The genetic contribution to phenotypic variation can be estimated.
- Twin studies provide an insight into the genetic contribution to phenotypic variance.
- Many multifactorial traits have social and cultural impacts.

Phenotypes can be discontinuous or continuous.

The problem with comparing the inheritance of height in pea plants with the inheritance of height in humans is that a single gene pair controls height in pea plants, whereas in humans height is a complex trait determined by several gene pairs and environmental interactions. The tall and short phenotypes in Mendel's pea plants are examples of **discontinuous variation** (Figure 5.1a). Interestingly, if Mendel had chosen to study height in tobacco plants, he would have encountered continuous phenotypic variation (Figure 5.1b). In humans, it is difficult to set up only two phenotypes for height. Instead, height in humans is an example of a phenotype with **continuous variation.** Unlike Mendel's pea plants, people are not either 18 inches or 84 inches tall; they fall into a series of overlapping phenotypic classes (Active Figure 5.2).

Understanding the distinction between discontinuous and continuous traits was an important advance in human genetics and led to the understanding that some traits are more complex than previously thought, that genes can interact with each other and with the environment, and that individual traits can be controlled by several genes.

What are complex traits?

Complex traits are determined by the cumulative effects of several genes and the influence of environmental factors. Traits controlled by two or more genes are called

Discontinuous variation Phenotypes that fall into two or more distinct, nonoverlapping classes.

Continuous variation A distribution of phenotypic characters that is distributed from one extreme to another in an overlapping, or continuous, fashion.

Complex traits Traits controlled by multiple genes, the interaction of genes with each other, and with environmental factors where the contributions of genes and environment are undefined.

FIGURE 5.1 A comparison of a trait (height) that shows discontinuous and continuous phenotypes in different plants. (a) Histograms show the percentage of plants with different heights in crosses between tall and dwarf strains of the pea plant. The F_1 generation has the tall phenotype, and the F_2 has two distinct phenotypic classes: 75% of the offspring are tall, and 25% are dwarf. (b) Histograms show the percentage of plants with different heights in crosses between tall and dwarf strains of tobacco plants carried to the F_2 generation. The F_1 generation is intermediate to the parents in height, and the F_2 shows a range of phenotypes from dwarf to tall. Most plants have a height intermediate to those of the P_1 generation. The differences between the pea plants and tobacco plants are explained by the fact that height in tobacco plants is controlled by two or more gene pairs, whereas height in peas is controlled by a single gene.

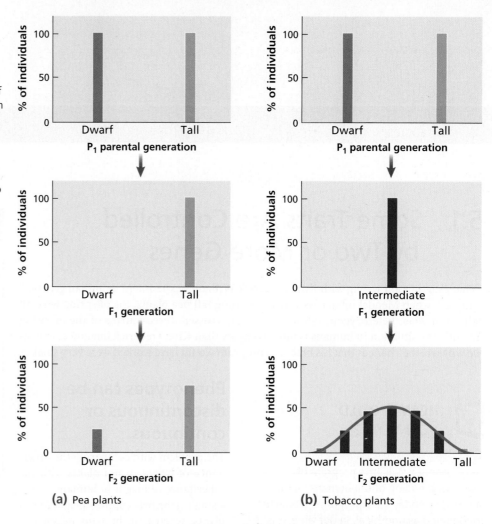

(a) Pea plants

(b) Tobacco plants

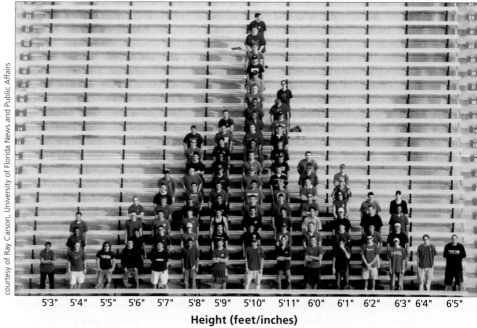

courtesy of Ray Carson, University of Florida News and Public Affairs

Height (feet/inches)

ACTIVE FIGURE 5.2 An example of continuous variation: biology students organized according to height.

CENGAGENOW Learn more about continuous variation by viewing the animation by logging on to *login.cengage.com/sso* and visiting CengageNOW's Study Tools.

polygenic traits; those controlled by two or more genes and that show significant environmental interactions are called **multifactorial traits.** Although each gene controlling multifactorial traits is inherited in Mendelian fashion, the interaction of genes with each other and with the environment produces variable phenotypes that often do not show clear-cut Mendelian ratios. Human height, for example, is a multifactorial trait; it is controlled by several genes, and environmental factors make significant contributions to variations in its expression.

Multifactorial inheritance underlies many human traits and diseases. Analysis of such traits is complicated by the fact that each gene contributes only a small amount to the phenotype and the environmental components can be hard to identify and measure. Complex traits can be fully understood only when all the genetic and environmental components are fully identified and their individual effects and interactions have been measured.

In this chapter, we examine traits controlled by two or more genes (polygenic traits) and traits controlled by two or more genes with significant environmental influences (multifactorial traits). In multifactorial inheritance, the degree to which genetics contributes to a trait can be estimated by measuring heritability. We consider this concept and the use of twins as a means of measuring the heritability, often using twin studies. In the last part of the chapter, we examine a number of human complex traits, some of which have been the subject of political and social controversy.

Polygenic traits Traits controlled by two or more genes.

Multifactorial traits Traits that result from the interaction of one or more environmental factors and two or more genes.

KEEP IN MIND

Many human diseases are controlled by the actions of several genes.

5.2 Polygenic Traits and Variation in Phenotype

In the early part of the twentieth century, interest in human genetics was largely centered on determining whether "social" traits, such as alcoholism, feeblemindedness, and criminal behavior, were inherited. Some geneticists constructed pedigrees and simply assumed that single genes controlled those traits. Other geneticists pointed out that those traits did not show the phenotypic ratios observed in experimental organisms and concluded that Mendelian inheritance might not apply to humans.

KEEP IN MIND

Environmental factors interact with genes to produce variations in phenotype.

Defining the genetics behind continuous phenotypic variation.

The controversy surrounding discontinuous versus continuous inheritance was resolved by crosses with experimental organisms. Work with these organisms showed that continuous phenotypic variations can be explained by Mendelian inheritance. In other words, traits determined by several genes can show a continuous distribution of phenotypes in the F2 generation, even though genotypic inheritance for each gene follows the rules of Mendelian inheritance.

In traits controlled by two or more genes, a small number of the offspring have phenotypes identical to the P1 generation (very short or very tall, for example). Most F2 offspring, however, have phenotypes between those extremes; their distribution follows

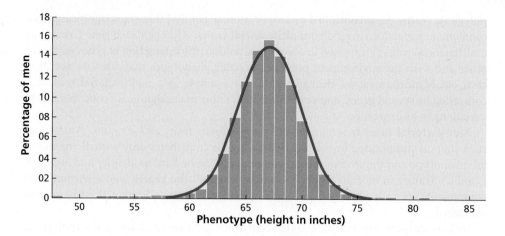

FIGURE 5.3 A bell-shaped, or "normal," curve shows the distribution of phenotypes for traits controlled by two or more genes. In a normal curve, few individuals are at the extremes of the phenotype, and most individuals are clustered around the average value. In this case, the phenotype is height measured in a population of human males.

a bell-shaped curve (Figure 5.3), with each gene adding a small but equal amount to the phenotype.

Polygenic inheritance has several distinguishing characteristics:

- Traits are usually quantified by measurement rather than by counting.
- Two or more genes contribute to the phenotype. Each gene contributes to the phenotype, and the effect of individual genes may be small.
- Phenotypic expression of polygenic inheritance varies across a wide range. This variation is best analyzed in populations rather than in individuals (Figure 5.4).

Polygenic inheritance is an important concept in human genetics. Traits such as height, weight, skin color, eye color, and intelligence are under polygenic control. In addition, congenital malformations such as neural tube defects, cleft palate, and clubfoot, as well as genetic disorders such as diabetes and hypertension, along with some behavioral disorders, are polygenic and/or multifactorial traits.

How many genes control a polygenic trait?

The distribution of phenotypes and F2 ratios in traits controlled by two, three, or four genes is shown in Figure 5.5. If two genes control a trait and each has a dominant and a recessive allele, there are five phenotypic classes in the F2, each of which is controlled by four, three, two, one, or zero dominant alleles. The F2 ratio of 1:4:6:4:1 results from the genotypic combinations that produce each phenotype. At one extreme is the homozygous dominant (AABB) genotype with four dominant alleles; at the other extreme is the homozygous recessive (aabb) genotype with no dominant alleles. The largest phenotypic class (6/16) has six genotypic combinations (AAbb, AaBb, aaBB, etc.). The five basic

FIGURE 5.4 Skin color is a polygenic trait controlled by three or four genes, producing a wide range of phenotypes. Environmental factors (exposure to the sun and weather) also contribute to the phenotypic variation, making this a multifactorial trait.

human eye colors (Figure 5.6) can be explained by a model using two genes (*A* and *B*), each of which has two alleles (*A* and *a*, *B* and *b*).

As the number of loci that controls a trait increases, the number of phenotypic classes increases. As the number of phenotypes increases, the difference between each class decreases. This means that there is a greater chance for environmental factors to override the small genotypic differences between classes, blending the phenotypes together to form a continuous distribution, or bell-shaped curve. For example, exposure to sunlight can alter skin color and obscure genotypic differences.

5.3 The Additive Model for Polygenic Inheritance

To explain how polygenes contribute to a trait and how genotypes contribute to variation in phenotypic expression, let's consider a simple model for polygenic inheritance. For this example, we will examine the model under the following conditions:

- The trait is controlled by three genes, each of which has two alleles (*A,a, B,b, C,c*).
- Each dominant allele makes an equal contribution to the phenotype, and recessive alleles make no contribution.
- The effect of each active (dominant) allele on the phenotype is small and additive.
- The genes controlling the trait are not linked; they assort independently.
- The environment acts equally on all genotypes.

For our example, let's look at King Frederick William's attempt to breed giants for his elite guard unit. We will assume that this breeding program used women who were at least 5 feet 9 inches tall. For simplicity, we'll also assume that the dominant alleles *A*, *B*, and *C* each add 3 inches above a base height of 5 feet 9 inches and that the recessive alleles *a*, *b*, and *c* add nothing above the base height. An individual with the genotype *aabbcc* would be 5 feet 9 inch tall, and an individual with the genotype *AABBCC* would be 7 feet 3 inches tall.

Suppose that a 6 foot 9 inch member of the guard with the genotype *AaBbCC* mates with a 6 foot 3 inch woman with the genotype *AaBbcc*. The possible genotypic and phenotypic outcomes are diagrammed in Figure 5.7. In this case, there are 4 paternal and 4 maternal gamete combinations, and 16 possible types of fertilizations with 5 phenotypic classes.

The frequency distribution of phenotypes from the possible offspring are diagrammed in Figure 5.8. As the king discovered to his frustration (after waiting about 18 years for them to grow up), most of the children tend toward the average height (6 feet 6 inches) between the two parents. In fact, 10 of the 16 genotypic combinations will result in children shorter than their fathers.

In this example the genotype represents the genetic potential for height. Full expression of the genotype depends on the environment.

2 genes

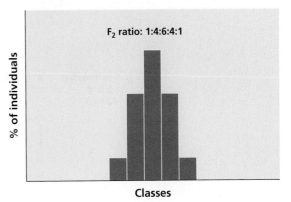

F$_2$ ratio: 1:4:6:4:1

3 genes

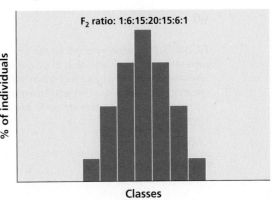

F$_2$ ratio: 1:6:15:20:15:6:1

4 genes

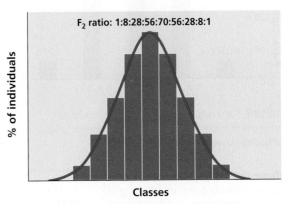

F$_2$ ratio: 1:8:28:56:70:56:28:8:1

FIGURE 5.5 The number of phenotypic classes in the F2 generation increases as the number of genes controlling the trait increases. This relationship allows geneticists to estimate the number of genes involved in expressing a polygenic trait. As the number of phenotypic classes increases, the distribution of phenotypes becomes a normal curve.

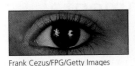

Frank Cezus/FPG/Getty Images

Frank Cezus/FPG/Getty Images

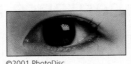

©2001 PhotoDisc

Ted Beaudin/FPG/Getty Images

Stan Sholik/FPG/Getty Images

FIGURE 5.6 Samples from the range of continuous variation in human eye color. Different alleles of more than one gene interact to synthesize and deposit melanin in the iris. Combinations of alleles result in small differences in eye color, making the distribution for eye color appear to be continuous over the range from light blue to black.

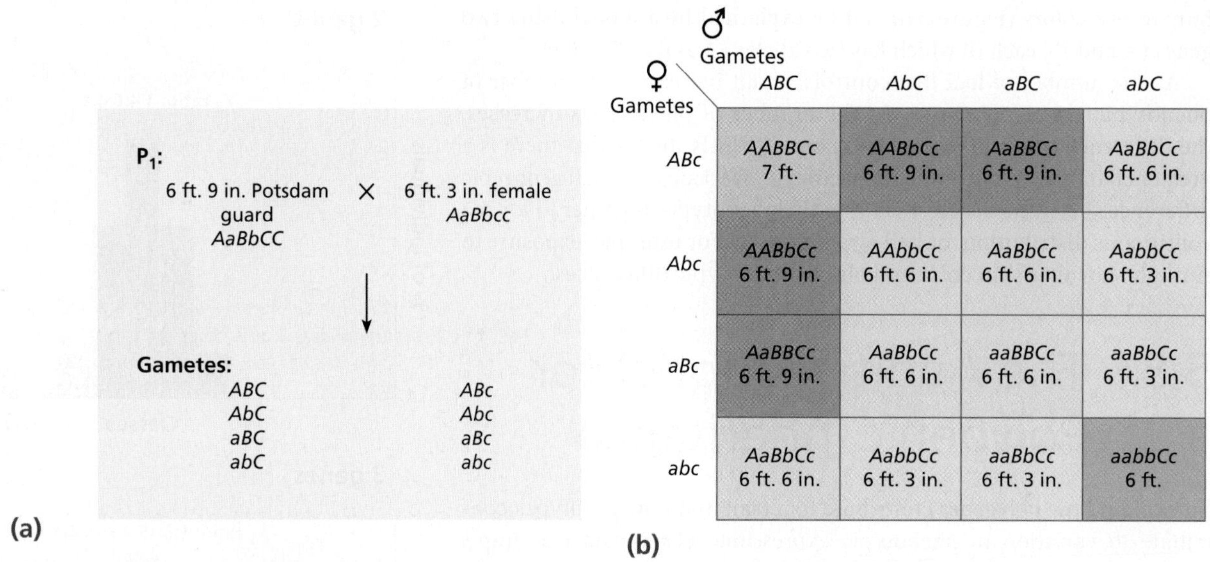

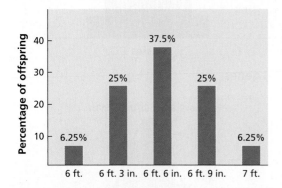

P₁:

6 ft. 9 in. Potsdam ✕ 6 ft. 3 in. female
guard   *AaBbcc*
AaBbCC

Gametes:

ABC	*ABc*
AbC	*Abc*
aBC	*aBc*
abC	*abc*

(a)

♀ Gametes \ ♂ Gametes	*ABC*	*AbC*	*aBC*	*abC*
ABc	*AABBCc* 7 ft.	*AABbCc* 6 ft. 9 in.	*AaBBCc* 6 ft. 9 in.	*AaBbCc* 6 ft. 6 in.
Abc	*AABBCc* 6 ft. 9 in.	*AAbbCc* 6 ft. 6 in.	*AaBbCc* 6 ft. 6 in.	*AabbCc* 6 ft. 3 in.
aBc	*AaBBCc* 6 ft. 9 in.	*AaBbCc* 6 ft. 6 in.	*aaBBCc* 6 ft. 6 in.	*aaBbCc* 6 ft. 3 in.
abc	*AaBbCc* 6 ft. 6 in.	*AabbCc* 6 ft. 3 in.	*aaBbCc* 6 ft. 3 in.	*aabbCc* 6 ft.

(b)

FIGURE 5.7 An additive model for the inheritance of height in the Potsdam Guards. In this example, the guards and their mates represent a subset of individuals in a population where height can range from 5 feet 9 inches (*aabbcc*) to 7 feet 3 inches (*AABBCC*). (a) Gametes produced by a 6 foot 9 inch male and a 6 foot 3 inch female. (b) Punnett square showing the 16 genotypic and 5 phenotypic combinations that result from fertilization of all combinations of gametes. The genotypes resulting in children who are as tall or taller than their father are noted. Most of the children will have a height intermediate to their parents, showing regression to the mean.

FIGURE 5.8 Frequency distribution of phenotypes from the possible offspring in Figure 5.7. Height of the offspring shows regression to the mean.

Regression to the mean In a polygenic system, the tendency of offspring of parents with extreme differences in phenotype to exhibit a phenotype that is the average of the two parental phenotypes.

Poor nutrition during childhood can prevent people from reaching their potential heights. On the other hand, optimal nutrition from birth to adulthood cannot make someone taller than the genotype dictates.

Averaging out the phenotype is called regression to the mean.

Sir Francis Galton, a cousin of Charles Darwin, studied the inheritance of many traits in humans. He noticed that children of tall parents were usually shorter than their parents and that children of short parents were usually taller than their parents. Children of very tall or very short parents are usually closer to the average height of the population rather than the average height of their parents. This important concept is called **regression to the mean** and explains why King Frederick William's attempt to breed giants for his elite guard unit failed. Using very tall parents (say, at least 5 feet 9 inches) results in more children with a height that is the average between the parents than it does tall children.

When you add in the fact that many of the Potsdam Grenadier Guards probably were tall because of environmental factors (endocrine malfunctions) and did not have the genotypes to produce tall offspring under any circumstances, it is easy to see why Frederick's program was a failure.

5.4 Multifactorial Traits: Polygenic Inheritance and Environmental Effects

In considering the interaction of polygenes and environmental factors, let's first review some basic concepts: (1) The genotype represents the genetic constitution of an individual. It is fixed at the moment of fertilization and, barring mutation, is unchanging. (2) The

Dissecting Genes and Environment in Spina Bifida

Spina bifida (SB) is one of the most common and most complex birth defects involving the nervous system. It occurs with a frequency of 1-2 per 1,000 births in the United States, but is higher in other populations. Spina bifida is one of a group of disorders called neural tube defects. The neural tube forms early in embryonic development and gives rise to the brain and the spinal cord. Neural tube defects occur during days 17 and 30 of development (the embryo is about the size of a grain of rice at this time) often before a woman realizes she is pregnant. Diagnosis usually occurs by ultrasound during week 15-17 of development. SB is a highly variable disorder, ranging from stillbirth to a form only discovered in apparently normal individuals by X-ray. Many affected individuals have nervous system problems that cause muscle imbalance resulting in crippling deformities and varying degrees of paralysis. In addition, most SB individuals have learning disabilities and may have bladder and bowel problems.

Family and twin studies show that SB has a significant genetic component. It is a multifactorial trait with significant environmental components; nutrition is a key factor. As with other complex traits, it is not clear whether SB is caused by a few rare genes (<10) each of which has a major effect, or is caused by many common genes (>100), each of which has a small effect. One gene associated with SB has been identified. This gene, VANGL1, normally controls the movement of cells during development. Mutations in this gene may cause abnormalities in neural tube formation, but how often this happens is not yet known.

Research has shown that nutrition, especially the amount of folate in the diet has a significant impact on the frequency of SB. Folate is a vitamin found in green, leafy vegetables, peas, and beans. It is essential for the formation of new cells, and is important for normal development. A diet rich in folate has been shown to reduce the incidence of SB by about 70% and reduces the severity of defects when they do occur. In the U.S., grain products have been fortified with folic acid since 1998. Women of childbearing age should eat a folate-rich diet to reduce the risk of having a child with spina bifida.

---

phenotype is the sum of the observable characteristics. It is variable and undergoes continuous change throughout the life of the organism. (3) The environment in which a gene exists and operates includes all other genes in the genotype and their effects and interactions, as well as all nongenetic factors, whether physical or social, that can interact with the genotype (see The Genetic Revolution: Dissecting Genes and Environment in Spina Bifida).

Multifactorial traits have several important characteristics:

- Traits are polygenic (controlled by several genes).
- Each gene controlling the trait contributes a small amount to the phenotype.
- Environmental factors interact with the genotype to produce the phenotype.

We can then ask what fraction of the total phenotypic variance is caused by genetic differences among individuals.

Several methods are used to study multifactorial traits.

Here we will briefly consider two ways of studying the genetic components of multifactorial traits: the threshold model and recurrence risk. Later in the chapter, we will briefly discuss the role of animal models in dissecting complex traits in humans and consider another method for studying multifactorial inheritance, called genome-wide association studies (GWAS).

Some multifactorial traits do not show a continuous phenotypic distribution; individuals are either affected or not. Congenital birth defects, such as clubfoot or cleft palate, are examples of traits that are distributed discontinuously but are, in fact,

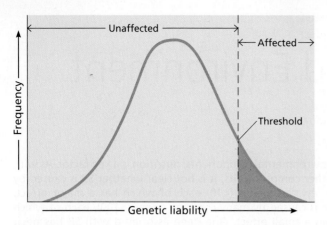

FIGURE 5.9 The threshold model explains the discontinuous distribution of some multifactorial traits. In this model, liability for a genetic disorder is distributed among individuals in a normal curve. This liability is caused by a number of genes, each acting additively. Only those individuals who have a genetic liability above a certain threshold are affected if exposed to certain environmental conditions. The severity of the disease usually increases as genetic liability moves away from the mean, and is affected by environmental factors.

multifactorial. In the threshold model, liability for a genetic disorder is distributed among individuals in a bell-shaped curve. This liability is in the form of genotypes, but only a limited number of genotypes express the phenotype (Figure 5.9). The predisposition is caused by a number of genes, each of which contributes to the liability in an additive way. Individuals with a liability above a certain genetic threshold will develop the genetic disorder if exposed to the proper environmental conditions. In other words, environmental conditions are most likely to have the greatest impact on those individuals with the highest level of genetic predisposition.

The threshold model is useful for explaining the frequency of certain disorders and congenital malformations. Evidence for a threshold in any specific disorder is indirect and comes mainly from family studies. To look for threshold effects in families, the frequency of the disorder among relatives of affected individuals is compared with the frequency of the disorder in the general population. In a family, first-degree relatives (parents-children) have one-half of their genes in common, second-degree relatives (grandparents-grandchildren) have one-fourth of their genes in common, and third-degree relatives (first cousins) have one-eighth of their genes in common. As the degree of relatedness declines, so does the probability that individuals will have the same combination of alleles for the genes that control the trait.

According to the threshold model, risk for a disorder should decrease as the degree of relatedness decreases. In fact, the distribution of risk for some congenital malformations, as shown in Table 5.1, declines as the degree of relatedness declines. The multifactorial threshold model provides only indirect evidence for the effect of genotype on traits and for the degree of interaction between the genotype and the environment. The model is helpful, however, in genetic counseling for predicting recurrence risks in families that have certain congenital malformations and multifactorial disorders.

In multifactorial disorders, the risk of recurrence depends on several factors. These include:

consanguinity—First-cousin parents have about a twofold higher risk than unrelated parents of having a child with a multifactorial disease because of the shared genes they carry.

previous affected child—If parents have two affected children, it means that their genotypes are probably close to the threshold, increasing the risk of recurrence.

severity of defect—A severely affected phenotype means that the affected child's genotype is well over the threshold and that the parental genotypes confer a higher recurrence risk on future children.

Table 5.1 Familial Risks for Multifactorial Threshold Traits

Multifactorial Trait	MZ Twins	Risk Relative to General Population		
		First-Degree Relatives	Second-Degree Relatives	Third-Degree Relatives
Clubfoot	300×	25×	5×	2.0×
Cleft lip	400×	40×	7×	3.0×
Congenital hip dislocation (females only)	200×	25×	3×	2.0×
Congenital pyloric stenosis (males only)	80×	10×	5×	1.5×

higher frequency in one sex—If the multifactorial disease is expressed more often in one sex than the other, the threshold in the less frequently affected sex is shifted to the right, and the recurrence risk for children of that sex is lower.

In multifactorial traits, the phenotype is triggered when a genetic predisposition is affected by environmental factors. Thus, for these disorders, it is important to assess the role of the environment as well as the interaction of the environment and the genotype. The question is: How can we measure this interaction? In the following sections, we will examine this question and survey some of the methods used to assess the genotypic contribution to the phenotypic variation.

> **KEEP IN MIND**
>
> The genetic contribution to phenotypic variation can be estimated.

5.5 Heritability Measures the Genetic Contribution to Phenotypic Variation

To measure the interaction between the genotype and the environment, we first must examine phenotypic variation in a population rather than individuals in the population. Phenotypic variation is derived from two sources: (1) different genotypes present in the population and (2) different environments in which identical genotypes are expressed.

Phenotypic variation caused by the presence of different genotypes in the population is known as **genetic variance**. Phenotypic variation among individuals with the same genotype is known as **environmental variance**.

The proportion of the total phenotypic variation that is due to genetic differences is called the **heritability** of a trait. Heritability uses a single number between 0 and 1 to express the fraction of phenotypic variation among individuals in a population that is due to their genotypes. In general, if heritability is high (it is 100% when $H = 1.0$), the observed variation in phenotypes is genetic, with little or no environmental contribution. If heritability is low (it is zero when $H = 0.0$), there is little or no genetic contribution to the observed phenotypic variation, and the environmental contribution is high.

Genetic variance The phenotypic variance of a trait in a population that is attributed to genotypic differences.

Environmental variance The phenotypic variance of a trait in a population that is attributed to differences in the environment.

Heritability An expression of how much of the observed variation in a phenotype is due to differences in genotype.

Correlation coefficients Measures the degree of interdependence of two or more variables.

Heritability estimates are based on known levels of genetic relatedness.

Heritability is calculated by using relatives because we know the fraction of genes shared by related individuals. These relationships are expressed as a **correlation coefficient**, or the fraction of genes shared by two relatives. A child receives half of his or her genes from each parent. The half set of genes received by a child from its parent corresponds to a correlation coefficient of 0.5. The genetic relatedness of identical twins is 100%, and the correlation coefficient therefore is 1.0. In such twins, all phenotypic differences may be due to environmental factors. Unless a mother and a father are related by descent, they should be genetically unrelated, and the correlation coefficient for this relationship is 0.0.

Using the genetic relatedness among population members expressed as a correlation coefficient and using the measured phenotypic variation expressed in quantitative units (inches, pounds, etc.), a heritability value can be calculated for a specific phenotype in a population. If the heritability value for a trait is 0.72, this means that 72% of the phenotypic variability seen in the population is caused by genetic differences in the population.

5.6 Twin Studies and Multifactorial Traits

Using correlation coefficients to measure the amount of observed phenotypic variability provides an estimate of heritability. This method, however, has one main problem: The closer the genetic relationship is, the more likely it is that the relatives have a common environment. In other words, how can we tell whether parents and children have similar phenotypes when it could be either because they have one-half of their genes in common or because they have a similar environment? Is there a way we can separate the effects of genotype on phenotypic variation from the effects of the environment?

To solve this problem, human geneticists look for situations in which genetic and environmental influences are clearly separated. One way to do this is to study twins (Figure 5.10). Identical twins have the same genotype. If identical twins are separated at birth and raised in different environments, the genotype is constant and the environments are different. To reverse the situation, geneticists compare traits in unrelated adopted children with those of natural children in the same family. In this situation, there is a similar environment and maximum genotypic differences. As a result, twin studies and adoption studies are important tools in measuring heritability in humans.

The biology of twins includes monozygotic and dizygotic twins.

Before examining the results of twin studies, let's briefly look at the biology of twinning. There are two types of twins: **monozygotic (MZ)** (identical) and **dizygotic (DZ)** (fraternal). Monozygotic twins originate from a single egg fertilized by a single sperm (Figure 5.11a). During an early stage of development, two separate embryos are formed. Additional splitting is also possible (see Exploring Genetics: Twins, Quintuplets, and Armadillos on page 106). Because they result from a single fertilization event, MZ twins have the same genotype, have the same sex, and carry the same genetic markers, such as blood types. Dizygotic twins originate from two separate fertilization events: Two eggs, ovulated in the same ovarian cycle, are fertilized independently by two different sperm (Figure 5.11b). DZ twins are no more related than are other pairs of siblings, have half of their genes in common, can differ in sex, and may have different genetic markers, such as blood types.

For heritability studies, it is essential to know whether a pair of twins is MZ or DZ. Comparison of many traits such as blood groups, sex, eye color, hair color, fingerprints, palm and sole prints, DNA fingerprinting, and analysis of DNA molecular markers are used to identify twins as MZ or DZ pairs.

Monozygotic (MZ) Twins derived from a single fertilization involving one egg and one sperm; such twins are genetically identical.

Dizygotic (DZ) Twins derived from two separate and nearly simultaneous fertilizations, each involving one egg and one sperm. Such twins share, on average, 50% of their genes.

FIGURE 5.10 Identical twins (monozygotic twins) have the same sex and share a single genotype.

Rosemarie Gearhart/iStockphoto

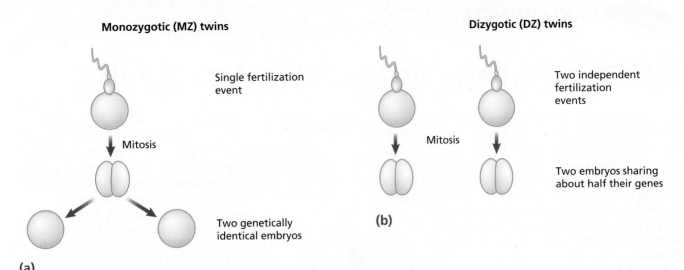

Monozygotic (MZ) twins

Single fertilization event

Mitosis

Two genetically identical embryos

(a)

Dizygotic (DZ) twins

Two independent fertilization events

Mitosis

Two embryos sharing about half their genes

(b)

FIGURE 5.11 (a) Monozygotic (MZ) twins result from the fertilization of a single egg by a single sperm. After one or more mitotic divisions, the embryo splits in two and forms two genetically identical individuals. (b) Dizygotic (DZ) twins result from the independent fertilization of two eggs by two sperm during the same ovulatory cycle. Although these two embryos simultaneously occupy the same uterine environment, they share only about half of their genes.

The study of heritability in twins makes several important assumptions: MZ twins share all their genes and their environment; DZ twins share half their genes and their environment. For a multifactorial trait such as height, if heritability is high (the variation in phenotype is largely genetic), then MZ twins will be closer in height than DZ twins. If heritability is low and variation in height is due mostly to environmental factors, then MZ twins should vary in height as much as DZ twins.

Concordance rates in twins.

To evaluate phenotypic differences between twins, traits are scored as present or absent rather than measured quantitatively. Twins show **concordance** if both have a trait and are discordant if only one twin has that trait. As was noted, MZ twins have 100% of their genes in common, whereas DZ twins on average have 50% of their genes in common. For a genetically determined trait, the correlation in MZ twins should be higher than that in DZ twins. If the trait is completely controlled by genes, concordance should be 1.0 in MZ twins and close to 0.5 in DZ twins.

The degree of difference in concordance between MZ and DZ twins is important; the greater the difference, the greater the heritability. Concordance values for several traits are listed in Table 5.2. The concordance value for cleft lip in MZ twins is higher than that for

Concordance Agreement between traits exhibited by both twins.

Table 5.2 Concordance Values in Monozygotic (MZ) and Dizygotic (DZ) Twins

Trait	Concordance Values (%)	
	MZ Twins	DZ Twins
Blood types	100	66
Eye color	99	28
Mental retardation	97	37
Hair color	89	22
Down syndrome	89	7
Handedness (left or right)	79	77
Epilepsy	72	15
Diabetes	65	18
Tuberculosis	56	22
Cleft lip	42	5

Twins, Quintuplets, and Armadillos

Monozygotic (MZ) twins are genetically identical because of the way in which they are formed. The process of embryo splitting that gives rise to MZ twins can be considered a form of human asexual reproduction. In fact, another mammal, the nine-banded armadillo, produces litters of genetically identical, same-sex offspring that result from embryo splitting. In armadillo reproduction, a single fertilized egg splits in two and daughter embryos can split again, resulting in litters of two to six genetically identical offspring.

Multiple births in humans occur rarely. About 1 in 7,500 births are triplets, and 1 in 658,000 births are quadruplets. In many cases, both embryo splitting and multiple fertilizations are responsible for naturally occurring multiple births. Triplets may be

Tetra Images/Getty Images

produced by the fertilization of two eggs in which one of them undergoes embryo splitting. The use of hormones to enhance fertility has slightly increased the frequency of multiple births. These drugs work by inducing the production of multiple eggs in a single menstrual cycle. The subsequent fertilizations have resulted in multiple births that have ranged from twins to septuplets.

Embryo splitting in naturally occurring births was documented in the Dionne quintuplets born in May 1934. That was the first case in which all five members of a set of quintuplets survived. Blood tests and physical similarities indicate that those quintuplets arose from a single fertilization followed by several embryo splits. From this, it seems that MZ twins, armadillos, and the Dionne quintuplets have something in common: embryo splitting.

DZ twins (42% versus 5%). Although this difference suggests a genetic component to that trait, the value is so far below 100% that environmental factors are obviously important in the majority of cases. As this example shows, concordance values must be interpreted cautiously.

Concordance values are converted to heritability values using a number of statistical methods. Some heritability values derived from concordance values for obesity are listed in the right column of Table 5.3. Obesity is measured by body mass index, a measure of weight in relation to height (BMI = weight in kilograms divided by the square of height in meters). Obesity is defined as a BMI equal to or greater than 30 (about 30 pounds overweight for a 5 foot 4 inch person).

KEEP IN MIND

Twin studies provide an insight into the genetic contribution to phenotypic variance.

We can study multifactorial traits such as obesity using twins and family studies.

Obesity is a trait that runs in families and is a rapidly worsening national health problem. In 1998, 42 states had obesity rates of less than 20%. In 2008, only 1 state had an obesity rate of less than 20%, 17 states had rates between 20% and 24%, 26 states had

Table 5.3 Heritability Estimates for Obesity in Twins (from Several Studies)

Condition	Heritability
Obesity in children	0.77–0.88
Obesity in adults (weight at age 45)	0.64
Obesity in adults (body mass index at age 20)	0.80
Obesity in adults (weight at induction into armed forces)	0.77
Obesity in twins reared together or apart	
Men	0.70
Women	0.66

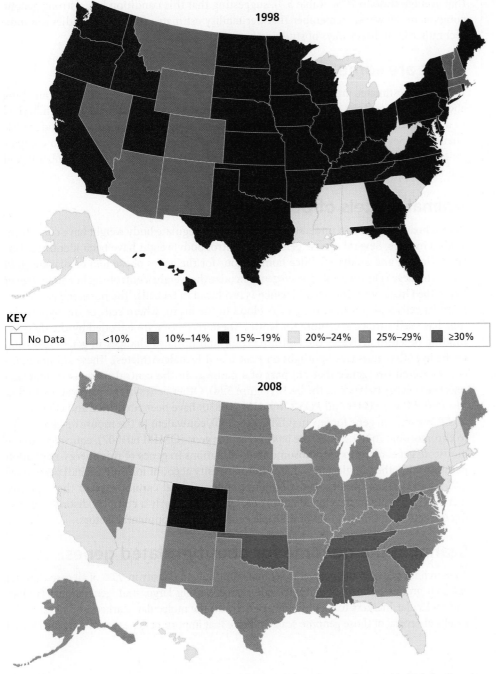

1998

KEY

☐ No Data	■ <10%	■ 10%–14%	■ 15%–19%	☐ 20%–24%	■ 25%–29%	■ ≥30%

2008

FIGURE 5.12 Fraction of obese individuals by state in 1998 and 2008. In 1998, 42 states had obesity rates below 20%, and no state had obesity rates above 25%. By 2008, only 1 state had an obesity rate below 20%, 17 had rates between 20–24%, 26 had rates between 25–29%, and 6 had rates over 30%.

rates between 25% and 29%, and 6 states had rates of more than 30% (Figure 5.12). As things now stand, about 68% of the adults in the United States are overweight and more than 33% are obese. All these individuals are at greatly increased risk for conditions such as high blood pressure, elevated blood levels of cholesterol, coronary artery disease, and adult-onset diabetes. Increases in the incidence of obesity have taken place in the last 30 to 40 years. It is unlikely that large-scale changes in our genetic makeup are responsible for this increase. Instead, we must look to changing environmental factors, including diet and physical activity interacting with our genotypes.

Twin, adoption, and family studies have been used to estimate the heritability of obesity. The results show high values of heritability for obesity, with heritability estimates

Leptin A hormone produced by fat cells that signals the brain and ovary. As fat levels become depleted, secretion of leptin slows and eventually stops.

FIGURE 5.13 The obese (*ob*) mouse mutant, shown on the left (a normal mouse is on the right), has provided many clues about how weight is controlled in humans.

that average close to 70% (Table 5.3), suggesting that this condition has a strong genetic component. However, remember that heritability estimates from twin studies are indirect rather than direct ways of studying multifactorial traits.

What are some genetic clues to obesity?

Heritability estimates are performed at the phenotypic level and cannot tell us anything about how many genes control the trait being studied; whether such genes are inherited in a dominant, recessive, or sex-linked fashion; or how such genes act to produce the phenotype. Several methods are being used to identify genes that contribute to complex traits (such as obesity) in humans. Two of these methods are animal models of obesity and genome-wide association studies.

Animal models of obesity.

Recent breakthroughs in understanding how genes regulate body weight have come from studies in mice. Several mouse genes that control body weight have been identified, isolated, cloned, and analyzed. Mice homozygous for the genes obese (*ob*) or diabetes (*db*) are both obese (Figure 5.13). The *ob* gene encodes the weight-controlling hormone **leptin** (from the Greek word for "thin"), which is produced in fat cells. The hormone is released from fat cells and travels through the blood to the brain, where cells of the hypothalamus have cell surface receptors for leptin. These receptors are encoded by the *db* gene. Binding of leptin activates the leptin receptor and initiates changes in gene expression in the hypothalamus (see Spotlight on Leptin and Female Athletes). These animal models uncovered two genes that are part of a pathway in the central nervous system that regulates energy balance in the body (Figure 5.14). Other genes in the pathway, including *MC4R* (OMIM 155541) and *POMC* (OMIM 176830) have been identified and cloned.

  LP, the human gene for leptin (OMIM 164160), equivalent to the mouse *ob* gene, maps to chromosome 7q31.1. *LEPR*, the leptin receptor gene (OMIM 601007), equivalent to the mouse *db* gene, maps to chromosome 1p31. Mutations in genes of this energy-regulation pathway and other single genes that result in obesity account for only a small percentage (5%) of all cases of obesity in the human population and cannot explain the explosive increase in obesity in developed countries. Obesity is clearly a complex disorder involving the action and interaction of multiple genes and environmental factors.

Scanning the genome for obesity-related genes.

To search for genes in the polygenic set that contribute to obesity in 95% of all cases, several research groups performed genome-wide searches using large, multigenerational families and molecular markers to identify linkage between the molecular markers and obesity. The results of several of those genome scans indicate that important genes for obesity are located

John Sholtis, The Rockefeller University, New York

on chromosomes 2, 3, 5, 6, 7, 10, 11, 17, and 20 (Figure 5.15). Further work may identify additional genes involved in obesity and provide a foundation for studying how these genes interact with environmental factors to cause obesity.

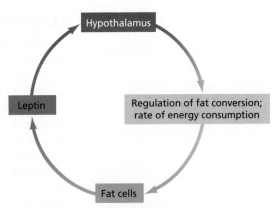

FIGURE 5.14 The hormone leptin is produced in fat cells, moves through the blood, and binds to receptors in the hypothalamus. Binding presumably activates a control mechanism (still unknown) that controls weight by regulating the conversion of food energy into fat and the rate of energy consumption.

5.7 Genetics of Height: A Closer Look

In an earlier section of the chapter, we examined a simple additive model for genes controlling height in humans. In this model, we assumed that height was controlled by a small number of genes and that some alleles made equal contributions to the phenotype, while others made no contribution. This model is useful for explaining how phenotypes result from the action of several genes, but research suggests that the number of alleles affecting a trait, their relative contributions to the phenotype, and the effect of environmental interactions on the phenotype is often difficult to measure accurately.

Let's consider some other possibilities. The phenotype of a complex trait might be controlled by a small number of genes (<10), each with a large effect. Or, at the other end of the spectrum, the trait might be controlled by a large number of genes (~100), each with very small effects. Between these extremes is the possibility that a phenotype is the product of a relatively small number of genes, each with different major effects on the phenotype, and a large number of genes, each having variable but small effects.

The development of new technology now allows researchers to use nucleotide variations (review SNPs in Chapter 1) to survey the genome to detect associations between common variations and a specific phenotype. This approach has opened new ground in the study of the genes controlling height in humans.

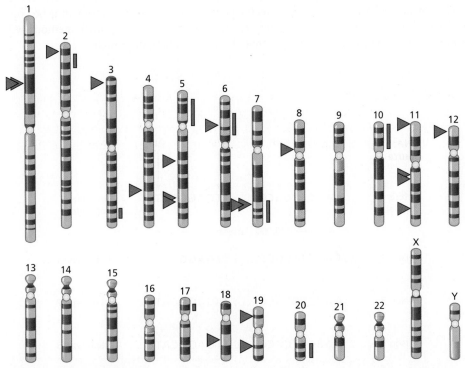

FIGURE 5.15 Some obesity genes in the human genome. The green triangles represent single genes in which obesity is the main phenotype with little environmental influence. The red triangles represent genes that are candidates for obesity genes. Mutations in these genes have an association with obesity. Red bars represent chromosome regions that may contain other genes associated with obesity. More than 70 genes associated with obesity have been identified.

Haplotypes and genome-wide association studies.

Genome-wide association (GWA) studies take advantage of the fact that the human genome contains more than 10 million (SNPs) and couples the use of a subset of 300,000 to 500,000 of these markers with technology that allows thousands of genomes to be analyzed, looking for an association between the SNP markers and a trait, such as height.

Recall from Chapter 1 that SNPs (single nucleotide polymorphisms) are single nucleotide differences in DNA sequence among individuals. Haplotypes (Figure 5.16) are combinations of SNPs (or other DNA sequence variants) located close together on a chromosome that are likely to be inherited together. In GWA studies, researchers use high-throughput technology employing up to 500,000 SNPs to scan tens of thousands of genomes to see whether a particular haplotype is found significantly more often in tall people than in short people. If so, a gene controlling height may be near the chromosome location of these SNPs, and candidate genes near the SNPs can be investigated.

Genes for human height: what have we learned so far?

GWA studies using SNPs have identified over 50 genes associated with height. Several conclusions can be drawn from these studies. First, height is clearly a polygenic trait, and alleles of many genes are involved. Second, some of the identified genes have mutant alleles involved with skeletal defects. For example, one of the first genes identified in these scans was *HMGA2* (OMIM 600698). Mice homozygous for a deletion of *HMGA2* are pygmy mice. A mutation that shortens this gene produces gigantic mice, and in humans, a chromosomal inversion that interrupts this gene is associated with an overgrowth syndrome. Third, several new pathways controlling growth have been identified, opening new areas of research into growth and growth disorders.

Perhaps the biggest surprise from these studies is that the genes discovered to date account for only about 5% of the total variation in adult height. Heritability studies indicate that about 80% of the variation in height is genetic, leaving most of the variation unexplained. GWA studies with larger sample sizes (more than 100,000 individuals) will probably identify additional genes that control adult height. However, other mechanisms such as gene-gene interactions, sex-specific gene expression, or variations in gene copy number may play a role in controlling growth. New studies using other methods will hopefully reveal more about the genes that play a role in human growth.

FIGURE 5.16 Single nucleotide polymorphisms (SNPs) are organized into blocks that are inherited together. These blocks, called haplotypes, are used as markers for specific regions on chromosomes. If a trait and a haplotype are inherited together, it means that a gene for the trait is located on the same chromosome as the haplotype and that the gene may be near the haplotype block.

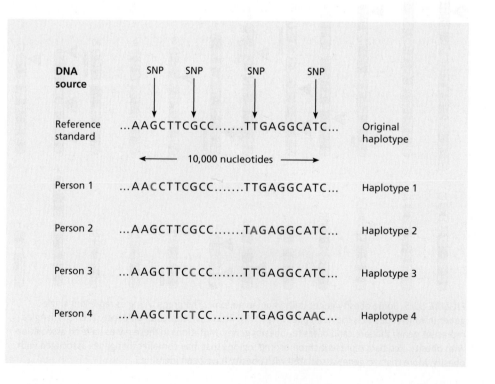

5.8 Skin Color and IQ Are Complex Traits

Questions about the inheritance of skin color inevitably raise other questions about races and the genetic differences between various populations. In Chapter 19, which examines the genetics of human populations, we will discuss whether or not there is a genetic basis for separating human populations into races. Here we will focus only on the multifactorial basis of skin color.

Skin color is a multifactorial trait.

One of the first investigations of the genetics of skin color, done between 1910 and 1914, studied black-white marriages in Bermuda and in the Caribbean. Skin color is controlled by genetic and environmental factors, making it a multifactorial trait.

The results illustrate several properties of polygenically controlled multifactorial traits. The F1 generation had skin colors intermediate to those of their parents. In the F2, a small number of children were as white as one grandparent, a small number were as black as the other grandparent, and most had skin color between those two extremes (Figure 5.17). Because F2 individuals could be grouped into five phenotypic classes, the investigators hypothesized that two gene pairs control skin color (see the distribution of genotypes for two loci in Figure 5.5). Each F2 phenotypic class represented a genotype produced by the segregation and assortment of two gene pairs. To help explain their results, let's suppose that these genes are *A* and *B*, respectively. Class 0 has the lightest skin color and represents the genotype *aabb*, Class 1 has the genotype *Aabb* or *aaBb*, and so forth, up to Class 4, which has the darkest skin color and represents the homozygous dominant (*AABB*) genotype.

Later work by other investigators using instruments that measure light reflected from the skin surface showed that skin color is actually controlled by more than two gene pairs. The data are most consistent with a model that involves three or four genes (Figure 5.18).

Although skin color is a highly variable phenotype among human populations, the question is whether this phenotypic diversity reflects an underlying level of genotypic divergence that justifies grouping human populations into racial groups. As we will see in Chapter 19, no genetic differences large enough to separate our species into clear-cut races have been discovered.

Intelligence and intelligence quotient (IQ): are they related?

The idea that intelligence can be measured quantitatively arose in the late eighteenth and early nineteenth centuries. Early on in the study of intelligence, phrenologists believed that the physical measurements of regions of the skull revealed how much intelligence,

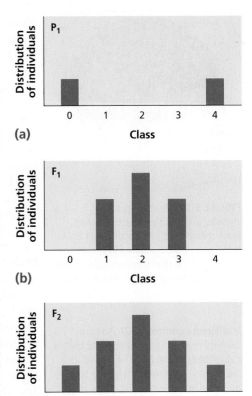

FIGURE 5.17 Frequency diagrams of skin colors. (a) Skin color distribution in the parents falls into two discontinuous classes. (b) Color values of seven children from the parents in *a* are intermediate to those of their parents. (c) Skin colors of 32 children of the parents in *b*. Color values range from one phenotypic extreme to the other, and most are clustered around a mean value. This normal distribution of phenotypes is characteristic of a polygenic trait.

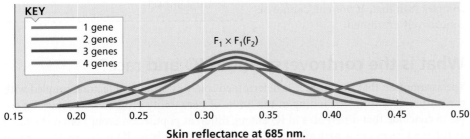

KEY
— 1 gene
— 2 genes
— 3 genes
— 4 genes

$F_1 \times F_1 (F_2)$

0.15   0.20   0.25   0.30   0.35   0.40   0.45   0.50

Skin reflectance at 685 nm.

FIGURE 5.18 Distribution of skin color as measured by a reflectometer at a wavelength of 685 nm. The results are shown for an additive model of skin color, with environmental effects, for one to four gene pairs. Distributions observed in several populations indicate that three or four gene pairs control skin color.

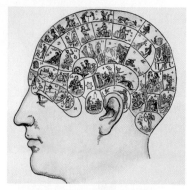

FIGURE 5.19 Phrenology model showing areas of the head overlaying brain regions that control different traits. Intelligence was estimated by measuring the area of the skull overlaying the region of the brain thought to control this trait.

Intelligence quotient (IQ) A score derived from standardized tests that is calculated by dividing the individual's mental age (determined by the test) by his or her chronological age and multiplying the quotient by 100.

courage, and so forth, an individual possessed (Figure 5.19). Later, these physical measurements gave way to overall brain size (craniometry). Large brains were associated with high intelligence and small brains with lower intelligence.

At the turn of the twentieth century, psychological rather than physical methods were used to measure intelligence. Alfred Binet, a French psychologist, developed a graded series of tasks related to basic mental processes such as comprehension, direction (sorting), and correction and tested children for their ability to perform those tasks. Each child began by performing the simplest tasks and progressed until the tasks became too difficult. The age assigned for the last task performed became the child's mental age, and the intellectual age was calculated by subtracting the mental age from the chronological age.

Wilhelm Stern, another psychologist, divided mental age by chronological age, and the number became known as the **intelligence quotient (IQ)**. If a child of 7 years (chronological age) was able to perform tasks for a 7-year-old but could not do tasks for an 8-year-old, a mental age of 7 would be assigned. To determine the IQ for this child, divide mental age by chronological age: mental age (7) divided by chronological age (7) = 1.0. The quotient is multiplied by 100 to eliminate the decimal point (1.0 × 100 = 100), obtaining an IQ of 100.

Both the physical and psychological methods of measuring intelligence assume that intelligence is a biological property that can be expressed quantitatively as a single number (see Spotlight on Building a Smarter Mouse). In fact, if anything, the use of IQ tests by governments and educational institutions has strengthened the assumption that IQ measures a fundamental, genetically determined physiological or biochemical property of the brain related to intelligence. The question is whether this assumption is correct and whether psychological methods (IQ tests) measure intelligence any more accurately than do the discredited physical methods of phrenology and craniometry.

The question can be answered only if intelligence is defined in such a way that it can be measured objectively, the way we measure height, weight, or fingerprint ridges. Intelligence is often thought of as abilities in abstract reasoning, mathematical skills, verbal expression, problem solving, and creativity. There is no evidence that any of these properties are measured directly by an IQ test, and there is at present no objective way to quantify such components of intelligence.

IQ values are heritable traits.

The values obtained in IQ measurements, however, do have significant heritable components. The evidence that IQ has genetic components comes from two areas: studies that estimate IQ heritability and comparison of IQs in individuals raised together (unrelated individuals, parents and children, siblings, and MZ and DZ twins) and individuals raised separately (unrelated individuals, siblings, and MZ twins). If we take IQ as a trait, heritability estimates range from 0.6 to 0.8. The high correlation observed for MZ twins raised together indicates that genetics plays a significant role in determining IQ (Figure 5.20). However, rearing MZ twins apart or raising siblings in different environments significantly reduces the correlation and provides evidence that the environment plays a substantial role in determining IQ. Heritability studies using twins cannot measure the effects of interactions between the genotype and the environment and may underestimate the impact of environment on IQ. As a result, intelligence is an example of a complex trait that, at present, cannot be conclusively examined.

> **KEEP IN MIND**
> Many multifactorial traits have social and cultural impacts.

What is the controversy about IQ and race?

The assumption that intelligence is determined solely by biological factors, coupled with the misuse or misunderstanding of the limits of heritability estimates, has misled people to conclude that differences in IQ among different population groups often classified as different races are genetically determined. On standardized IQ tests, blacks score an

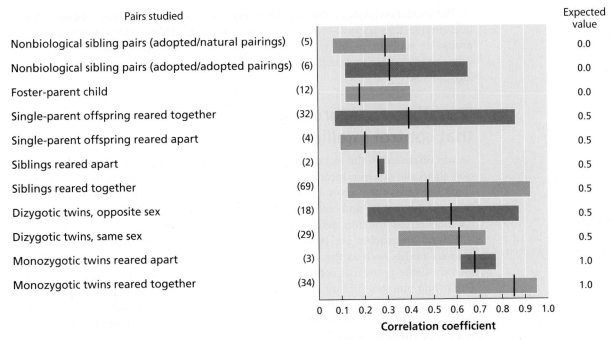

Pairs studied

		Expected value
Nonbiological sibling pairs (adopted/natural pairings)	(5)	0.0
Nonbiological sibling pairs (adopted/adopted pairings)	(6)	0.0
Foster-parent child	(12)	0.0
Single-parent offspring reared together	(32)	0.5
Single-parent offspring reared apart	(4)	0.5
Siblings reared apart	(2)	0.5
Siblings reared together	(69)	0.5
Dizygotic twins, opposite sex	(18)	0.5
Dizygotic twins, same sex	(29)	0.5
Monozygotic twins reared apart	(3)	1.0
Monozygotic twins reared together	(34)	1.0

0   0.1 0.2 0.3 0.4 0.5 0.6 0.7 0.8 0.9 1.0
Correlation coefficient

FIGURE 5.20 A graphical representation of correlations in IQ measurements in different sets of individuals. The expected correlation coefficients are determined by the degree of genetic relatedness in each set of individuals. The vertical line represents the median correlation coefficient in each case.

average of 15 points lower than the average score of 100 by whites, and Asians score significantly above that average. The controversy is over what causes the differences. Are such differences genetic in origin, do they reflect environmental differences, or are both factors at work? If both, to what degree does inheritance contribute to the differences? The debate about these questions was renewed by the 1994 publication of *The Bell Curve* by Herrnstein and Murray.

The results of heritability studies have been used to support the argument that intelligence is mainly innate and inherited, citing heritability values of 0.8 for intelligence. In most cases, however, the reasoning used to support this argument misuses the concept of heritability. Recall that a measured heritability of 0.8 (for example) means that 80% of the phenotypic variation observed is due to genetic differences *within that* population. Heritability differences *between* two populations cannot be compared, because heritability measures only variation within a population at the time of measurement. By definition, it cannot be used to estimate genetic variation between populations. In other words, we cannot use heritability differences between groups to conclude that there are genetic differences between those groups. As we will see in Chapter 19, genetic variation within any single population is much greater than the amount of genetic variation between any two populations. In fact, the amount of genetic variation within a population is so great that it swamps the genetic differences between populations, invalidating the idea that human populations can be sorted into racial groups.

It is quite evident that both genetic and environmental factors make important contributions to intelligence. Clearly, the relative amount that each contributes cannot be measured accurately at this time. Several points about this debate should be kept in mind. First, IQ test scores cannot be equated with intelligence. Second, IQ scores are not fixed and can be changed significantly by training in problem solving and, in fact—like many other phenotypes—change somewhat throughout the life of an individual. Variation in IQ scores is quite wide, and values measured in one racial or ethnic group greatly overlap those of other groups, making comparisons more difficult. In addition, worldwide IQ scores have been rising significantly for over 50 years. An IQ score regarded as average in 1944 would be well below average in 2008, emphasizing once again the role of environment in IQ scores.

Spotlight on...

Building a Smarter Mouse

During the learning process, chemical changes occur at synapses, the gaps between two nerve cells where signals are transferred between nerves. If both nerve cells are active at the same time, learning and memory are enhanced. By creating mice that overexpress or underexpress genes encoding proteins that transmit signals across the gap in the synapse, researchers can test the effects of overexpression or underexpression of these genes on learning and memory. One protein, called the NMDA receptor, has two subunits: NR1 and NR2. Young mice have higher levels of NR2 and learn better than do adult mice. Researchers genetically engineered mice to overexpress the NR2 subunits and found that when those mice were adults, they learned faster and had better memories than other adult mice. This work shows that genetic enhancement of intelligence and memory is possible and has identified a key gene in learning and memory.

The problem with discussing the differences in IQ scores between groups arises when quantitative differences in scores are converted into qualitative judgments used to rank groups as superior or inferior. Genetics, like all sciences, progresses by formulating hypotheses that can be tested rigorously and objectively, but it is often misused for ideological ends.

Scientists are searching for genes that control intelligence.

As was discussed in the section on obesity, heritability studies cannot provide information about the number, location, or identity of genes involved in intelligence. To learn about the genes themselves, scientists are using an expanded definition of intelligence that goes beyond IQ and recognizes that many genes are involved in normal cognitive and intellectual function, defined as **general cognitive ability.**

Several approaches are being used to identify genes associated with intelligence. Animal models such as the fruit fly, *Drosophila*, and the mouse are being studied to identify single genes that control aspects of learning, memory, and spatial perception. *Drosophila* mutants such as *dunce*, *rutabaga*, and *amnesiac*, among others, have been used to identify biochemical pathways in the nervous system that play important roles in learning and memory. Human brains have the same pathways, and they may play similar roles in our nervous system.

Another approach uses genomics to identify genes that affect specific polygenic traits, such as reading ability and IQ. This work uses SNPs and other DNA markers to identify chromosome regions that contain genes related to these traits. Identifying the number of these regions associated with variation in a specific phenotype is an indirect way of determining whether that phenotype is controlled by many genes, each having a small effect, or by a small number of genes, each with a major effect on the phenotype. This searching has uncovered genes associated with reading disability (developmental dyslexia) on chromosome 6 (OMIM 600202) and chromosome 15. More recently, a gene associated with cognitive ability has been identified on chromosome 4 (OMIM 603783). With the use of approaches that combine behavioral genetics, twin studies, and molecular genetics, more genes controlling aspects of cognitive ability are likely to be identified in the near future. As results from the Human Genome Project are analyzed, it will become easier to define the number and actions of genes involved in higher mental processes and provide insight into the genetics of intelligence.

General cognitive ability
Characteristics that include verbal and spatial abilities, memory, speed of perception, and reasoning.

Genetics in Practice

Genetics in Practice case studies are critical-thinking exercises that allow you to apply your new knowledge of human genetics to real-life problems. You can find these case studies and links to relevant websites at *www.cengage.com/biology/cummings*

CASE 1

Sue and Tim were referred for genetic counseling after they inquired about the risk of having a child with a cleft lip. Tim was born with a mild cleft lip that was surgically repaired. He expressed concern that his future children could be at risk for a more severe form of clefting. Sue was in her 12th week of pregnancy, and both were anxious about the pregnancy because Sue had had a difficult time conceiving. The couple stated that they would not consider terminating the pregnancy for any reason but wanted to be prepared for the possibility of having a child with a birth defect. The genetic counselor took a three-generation family history from both Sue and Tim and found that Tim was the only person to have had a cleft lip. Sue's family history showed no cases of cleft lip. Tim and Sue had several misconceptions about clefting, and the genetic counselor spent time explaining how cleft lips occur and some of the known causes of this birth defect. The following list summarizes the counselor's discussion with the couple.

- Fathers, as well as mothers, can pass on genes that cause clefting.
- Some clefts are caused by environmental factors, meaning that the condition didn't come from the father or the mother.
- One child in 33 is born with some sort of birth defect.
- One in 700 is born with a cleft-related birth defect.
- Most clefts occur in boys; however, a girl can be born with a cleft.
- If a person (male or female) is born with a cleft, the chances of that person having a child with a cleft, given no other obvious factor, is 7 in 100.
- Some clefts are related to identifiable syndromes. Of those, some are autosomal dominant. A person with an autosomal dominant gene has a 50% probability of passing the gene to an offspring.
- Many clefts run in families even when there does not seem to be any identifiable syndrome present.
- Clefting seems to be related to ethnicity, occurring most often among Asians, Latinos, and Native Americans (1:500); next most often among persons of European ethnicity (1:700); and least often among persons of African origin (1:1,000).
- A cleft condition develops during the fourth to the eighth week of pregnancy. After that critical period, nothing the mother does can cause a cleft. Sometimes a cleft develops even before the mother is aware that she is pregnant.
- Women who smoke are twice as likely to give birth to a child with a cleft.
- Women who ingest large quantities of vitamin A or low quantities of folic acid are more likely to have children with a cleft.
- In about 70% of cases, the fetal face is clearly visible using ultrasound. Facial disorders have been detected at the 15th gestational week of pregnancy. Ultrasound can be precise and reliable in diagnosing fetal craniofacial conditions.

1. After hearing this information, should Sue and Tim feel that their chances of having a child with a cleft lip are increased over that of the general population?

2. Can cleft lip be surgically corrected?

3. If the child showed a cleft lip through ultrasound analysis and the parents then started blaming each other (because Sue is a smoker and Tim was born with the defect), how would you counsel them?

CASE 2

Louise was an active 27-year-old gymnastics instructor at a local YMCA. She was recently accepted into law school and worked part time in the evenings at the local Gap clothing store. She was very busy and therefore thought nothing of her increasing fatigue until it started to affect her daily activities. She also complained of occasional dizziness, difficulty hearing, constipation, and problems controlling her bladder. Her general practitioner conducted a series of blood tests to determine what was wrong. All tests showed that she was in perfect health, but her symptoms were becoming progressively worse. Over a period of months, she was forced to quit teaching gymnastics and was soon confined to a wheelchair. Her doctor finally referred her for a genetic evaluation. The clinical geneticist immediately recognized Louise's symptoms as signs of multiple sclerosis (MS). The geneticist explained that fatigue, almost to the point of being disabling, is the most common symptom of MS. Some medications could help her fatigue, but their effect might not last. The best cure for fatigue would be to listen to her body and just rest whenever possible. The geneticist warned her of other possible symptoms of MS, including (1) numbness, which is most likely a direct result of the destruction of myelin in the nerves; (2) tingling or a pins-and-needles sensation in her feet (the same sensation you get when your foot "falls asleep"), for which little can be done other than massaging the affected area and resting; (3) tremors, usually of the hands, also due to myelin destruction; (4) muscle spasms (sustained or temporary muscle contractions) in the arms, legs, abdomen, back—just about anywhere; (5) depression and mood swings; (6) memory problems; (7) loss of balance with or without dizziness; and (8) bowel and bladder problems. MS is the most common autoimmune disease involving the nervous system, affecting approximately 250,000 individuals in the United States. The cause of MS is unknown; however, some studies have suggested that the risk to a first-degree relative (sibling, parent, or child) of a patient with MS is at least 15 times that for a member of the general population. Unfortunately, no definite genetic pattern is discernible.

1. What could explain why a first-degree relative of an MS patient is 15 times more likely to have MS than is the general population? Does this mean that MS has a genetic component?

2. What is an autoimmune disease? What kinds of genes could be involved in this process?

3. What kinds of questions should medical researchers ask in the study of MS and possible treatments?

Summary

5.1 Some Traits Are Controlled by Two or More Genes

- The pattern of inheritance that controls traits that can be measured quantitatively is called polygenic inheritance, because two or more genes are usually involved. Polygenic traits with significant environmental interaction are called multifactorial traits.

5.2 Polygenic Traits and Variation in Phenotype

- The distribution of polygenic traits through the population follows a bell-shaped, or normal, curve. Parents whose phenotypes are near the extremes of this curve usually have children whose phenotypes are less extreme than those of the parents and are closer to the population mean. This phenomenon is known as regression to the mean.

5.3 The Additive Model for Polygenic Inheritance

- In a simple additive model, a number of alleles each make an equal contribution to the trait, while other alleles make no contribution. In an additive model, the phenotypes are distributed in a continuous distribution. The action of environmental factors on the phenotype further smoothes out the distribution, reducing distinctions between phenotypic classes.

5.4 Multifactorial Traits: Polygenic Inheritance and Environmental Effects

- Variations in the expression of polygenic traits are often due to the action of environmental factors. Polygenic traits with a strong environmental component are called multifactorial traits or complex traits. The impact of environment on genotype can cause genetically susceptible individuals to exhibit a trait discontinuously even though there is an underlying continuous distribution of genotypes for the trait.

5.5 Heritability Measures the Genetic Contribution to Phenotypic Variation

- The degree of phenotypic variation produced by a genotype in a specific population can be estimated by calculating the heritability of a trait. Heritability is estimated by observing the amount of variation among relatives who have a known fraction of genes in common. MZ twins have 100% of their genes in common and, when raised in separate environments, provide an estimate of the degree of environmental influence on gene expression. Heritability is a variable that is calculated validly only for the population under study and the environmental condition in effect at the time of the study.

5.6 Twin Studies and Multifactorial Traits

- In twin studies, the degree of concordance for a trait is compared in MZ and DZ twins reared together or apart. MZ twins result from the splitting of an embryo produced by a single fertilization, whereas DZ twins are the products of multiple fertilizations. Although twin studies can be useful in determining whether a trait is inherited, they cannot provide any information about the mode of inheritance or the number of genes involved.

5.7 Genetics of Height: A Closer Look

- Although height in humans has long been regarded as a classic multifactorial trait, new techniques reveal that it is more complex than previously thought. Genome-wide association studies have identified a large number of genes that contribute to height. However, genes identified to date account for only a small percentage of the observed heritability. Gene-gene interactions, genome variations including differences in gene copy number, and other factors are being explored to uncover other genes related to height.

5.8 Skin Color and IQ Are Complex Traits

- Many human traits are multifactorial, combining polygenes and action by environmental factors. Some of these include skin color and IQ. The genetics of some of these traits has often been misused and misrepresented for ideological or political ends. New genetic approaches using recombinant DNA methods are helping identify genes involved in those traits.

Questions and Problems

Some Traits Are Controlled by Two or More Genes

1. Describe why continuous variation is common in humans and provide examples of such traits.

2. The text outlines some of the problems Frederick William I encountered in his attempt to breed tall Potsdam Guards.
   a. Why were the results he obtained so different from those obtained by Mendel with short and tall pea plants?
   b. Why were most of the children shorter than their tall parents?

3. What role might environment have played in causing Frederick William's problems, especially at a time when nutrition varied greatly from town to town and from family to family?

4. Do you think Frederick William's experiment would have worked better if he had ordered brother-sister marriages within tall families instead of just choosing the tallest individuals from throughout the country?

5. As it turned out, one of the tallest Potsdam Guards had an unquenchable attraction to short women. During his tenure as guard, he had numerous clandestine affairs. In each case, children resulted. Subsequently, some of the children—who had no way of knowing that they were related—married and had children of their own. Assume that two pairs of genes determine height. The genotype of the 7-foot-tall Potsdam Guard was $A'A'B'B'$, and the genotype of all of his 5-foot clandestine lovers was $AABB$. An A' or B' allele in the offspring each adds 6 inches to the base height of 5 feet conferred by the $AABB$ genotype.
   a. What were the genotypes and phenotypes of all the F1 children?
   b. Diagram the cross between the F1 offspring, and give all possible genotypes and phenotypes of the F2 progeny.

Polygenic Traits and Variation in Phenotype

6. Describe why there is a fundamental difference between the expression of a trait that is determined by polygenes and the expression of a trait that is determined monogenetically.

The Additive Model for Polygenic Inheritance

7. Sunflowers with flowers 10 cm in diameter are crossed with a plant that has 22 cm flowers. The F1 plants have flowers 15 cm in diameter. In the F2 generation, 4 flowers are 10 cm in diameter and 4 are 20 cm in diameter. Between these are 5 phenotypic classes with diameters intermediate to those at the extremes.

   a. Assuming that the alleles that contribute to flower diameter act additively, how many genes control flower size in this strain of sunflowers?
   b. How much does each additive allele contribute to flower diameter?
   c. What size flower makes up the largest phenotypic class?

Multifactorial Traits: Polygenic Inheritance and Environmental Effects

8. Clubfoot is a common congenital birth defect. This defect is caused by a number of genes but appears to be phenotypically distributed in a noncontinuous fashion. Geneticists use the multifactorial threshold model to explain the occurrence of this defect. Explain this model. Explain predisposition to the defect in an individual who has a genotypic liability above the threshold versus an individual who has a liability below the threshold.

Heritability Measures the Genetic Contribution to Phenotypic Variation

9. Define genetic variance.
10. Define environmental variance.
11. How is heritability related to genetic and environmental variance?
12. Why are relatives used in the calculation of heritability?
13. If there is no genetic variation within a population for a given trait, what is the heritability for the trait in the population?

Twin Studies and Multifactorial Traits

14. Can conjoined (Siamese) twins be dizygotic twins in light of the theory that conjoined twins result from incomplete division of the embryo?

15. Dizygotic twins:
   a. are as closely related as monozygotic twins
   b. are as closely related as non-twin siblings
   c. share 100% of their genetic material
   d. share 25% of their genetic material
   e. none of the above

16. Why are monozygotic twins who are reared apart so useful in the calculation of heritability?

17. Monozygotic (MZ) twins have a concordance value of 44% for a specific trait, whereas dizygotic twins have a concordance value of less than 5% for the same trait. What could explain why the value for MZ twins is significantly less than 100%?

18. If monozygotic twins show complete concordance for a trait, whether they are reared together or apart, what does this suggest about the heritability of the trait?

19. Researchers set up an obesity study in which MZ and DZ twins who served in the armed forces were studied at induction into the military and 25 years later. Results indicated that obesity has a strong genetic component.
    a. What are some of the problems with this study?
    b. Design a better study to test whether obesity has a genetic component.
20. What does the *ob* gene code for? How does it work? Is this a gene found only in animals, or do humans have it also?
21. What is the importance of the comparison of traits between adopted and natural children in determining heritability?

Genetics of Height: A Closer Look

22. Height in humans is controlled by the additive action of genes and the action of environmental factors. For the purposes of this problem, assume that height is controlled by four genes—A, B, C, and D—and that there are no environmental effects. Assume further that additive alleles contribute two units of height and partially additive alleles contribute one unit of height.
    a. Given these assumptions, can two individuals of moderate height produce offspring that are much taller and shorter than either parent? If so, how can this happen?
    b. Can someone of minimum height and someone of intermediate height have children taller than the parent of intermediate height? Why or why not?

Skin Color and IQ Are Complex Traits

23. If diseases such as cardiovascular disease (hypertension and atherosclerosis) are familial, is this an indication that there is a genetic contribution to these traits? What would you do to confirm that genetics is involved in this condition?
24. Discuss the difficulties in attempting to determine whether intelligence is genetically based.
25. At the age of 9 years, your genetics instructor was able to perform the mental tasks of an 11-year-old. According to Wilhelm Stern's method, calculate his or her IQ.
26. Suppose that a team of researchers analyzes the heritability of high SAT scores and assigns a heritability of 0.75 for this ability. The team also determines that a certain ethnic group has a heritability value that is 0.12 lower compared with that of other ethnic groups. The group concludes that there must be a genetic explanation for the differences in scores. Why is this an invalid conclusion?

Internet Activities

Internet Activities are critical-thinking exercises using the resources of the World Wide Web to enhance the principles and issues covered in this chapter. For a full set of links and questions investigating the topics described below, visit *www.cengage.com/biology/cummings*

1. *Twin Studies.* Access the website of the International Society for Twin Studies. At this site, click on the link to the Genetic Epidemiology Group of the Queensland Institute of Medical Research. Once there, click on "Studies" to view information on several long-term twin study projects. The project overview mentions several different ongoing studies. This website also has a link to the International Society for Twin Studies. At the ISTS website, open and read the link to the "Declaration of Rights and Statement of Needs of Twins and Higher Order Multiples."
2. *A Multifaceted Look at a Multifactorial Trait.* Body weight is an example of a multifactorial trait that results from the interaction between multiple genes and the environment. At the same time, the issue of body weight has taken on different connotations in different societies. The PBS series *Frontline* has a website that addresses both the question of what makes people "fat" and how fatness is perceived by our society.
    a. At the website, click on "What the Experts Said in This Report" regarding body weight, especially those comments regarding the various causes—cultural, societal, and genetic—of fatness in people.
    b. Follow the link to the chapter of Richard Klein's book *Eat Fat.* Klein discusses how fatness and even the word *fat* once had very positive connotations; however, this is generally untrue today, especially in the industrialized Western world.

The idea of selectively breeding humans, as King Frederick William of Prussia attempted to do with the Potsdam Guards, is generally considered highly unethical. However, with increased knowledge of genetics and advances in reproductive technology, it is possible in our time to test embryos for their genetic characteristics—such as genetic markers associated with higher IQ levels—before they are implanted in a mother's womb. Now that you know more about multifactorial traits that involve polygenic inheritance and the effects of environmental factors, what do you think? Would you consider having such tests done and implanting only those embryos carrying the desired markers? Visit the *Human Heredity* companion website at *www.cengage.com/biology/cummings* to find out more on the issue; then cast your vote online.

6 Cytogenetics: Karyotypes and Chromosome Aberrations

Tierney's home pregnancy test confirmed the good news: She and her husband, Greg, were having a baby. As the pregnancy progressed, they went to their obstetrician for routine exams. After an ultrasound test, they were told the fetus had a heart defect. Worse was the news that the nature of the defect made the doctor suspect it was caused by Down syndrome. Further testing confirmed that diagnosis.

Now Greg and Tierney Fairchild were struggling with an array of emotional conflicts and difficult decisions. First, they faced the guilt and fear that accompany the news that a fetus is not normal. Next, they had to decide whether to terminate the pregnancy or have a child who would face major surgery, a lifetime of retardation, and medical risks such as hypothyroidism and leukemia. The Fairchilds had made tough decisions before. As an interracial couple, they had faced the specter of discrimination and intolerance with their decision to marry. The current decision would involve not only another person but one burdened by both racial discrimination and mental retardation. Before making their decision about the pregnancy, they sought information, advice, and help.

Unfortunately, there is no way to predict how retarded a Down syndrome child will be. About 10% are severely retarded—unable to dress, eat, or use the toilet by themselves. Others attend school, graduate, have jobs, and live in group homes or supervised settings as adults. After much anguish and conflicting advice, they decided to continue the pregnancy and had a daughter, Naia. She underwent successful heart surgery, and both parents are now deeply involved in her education at home and at school. Despite her developmental challenges, at age 10 Naia is reading at grade level, she has many friends, and her outside interests include riding horses. Early on, after the prenatal diagnosis, the Fairchilds opened their lives to Mitchell Zukoff, a reporter from the *Boston Globe*, who wrote a series of articles and later a book, *Choosing Naia*, about the choices and decisions the Fairchilds faced during and after this pregnancy.

© moodboard/Alamy

A young girl with Down syndrome in her computer classroom.

6.1 The Human Chromosome Set

The number of chromosomes in the nucleus of an organism is characteristic for a species: Cells of the fruit fly *Drosophila melanogaster* have 8 chromosomes, corn plants have 20 chromosomes, and humans have 46 chromosomes. The chromosome numbers for several species of plants and animals are given in Table 6.1. Recall from Chapter 2 that human chromosomes exist in pairs, with most cells having 23 homologous pairs, or 46 chromosomes. This is the diploid, or *2n*, number of chromosomes. One member of each homologous pair is contributed by each parent. Certain cells, such as eggs and sperm (gametes), contain only one copy of each chromosome. These cells have 23 chromosomes, which is the haploid, or *n*, number of chromosomes.

As chromosomes condense and become visible in the early stages of cell division, certain structural features can be recognized. Each chromosome contains a specialized region known as the **centromere**, which divides the chromosome into two arms. The location of the centromere is characteristic for each chromosome (Figure 6.1). Chromosomes with centrally located

KEEP IN MIND AS YOU READ

- Karyotype construction and analysis are used to identify chromosome abnormalities.
- Polyploidy results when there are more than two complete sets of chromosomes.
- Monosomy and trisomy involve the loss and gain of a single chromosome to a diploid genome.
- Age of the mother is the best-known risk factor for trisomy.
- Changes in the number of sex chromosomes have less impact than changes in autosomes.
- Chromosomes can lose, gain, or rearrange segments.
- Some fragile sites are associated with mental retardation.

Centromere A region of a chromosome to which spindle fibers attach during cell division. The location of a centromere gives a chromosome its characteristic shape.

Table 6.1   Chromosome Number in Selected Organisms

Organism	Diploid Number (2n)	Haploid Number (n)
Human (*Homo sapiens*)	46	23
Chimpanzee (*Pan troglodytes*)	48	24
Gorilla (*Gorilla gorilla*)	48	24
Dog (*Canis familiaris*)	78	39
Housefly (*Musca domestica*)	12	6
Corn (*Zea mays*)	20	10
Mouse (*Mus musculus*)	40	20
Fruit fly (*Drosophila melanogaster*)	8	4
Nematode (*Caenorhabditis elegans*)	12	6

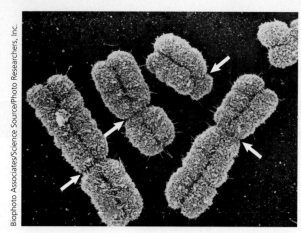

FIGURE 6.1 Human chromosomes as seen at metaphase of mitosis in the scanning electron microscope. The replicated chromosomes appear as double structures, consisting of sister chromatids joined by a single centromere (arrows).

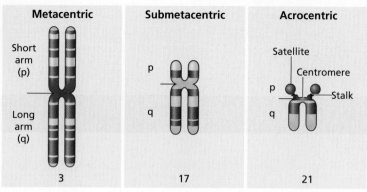

FIGURE 6.2 Human mitotic metaphase chromosomes are identified by size, centromere location, and banding pattern. The relative size, centromere locations, and banding patterns for three representative human chromosomes are shown. Chromosome 3 is one of the largest human chromosomes and, because the centromere is centrally located, is a metacentric chromosome. Chromosome 17 is a submetacentric chromosome because the centromere divides the chromosome into two arms of unequal size. Chromosome 21 has a centromere placed very close to one end and is called an acrocentric chromosome. In humans, the short arm of each chromosome is called the p arm, and the long arm is called the q arm. The symbol p was chosen for the short arm because it stands for the word "petit," which in French means small. The symbol q was chosen for the long arm because it is the next letter in the alphabet.

Metacentric Describes a chromosome that has a centrally placed centromere.

Submetacentric Describes a chromosome whose centromere is placed closer to one end than the other.

Acrocentric Describes a chromosome whose centromere is placed very close to, but not at, one end.

Sex chromosomes In humans, the X and Y chromosomes that are involved in sex determination.

Autosomes Chromosomes other than the sex chromosomes. In humans, chromosomes 1 to 22 are autosomes.

Karyotype A complete set of chromosomes from a cell that has been photographed during cell division and arranged in a standard sequence.

centromeres have arms of equal length and are known as **metacentric** chromosomes (Figure 6.2). If the centromere is located at a distance from the center, the arms are unequal in length, and the chromosome is called a **submetacentric** chromosome. If the centromere is located very close to one end, the chromosome is called an **acrocentric** chromosome.

Humans (and other animal species) have one pair of **sex chromosomes** that are not completely homologous. Females have two homologous X chromosomes, and males have a nonhomologous pair consisting of one X chromosome and one Y chromosome. The nonsex human chromosomes (chromosomes 1 to 22) are called **autosomes**.

Chromosomes are usually studied and photographed while they are in metaphase of mitosis. For convenience, the chromosome images are arranged in pairs according to size and centromere location to form a **karyotype** (Figure 6.3).

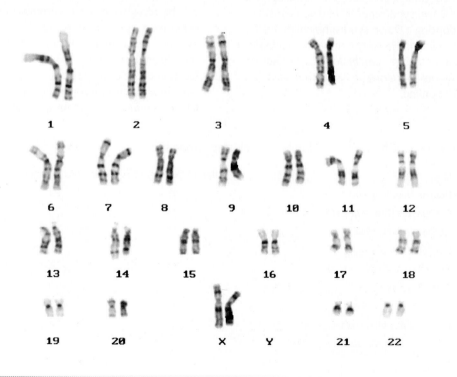

FIGURE 6.3 A human karyotype showing replicated chromosomes from a cell in metaphase of mitosis. This female has 46 chromosomes, including two X chromosomes.

Stains and dyes are used to produce chromosome banding patterns and to identify specific regions on each chromosome. The standardized banding pattern for the human chromosome set is shown in Figure 6.4. By convention, the short arm of each chromosome is designated the p arm, and the long arm the q arm (in metacentric chromosomes, the short arm was designated by an international committee). Each arm is subdivided into numbered regions from the centromere to a region at the end of the chromosome called the **telomere**. Within each region, bands are identified by number. Thus, any region in the human karyotype can be identified by a descriptive address, such as 1q2.4. This address includes the chromosome number (1), the arm (q), the region (2), and the band (4) (Figure 6.5). Karyotypic analysis of banded chromosomes is a powerful tool in chromosomal studies and is one of the basic techniques in human genetics.

Telomere Short repeated DNA sequences located at each end of chromosomes.

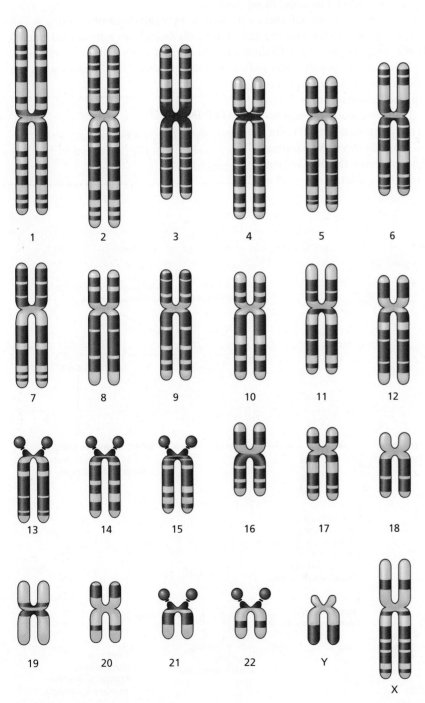

FIGURE 6.4 A karyogram of the human chromosome set, showing the distinctive banding pattern of each chromsome.

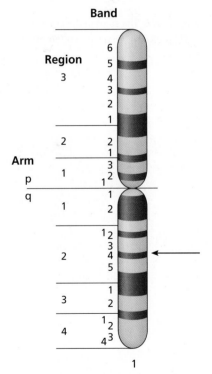

FIGURE 6.5 The system of naming chromosome bands. Each autosome is numbered from 1 to 22. The sex chromosomes are X and Y. Within a chromosome, the short arm is the p arm and the long arm is the q arm. Each arm is divided into numbered regions. Within each region, the bands are designated by numbers. The area marked with an arrow is 1q2.4 (chromosome 1), q (long arm), 2 (region), 4 (band).

6.2 Making a Karyotype

Karyotypes can be constructed using cells from a number of sources, including white blood cells (lymphocytes), skin cells (fibroblasts), amniotic fluid cells (amniocytes), and chorionic villus cells (placental cells). One of the most common methods of preparing lymphocytes for karyotype analysis begins with a blood sample and is shown in Figure 6.6. A few drops of the blood are added to a flask containing a nutrient growth medium. Because lymphocytes normally do not divide, a mitosis-inducing chemical such as phytohemagglutinin is added to the flask, and the cells are grown for 2 or 3 days at body temperature (37°C) in an incubator. Then a drug such as Colcemid is added to stop mitosis at metaphase. Over a period of about 2 hours after the drug is added, all cells entering mitosis stop dividing when they enter metaphase.

The treated cells are collected and concentrated by centrifugation; adding a salt solution breaks open and destroys the red blood cells (which do not contain nuclei) and swells the lymphocytes. After fixation in a mixture of methanol and acetic acid, the swollen lymphocytes are dropped onto a microscope slide. The impact causes the fragile cells to break open, spreading metaphase chromosomes onto the slide. The chromosome preparation is partially digested with trypsin, an enzyme that enhances the banding pattern. After staining, the preparation is examined with a microscope, and a

> **KEEP IN MIND**
>
> Karyotype construction and analysis are used to identify chromosome abnormalities.

FIGURE 6.6 The steps in the process of creating a karyotype for chromosome analysis.

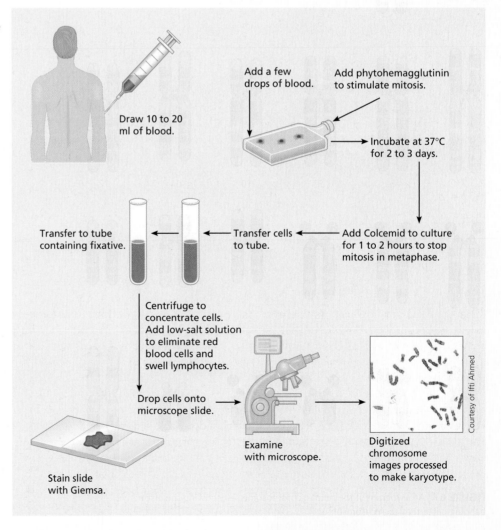

Draw 10 to 20 ml of blood.

Add a few drops of blood.

Add phytohemagglutinin to stimulate mitosis.

Incubate at 37°C for 2 to 3 days.

Add Colcemid to culture for 1 to 2 hours to stop mitosis in metaphase.

Transfer cells to tube.

Transfer to tube containing fixative.

Centrifuge to concentrate cells. Add low-salt solution to eliminate red blood cells and swell lymphocytes.

Drop cells onto microscope slide.

Stain slide with Giemsa.

Examine with microscope.

Digitized chromosome images processed to make karyotype.

Courtesy of Ifti Ahmed

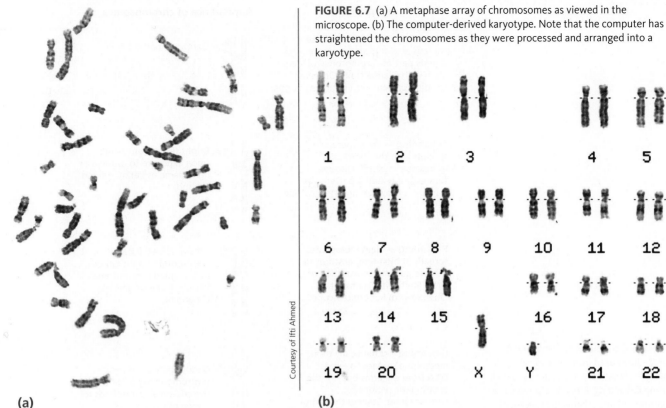

FIGURE 6.7 (a) A metaphase array of chromosomes as viewed in the microscope. (b) The computer-derived karyotype. Note that the computer has straightened the chromosomes as they were processed and arranged into a karyotype.

Courtesy of Ifti Ahmed

(a)

(b)

cluster of metaphase chromosomes is located on the slide (Figure 6.7a). A video camera attached to the microscope transmits the chromosome images to a computer, where they are recorded, digitized, and processed to make a karyotype. The metaphase chromosomes shown in Figure 6.7a were recorded by this method. Figure 6.7b shows the computer-derived karyotype.

6.3 Constructing and Analyzing Karyotypes

Stains and dyes are used to produce a pattern of bands specific to each chromosome (although homologous chromosomes usually have the same pattern). Metaphase chromosomes have a total of about 550 bands. More bands can be produced by using cells in early metaphase or late prophase. In those stages, chromosomes are less condensed, and up to 2,000 bands can be identified in the normal human karyotype. Some commonly used banding methods are reviewed in Figure 6.8.

A karyotype provides several kinds of information: (1) the number of chromosomes present, (2) number and type of sex chromosomes, (3) the presence or absence of individual chromosomes, and (4) the nature and extent of detectable structural abnormalities. The symbols for structural abnormalities detected in karyotypes include t for a translocation, dup for duplication, and del for deletion. These structural abnormalities are discussed later in this chapter. If a male carries a deletion in the short arm of chromosome 5 but otherwise is chromosomally normal, this is written as 46,XY,del(5p). The number of chromosomes is 46, the sex-chromosome set is XY, and there is a deletion (del) on the

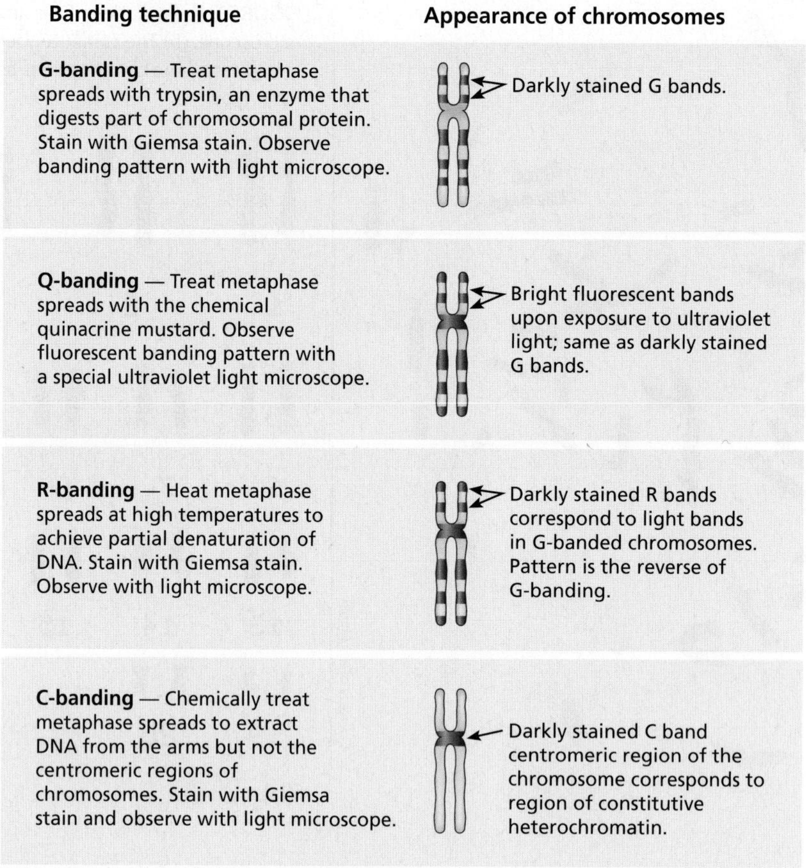

Banding technique	Appearance of chromosomes
G-banding — Treat metaphase spreads with trypsin, an enzyme that digests part of chromosomal protein. Stain with Giemsa stain. Observe banding pattern with light microscope.	Darkly stained G bands.
Q-banding — Treat metaphase spreads with the chemical quinacrine mustard. Observe fluorescent banding pattern with a special ultraviolet light microscope.	Bright fluorescent bands upon exposure to ultraviolet light; same as darkly stained G bands.
R-banding — Heat metaphase spreads at high temperatures to achieve partial denaturation of DNA. Stain with Giemsa stain. Observe with light microscope.	Darkly stained R bands correspond to light bands in G-banded chromosomes. Pattern is the reverse of G-banding.
C-banding — Chemically treat metaphase spreads to extract DNA from the arms but not the centromeric regions of chromosomes. Stain with Giemsa stain and observe with light microscope.	Darkly stained C band centromeric region of the chromosome corresponds to region of constitutive heterochromatin.

FIGURE 6.8 Four common staining procedures used in chromosomal analysis. Most karyotypes are prepared using G-banding. R-banding produces a pattern of bands that is the reverse of those in G-banded chromosomes.

short arm of chromosome 5 (5p). Table 6.2 includes descriptions of some chromosomal aberrations, using this system.

Chromosome analysis is a painstaking procedure; to make it easier to spot abnormalities, cytogeneticists use a technique called chromosome painting, which uses chromosome-specific DNA sequences attached to fluorescent dyes. The labeled DNA sequences attach to their target chromosomes, painting them distinctive colors. Combinations of different DNA sequences and fluorescent dyes produce a unique pattern for each human chromosome (Figure 6.9).

What cells are obtained for chromosome studies?

In adults, several cell types, including white blood cells (lymphocytes), skin cells (fibroblasts), and cells from biopsies or surgically removed tumor cells are used for karyotype preparations.

In a fetus, cells are collected for chromosomal analysis using two methods: amniocentesis and chorionic villus sampling. A less invasive method now under development involves the collection of fetal cells or fetal DNA from the mother's circulatory system (see Exploring Genetics: Noninvasive Prenatal Diagnosis on page 129).

Table 6.2 Chromosomal Aberrations

Chromosomal Abnormality		Syndrome Phenotype
46,del(5p)	Cri du chat syndrome	Small head; round face; low-set ears; weak, catlike cry; low, broad nasal ridge; mental retardation
46,t(9;22) (q34a11)	CML (chronic myelogenous leukemia)	Enlargement of liver and spleen; anemia; excessive, unrestrained growth of white cells (granulocytes) in the bone marrow
46,dup(17p12)	Charcot-Marie-Tooth syndrome	Loss of sensation and muscle atrophy in the feet and legs that spreads to the arms and hands as the disease progresses

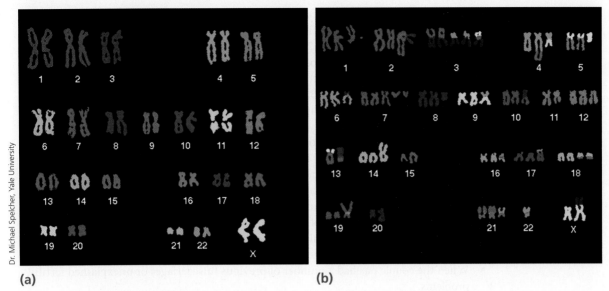

FIGURE 6.9 (a) Chromosome painting with five different-colored markers in a normal cell. (b) Painting produces a pattern that highlights the translocations, duplications, deletions, and aneuploidies found in cancer cells.

Amniocentesis collects cells from the fluid surrounding the fetus.

In **amniocentesis**, the fetus and placenta are located by ultrasound, and a needle is inserted through the abdominal and uterine walls (avoiding the placenta and fetus) into the amniotic sac surrounding the fetus. Approximately 10 to 30 ml of fluid is withdrawn by syringe. Amniotic fluid contains cells shed from the skin, respiratory tract, and urinary tract of the fetus. Cells are collected by centrifuging the fluid (Active Figure 6.10) and then grown in the laboratory before use. Amniotic cells can be used to detect biochemical disorders or chromosome abnormalities. Karyotype analysis makes it possible to diagnose the sex of the fetus and identify many chromosomal abnormalities.

Amniocentesis A method of sampling the fluid surrounding the developing fetus by inserting a hollow needle and withdrawing suspended fetal cells and fluid; used in diagnosing fetal genetic and developmental disorders; usually performed in the sixteenth week of pregnancy.

Removal of about 20 ml of amniotic fluid containing suspended cells that were sloughed off from the fetus

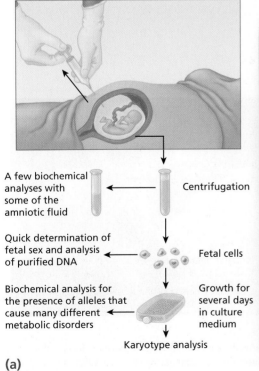

(a)

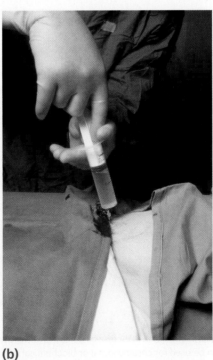

(b)

ACTIVE FIGURE 6.10 (a) In amniocentesis, a syringe needle is inserted through the walls of the abdomen and uterus to collect a sample of amniotic fluid. The fluid contains fetal cells that can be used for prenatal chromosomal or biochemical analysis. (b) A woman undergoing amniocentesis.

CENGAGENOW Learn more about amniocentesis by viewing the animation by logging on to *login.cengage.com/sso* and visiting CengageNOW's Study Tools.

Amniocentesis is usually performed around the 16th week of pregnancy. Before that time, there is not enough amniotic fluid, and the presence of maternal cells in the fluid is a potential problem. There is a 0.2% (1 in 500) to 0.3% (1 in 300) risk of miscarriage with amniocentesis and a small risk of maternal infection. Because of these risks, the use of amniocentesis is restricted to certain circumstances:

- When the mother is over age 35. The risk of having children with chromosome abnormalities increases dramatically after this age, so amniocentesis is recommended for pregnant women who are 35 years or older. The majority of all amniocentesis procedures are performed because of advanced maternal age.
- When the mother has already had a child with a chromosomal aberration. The recurrence risk in such cases is 1% to 2%.
- When either parent carries one or more structurally abnormal chromosomes. This situation may cause an abnormal number of chromosomes in a child.
- When the mother is a carrier of a disorder caused by a gene on the X chromosome. If the mother is a carrier of an X-linked biochemical disorder that cannot otherwise be diagnosed prenatally and is willing to abort if the fetus is male, amniocentesis is recommended.
- When the couple has had a number of previous miscarriages or unexplained fertililty problems.

Chorionic villus sampling retrieves fetal tissue from the placenta.

Chorionic villus sampling (CVS) A method of sampling fetal chorionic cells by inserting a catheter through the vagina or abdominal wall into the uterus. Used in diagnosing biochemical and cytogenetic defects in the embryo. Usually performed in the eighth or ninth week of pregnancy.

For **chorionic villus sampling (CVS)**, a flexible catheter is inserted through the vagina or abdomen into the uterus, guided by ultrasound images. Some chorionic villi (fetal tissue that forms part of the placenta) are removed by suction (Figure 6.11) for karyotype preparation and analysis. CVS has several advantages over amniocentesis; it can be performed earlier in the pregnancy (8 to 10 weeks, compared with 16 weeks for amniocentesis). Because placental cells are already dividing, karyotypes are available within a day or two. This gives CVS a distinct advantage over amniocentesis, because the results are available in the first trimester.

Both amniocentesis and CVS can be coupled with genomic DNA testing for prenatal diagnosis of mutant alleles. The role of DNA technology in prenatal testing is discussed in Chapter 16.

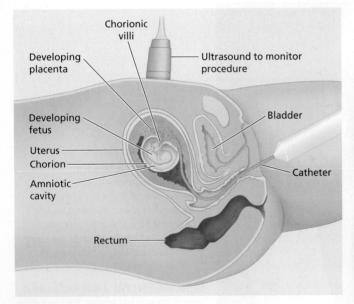

Chorionic villi

Developing placenta

Ultrasound to monitor procedure

Developing fetus

Bladder

Uterus

Chorion

Catheter

Amniotic cavity

Rectum

(a)

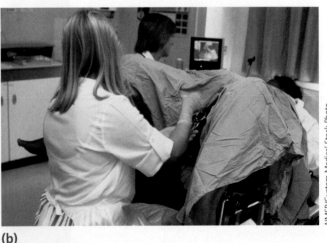

NMSB/Custom Medical Stock Photo

(b)

FIGURE 6.11 The chorionic villus sampling technique. (a) A catheter is inserted into the uterus through the vagina to remove a sample of fetal tissue from the chorion. Cells in the tissue can be used for chromosomal or biochemical analysis. (b) A woman having chorionic villus sampling.

Noninvasive Prenatal Diagnosis

In 1969, cytogeneticists observed cells with a Y chromosome in the blood of women who later gave birth to male infants. In the decades following that discovery, research has been directed at finding ways to detect fetal genetic disorders (e.g., sickle cell anemia, cystic fibrosis, etc.) and fetal chromosomal abnormalities (e.g., Down syndrome) using fetal cells recovered from maternal blood samples without using the invasive procedures of amniocentesis and chorionic villus sampling (CVS), which carry a risk to the fetus.

Two different approaches to noninvasive prenatal diagnosis have developed: (1) isolation and analysis of fetal cells present in the maternal circulation, and (2) the recovery and analysis of cell-free fetal DNA (cff DNA) from maternal blood plasma.

Several types of fetal cells enter the maternal circulation, but because there are only one to two fetal cells per milliliter of maternal blood, collecting enough fetal cells for analysis of single-gene disorders (cystic fibrosis, sickle cell anemia, etc.) or chromosomal abnormalities (trisomy 21) is a major challenge facing those working to develop this technique.

cff DNA was discovered in the maternal circulation in 1997 and can represent up to 10% of the non-cellular DNA in the mother's blood. Several diagnostic methods using cff DNA have been developed, including determination of fetal sex, analysis of Rh blood group, and the presence of paternal mutant alleles in autosomal dominant disorders. Future advances in noninvasive fetal testing will depend on improving the methods for separating fetal cells from maternal cells, the development of methods for enriching fetal DNA from the maternal circulation, and the application of genomic techniques such as protein analysis and DNA sequencing. If these advancements can be made, fetal blood sampling may gradually replace amniocentesis and CVS for prenatal chromosome analysis.

Image Source Black/Getty Images

CVS is used less frequently than amniocentesis. Early studies indicated that CVS posed a higher risk to mother and fetus than amniocentesis, but recent studies have concluded that when CVS is performed at a major medical center, the risk (1 in 300 to 1 in 500) is about the same as for amniocentesis. CVS offers early diagnosis of genetic diseases, and if termination of pregnancy is elected, maternal risks are lower at 9 to 12 weeks than at 16 weeks.

6.4 Variations in Chromosome Number

Anxious parents usually have two questions about a newborn: Is it a boy or a girl, and does the baby have any birth defects? The causes of such defects, of course, are both genetic and environmental. Among the genetic causes, we discussed single-gene disorders such as sickle cell anemia and Marfan syndrome in Chapter 4; and in Chapter 5, we considered complex disorders such as cleft palate. Changes in chromosome number or changes in chromosome structure can also cause genetic disorders. Those changes may involve entire chromosome sets, individual chromosomes, or alterations within individual chromosomes.

The diploid ($2n$) and haploid (n) chromosome numbers in human cells are the normal, or euploid, condition. An increase in the number of chromosome sets in a cell is called **polyploidy**. Cells with three sets of chromosomes are triploid; those with four sets are tetraploid, and so forth. Changes in chromosome number that involve less than a whole chromosome set results in **aneuploidy**. In its simplest form, aneuploidy involves the gain or loss of a single chromosome. Loss of a single chromosome is known as **monosomy** ($2n - 1$), and the gain of a single chromosome is known as **trisomy** ($2n + 1$).

Polyploidy A chromosomal number that is a multiple of the normal haploid chromosomal set.

Aneuploidy A chromosomal number that is not an exact multiple of the haploid set.

Monosomy A condition in which one member of a chromosomal pair is missing; having one less than the diploid number ($2n - 1$).

Trisomy A condition in which one chromosome is present in three copies, whereas all others are diploid; having one more than the diploid number ($2n + 1$).

Chromosome abnormalities in humans are common.

Chromosome abnormalities are fairly common in humans and are a major cause of reproductive failure. To understand the impact of these abnormalities, let's survey the chromosome abnormalities in a random selection of 10,000 early embryos. In this sample, there will be close to 800 embryos with chromosome abnormalities. Of these 800:

- More than 110 will have an extra copy of chromosome 16.
- More than 20 will have an extra copy of chromosome 18.
- More than 40 will have an extra copy of chromosome 21.
- About 140 will be missing an X or a Y chromosome.
- The rest of the 800 embryos will have a variety of chromosome abnormalities.

About 750 of the 800 chromosomally abnormal embryos will be lost through miscarriage. The remaining 50 will survive and be born. Of these there will be about:

- 1 with an extra chromosome 18 (trisomy 18)
- 10 with an extra chromosome 21 (Down syndrome)
- 1 with a missing X or Y chromosome (Turner syndrome)
- 15 with an extra X or Y chromosome
- 23 with other chromosome abnormalities

Close to 1 in every 170 live births have a chromosome abnormality, and from 5% to 7% of all early childhood deaths are related to chromosome abnormalities, including aneuploidy. Humans have a rate of aneuploidy that is up to 10 times higher than that of other mammals. Understanding the causes of aneuploidy in humans remains one of the great challenges in human genetics.

We will first review the causes and effects of polyploidy and then discuss several aspects of aneuploidy.

Polyploidy changes the number of chromosomal sets.

Abnormalities in the number of chromosomal sets can arise in several ways:

- errors in meiosis during gamete formation
- events at fertilization
- errors in mitosis after fertilization

If homologous chromosomes fail to separate during meiosis I, the division in meiosis II will produce diploid gametes. Fusion of this diploid gamete with a normal haploid gamete will produce a triploid zygote (Figure 6.12).

FIGURE 6.12 The karyotype of a triploid individual contains three copies of each chromosome.

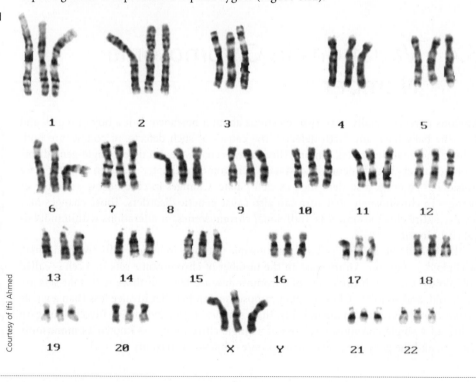

Courtesy of Ifti Ahmed

Polyploidy can also also be produced at fertilization by the simultaneous fusion of a haploid egg with two haploid sperm (dispermy). The resulting zygote contains three haploid chromosome sets and is triploid.

Triploidy.

The most common form of polyploidy in humans is **triploidy**, which is found in 15% to 18% of all miscarriages. Three types of triploid chromosome sets are observed: 69,XXY, 69,XXX, and 69,XYY. Approximately 75% of all cases of triploidy are 69,XYY and have two sets of paternal chromosomes. Meiotic accidents do not occur this often in males, so most triploid zygotes probably arise through dispermy. Although biochemical changes that accompany fertilization normally prevent such fertilizations, this system is not fail-safe.

Almost 1% of all conceptions are triploid, but over 99% of those result in miscarriage; only 1 in 10,000 live births is triploid, but most of these infants die within a month. Triploid newborns have multiple abnormalities, including an enlarged head (Figure 6.13), fused fingers and toes, and malformations of the mouth, eyes, and genitals. The high rate of embryonic death and failure to survive after birth indicates that triploidy is a lethal condition.

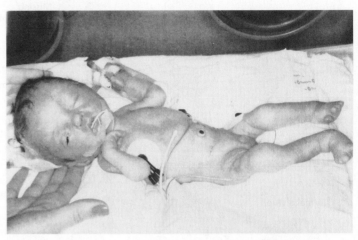

Reproduced by permission of *Pediatrics*, Vol. 74, p. 296 ©1984
Falix et al. *Pediatrics* 74:296–299, 1984, Figure 29.1

FIGURE 6.13 An infant with triploidy, showing the characteristic enlarged head.

Tetraploidy.

Tetraploidy is found in about 5% of all miscarriages but is extremely rare in live births. Tetraploidy can result from a failure of cytokinesis in the first mitotic division after fertilization. If tetraploidy arises at some point after the first mitotic division, two different cell types will be present in the embryo: normal diploid cells and tetraploid cells. These mosaic individuals survive somewhat longer than do full tetraploids, but the condition is still life threatening.

In summary, polyploidy in humans has at least two different causes: errors in cell division and accidents at fertilization. In either case, it is inevitably lethal. Polyploidy does not involve the mutation of any genes, only changes in the number of gene copies. How this quantitative change in gene number is related to lethality in development is unknown.

Triploidy A chromosomal number that is three times the haploid number, having three copies of all autosomes and three sex chromosomes.

Tetraploidy A chromosomal number that is four times the haploid number, having four copies of all autosomes and four sex chromosomes.

Aneuploidy changes the number of individual chromosomes.

As defined earlier, aneuploidy is the addition or deletion of individual chromosomes from the normal diploid set of 46. The most common cause of aneuploidy is **nondisjunction**: the failure of chromosomes to separate at anaphase (Active Figure 6.14). Although this failure can occur in either meiosis or mitosis, nondisjunction in meiosis is the leading cause of aneuploidy in humans. There are two cell divisions in meiosis, and nondisjunction can occur in either the first or the second division, with different genetic consequences.

If nondisjunction occurs in meiosis I (Active Figure 6.14a), all gametes will be abnormal and carry either both members of a chromosomal pair or neither member of the pair. Nondisjunction in meiosis II (Active Figure 6.14b) produces two normal haploid cells and two abnormal cells, one with an extra copy of a chromosome and one missing a chromosome.

Nondisjunction The failure of homologous chromosomes to separate properly during meiosis or mitosis.

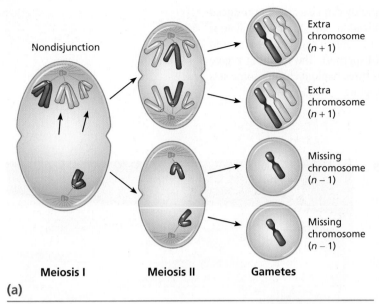

Nondisjunction

Meiosis I   Meiosis II   Gametes

Extra chromosome (*n* + 1)

Extra chromosome (*n* + 1)

Missing chromosome (*n* − 1)

Missing chromosome (*n* − 1)

(a)

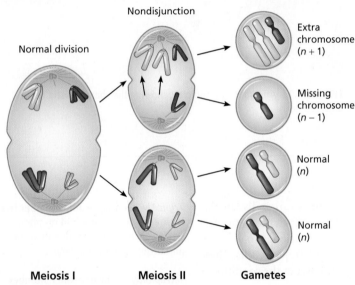

Nondisjunction

Normal division

Meiosis I   Meiosis II   Gametes

Extra chromosome (*n* + 1)

Missing chromosome (*n* − 1)

Normal (*n*)

Normal (*n*)

(b)

ACTIVE FIGURE 6.14 (a) In nondisjunction in meiosis I, one pair of chromosomes fails to separate properly at anaphase I of meiosis. Two gametes carry both members of a chromosome pair (*n* + 1), and two are missing one chromosome (*n* − 1). (b) In meiosis II, nondisjunction occurs in one cell, producing two cells with the normal haploid number, one cell with an extra chromosome, and one cell with a missing chromosome.

CENGAGENOW Learn more about nondisjunction by viewing the animation by logging on to *login.cengage.com/sso* and visiting CengageNOW's Study Tools.

Trisomy 21 Aneuploidy involving the presence of an extra copy of chromosome 21, resulting in Down syndrome.

Gametes with an extra copy of a chromosome will result in trisomy, and those missing a copy of a specific chromosome will result in monosomy. The phenotypic effects of aneuploidy range from minor physical symptoms to devastating and lethal conditions. Among survivors, phenotypic effects often include behavioral deficits and mental retardation. In the following section, we look at some of the important features of autosomal aneuploid phenotypes. Then we consider the phenotypic effects of sex-chromosome aneuploidy.

Autosomal monosomy is a lethal condition.

Aneuploidy during gamete formation produces equal numbers of monosomic and trisomic gametes. However, autosomal monosomies are observed only rarely in miscarriages and live births, and is regarded as universally lethal. The likely explanation is that the majority of autosomal monosomic embryos are lost very early, even before pregnancy is recognized.

Autosomal trisomy is relatively common.

Most autosomal trisomies are lethal during prenatal development and account for up to 50% of the chromosomal abnormalities seen in miscarriages. Karyotypic analysis of these cases reveals that autosomes are differentially involved in trisomy (Figure 6.15). Trisomies for chromosomes 1, 3, 12, and 19 are rarely observed, whereas trisomy for chromosome 16 accounts for almost one-third of all cases. As a group, the acrocentric chromosomes (13–15, 21, and 22) are present in 40% of all miscarriages. Only a few autosomal trisomies result in live births (trisomy 8, 13, and 18). **Trisomy 21** (Down syndrome) is the only autosomal trisomy that allows survival into adulthood. In the following section, we will review the characteristics of some autosomal trisomies.

Trisomy 13: Patau syndrome (47,+13).

Only 1 in 10,000 live births have trisomy 13 (47,+13), and the condition is lethal. Half of all affected individuals die in the first month. The phenotype of trisomy 13 includes facial malformations (Figure 6.16), eye defects, extra fingers or toes, and feet with large protruding heels. Internally, there are usually severe malformations of the brain and nervous system, as well as congenital heart defects.

The involvement of so many organ systems indicates that abnormalities form early in embryonic development—perhaps as early as the sixth week. Parental age is

> **KEEP IN MIND**
>
> Monosomy and trisomy involve the loss and gain of a single chromosome to a diploid genome.

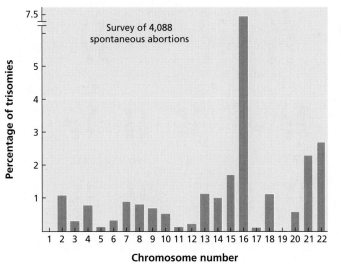

Courtesy of Dr. Ira Rosenthal, University of Chicago Children's Hospital

Survey of 4,088 spontaneous abortions

FIGURE 6.15 The results of a cytogenetic survey of over 4,000 miscarriages show a wide variation in how often specific chromosomes are involved in aneuploidy.

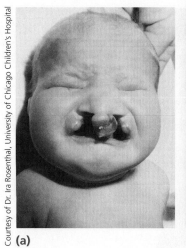
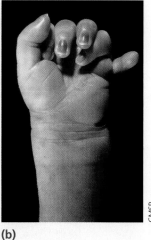

(a) **(b)**

CMSP

FIGURE 6.16 (a) An infant with trisomy 13, showing a cleft lip and palate (the roof of the mouth). (b) The hand of a trisomy 13 infant, showing polydactyly (extra fingers).

the only factor known to be related to trisomy 13. Parents of children with trisomy 13 are older (averaging about 32 years) than parents who have normal children. The relationship between parental age and aneuploidy is discussed later in this chapter.

Trisomy 18: Edwards syndrome (47,+18).

Infants with trisomy 18 are small at birth, grow very slowly, and are mentally retarded. For reasons still unknown, 80% of all trisomy 18 births are female. Clenched fists, with the second and fifth fingers overlapping the third and fourth fingers (Figure 6.17), and malformed feet are also characteristic. Heart malformations are almost always present, and heart failure or pneumonia are the usual causes of death. Trisomy 18 occurs with a frequency of 1 in 11,000 live births, and the average survival time is 2 to 4 months. As in trisomy 13, the age of the mother is a factor in trisomy 18.

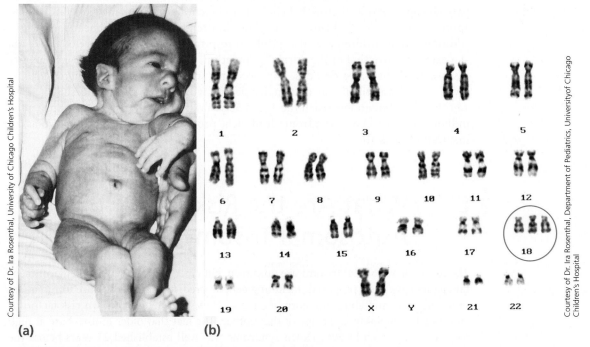

Courtesy of Dr. Ira Rosenthal, University of Chicago Children's Hospital

Courtesy of Dr. Ira Rosenthal, Department of Pediatrics, University of Chicago Children's Hospital

(a) **(b)**

FIGURE 6.17 (a) An infant with trisomy 18. (b) A trisomy 18 karyotype.

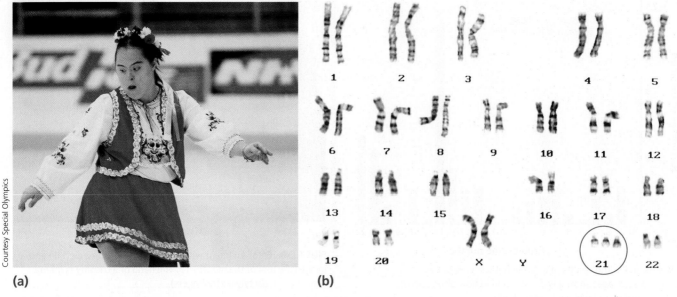

FIGURE 6.18 (a) A child with trisomy 21. (b) A karyotype shows this child has three copies of chromosome 21.

Trisomy 21: Down syndrome (47,+21).

The features of trisomy 21 (OMIM 190685) were first described by John Langdon Down in 1866. He called the condition "mongolism" because of the distinctive skin fold—known as an epicanthic fold—in the corner of the eye (Figure 6.18). To remove the racist implications inherent in the term, Lionel Penrose and others changed the name to Down syndrome. Trisomy 21 was the first chromosomal abnormality discovered in humans. In 1959, Jerome Lejeune and his colleagues discovered that the presence of an extra copy of chromosome 21 is the underlying cause of Down syndrome.

In the United States, Down syndrome occurs in about 1 in 800 live births and is a leading cause of childhood mental retardation and heart defects. Affected individuals usually have a wide, flat skull, folds in the corners of the eyelids, and spots on the irises. They may also have large, furrowed tongues that cause the mouth to remain partially open. Physical growth, behavior, and mental development are retarded, and approximately 40% of all Down syndrome children have congenital heart defects. In addition, these children are susceptible to respiratory infections, develop leukemia at a rate far above that of the normal population, and are at high risk for Alzheimer disease. In the last two decades, improvements in medical care have increased survival rates dramatically so that many people with Down syndrome survive into adulthood, although few reach the age of 50 years. In spite of their handicaps, many individuals with Down syndrome lead rich, productive lives and can serve as an inspiration to us all.

6.5 What Are the Risks for Autosomal Trisomy?

The causes of autosomal trisomy are unknown, but a variety of genetic and environmental factors have been proposed, including genetic predisposition, exposure to radiation, viral infection, and abnormal hormone levels. However, the only known risk factor for autosomal aneuploidy is the age of the mother. The fact that older mothers are at high risk for having a child with Down syndrome was well established 25 years before the chromosomal basis for the condition was discovered.

Maternal age is the leading risk factor for trisomy.

Young mothers have a low probability of having children with trisomy 21, but the risk increases rapidly after age 35 years. At age 20, the risk of having a child with Down syndrome is 1 in 2,000 (0.05%); by age 35, the risk has climbed to 1 in 111 (0.9%); and at age 45, 1 in 33 (3%) of all newborns have trisomy 21 (Figure 6.19a). Maternal age is also a risk factor for other autosomal aneuploidies (Figure 6.19b). Although paternal age has been proposed as a factor in autosomal trisomy, no clear-cut link has been demonstrated.

The age of the mother as a risk factor is supported by studies on the parental origin of chromosomes in trisomies. Occasionally, members of a chromosome pair have minor differences in their banding patterns. Examination of chromosomes from a trisomic child and the parents often allows the nondisjunction event to be traced to one parent. For trisomy 21, nondisjunction occurs about 94% of the time in the mother, and the great majority of these meiotic accidents take place at meiosis I.

Why is maternal age a risk factor?

The reasons why maternal age is a risk factor for nondisjunction are unknown, but at least two possible explanations are being studied. One of these focuses on the duration of meiosis in females. Recall from Chapter 2 that primary oocytes are formed early in embryonic development and enter meiotic prophase I well before birth. However, meiosis I stops and is not completed until ovulation, so oocytes produced at age 40 have been in meiosis I for over 40 years. During this time, intracellular events or environmental agents may cause damage that results in aneuploidy when meiosis resumes.

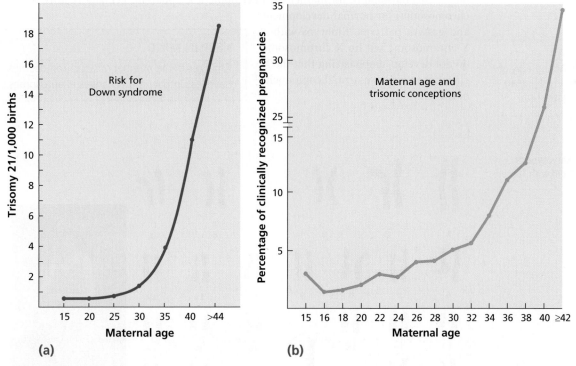

(a) (b)

FIGURE 6.19 (a) The relationship between maternal age and the frequency of trisomy 21 (Down syndrome). The risk increases rapidly after 35 years of age. (b) Maternal age is the major risk factor for autosomal trisomies of all types. By age 42, about one in three identified pregnancies is trisomic.

A second possible explanation centers on interactions between the implanting embryo and the uterus. According to this idea, embryo–uterine interactions normally result in the miscarriage of chromosomally abnormal embryos—a process called maternal selection. If maternal selection becomes less effective as women age, this would allow more chromosomally abnormal embryos to implant and develop. Other as yet unidentified factors may also play a role in the relationship between maternal age and autosomal aneuploidy, and more research is needed to clarify the underlying mechanisms.

6.6 Aneuploidy of the Sex Chromosomes

Aneuploidy involving the X and Y chromosomes is more common than autosomal aneuploidy. The overall incidence of sex-chromosome anomalies in live births is 1 in 400 for males and 1 in 650 for females. These abnormalities include both monosomy and trisomy.

Turner syndrome (45,X).

Females with monosomy of the X chromosome (**Turner syndrome**) are typically short, wide-chested, with rudimentary ovaries and puffiness of the hands and feet (Figure 6.20). They may have an aortic constriction, but there is no mental retardation associated with this condition. Turner syndrome occurs with a frequency of 1 in 10,000 female births.

The phenotypic impact of X chromosome monosomy in Turner syndrome is strikingly illustrated in a case of identical twins, one of them 46,XX and the other 45,X (Figure 6.21). Despite being identical twins, they have significant differences in height, sexual development, hearing, and other physical characteristics. Although environmental factors may contribute to these differences, the major role of the second X chromosome in normal female development is apparent. Females require two X chromosomes for normal development and growth patterns. Embryos with a Y chromosome but no X chromosome do not develop, emphasizing that the X chromosome is an essential component of the karyotype.

Turner syndrome A monosomy of the X chromosome (45,X) that results in female sterility.

Klinefelter syndrome Aneuploidy of the sex chromosomes involving an XXY chromosomal constitution.

> **KEEP IN MIND**
>
> Changes in the number of sex chromosomes have less impact than changes in autosomes.

FIGURE 6.20 (a) The karyotype of Turner syndrome. (b) A girl with Turner syndrome.

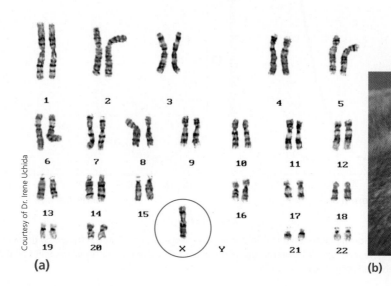

Courtesy of Dr. Irene Uchida

Wellcome Library/CMSP

(a)

(b)

Klinefelter syndrome (47,XXY).

Most males with an extra X chromosome (XXY, **Klinefelter syndrome**) have fertility problems but few other symptoms (Figure 6.22). Klinefelter syndrome occurs in approximately 1 in 1,000 male births.

Overall, about 60% of the cases result from maternal nondisjunction, and advanced maternal age is known to increase the risk of having affected sons. Other forms of Klinefelter have XXYY, XXXY, and XXXXY sex-chromosome sets. Additional X chromosomes in these karyotypes increase the severity of the phenotypic symptoms and bring on clear-cut mental retardation. A significant number of Klinefelter males are mosaics, with some of their cells having an XY chromosome set and others carrying an XXY set. In these cases, nondisjunction occurred during mitosis in embryos.

XYY syndrome (47,XYY).

In 1965, a cytogenetic survey of 197 males imprisoned for violent and dangerous antisocial behavior aroused a great deal of interest in the scientific community and the popular press. The survey results showed that 9 of those males (about 4.5% of the total) were XYY

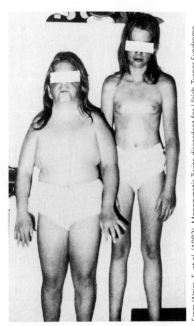

From Weiss, E. et al. (1982), Monozygotic Twins discordant for Ulrich-Turner Syndrome. Am. J. Med. Genet. 13:389–399

FIGURE 6.21 Monozygotic twins, one of which has Turner syndrome. The twin with Turner syndrome (*left*) is 45,X; the other twin (*right*) is 46,XX. In this case, the condition arose through mitotic nondisjunction, which occurred after fertilization.

Courtesy of Dr. Irene Uchida

Stefan Schwarz

Stefan Schwarz

FIGURE 6.22 (a) The characteristic karyotype of Klinefelter syndrome. (b) The young man (*left*) in these photos has Klinefelter syndrome.

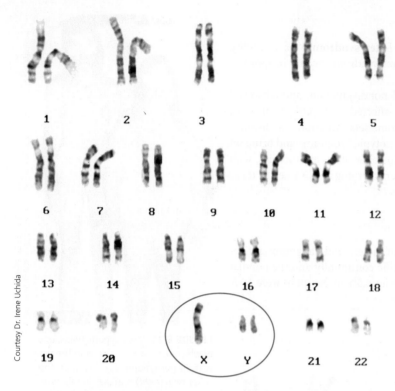

1 2 3 4 5

6 7 8 9 10 11 12

13 14 15 16 17 18

19 20 X Y 21 22

FIGURE 6.23 The karyotype of an XYY male. Affected individuals are usually taller than normal, and some, but not all, have personality disorders.

XYY karyotype Aneuploidy of the sex chromosomes involving an XYY chromosomal constitution.

(Figure 6.23). Phenotypically, most were above average in height, had personality disorders, and seven of the nine were of subnormal intelligence. Subsequent work showed that the frequency of XYY males in the general population is 1 in 1,000 male births (about 0.1% of the males in the general population) and that the frequency of XYY individuals in penal and mental institutions is significantly higher than it is in the population at large.

Early investigators associated the tendency to violent criminal behavior with the presence of an extra Y chromosome. If this were true, this would mean that some tendencies towards violent behavior are genetically determined. In fact, some defendants have attempted to use their **XYY karyotype** as a legal defense (unsuccessfully, so far) in criminal trials. The question is this: Is there really a connection between the XYY condition and criminal behavior? There is no evidence of a direct link between this karyotype and antisocial behavior, nor any evidence that an extra Y chromosome has a substantial phenotypic consequence. In fact, the vast majority of XYY males lead socially normal lives. In the United States, long-term studies of the relationship between antisocial behavior and the 47,XYY karyotype were discontinued. Researchers feared that identifying children with potential behavioral problems might lead parents to treat them differently and result in behavioral problems as a self-fulfilling prophecy.

What can we conclude about sex-chromosome aneuploidy?

Several conclusions can be drawn from the study of sex-chromosome disorders. First, at least one copy of an X chromosome is essential for survival. Embryos without an X chromosome (44,–XX and 45,Y) are not observed in miscarriages; these must be eliminated early in development—perhaps even before pregnancy is recognized. Second, adding more copies of the X or the Y chromosome interferes with normal development and can result in both physical and mental problems. As the number of sex chromosomes in the karyotype increases, the phenotype becomes more severe, indicating that a balance of sex chromosomes is essential to normal development in both males and females.

6.7 Structural Changes Within Chromosomes

Now that we have discussed changes in chromosome number, let's examine the phenotypic impact of structural changes within and between chromosomes. Such changes can involve one or more chromosomes. One type of change, chromosome breaks, can occur spontaneously through errors in replication or recombination. Environmental agents such as ultraviolet light, radiation, viruses, and chemicals can also produce breaks. The resulting structural changes can include duplications (extra copies of a chromosome part), translocations (transfer of a chromosome part to another, nonhomologous chromosome), and deletions (loss of a chromosome part). These changes

KEEP IN MIND
Chromosomes can lose, gain, or rearrange segments.

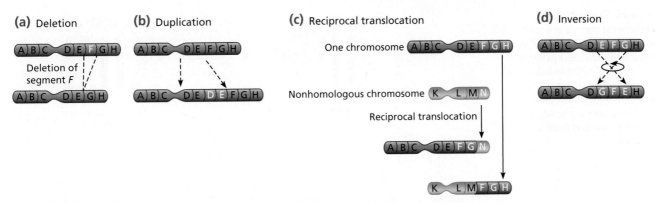

(a) Deletion

A B C D E F G H

Deletion of segment F

A B C D E G H

(b) Duplication

A B C D E F G H

A B C D E D E F G H

(c) Reciprocal translocation

One chromosome A B C D E F G H

Nonhomologous chromosome K L M N

Reciprocal translocation

A B C D E F G N

K L M F G H

(d) Inversion

A B C D E F G H

A B C D G F E H

FIGURE 6.24 Some of the common structural abnormalities seen in chromosomes. (a) In a deletion, part of a chromosome is lost (here segment F is lost). (b) Duplications have a chromosomal segment repeated (here segment DE is duplicated). (c) In a translocation, parts are exchanged between chromosomes. In this example, segments of nonhomologous chromosomes are exchanged. (d) In an inversion, the order of chromosome segments is reversed (here segment EFG is inverted).

are summarized in Figure 6.24. Rather than considering how such aberrations are produced, we'll look at the phenotypic effects of these alterations and what they can tell us about the location and action of genes.

Deletions involve loss of chromosomal material.

Deletion of a chromosome region causes developmental abnormalities; deletion of an entire autosome is lethal. Consequently, only a few viable conditions are associated with large-scale deletions. Some of these conditions are listed in Table 6.3.

Cri du chat syndrome (OMIM 123450) is caused by a deletion in the short arm of chromosome 5 and occurs in 1 in 20,000 to 1 in 50,000 births. Affected infants are mentally retarded, with defects in facial development and an abnormal larynx. They have a cry that sounds like a cat meowing—hence the name cri du chat. The deletion affects the motor and mental development of affected individuals but does not seem to be life threatening.

By comparing phenotypes with chromosomal breakpoints, two distinct regions have been identified on the short arm of chromosome 5 (Figure 6.25). Deletion of one region (5p15.3) causes abnormal larynx development; deletion of a neighboring region (5p15.2) is associated with mental retardation and other characteristics of cri du chat syndrome. One of the genes in this region is CTNND2 (OMIM 604275), and deletion of one copy of this gene may cause abnormal migration of nerve cells during development, resulting in mental retardation. These results indicate that genes controlling larynx development may be located in 5p15.3 and that two copies of all genes in 5p15.2 must be present for normal development of the nervous system.

Translocations involve exchange of chromosomal parts.

Translocations move part of a chromosome to another, nonhomologous chromosome. There are two major types of translocations: reciprocal translocations and Robertsonian translocations. In a reciprocal translocation, two nonhomologous chromosomes exchange parts. No genetic information is gained or lost from the cell in the exchange, but genes are moved to new chromosomal locations. In some cases, there are no phenotypic effects, and the translocation is passed through a family for generations. Robertsonian

p

15.3 — Larynx development
15.2 — Nervous system; CTNND2 gene
15.1

14

q

5

FIGURE 6.25 A deletion of part of chromosome 5 is associated with cri du chat syndrome. By comparing the region deleted with its associated phenotype, investigators have identified regions of the chromosome that carry genes involved in developing the larynx.

Table 6.3 Chromosomal Deletions

Deletion	Syndrome	Phenotype
5p−	Cri du chat syndrome	Infants have catlike cry, some facial anomalies, severe mental retardation
11q−	Wilms tumor	Kidney tumors, genital and urinary tract abnormalities
13q−	Retinoblastoma	Cancer of eye, increased risk of other cancers
15q−	Prader-Willi syndrome	Infants: weak, slow growth; children and adults: obesity, compulsive eating

Cri du chat syndrome A deletion of the short arm of chromosome 5 associated with an array of congenital malformations, the most characteristic of which is an infant cry that resembles a meowing cat.

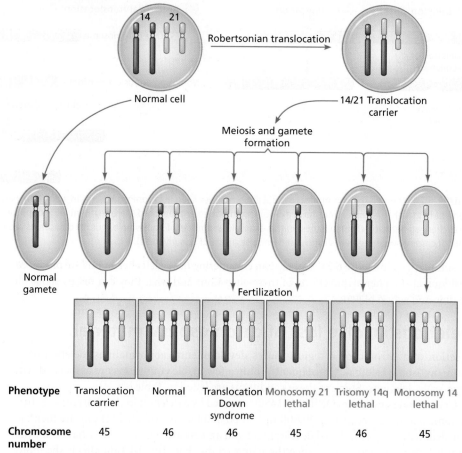

FIGURE 6.26 Segregation of chromosomes at meiosis in a 14/21 translocation carrier. Six types of gametes are produced. When these gametes fuse with those of a normal individual, six types of zygotes are produced. Of these, two (translocational carrier and normal) have a normal phenotype, one is Down syndrome, and three are lethal combinations.

| | | 14 | 21 |
Robertsonian translocation
Normal cell
14/21 Translocation carrier
Meiosis and gamete formation
Normal gamete
Fertilization

Phenotype	Translocation carrier	Normal	Translocation Down syndrome	Monosomy 21 lethal	Trisomy 14q lethal	Monosomy 14 lethal
Chromosome number	45	46	46	45	46	45

translocations can produce genetically unbalanced gametes with duplicated or deleted chromosomal segments that can result in embryonic death or abnormal offspring.

About 5% of all Down syndrome cases involve a Robertsonian translocation, most often between chromosomes 21 and 14. In these translocations, the centromeres of the two chromosomes fuse, and the short arms of both chromosomes are lost. Carriers of this translocation are phenotypically normal, even though they are actually aneuploid (they have only 45 chromosomes) and are missing the short arms of both chromosomes (no protein-coding genes are known to be on the short arms of these chromosomes). Meiosis in a carrier produces six types of gametes in equal proportions (Figure 6.26). Following fertilization, three of these result in lethal conditions caused by aneuploidy. Of the remaining three, one will produce a Down syndrome child, one will produce a translocation carrier, and one will result in a chromosomally normal individual.

Although it might seem that translocation heterozygotes have a 33% risk of having a Down syndrome child, the observed frequency is somewhat lower. It is important to remember that this risk is independent of maternal age. In addition, there is a one in three risk of having a child who is a translocation carrier who will be at risk of having children with Down syndrome. For this reason it is important to do a karyotypic analysis on a Down syndrome child and the parents to determine whether a translocation is involved. This information is essential in counseling parents about future reproductive risks.

6.8 What Are Some Consequences of Aneuploidy?

Aneuploidy is the most common chromosomal abnormality in humans and has several important consequences. Aneuploidy is a major cause of miscarriages (see Figure 6.15). Table 6.4 summarizes some of the major chromosomal abnormalities found in

Table 6.4 Chromosomal Abnormalities in Miscarriages

Abnormality	Frequency (%)
Trisomy 16	15
Trisomies 13, 18, 21	9
XXX, XXY, XYY	1
45,X	18
Triploidy	17
Tetraploidy	6

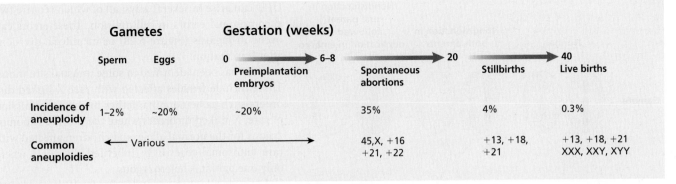

	Gametes		Gestation (weeks)			
	Sperm	Eggs	0 ──────► 6–8 Preimplantation embryos	──────► 20 Spontaneous abortions	Stillbirths	40 Live births
Incidence of aneuploidy	1–2%	~20%	~20%	35%	4%	0.3%
Common aneuploidies	◄─ Various ─────────────────►			45,X, +16 +21, +22	+13, +18, +21	+13, +18, +21 XXX, XXY, XYY

FIGURE 6.27 The frequency of aneuploidy changes dramatically over developmental time. Between 6 to 8 weeks and 20 weeks, about 35% of spontaneous abortions are aneuploid. Around 20 weeks, the frequency falls by an order of magnitude to about 4% stillbirths. The frequency decreases again by an order of magnitude, with about 0.3% of newborns being aneuploid.

miscarriages. These include triploidy, monosomy for the X chromosome (45,X), and trisomy 16. It is interesting to compare the frequency of chromosomal abnormalities found in miscarriages with those in live births. Triploidy is found in 17 of every 100 miscarriages but in only about 1 in 10,000 live births; 45,X is found in 18% of chromosomally abnormal miscarriages but in only 1 in 7,000 to 10,000 live births.

Comparison of the number of chromosomal abnormalities detected by CVS (performed at 8 to 10 weeks of gestation) versus amniocentesis (at 16 weeks of gestation) shows that the abnormalities detected by CVS are two to five times more common than those detected by amniocentesis, which in turn are about two times more common than those found in newborns. This decrease in the frequency of chromosomal abnormalities during pregnancy provides evidence that almost all chromosomally abnormal embryos and fetuses are eliminated as pregnancy progresses (Figure 6.27).

Birth defects are another consequence of chromosomal abnormalities. Only trisomy 13, 18, and 21 occur with any frequency in live births. Trisomy 21 occurs with a frequency of about 1 in 800 births, but cytogenetic surveys indicate that about two-thirds of such conceptions end in miscarriage. Overall, the high rate of nondisjunction in humans means that there is a significant reproductive risk for chromosomal abnormalities. About 0.5% of all newborns are affected with an abnormal karyotype.

A significant number of cancers, especially leukemia, are associated with specific chromosomal translocations. Solid tumors have a wide range of chromosomal abnormalities, including aneuploidy, translocations, and duplications. Evidence suggests that these abnormalities may arise during a period of genomic instability that precedes or accompanies the transition of a normal cell into a malignant cell. The chromosomal changes that accompany the development of cancer are discussed in Chapter 12.

6.9 Other Forms of Chromosome Changes

In some cases, the karyotype and individual chromosomes appear to be normal, but the phenotype is abnormal, and careful analysis reveals a subtle chromosome change. These situations include uniparental disomy and copy number variations. Fragile sites are another rare form of chromosome abnormality, which can be observed only when cells are grown in the laboratory and certain chemicals are added to the growth medium.

Uniparental disomy.

Normally, everyone inherits one member of each chromosomal pair from their mother and the other from their father. On rare occasions, however, a child gets both copies of a chromosome from one parent—a condition known as **uniparental disomy (UPD)**.

Uniparental disomy (UPD) A condition in which both copies of a chromosome are inherited from one parent.

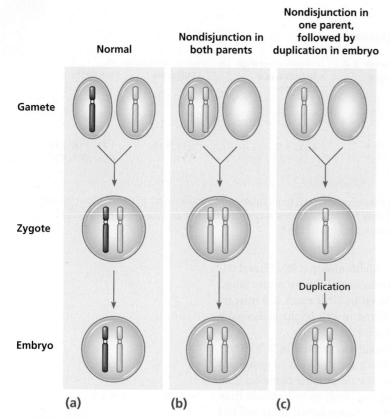

Gamete

Zygote

Embryo

Normal

Nondisjunction in both parents

Nondisjunction in one parent, followed by duplication in embryo

Duplication

(a)   (b)   (c)

FIGURE 6.28 Uniparental disomy can be produced by several mechanisms involving nondisjunction in meiosis, nondisjunction in the zygote, or early embryo. (a) Normally, gametes contain one copy of each chromosome, and fertilization produces a zygote carrying two copies of a chromosome—one derived from each parent. (b) Nondisjunction in both parents, in which one gamete carries both copies of a chromosome and the other gamete is missing a copy of that chromosome. Fertilization produces a diploid zygote, but both copies of one chromosome are inherited from a single parent. (c) Nondisjunction in one parent, resulting in the loss of a chromosome. This gamete fuses with a normal gamete to produce a zygote monosomic for a chromosome. An error in the first mitotic division results in duplication of the monosomic chromosome, producing uniparental disomy.

UPD can arise in several ways, all of which require two chromosomal errors in cell division. These errors can occur in meiosis (Figure 6.28) or in mitotic divisions after fertilization.

UPD has been identified in some unusual situations. These include females affected with rare X-linked disorders such as hemophilia; father-to-son transmission of rare, X-linked disorders where the mother is homozygous for the normal allele; and children affected with rare autosomal recessively inherited disorders where only one parent is heterozygous.

Let's consider two examples. Prader-Willi syndrome (OMIM 176270) and Angelman syndrome (OMIM 105830) can be caused by deletions in the long arm of chromosome 15 or by UPD. If both copies of chromosome 15 are inherited from the mother, the child will have Prader-Willi syndrome; if both copies of chromosome 15 are inherited from the father, the child will have Angelman syndrome. The origin of these disorders by UPD is discussed in detail in Chapter 11.

Copy number variation.

Earlier, we discussed how large-scale structural alterations within and between chromosomes play a role in genetic disorders, including cri du chat syndrome (deletions) and Down syndrome (translocations). The discovery that very small duplications and deletions of chromosomal regions are associated with genetic disoders has spurred an effort to identify and map structural changes in the human genome at and below the level of microscopic detection.

Structural abnormalities on the short arm of chromosome 17 illustrate the role of small duplications and deletions in genetic disorders. A duplication in 17p12 contains an extra copy of a gene (*PMP22*; OMIM 601097) that encodes a protein involved in making a sheath surrounding nerve cells (Figure 6.29). The duplication disrupts the production of this protective sheath and causes an autosomal dominant disease, Charcot-Marie-Tooth Type 1A syndrome (CMT 1A; OMIM 118220). Normal individuals have two copies of this region—one on each homologue—but those with CMT have three copies of this chromosome region, two on one homologue, and one on the other.

Deletions of one copy of *PMP22* cause another genetic disorder called hereditary neuropathy with liability to pressure palsies (HNPP; OMIM 162500). Affected individuals have only one copy of this chromosome region.

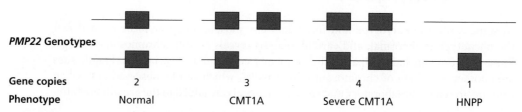

PMP22 Genotypes

Gene copies	2	3	4	1
Phenotype	Normal	CMT1A	Severe CMT1A	HNPP

FIGURE 6.29 The relationship between copy number variation (CNV) in the *PMP22* gene and the symptoms of Charcot-Marie-Tooth (CMT) syndrome. Having more than two copies of this gene causes (CMT). As the copy number increases, the symptoms become more severe.

Source: Lee, J. A. and Lupski, J. R. 2006. Genomic rearrangements and gene-copy number alterations as a cause of nervous system disorders. *Neuron* 52: 103–121.

Table 6.5 Human Diseases Associated with Copy Number Variants

Disease	Gene	Phenotype
Alzheimer disease	*APP*	Buildup of amyloid protein precursor, death of certain brain cells
Autsomal dominant adrenolukodystrophy	*LMNB1*	Abnormalities of white matter in brain, breakdown of myelin sheath surrounding nerves
Charcot-Marie-Tooth Type 1A	*PMP22*	Numbness in arms, legs, breakdown of myelin sheath surrounding nerves
Drug metabolism	*CYP2D6*	Increase or decrease in rate of drug metabolism, causing side effects and variation in effectiveness of drug
HIV infection/AIDS	*CCL3L1*	Increased susceptibility to HIV infection and AIDS
Lupus	*FCGR3B*	Increased susceptibility to kidney failure
Parkinson's disease	*SNCA*	Death of certain brain cells, tremors, increasing rigidity of body
Smith-Magenis syndrome	*RAI1*	Mental retardation
X-linked hypopituitarism	*SOX3*	Short stature, mild mental retardation; affects mostly males

Source: Cohen, J. 2009. DNA duplications and deletions help determine health. *Genomic Structural Variation in Human Diversity*; pp. 4–7. Washington D.C.: American Association for the Advancement of Science.

In a nearby region on chromosome 17, a deletion that includes the *RAI1* gene (OMIM 607642) causes Smith-Magenis syndrome (OMIM 182290), resulting in neurobehavioral abnormalities. A duplication of *RAI1* causes Potocki-Lupski syndrome (OMIM 610883), a disorder associated with cardiovascular abnormalities and autistic spectrum disorder.

Structural variations in chromosomes caused by changes in the number of copies of DNA segments are called **copy number variants (CNVs)**. Not all CNVs produce visible changes in chromosomes. In fact, most are submicroscopic changes that are detected using genomic rather than microscopic techniques. Copy number variants are defined as deletion or duplication of DNA segments from about 1,000 base pairs (1 kilobase) up to about 5 million base pairs (5 megabases). Chromosome changes of 3 to 5 megabases are just at the limit of what can be seen with a light microscope. CNVs change the structure of the genome and, along with submicroscopic inversions and translocations, are all forms of genome structural variation. We will explore genome variation in more detail in Chapter 15.

Thousands of CNVs have been discovered, some of which are associated with complex diseases—including autism, Alzheimer disease, Parkinson's disease, and schizophrenia—as well as some that affect drug metabolism in the body (Table 6.5). Geneticists are now identifying and mapping CNVs across the genome and establishing their association with human genetic disorders.

Copy number variation (CNV) A DNA segment at least 1,000 base pairs long with a variable copy number in the genome.

Fragile X syndrome An X chromosome that carries a gap, or break, at band q27; associated with mental retardation in males.

Fragile sites appear as gaps or breaks in chromosomes.

Fragile sites appear as gaps or breaks at specific sites on a chromosome and are inherited as codominant traits. Over 100 fragile sites have been identified in the human genome. Spontaneous chromosome breaks often occur at fragile sites, producing chromosome fragments, deletions, and other aberrations. The molecular nature of most fragile sites is unknown but is of great interest because those sites represent regions susceptible to breakage. Two fragile sites on the X chromosome, *FRAX E* and *FRAX A* are associated with genetic disorders

- -

KEEP IN MIND

Some fragile sites are associated with mental retardation.

- -

(Figure 6.30). The *FRAX A* site near the tip of the long arm of the X chromosome is associated with an X-linked form of mental retardation known as Martin-Bell syndrome, or **fragile-X syndrome**. The fragile-X syndrome (OMIM 309500) is caused by mutation in the *FMR-1* gene and is discussed in Chapter 11.

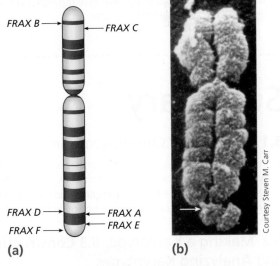

FRAX B
FRAX C
FRAX D
FRAX A
FRAX F
FRAX E

(a)

(b)

Courtesy Steven M. Carr

FIGURE 6.30 (a) The fragile sites on the human X chromosome. Sites B, C, and D are common sites and are found on almost all copies of the X chromosome. A, E, and F are rare sites; expression of A is associated with fragile-X syndrome. (b) A photograph of an X chromosome showing a fragile site (arrow).

Genetics in Practice

Genetics in Practice case studies are critical-thinking exercises that allow you to apply your new knowledge of human genetics to real-life problems. You can find these case studies and links to relevant websites at *www.cengage.com/biology/cummings*

CASE 1

Michelle was a 42-year-old Caucasian woman who had declined counseling and amniocentesis at 16 weeks of pregnancy but was referred for genetic counseling after an abnormal ultrasound at 20 weeks gestation. After the ultrasound, a number of findings suggested a possible chromosome abnormality in the fetus. The ultrasound showed swelling under the skin at the back of the fetus's neck; shortness of the femur, humerus, and ear length; and underdevelopment of the middle section of the fifth finger. Michelle's physician performed an amniocentesis and referred her to the genetics program. Michelle and her husband did not want genetic counseling before receiving the results of the cytogenetic analysis.

This was Michelle's third pregnancy; she and her husband, Mike, had a 6-year-old daughter and a 3-year-old son. At their next session, the counselor informed the couple that the results revealed trisomy 21, explored their understanding of Down syndrome, and elicited their experiences with people with disabilities. She also reviewed the clinical concerns revealed by the ultrasound and associated anomalies (mild to severe mental retardation, cardiac defects, and kidney problems). The options available to the couple were outlined. They were provided with a booklet written for parents making choices after the prenatal diagnosis of Down syndrome. After a week of careful deliberation with their family, friends, and clergy, they elected to terminate the pregnancy.

1. Do you think that this couple had the right to terminate the pregnancy in light of the prenatal diagnosis? If not, under what circumstance would a couple have this right? What other options were available to the couple?

2. Should physicians discourage a 42-year-old woman from having children, because of an increased chance of a chromosomal abnormality?

CASE 2

A genetic counselor was called to the nursery for consultation on a newborn described as "floppy with a weak cry." The counselor noted that the newborn's chart indicated that he was having feeding problems and had not gained weight since his delivery 15 days earlier. The counselor noted several other findings during her evaluation. The infant had almond-shaped eyes, a small mouth with a thin upper lip, downturned corners of the mouth, and a narrow face. He was born with undescended testes and a small penis. The counselor suspected that this child had Prader-Willi syndrome.

Prader-Willi syndrome is caused by a deletion or a mutation on the long arm of chromosome 15 or by UPD. Prader-Willi can happen three ways: a deletion on the father's copy of chromosome 15, a mutation in the Prader-Willi gene on the father's copy of chromosome 15, or maternal UPD, where both copies of chromosome 15 are from the mother and none are contributed by the father.

The child and his parents were tested for a deletion in the long arm of chromosome 15 (15q11–q13) by fluorescence *in situ* hybridization (FISH) and for uniparental disomy 15 by polymerase chain reaction (PCR). In this case, maternal disomy—which is the cause of Prader-Willi syndrome in about 30% of the cases—was detected by PCR.

1. Why is a copy of the paternal chromosome 15 needed to prevent Prader-Willi syndrome?

2. Are there any treatments for Prader-Willi syndrome? What steps should the family now take to cope with the diagnosis?

3. Explain to the parents how maternal disomy happens during gamete formation and/or in mitosis after fertilization.

Summary

6.1 The Human Chromosome Set

- Human chromosomes are analyzed by the construction of karyotypes. A system of identifying chromosome regions allows any region to be identified by a descriptive address. Chromosome analysis is a powerful and useful technique in human genetics.

6.2 Making a Karyotype, 6.3 Constructing and Analyzing Karyotypes

- The study of variations in chromosomal structure and number began in 1959 with the discovery that Down syndrome is caused by the presence of an extra copy of chromosome 21. Since that time, the number of genetic diseases related to chromosomal aberrations has steadily increased. The development of chromosome banding and techniques for identifying small changes in chromosomal structure has contributed greatly to the information that is now available.

6.4 Variations in Chromosome Number

- There are two major types of chromosomal changes: a change in chromosomal number and a change in chromosomal arrangement. Polyploidy and aneuploidy are major causes of reproductive failure in humans. Polyploidy is seen only rarely in live births, but the rate of aneuploidy in humans is reported to be more than tenfold higher than in other primates and mammals. The reasons for the difference are unknown, but this represents an area of intense scientific interest.

6.5 What Are the Risks for Autosomal Trisomy?

▪ The loss of a single chromosome creates a monosomic condition, and the gain of a single chromosome is called a trisomic condition. Autosomal monosomy is eliminated early in development. Autosomal trisomy is selected against less stringently, and cases of partial development and live births of trisomic individuals are observed. Most cases of autosomal trisomy greatly shorten life expectancy, and only individuals who have trisomy 21 survive into adulthood.

6.6 Aneuploidy of the Sex Chromosomes

▪ Aneuploidy of sex chromosomes involves both the X and Y chromosomes. Studies of sex-chromosome aneuploidies indicate that at least one copy of the X chromosome is required for development. Increasing the number of copies of the X or Y chromosome above the normal range causes progressively greater disturbances in phenotype and behavior, indicating the need for a balance in gene products for normal development.

6.7 Structural Changes Within Chromosomes

▪ Changes in the arrangement of chromosomes include duplications, inversions, translocations, and deletions. Deletions of chromosomal segments are associated with several genetic disorders, including cri du chat and Prader-Willi syndromes. Translocations often produce no overt phenotypic effects but can result in genetically imbalanced and aneuploid gametes. We discussed a translocation resulting in Down syndrome that in effect makes Down syndrome a heritable genetic disease, potentially present in one in three offspring.

6.8 What Are Some Consequences of Aneuploidy?

▪ Aneuploidy is the leading cause of reproductive failure in humans, resulting in spontaneous abortions and birth defects. In addition, aneuploidy is associated with most cancers.

6.9 Other Forms of Chromosome Changes

▪ Uniparental disomy (UPD) is a condition in which both copies of a chromosome are inherited from a single parent. UPD is associated with several genetic diseases. Copy number variations (CNVs) are deletions or duplications that change gene dosage and are associated with many common genetic disorders. Fragile sites appear as gaps, or breaks, in chromosome-specific locations. One of these fragile sites on the X chromosome is associated with a common form of mental retardation that affects a significant number of males.

Questions and Problems

CENGAGENOW Preparing for an exam? Assess your understanding of this chapter's topics with a pre-test, a personalized learning plan, and a post-test by logging on to *login.cengage.com/sso* and visiting CengageNOW's Study Tools.

Constructing and Analyzing Karyotypes

1. Originally, karyotypic analysis relied on size and centromere placement to identify chromosomes. Because many chromosomes are similar in size and centromere placement, the identification of individual chromosomes was difficult, and chromosomes were placed into eight groups, identified by the letters A to G. Today, each human chromosome can be readily identified.
   a. What technical advances led to this improvement in chromosome identification?
   b. List two ways this improvement can be implemented.
   c. What clinical information does a karyotype provide?
2. Given the karyotype on the right, is this a male or a female? Normal or abnormal? What would the phenotype of this individual be?

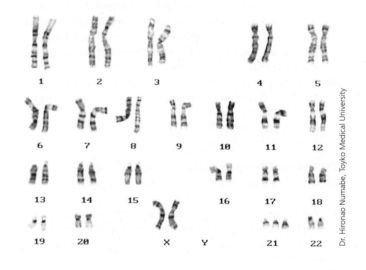

Dr. Hironao Numabe, Toyko Medical University

3. A colleague e-mails you saying that she has identified an interesting chromosome variation at 21q13. In discussing this discovery with a friend who is not a cytogeneticist, explain how you would describe this location, defining each term in the chromosome address 21q13.

4. What are the two most commonly used methods of prenatal diagnosis? Which technique can be performed earlier, and why is this an advantage?

5. What are some conditions that warrant prenatal diagnosis?

Variations in Chromosome Number—Polyploidy

6. Discuss the following sets of terms:
   a. trisomy and triploidy
   b. aneuploidy and polyploidy

7. What chromosomal abnormality can result from dispermy?

8. Tetraploidy may result from:
   a. lack of cytokinesis in meiosis II
   b. nondisjunction in meiosis I
   c. lack of cytokinesis in mitosis
   d. nondisjunction in mitosis in the early embryo
   e. none of the above

9. A cytology student believes he has identified an individual with monoploidy. The instructor views the dividing cells under the microscope and correctly dismisses the claim. Why was the claim dismissed? What types of cells were being viewed?

10. An individual is found to have some tetraploid liver cells but diploid kidney cells. Be specific in explaining how this condition might arise.

11. A spermatogonial cell undergoes mitosis before entering the meiotic cell cycle en route to the production of sperm. However, during mitosis the cytoplasm fails to divide, and only one daughter cell is produced. A resultant sperm eventually fertilizes a normal ovum. What is the chromosomal complement of the embryo?

12. A teratogen is an agent that produces nongenetic abnormalities during embryonic or fetal development. Suppose a teratogen is present at conception. As a result, during the first mitotic division the centromeres fail to divide. The teratogen then loses its potency and has no further effect on the embryo. What is the chromosomal complement of this embryo?

13. As a physician, you deliver a baby with protruding heels and clenched fists with the second and fifth fingers overlapping the third and fourth fingers.
   a. What genetic disorder do you suspect the baby has?
   b. How do you confirm your suspicion?

Variations in Chromosome Number—Aneuploidy

14. Describe the process of nondisjunction and explain when it takes place during cell division.

15. A woman gives birth to monozygotic twins. One boy has a normal genotype (46,XY), but the other boy has trisomy 13 (47,+13). What events—and in what sequence—led to this situation?

16. Assume that a meiotic-nondisjunction event causes trisomy 8 in a newborn. If two of the three copies of chromosome 8 are absolutely identical, at what point during meiosis did the nondisjunction event take place?

17. Two hypothetical human conditions have been found to have a genetic basis. Suppose that a hypothetical genetic disorder responsible for condition 1 is similar to Marfan syndrome. The defect responsible for condition 2 resembles Edwards syndrome. One of the two conditions results in more severe defects, and death occurs in infancy. The other condition produces a mild phenotypic abnormality and is not lethal. Which condition is most likely lethal, and why?

18. What is the genetic basis and phenotype for each of the following disorders (use proper genetic notation)?
   a. Edwards syndrome
   b. Patau syndrome
   c. Klinefelter syndrome
   d. Down syndrome

19. The majority of nondisjunction events leading to Down syndrome are maternal in origin. Based on the duration of meiosis in females, speculate on the possible reasons for females contributing aneuploid gametes more frequently than males do.

20. Name and describe the theory that deals with embryo–uterus interaction that explains the relationship between advanced maternal age and the increased frequency of aneuploid offspring.

21. If all the nondisjunction events leading to Turner syndrome were paternal in origin, what trisomic condition might be expected to occur at least as frequently?

Structural Changes Within Chromosomes

22. Identify the type of chromosomal aberration described in each of the following cases:
   a. Loss of a chromosome segment
   b. Extra copies of a chromosome segment
   c. Reversal in the order of a chromosome segment
   d. Movement of a chromosome segment to another, nonhomologous chromosome

23. Describe the chromosomal alterations and phenotype of cri du chat syndrome and Prader-Willi syndrome.

24. A geneticist discovers that a girl with Down syndrome has a Robertsonian translocation involving chromosomes 14 and 21. If she has an older brother who is phenotypically normal, what are the chances that he is a translocation carrier?

25. Albinism is caused by an autosomal recessive allele of a single gene. An albino child is born to phenotypically normal parents. However, the paternal grandfather is albino. Exhaustive analysis suggests that neither the mother nor her ancestors carry the allele for albinism. Suggest a mechanism to explain this situation.

Other Forms of Chromosome Changes

26. Fragile-X syndrome causes the most common form of inherited mental retardation. What is the chromosomal abnormality associated with this disorder? What is the phenotype of this disorder?

Internet Activities

Internet Activities are critical-thinking exercises using the resources of the World Wide Web to enhance the principles and issues covered in this chapter. For a full set of links and questions investigating the topics described below, visit *www.cengage.com/biology/cummings*

1. *Identifying Chromosomes.* The University of Arizona's Biology Project provides a chromosome karyotyping activity. In this exercise, you have the opportunity to create part of a human karyotype. In the first part of the activity, you will be arranging chromosomes onto a karyotyping sheet; once you have completed the karyotype, you will interpret the results of your efforts. Read the introductory material and then proceed to "Patient Histories."
   Further Exploration. To read more about the latest high-tech methods in karyotyping, go to The Biology Project's "New Methods for Karyotyping" Web page.

2. *Exploring a Chromosomal Defect.* The chromosomal abnormality called fragile-X syndrome, discussed in this chapter, is a leading genetic cause of mental retardation. Go to the *Your Genes, Your Health* website maintained by the Dolan DNA Learning Center at Cold Spring Harbor Laboratory and click on the "Fragile X Syndrome" link. (If you want to find out about hemophilia or Marfan syndrome, there are links at this site.) For this exercise, you should choose the "What causes it?" link. We'll continue to discuss various aspects of fragile-X syndrome in later chapters of this text. If you would like to investigate some of this information now, go to the fragile-X Internet Activities for Chapters 7 and 11.
   Further Exploration. To find out more about general aspects of fragile-X syndrome, from current research to how to get involved with support groups, go to the *FRAXA* (Fragile X Research Foundation) website.

HOW WOULD YOU VOTE NOW?

The most common chromosomal disorder in humans is Down syndrome, which occurs in about 1 in every 800 births. The symptoms of Down syndrome are variable and cannot be predicted accurately before birth. Prenatal diagnostic testing can reveal whether a fetus has Down syndrome. More than 90% of couples learning of such a diagnosis elect to terminate the pregnancy. The Fairchild family, discussed in this chapter's opening story, chose to continue the pregnancy of their Down syndrome child, Naia, who is now a loving child and an integral part of their family. Now that you know more about chromosomal abnormalities, risk factors, and outcomes, what do you think? Would you elect to terminate or continue a pregnancy after a diagnosis of Down syndrome? Would you consider adopting a Down syndrome child? Visit the *Human Heredity* companion website at *www.cengage.com/biology/cummings* to find out more on the issue; then cast your vote online.

7 Development and Sex Determination

n the summer of 1965, Janet and Ron Reimer had twin sons, Bruce and Brian. A few months later, the boys were circumcised. For Bruce, the procedure went terribly wrong, and most of his penis was burned so badly that it could not be repaired. When he was 21 months of age, he was examined at a clinic in the United States, and his parents were advised to have reconstructive surgery done and raise Bruce as a female. Bruce underwent sex reassignment surgery and went home with a new name: Brenda. His parents had instructions not to tell Brenda the truth and to raise him as a girl.

During Brenda's childhood as a girl, this case was hailed as proof that children are psychosexually neutral at birth and that nurture has more to do with sexual roles than nature. Although Bruce's case started with an accident during circumcision, the apparent success of his transformation from male to female was used as a guideline in treating the 1 in 1,500 to 1 in 2,000 children born every year with genital structures that are not fully male or fully female, a condition known as ambiguous genitalia.

The treatment became focused on what was surgically possible, and little attention was given to the psychological, social, or ethical consequences of these decisions. In general, males with a small or malformed penis were surgically altered into females because in reconstructive genital surgery it is easier to make a vagina than a penis.

In spite of the glowing reports about Bruce and his progress as Brenda, the reality was much different. As a child, Brenda refused to wear dresses and preferred to play with boys. As a young teen, Brenda rebelled at having further surgery to construct a vagina and threatened suicide. One day, on the way home after a counseling session, Brenda's father told him the truth. Within weeks, Brenda demanded sex-change surgery and changed his name to David. After surgery to reconstruct a penis, he married and, through adoption, became the father of three children. Unfortunately,

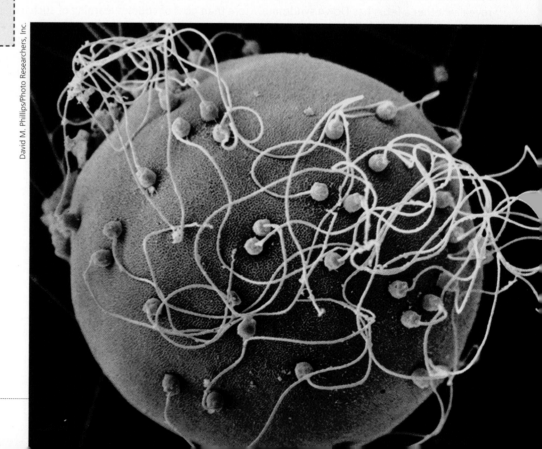

A mammalian egg surrounded by many sperm.

David M. Phillips/Photo Researchers, Inc.

overwhelmed by his problems, David committed suicide in 2004. The story of his life is narrated in the book *As Nature Made Him: The Story of a Boy Who Was Raised As a Girl*, by John Calpitano.

In an important follow-up, investigators concluded that it is wrong to assume that sexual identity is neutral at birth and that it can be shaped by the environment. This conclusion was confirmed by studies of children born as males with ambiguous genitals and surgically reassigned as females.

In this chapter, we will review the stages of human development and discuss the genetic and environmental factors that interact during prenatal sexual differentiation. We will also consider how gene dosage differences between males and females are adjusted and how the same gene can produce different phenotypes in males and females.

7.1 The Human Reproductive System

We all begin as a single cell, the **zygote**, produced by the fusion of a **sperm** and an **oocyte**. The sperm (from the father) and the oocyte (from the mother) are **gametes**. Males and females produce gametes in their **gonads**: paired organs with associated ducts and accessory glands. The **testes** in males produce spermatozoa and male sex hormones called androgens, and the **ovaries** of females produce oocytes and female sex hormones called estrogens. Within the gonads, cells produced by meiosis mature into gametes, and by fertilization, gametes from two parents unite to form a zygote, from which a new individual develops.

The male reproductive system.

Testes form in the abdominal cavity during male embryonic development; before birth, they descend into the **scrotum**, a pouch of skin outside the body cavity. In addition, the male reproductive system (Active Figure 7.1) includes:

- a duct system that transports sperm out of the body
- three sets of glands that secrete fluids to maintain sperm viability and motility
- the penis

Zygote The fertilized egg that develops into a new individual.

Sperm Male gamete.

Oocyte A cell from which an ovum develops by meiosis.

Gametes Unfertilized germ cells.

Gonads Organs where gametes are produced.

Testes Male gonads that produce spermatozoa and sex hormones.

Ovaries Female gonads that produce oocytes and female sex hormones.

Scrotum A pouch of skin outside the male body that contains the testes.

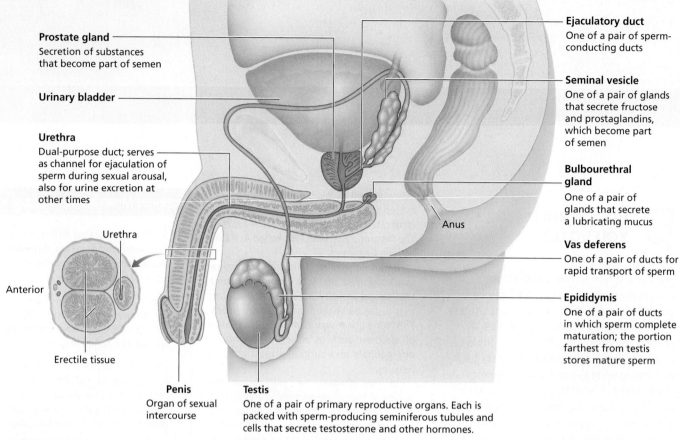

Prostate gland
Secretion of substances that become part of semen

Urinary bladder

Urethra
Dual-purpose duct; serves as channel for ejaculation of sperm during sexual arousal, also for urine excretion at other times

Urethra

Anterior

Erectile tissue

Penis
Organ of sexual intercourse

Testis
One of a pair of primary reproductive organs. Each is packed with sperm-producing seminiferous tubules and cells that secrete testosterone and other hormones.

Ejaculatory duct
One of a pair of sperm-conducting ducts

Seminal vesicle
One of a pair of glands that secrete fructose and prostaglandins, which become part of semen

Bulbourethral gland
One of a pair of glands that secrete a lubricating mucus

Anus

Vas deferens
One of a pair of ducts for rapid transport of sperm

Epididymis
One of a pair of ducts in which sperm complete maturation; the portion farthest from testis stores mature sperm

ACTIVE FIGURE 7.1 The anatomy of the male reproductive system and the functions of its components.

CENGAGENOW Learn more about the male reproductive system by viewing the animation by logging on to *login.cengage.com/sso* and visiting CengageNOW's Study Tools.

Seminiferous tubules Small, tightly coiled tubes inside the testes where sperm are produced.

Spermatogenesis The process of sperm production.

Spermatocytes Diploid cells that undergo meiosis to form haploid spermatids.

Epididymis A part of the male reproductive system where sperm are stored.

Vas deferens A duct connected to the epididymis, which sperm travels through.

Ejaculatory duct In males, a short connector from the vas deferens to the urethra.

Urethra A tube that passes from the bladder and opens to the outside. It functions in urine transport and, in males, also carries sperm.

Seminal vesicles Glands in males that secrete fructose and prostaglandins into the semen.

Prostaglandins Locally acting chemical messengers that stimulate contraction of the female reproductive system to assist in sperm movement.

Inside the testis are tightly coiled lengths of **seminiferous tubules**, where sperm are produced in a process called **spermatogenesis** (Active Figure 7.2). In the seminiferous tubules, cells called **spermatocytes** divide by meiosis to produce four haploid spermatids, which in turn differentiate to form mature sperm. Spermatogenesis begins at puberty and continues throughout life; each day, several hundred million sperm are in various stages of maturation. Immature sperm move from the seminiferous tubules to the **epididymis**, where they mature and are stored.

Sperm move through the male reproductive system in stages. In the first stage, when a male is sexually aroused, sperm move from the epididymis into the **vas deferens**, a duct lined with muscles, which contract rhythmically to move sperm forward. The vas deferens from each testis joins to form a short **ejaculatory duct** that connects to the **urethra**. The urethra (which also functions in urine transport) passes through the penis and opens to the outside. In the second stage, sperm are propelled by the muscular contractions that accompany orgasm from the vas deferens through the urethra and are expelled from the body.

As sperm are transported in the first stage, secretions are added from three sets of glands. The **seminal vesicles** secrete fructose, a sugar that serves as an energy source for the sperm, and **prostaglandins**, locally acting chemical messengers that stimulate contraction of the female reproductive system to assist in sperm movement. The **prostate gland** secretes a milky, alkaline fluid that neutralizes acidic vaginal secretions and enhances sperm viability. The **bulbourethral glands** secrete a mucus-like substance that provides lubrication for intercourse. Together, the sperm and these various glandular

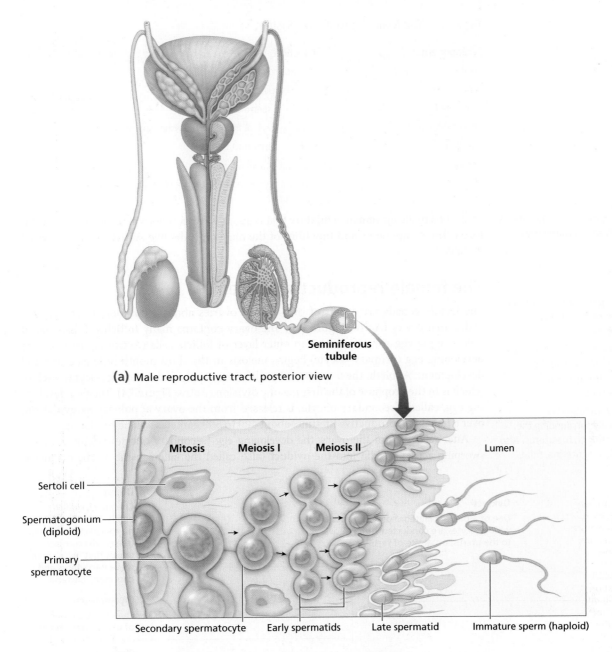

(a) Male reproductive tract, posterior view

Seminiferous tubule

Mitosis Meiosis I Meiosis II Lumen

Sertoli cell

Spermatogonium (diploid)

Primary spermatocyte

Secondary spermatocyte Early spermatids Late spermatid Immature sperm (haploid)

(b) Part of a cross-section through a seminiferous tubule

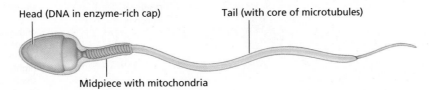

Head (DNA in enzyme-rich cap) Tail (with core of microtubules)

Midpiece with mitochondria

(c) Structure of a mature human sperm

ACTIVE FIGURE 7.2 (a) The male reproductive tract. (b) Cross section of the seminiferous tubule showing the process of sperm formation. Mitosis, meiosis, and incomplete cytokinesis produce haploid cells that differentiate into mature sperm. (c) A mature sperm and its components.

CENGAGENOW Learn more about sperm production by viewing the animation by logging on to *login.cengage.com/sso* and visiting CengageNOW's Study Tools.

Prostate gland A gland that secretes a milky, alkaline fluid that neutralizes acidic vaginal secretions and enhances sperm viability.

Bulbourethral glands Glands in the male that secrete a mucus-like substance that provides lubrication for intercourse.

Table 7.1   The Male Reproductive System

Component	Function
Testes	Produce sperm and male sex steroids
Epididymis	Site of sperm maturation and storage
Vas deferens	Conducts sperm to urethra
Accessory glands	Produce seminal fluid that nourishes sperm
Urethra	Conducts sperm to outside
Penis	Organ of sexual intercourse
Scrotum	Provides proper temperature for sperm formation by testes

Semen A mixture of sperm and various glandular secretions containing 5% spermatozoa.

secretions make up **semen**, a mixture that is about 95% secretions and about 5% spermatozoa. The components and functions of the male reproductive system are summarized in Table 7.1.

The female reproductive system.

Follicle A developing egg surrounded by an outer layer of follicle cells, contained in the ovary.

Ovulation The release of a secondary oocyte from the follicle; usually occurs monthly during a female's reproductive lifetime.

Oviduct A duct with fingerlike projections partially surrounding the ovary and connecting to the uterus. Also called the fallopian, or uterine, tube.

The female gonads are a pair of oval-shaped ovaries about 3 cm long, located in the abdominal cavity (Active Figure 7.3). The ovary contains many **follicles**, consisting of a developing egg surrounded by an outer layer of follicle cells (Active Figure 7.4). The developing egg (primary oocyte) begins meiosis in the third month of female prenatal development. At birth, the female carries a lifetime supply of developing oocytes, each of which is in the prophase of the first meiotic division (Active Figure 7.4). The first developing egg, called a secondary oocyte, is released from the ovary at puberty by **ovulation**; over a female's reproductive lifetime, about 400 to 500 gametes will be produced.

After release from the ovary, the developing egg (secondary oocyte) is moved by the sweeping action of cilia into the **oviduct** (also called the fallopian tube). The oviduct is

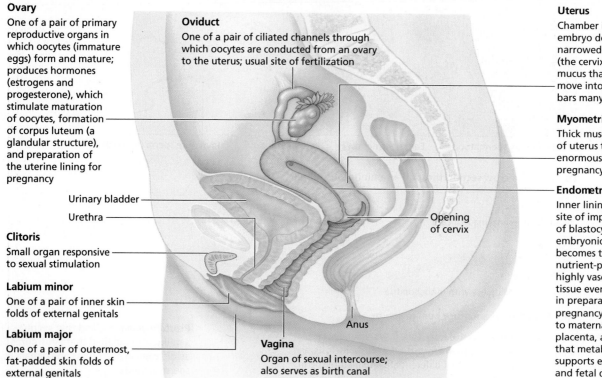

Ovary

One of a pair of primary reproductive organs in which oocytes (immature eggs) form and mature; produces hormones (estrogens and progesterone), which stimulate maturation of oocytes, formation of corpus luteum (a glandular structure), and preparation of the uterine lining for pregnancy

Oviduct

One of a pair of ciliated channels through which oocytes are conducted from an ovary to the uterus; usual site of fertilization

Urinary bladder

Urethra

Clitoris

Small organ responsive to sexual stimulation

Labium minor

One of a pair of inner skin folds of external genitals

Labium major

One of a pair of outermost, fat-padded skin folds of external genitals

Opening of cervix

Anus

Vagina

Organ of sexual intercourse; also serves as birth canal

Uterus

Chamber in which embryo develops; its narrowed-down portion (the cervix) secretes mucus that helps sperm move into uterus and bars many bacteria

Myometrium

Thick muscle layers of uterus that stretch enormously during pregnancy

Endometrium

Inner lining of uterus; site of implantation of blastocyst (early embryonic stage); becomes thickened, nutrient-packed, highly vascularized tissue every month in preparation for a pregnancy; gives rise to maternal portion of placenta, an organ that metabolically supports embryonic and fetal development

ACTIVE FIGURE 7.3 The anatomy of the female reproductive system and the functions of its components.

CENGAGENOW Learn more about the female reproductive system by viewing the animation by logging on to *login.cengage.com/sso* and visiting CengageNOW's Study Tools.

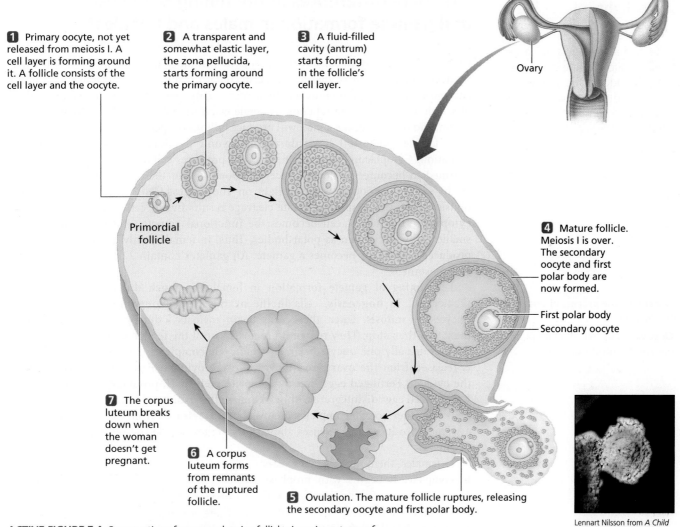

1 Primary oocyte, not yet released from meiosis I. A cell layer is forming around it. A follicle consists of the cell layer and the oocyte.

2 A transparent and somewhat elastic layer, the zona pellucida, starts forming around the primary oocyte.

3 A fluid-filled cavity (antrum) starts forming in the follicle's cell layer.

Ovary

Primordial follicle

4 Mature follicle. Meiosis I is over. The secondary oocyte and first polar body are now formed.

First polar body
Secondary oocyte

7 The corpus luteum breaks down when the woman doesn't get pregnant.

6 A corpus luteum forms from remnants of the ruptured follicle.

5 Ovulation. The mature follicle ruptures, releasing the secondary oocyte and first polar body.

Lennart Nilsson from *A Child Is Born* © 1966, 1977, Dell Publishing Company

ACTIVE FIGURE 7.4 Cross section of an ovary showing follicles in various stages of development. The photomicrograph at the right shows a secondary oocyte being released from the surface of the ovary. This oocyte will enter the fallopian tube and move toward the uterus.

CENGAGENOW Learn more about the development of oocytes by viewing the animation by logging on to *login.cengage.com/sso* and visiting CengageNOW's Study Tools.

connected to the **uterus**, a hollow, pear-shaped muscular organ about 7.5 cm (3 in.) long and 5 cm (2 in.) wide. The uterus consists of a thick, muscular outer layer called the myometrium and an inner membrane called the **endometrium**. This blood-rich inner lining is shed at menstruation if fertilization has not occurred. The lower neck of the uterus, the **cervix**, opens into the **vagina**. The vagina receives the penis during intercourse and also serves as the birth canal. The vagina opens to the outside of the body behind the urethra. The components and functions of the female reproductive system are summarized in Table 7.2.

Uterus A hollow, pear-shaped muscular organ where an early embryo will implant and develop throughout pregnancy.

Endometrium The inner lining of the uterus that is shed at menstruation if fertilization has not occurred.

Cervix The lower neck of the uterus, opening into the vagina.

Vagina The opening that receives the penis during intercourse and also serves as the birth canal.

Table 7.2 The Female Reproductive System

Component	Function
Ovaries	Produce ova (eggs) and female sex hormones
Oviducts	Transport sperm to ova; transport fertilized ova to uterus
Uterus	Nourishes and protects embryo and fetus
Vagina	Receptacle for sperm; birth canal

Oogenesis The process of oocyte production.

Oogonia Cells that produce primary oocytes by mitotic division.

Are there differences in the timing of meiosis and gamete formation in males and females?

In males, spermatogenesis begins at puberty and takes about 48 days: 16 for meiosis I, 16 for meiosis II, and 16 to convert the spermatid into the mature sperm. Each of the four products of meiosis forms sperm. All sperm contain 22 autosomes and either an X or a Y chromosome. The seminiferous tubules contain many spermatocytes, and large numbers of sperm are always in production. A single ejaculate may contain 200 to 400 million sperm, and over a lifetime a man will produce billions of sperm.

In females, meiosis I starts during embryonic development and is completed at ovulation. Cytoplasmic cleavage in meiosis I produces cells of unequal size. One cell, destined to become the oocyte, receives about 95% of the cytoplasm and is known as the secondary oocyte (see Spotlight on The Largest Cell). In the second meiotic division, the same disproportionate cleavage results in one cell retaining most of the cytoplasm. The largest cell becomes the functional gamete, and the nonfunctional smaller cells are known as polar bodies. Thus, in females, only one of the four cells produced by meiosis becomes a gamete. All gametes contain 22 autosomes and an X chromosome.

The timing of gamete formation in females is much different than in males (Table 7.3). In **oogenesis**, cells in the ovaries (called **oogonia**) produce primary oocytes by mitosis. Later, these cells begin meiosis I during embryonic development and then stop. They remain in meiosis I until the female undergoes puberty. At puberty, usually one oocyte per menstrual cycle completes the first meiotic division, is released from the ovary, and moves into the oviduct. Fertilization takes place in the oviduct. Fertilized eggs quickly complete meiosis II, producing a diploid zygote. Unfertilized eggs disintegrate within 24 hours after ovulation. Each month until menopause, another oocyte completes meiosis I and is released from the ovary. Altogether, a woman produces and releases about 450 oocytes during the reproductive phase of her life.

In females, then, meiosis takes years to complete. It begins with prophase I, while she is still an embryo, and continues until the completion of meiosis II after fertilization. Depending on the time of ovulation, meiosis can take from 12 to 50 years in human females.

KEEP IN MIND

There are important differences in the timing and duration of meiosis and gamete formation between males and females.

Table 7.3   A Comparison of the Duration of Meiosis in Males and Females

Spermatogenesis		Oogenesis	
Begins at Puberty		**Begins During Embryogenesis**	
Spermatogonium ↓	}	Oogonium ↓	} Forms at 2 to 3 months after conception
Primary spermatocyte ↓	} 16 days	Primary oocyte ↓	} Forms at 2 to 3 months of gestation. Remains in meiosis I until ovulation, 12 to 50 years after formation.
Secondary spermatocyte ↓	} 16 days	Secondary oocyte ↓	
Spermatid ↓	} 16 days	Ootid	} Less than 1 day, when fertilization occurs
Mature sperm Total time	48 days	Mature egg-zygote Total time	12 to 50 years

7.2 A Survey of Human Development from Fertilization to Birth

Fertilization, the fusion of male and female gametes, usually occurs in the upper third of the oviduct (Active Figure 7.5). Sperm deposited in the vagina swim through the cervix, up the uterus, and into the oviduct. About 30 minutes after ejaculation, sperm are present in the oviduct. Sperm travel this distance (about 7 inches) by swimming, using whip-like contractions of their tails, and are assisted by muscular contractions of the uterus.

Usually only one sperm fertilizes the egg, but many other sperm assist (Active Figure 7.5) by helping to trigger chemical changes near the surface of the egg. During fertilization, a sperm binds to receptors on the surface of the egg (technically, a secondary oocyte) and fuses with the cell's outer membrane. This fusion triggers a series of chemical changes in the membrane and prevents any other sperm from entering the oocyte. As a sperm enters the cytoplasm, its presence helps initiate the second meiotic division. After meiosis, the haploid oocyte nucleus fuses with the haploid sperm nucleus, forming a diploid zygote.

After fertilization, the zygote is swept along by cilia lining the walls of the oviduct and travels down the oviduct to the uterus over the next 3 to 4 days. Mitosis and embryogenesis begin while the zygote is in the oviduct. The embryo, consisting of a small number of cells, descends into the uterus and floats unattached in the uterine interior for a few days, drawing nutrients from the uterine fluids. Cell division continues during this time, and the embryo enters a new stage of development; it is now called a **blastocyst**

Fertilization The fusion of two gametes to produce a zygote.

Blastocyst The developmental stage at which the embryo implants into the uterine wall.

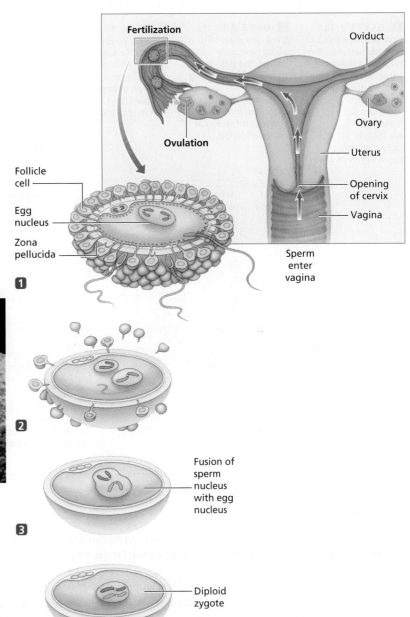

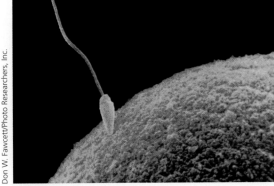

ACTIVE FIGURE 7.5 (1–2) In fertilization, many sperm surround the secondary oocyte and secrete enzymes that dissolve the outer barriers surrounding the oocyte. Only one sperm enters the egg. Penetration stimulates the oocyte to begin meiosis II. (3) The sperm tail degenerates, and its nucleus enlarges and fuses with the oocyte nucleus after meiosis II. (4) After fertilization, a zygote has formed.

CENGAGENOW Learn more about fertilization by viewing the animation by logging on to *login.cengage.com/sso* and visiting CengageNOW's Study Tools.

Don W. Fawcett/Photo Researchers, Inc.

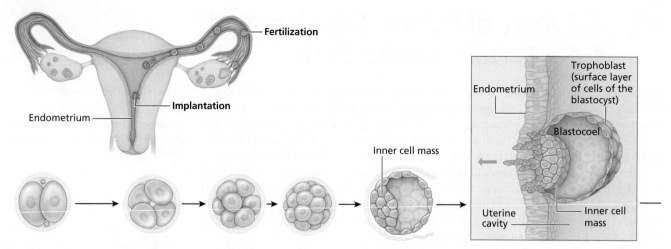

1 **DAYS 1–2.** The first cleavage furrow extends between the two polar bodies. Later cleavage furrows are angled, so cells become asymmetrically arranged. They are loosely organized with space between them.

2 **DAY 3.** After the third cleavage, cells abruptly huddle into a compacted ball, and tight junctions among the outer cells stabilize. Gap junctions formed along the interior cells enhance intercellular communication.

3 **DAY 4.** By 96 hours, the embryo is a solid ball of cells called a morula. Cells of the surface layer will function in implantation and give rise to a membrane, the chorion.

4 **DAY 5.** A fluid-filled cavity called the blastocoel forms in the morula and the inner cell mass forms. Differentiation occurs in the inner cell mass and gives rise to the embryo proper. This embryonic stage is the blastocyst.

5 **DAYS 6–7.** Some of the blastocyst's surface cells attach themselves to the endometrium and start to burrow into it. Implantation has started.

Actual size

6 **DAYS 10–11.** The yolk sac, embryonic disk, and amniotic cavity have started to form from the blastocyst.

Actual size

7 **DAY 12.** Blood-filled spaces form in maternal tissue. The chorionic cavity starts to form.

Actual size

8 **DAY 14.** A connecting stalk has formed between the embryonic disk and the chorion. Chorionic villi, which will be features of a placenta, start to form.

Actual size

ACTIVE FIGURE 7.6 Development from fertilization through implantation. A blastocyst forms, and its inner cell mass gives rise to a disc-shaped early embryo. As the blastocyst implants into the uterus, cords of chorionic cells start to form. When implantation is complete, the blastocyst is buried in the endometrium.

CENGAGENOW Learn more about early development and implantation by viewing the animation by logging on to *login.cengage.com/sso* and visiting CengageNOW's Study Tools.

Inner cell mass A cluster of cells in the blastocyst that gives rise to the embryonic body. The inner cell mass contains the embryonic stem cells.

Trophoblast The outer layer of cells in the blastocyst that gives rise to the membranes surrounding the embryo.

(Active Figure 7.6). A blastocyst, made up of about 100 cells, has several parts: the **inner cell mass**, an internal cavity, and an outer layer of cells (the **trophoblast**). The inner cell mass contains human embryonic stem cells (hESC), which will eventually form all the cells of the body. There has been a great deal of controversy about the isolation and use of human embryonic stem cells to develop treatments for many diseases and disorders. We will discuss the therapeutic uses of stem cells in later chapters, as well as the ethical and legal issues surrounding their use.

While the blastocyst is forming, the cells lining the uterus (called the endometrium) enlarge and differentiate, preparing for the attachment of the embryo. During the week-long process of implantation, the embryo's trophoblast attaches to the endometrium and releases enzymes that dissolve endometrial cells, allowing fingerlike growths of trophoblasts to lock the embryo into place (Active Figure 7.6).

By about 12 days after fertilization, the embryo is firmly implanted in the uterine wall and has formed a two-layered structure called the **chorion**. The chorion makes and releases a hormone called human chorionic gonadotropin (hCG), which prevents breakdown of the uterine lining and stimulates endometrial cells to release hormones that help maintain the pregnancy. Excess hCG is eliminated in the urine. Home pregnancy tests work by detecting elevated hCG levels as early as the first day of a missed menstrual period.

Chorion A two-layered structure formed during embryonic development from the trophoblast.

As the chorion grows, it forms fingerlike villi that extend into endometrial cavities filled with maternal blood. Capillaries from the embryo's developing circulatory system extend into the villi. As a result, the embryonic and maternal circulatory systems are separated from each other only by a thin layer of cells. Food molecules and oxygen cross easily from the mother's blood into the embryo, and waste molecules and carbon dioxide move from the embryo into the mother's blood. The chorionic villi eventually form the placenta, a disc-shaped structure that will nourish the embryo throughout prenatal development. Membranes connecting the embryo to the placenta form the umbilical cord, which contains two umbilical arteries and a single umbilical vein as extensions of the embryo's circulatory system.

Development is divided into three trimesters.

Development between fertilization and birth is divided into three trimesters, each of which lasts about 12 to 13 weeks. During the 36 to 39 weeks of development, the single-celled zygote undergoes 40 to 44 rounds of mitosis, producing trillions of cells that become organized into the tissues and organs of the fully developed fetus.

Organ formation occurs in the first trimester.

The first trimester is a period of radical change in the size, shape, and complexity of the embryo (Figure 7.7). Soon after implantation, organ systems begin to take shape. At 4 weeks, the embryo is about 5 mm long (about one-fifth of an inch), and much of the body is composed of paired segments.

During the second month, the embryo grows to a length of about 3 cm (about 1.12 in.) and undergoes a 500-fold increase in size. Most of the major organ systems, including the heart, are formed. Limb buds develop into arms and legs, complete with fingers and toes. The head is very large in relation to the rest of the body because of the rapid development of the nervous system.

By about 8 to 9 weeks, the embryo is called a fetus. Although chromosomal sex (XX in females and XY in males) is determined at the time of fertilization, the fetus is neither male nor female at the beginning of the third month. Soon after, specific gene sets are activated and sexual development is initiated. External sex organs can be seen in ultrasound scans between the 12th and 15th weeks. Sex differentiation is discussed later in this chapter.

By the end of the first trimester, the fetus is about 9 cm (about 3.5 in.) long and weighs about 15 g (about half an ounce). All the major organ systems have formed and are functional.

KEEP IN MIND

Most of the important events in human development occur in the first trimester. The remaining months are mainly a period of growth.

The second trimester is a period of organ maturation.

In the second trimester, bony parts of the skeleton begin to form, and the heartbeat can be heard with a stethoscope. Fetal movements begin in the third month, and by the fourth month the mother can feel movements of the fetus's arms and legs. At the end of the second trimester, the fetus weighs about 700 g (27 oz.) and is 30 to 40 cm (about 13 in.) long. It has a well-formed face, its eyes can open, and it has fingernails and toenails.

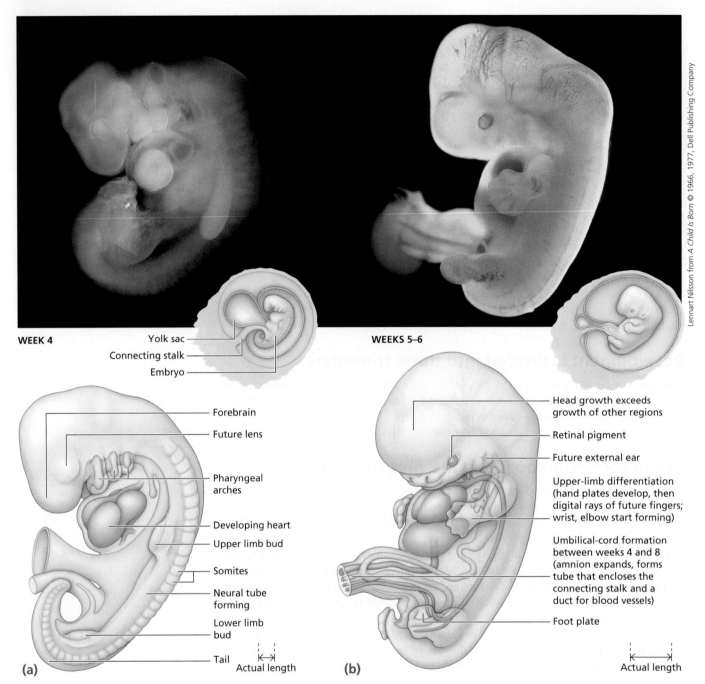

WEEK 4

Yolk sac
Connecting stalk
Embryo

WEEKS 5–6

Forebrain

Future lens

Pharyngeal arches

Developing heart

Upper limb bud

Somites

Neural tube forming

Lower limb bud

Tail

(a)

Actual length

Head growth exceeds growth of other regions

Retinal pigment

Future external ear

Upper-limb differentiation (hand plates develop, then digital rays of future fingers; wrist, elbow start forming)

Umbilical-cord formation between weeks 4 and 8 (amnion expands, forms tube that encloses the connecting stalk and a duct for blood vessels)

Foot plate

(b)

Actual length

FIGURE 7.7 Stages of human development. (a) Human embryo 4 weeks after fertilization. (b) Embryo at 4 to 5 weeks of development. (c) Embryo at week 8, during transition to the fetal stage of development. (d) Fetus at 16 weeks of development.

Rapid growth takes place in the third trimester.

The fetus grows rapidly in the third trimester, and the circulatory and respiratory systems mature to prepare for breathing air. During this period of rapid growth, maternal nutrition is important because most of the protein eaten by the mother will be used for growth and development of the fetal brain and nervous system. Similarly, much of the calcium in the mother's diet is used to develop the fetal skeletal system.

In the two months before birth, the fetus doubles in size, and chances for survival outside the uterus increase rapidly during that time. In the last month, antibodies pass from the mother to the fetus, giving the fetus a temporary immune system. At the end of the third trimester, the fetus is about 50 cm (19 in.) long and weighs from 2.5 to 4.8 kg (5.5 to 10.5 lb.).

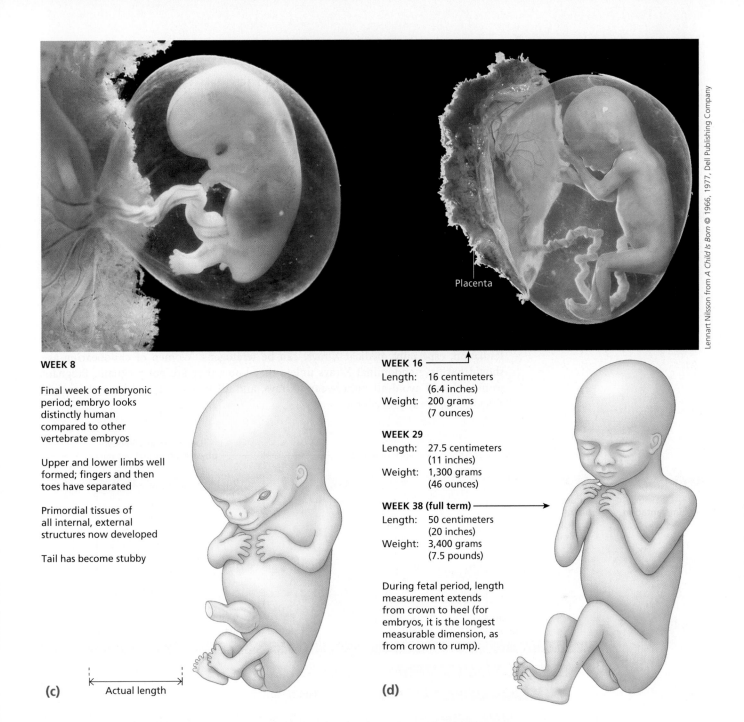

Lennart Nilsson from *A Child Is Born* © 1966, 1977, Dell Publishing Company

Placenta

WEEK 8

Final week of embryonic period; embryo looks distinctly human compared to other vertebrate embryos

Upper and lower limbs well formed; fingers and then toes have separated

Primordial tissues of all internal, external structures now developed

Tail has become stubby

(c) |← Actual length →|

WEEK 16

Length: 16 centimeters (6.4 inches)
Weight: 200 grams (7 ounces)

WEEK 29

Length: 27.5 centimeters (11 inches)
Weight: 1,300 grams (46 ounces)

WEEK 38 (full term)

Length: 50 centimeters (20 inches)
Weight: 3,400 grams (7.5 pounds)

During fetal period, length measurement extends from crown to heel (for embryos, it is the longest measurable dimension, as from crown to rump).

(d)

Birth is hormonally induced.

Birth is a hormonally induced process. During the last trimester, the cervix softens and the fetus shifts downward, usually with its head pressed against the cervix. Hormone-induced mild uterine contractions start during the third trimester, but at the start of the birth process, they become more frequent and intense. The hormone oxytocin, released from the pituitary gland, helps stimulate uterine contractions. During labor, the cervical opening dilates in stages to allow passage of the fetus, and uterine contractions expel the fetus. The head usually emerges first. If another body part enters the birth canal first, the result is called a breech birth. A short time after delivery, a second round of uterine contractions begins the expulsion of the placenta. These contractions separate the placenta from the lining of the uterus, and the placenta is expelled through the vagina.

7.3 Teratogens Are a Risk to the Developing Fetus

Although about 97% of all babies are normal at birth, birth defects can be produced by developmental accidents, genetic disorders, or exposure to environmental agents (Active Figure 7.8). Most birth defects are caused by disruptions of embryonic development, but the brain and nervous system can be damaged at any time during development, leading to conditions such as learning disabilities and mental retardation.

Chemicals and other agents that produce embryonic and/or fetal abnormalities are called **teratogens**. Defects produced by teratogens are nongenetic and are not passed on to the following generations. In 1960, only four or five agents were known to be teratogens. The discovery that thalidomide, a tranquilizer prescribed to stop morning sickness, caused limb defects in unborn children helped focus attention on environmental factors that produce birth defects. Today, we know that 30 to 40 agents are teratogens. Another 10 to 12 chemicals are strongly suspected of causing birth defects.

Teratogen Any physical or chemical agent that brings about an increase in congenital malformations.

Radiation, viruses, and chemicals can be teratogens.

Radiation, especially medical X-rays, can be teratogens. Women of childbearing age should not have abdominal X-rays unless they know they are not pregnant. Pregnant women should avoid all unnecessary X-rays, and all females should have abdominal shielding for X-ray procedures.

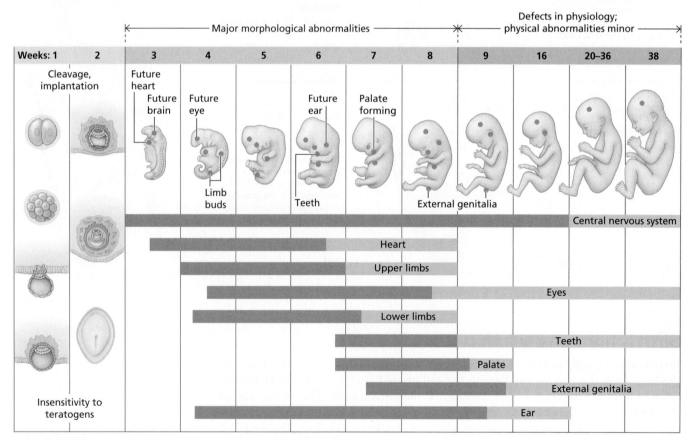

ACTIVE FIGURE 7.8 Teratogens are chemical and physical agents that can produce deformities in the embryo and the fetus. The effects of most teratogens begin after 3 weeks of development. Dark blue represents periods of high sensitivity; light blue shows periods of development with less sensitivity to teratogens.

CENGAGENOW Learn more about the action of teratogens by viewing the animation by logging on to *login.cengage.com/sso* and visiting CengageNOW's Study Tools.

Some viruses are teratogens. They include HIV, the measles virus, the German measles virus (rubella), herpes simplex (the virus that causes genital herpes), and HBV (the virus that causes hepatitis B). Fetuses infected with HIV are at risk for being stillborn or born prematurely and with low birth weight. The other viruses can cause a variety of effects, including severe brain damage or mental retardation in a developing fetus, or cause risks for diseases as adults. Some infectious organisms, such as *Toxoplasma gondii*, which is transmitted to humans by cats, raw meat, or fecal contamination of hands, are teratogenic and can result in a stillborn child or a child with mental retardation or other disorders.

Many chemicals, including medications such as the antibiotic tetracycline, are teratogens. Case 1 at the end of this chapter discusses drugs with teratogenic effects.

Fetal alcohol syndrome is a preventable tragedy.

A fetus's exposure to alcohol is one of the most serious teratogenic problems and is the most widespread of these; it is also the leading preventable cause of birth defects. Alcohol consumption during pregnancy can result in miscarriage, growth retardation, facial abnormalities (Figure 7.9), mental retardation, and learning disabilities. This collection of defects is known as **fetal alcohol syndrome (FAS)**. Related but milder forms of this disorder are called alcohol-related birth defects (ARBD) and alcohol-related developmental disorder (ARDD). The incidence of FAS is between 0.5 and 2.0 affected infants per 1,000 births, and the incidence for ARBD and ARDD is about 3.5 affected infants per 1,000 births. Together, these defects may affect 1 in every 100 (1%) births in the United States.

The teratogenic effects of alcohol can occur at any time during pregnancy, but weeks 8 to 12 are particularly sensitive periods. Even in the third trimester, alcohol can seriously impair fetal growth. Most studies show that the consumption of one or more drinks per day is associated with an increased risk of having a child with growth retardation. However, because fetal damage is related to blood alcohol levels, thinking about averages can be misleading. Having six drinks in a day and no drinks the rest of the week may pose a greater risk to the fetus than having one drink each day of the week. To emphasize the risks, the U.S. Surgeon General requires that all alcohol containers carry this warning:

> Drinking during pregnancy may cause mental retardation and other birth defects. Avoid alcohol during pregnancy.

The American Academy of Pediatrics has issued this policy statement: "Because there is no known safe amount of alcohol consumption during pregnancy, the Academy recommends abstinence from alcohol for women who are pregnant or who are planning a pregnancy."

The economic cost of FAS is enormous. The lifetime cost of caring for one child with FAS exceeds $1.4 million, and the overall costs to society range into the billions of dollars. The mental retardation associated with FAS accounts for 11% of the cost of treating all institutionalized, mentally retarded individuals. The emotional costs and social effects are difficult to estimate. Insight into the struggles of a family with an FAS child is recorded by Michael Dorris in his book *The Broken Cord*.

Aside from the well-known effects of alcohol, more work is needed to resolve the degree of risk associated with other chemicals and substances that are suspected teratogens and to identify new teratogens among the thousands of chemicals currently used. More importantly, research is needed to investigate the genetic basis for susceptibility to teratogenic agents and to develop tests to identify those who are susceptible to teratogens.

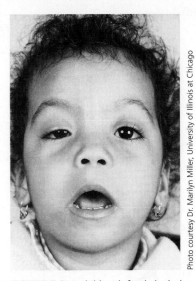

FIGURE 7.9 A child with fetal alcohol syndrome. The misshapen eyes, flat nose, and distinctive facial features are hallmarks of this condition.

Photo courtesy Dr. Marilyn Miller, University of Illinois at Chicago

Fetal alcohol syndrome (FAS) A constellation of birth defects caused by maternal alcohol consumption during pregnancy.

7.4 How Is Sex Determined?

In humans, as in many other species, we can see obvious differences between males and females. In humans, secondary sex characteristics such as the development of a penis or vagina, body size, muscle mass, patterns of fat distribution, and amounts and distribution of body hair emphasize the differences between the sexes. These differences are the outcome of a long chain of events that begins early in embryonic development and involves a network of interactions between gene expression and the environment.

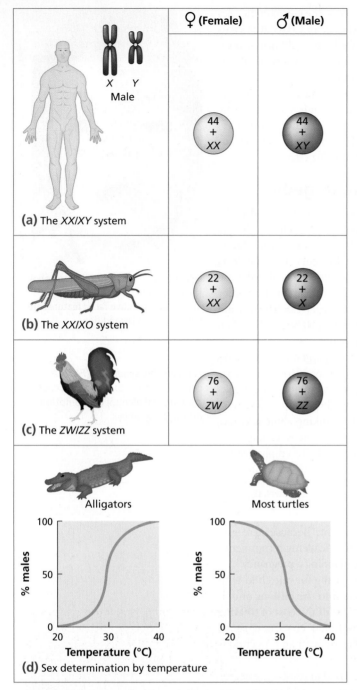

	♀ (Female)	♂ (Male)
(a) The *XX/XY* system	44 + XX	44 + XY
(b) The *XX/XO* system	22 + XX	22 + X
(c) The *ZW/ZZ* system	76 + ZW	76 + ZZ

Male

X Y

Alligators Most turtles

(d) Sex determination by temperature

FIGURE 7.10 (a–c) Animals have several mechanisms of sex determinations that involve chromosomes. (d) In some reptiles, the temperature at which the fertilized egg is incubated determines the sex of the offspring.

Environmental interactions can help determine sex.

Maleness and femaleness are determined by complex interactions between genes and the environment. To emphasize the role of environment, let's look at sex determination in some reptiles (Figure 7.10). These animals dig shallow holes, lay eggs, and then cover the nest as protection against predators. The internal temperature of the nest in which the eggs develop determines the sex of the offspring. In some species, higher temperatures produce females and lower temperatures produce males. In other species, the opposite is true.

Chromosomes can help determine sex.

In humans, maleness or femaleness is determined in stages beginning at fertilization, when the sex chromosomes carried by the gametes combine in the zygote. As was discussed in Chapter 2, females have two X chromosomes (XX) and males have an X chromosome and a Y chromosome.

Although saying that females are XX and males are XY seems straightforward, it does not provide all the answers to the question of what determines maleness and femaleness. Is a male a male because he has a Y chromosome or because he does *not* have two X chromosomes? Can someone be XY and develop as a female? Can someone be XX and develop as a male? While some of these questions have been partially answered, much more remains to be discovered. We do know that those with only one X chromosome (45,X) are female. We also know that those who carry two X chromosomes along with a Y chromosome (47,XXY) are males. From the study of people with abnormal numbers of sex chromosomes, it is clear that some females have only one X chromosome and some males can have more than one X chromosome. Furthermore, anyone with a Y chromosome is almost always male, no matter how many X chromosomes he may have. However, having an XX or XY chromosome set does not always mean someone develops as a female or male. The outcome depends on the distribution of genes on the X and Y chromosomes and interactions between genes on these chromosomes with many different environmental factors.

> **KEEP IN MIND**
>
> Chromosomal sex is determined at fertilization. Sexual differentiation begins in the seventh week and is influenced by a combination of genetic and environmental factors.

The human sex ratio changes with stages of life.

All eggs produced by females carry an X chromosome; in males, about half the mature sperm carry an X chromosome and half carry a Y chromosome. An egg fertilized by an X-bearing sperm results in an XX zygote that will develop as a female. Fertilization by a Y-bearing sperm will produce an XY, or male, zygote (Active Figure 7.11).

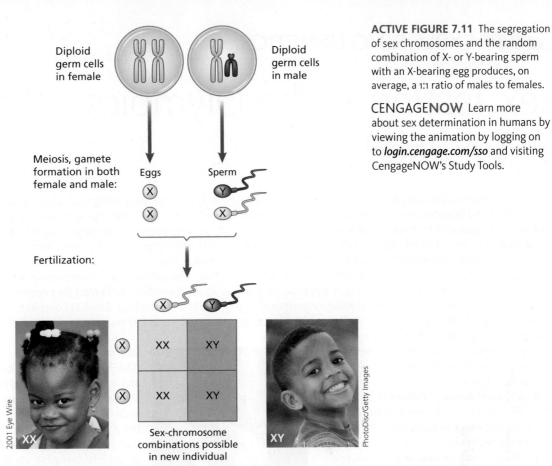

Diploid germ cells in female

Diploid germ cells in male

Meiosis, gamete formation in both female and male:

Eggs
X
X

Sperm
Y
X

Fertilization:

X Y

	X	Y
X	XX	XY
X	XX	XY

XX

XY

Sex-chromosome combinations possible in new individual

2001 Eye Wire

PhotoDisc/Getty Images

ACTIVE FIGURE 7.11 The segregation of sex chromosomes and the random combination of X- or Y-bearing sperm with an X-bearing egg produces, on average, a 1:1 ratio of males to females.

CENGAGENOW Learn more about sex determination in humans by viewing the animation by logging on to *login.cengage.com/sso* and visiting CengageNOW's Study Tools.

Because males produce approximately equal numbers of X- and Y-bearing sperm, males and females should be produced in equal proportions (Active Figure 7.11). This proportion, known as the **sex ratio**, changes throughout life. At fertilization, the sex ratio (known as the primary sex ratio) should be 1:1. Although direct determinations are impossible, estimates indicate that more males than females are conceived. At birth, the sex ratio (the secondary sex ratio) is about 1.05 (105 males for every 100 females). The tertiary sex ratio is measured in adults. Between the ages of 20 and 25, the ratio is close to 1:1. After that, females outnumber males in ever-increasing proportions. Genetic and environmental factors are responsible for the higher death rate among males. The expression of deleterious X-linked recessive genes is one cause of male death in both prenatal and postnatal stages of life. Between the ages of 15 and 35, accidents are the leading cause of death in males.

Sex ratio The proportion of males to females, which changes throughout the life cycle. The ratio is close to 1:1 at fertilization, but the ratio of females to males increases as a population ages.

7.5 Defining Sex in Stages: Chromosomes, Gonads, and Hormones

Having XX females and XY males provides a genetic framework for developmental events that guide the embryo toward the male or female phenotype (Figure 7.12). The formation of male or female reproductive structures depends on several factors, including gene action, interactions within the embryo, interaction with other embryos that may be in the uterus, and interactions with the maternal environment. As a result, the chromosomal sex (XX or XY) of an individual may differ from their phenotypic sex. The result can be a phenotype

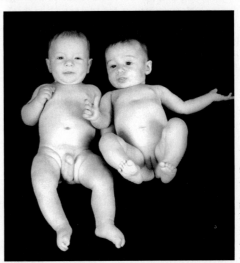

Sandra Wavick/Photonica/Getty Images

FIGURE 7.12 A cascade of gene action that begins in the seventh week of gestation results in the development of the male and female sexual phenotypes.

Sex Testing in the Olympics—Biology and a Bad Idea

Success in athletics, including the Olympics, is often a prelude to a career and financial rewards and acclaim that come with being a professional athlete. Because the stakes are so high, several methods are used to guard against cheating in competition. Competitors in many international events are required to submit urine samples (collected while someone watches) for drug testing. In other cases, urine testing is done at random in an attempt to detect and thus eliminate the use of steroids or performance-enhancing drugs. In the 1960s, rumors about males attempting to compete as females led the International Olympic Committee (IOC) to require sex testing of all female athletes, beginning with the 1968 Olympic Games.

The IOC's test involved analysis of Barr bodies in cells collected by scraping the inside of the mouth. In genetic females (XX), the inactivated X chromosome forms a Barr body, which can be stained and viewed in a microscope. Genetic males (XY) do not have a Barr body. The procedure is noninvasive, and females were not required to submit to a physical examination of their genitals. If sexual identity was called into question as a result of the test, a karyotype was required, and if necessary, a gynecological examination followed.

In both theory and practice, the IOC's test was a bad idea for several reasons. Barr-body testing is unreliable and leads to both false positive and false negative results. It fails to take into account phenotypic females who are XY with androgen insensitivity and other conditions that result in a discrepancy between chromosomal sex and phenotypic sex. In addition, the test does not take into

Stockbyte/Getty Images RF

account the psychological, social, and cultural factors that enter into one's identity as a male or a female. Ironically, no men attempting to compete as women were identified, but the test unfairly prevented females from competition. Of the more than 6,000 women athletes tested, 1 in 500 had to withdraw from competition as a consequence of failing the sex test. The Spanish hurdler Maria Martinez Patino led a courageous fight against sex testing. She has complete androgen insensitivity, was raised as a female, and competed as a female.

In response to criticism, the IOC and the International Amateur Athletic Federation (IAAF) reconsidered the question of sex testing and instituted a new test, based on recombinant DNA technology, to detect the presence of the male-determining gene *SRY*, which is carried on the Y chromosome. This test was instituted at the 1992 Winter Olympics. A positive test makes an athlete ineligible to compete as a female. However, the test was again flawed because it fails to recognize several chromosomal combinations that result in a female phenotype even though an *SRY* gene is present. At the 1996 Summer Olympic Games in Atlanta, eight of 3,387 females were *SRY* positive; seven of the eight had partial or complete androgen insensitivity. Again, no males attempting to compete as females were identified.

Finally, in the face of criticism from medical professionals and athletes, in 1999 the IOC decided to abandon the use of genetic screening of female athletes at the 2000 Olympic Games in Australia. However, the IAAF still retains the option of testing a competitor should the question of sexual identity arise.

opposite to the chromosomal sex, a phenotype intermediate to those of the two sexes, or a phenotype with characteristics and genitalia of both sexes. The sex of an individual can be defined at several levels: chromosomal sex, gonadal sex, and phenotypic sex. In most cases, all these definitions are consistent, but in others they are not (see Exploring Genetics: Sex Testing in the Olympics—Biology and a Bad Idea). To understand these variations and the interactions of genes with the environment, let's first consider what happens during normal sexual differentiation.

Sex differentiation begins in the embryo.

The first step in sex differentiation—establishment of chromosomal sex—occurs at fertilization with the formation of a diploid zygote with an XX or XY chromosome pair in addition to the 22 pairs of autosomes. Although the chromosomal sex of the zygote is established at fertilization, the external genitalia of early embryos are neither male nor female until about the start of the third month of development. Before this time, two undifferentiated gonads are present, along with sets of both male and female reproductive duct systems (Figure 7.13a). The two internal duct systems are the Wolffian (male) and the Müllerian (female) ducts (Figure 7.13b). The second step begins around 8 to 9 weeks when gene expression activates different developmental pathways and causes the undifferentiated gonads to develop as testes or ovaries. This step—gonadal sex differentiation—takes place over the next 4 to 6 weeks. Although it is convenient to think of only two pathways—one leading to males and the other to females—there are many alternative pathways that produce intermediate outcomes in gonadal sex and sexual phenotypes, some of which we consider in the following paragraphs.

If a Y chromosome is present, expression of genes on the Y chromosome causes the indifferent gonad to develop as a testis. A gene called **SRY**, the sex-determining region of the Y (OMIM 480000) located on the short arm of the Y chromosome, activates the expression of other genes and plays a major role in testis development. Other genes on the Y chromosome and on autosomes also play important roles at this time.

Once testis development is initiated by action of the *SRY* gene, cells in the testis secrete two hormones: **testosterone** and **anti-Müllerian hormone (AMH)**. Testosterone stimulates the Wolffian ducts to form the male internal duct system that will carry sperm. These ducts include the epididymis, seminal vesicles, and vas deferens. AMH secreted by the developing testis stops further development of female duct structures and causes the Müllerian ducts to degenerate (Figure 7.13b).

In embryos with two X chromosomes, the absence of the Y chromosome (and its *SRY* gene) and the presence of the second X chromosome cause the embryonic gonad to develop as an ovary. Ovarian development begins as cells along the outer edge of the gonad divide and push into the interior, forming an ovary. Because the ovary does not produce testosterone, the Wolffian duct system degenerates (Figure 7.13b), and because no AMH is produced, the Müllerian duct system develops to form the fallopian tubes, the uterus, and parts of the vagina.

Hormones help shape male and female phenotypes.

After gonadal sex has been established, the third phase of sexual differentiation—the development of the sexual phenotype—begins (Figure 7.13c). In males, testosterone is converted into the hormone dihydrotestosterone (DHT), which directs the formation of the external genitalia and influences brain development and organ size. Under the influence of DHT (mediated by receptors on the surface of cells), the genital folds and genital tubercle develop into the penis, and the labioscrotal swelling forms the scrotum. In females, no DHT is present, and the genital tubercle develops into the clitoris, the genital folds form the labia minora, and the labioscrotal swellings form the labia majora (Figure 7.13c).

In terms of gene action, it is important to remember that gonadal sex and sexual phenotype in males and females are produced by separate developmental pathways

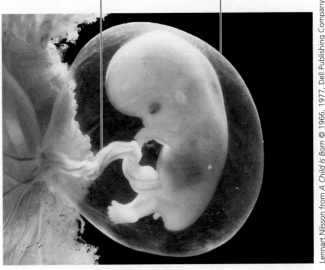

Umbilical cord (lifeline between the embryo and the mother's tissues)

Amnion (a protective, fluid-filled sac surrounding and cushioning the embryo)

Lennart Nilsson from *A Child Is Born* © 1966, 1977, Dell Publishing Company

(a)

FIGURE 7.13 (continued on next page) (a) A human embryo at 8 weeks, about the time sex differentiation begins.

SRY A gene, called the sex-determining region of the Y, located near the end of the short arm of the Y chromosome that plays a major role in causing the undifferentiated gonad to develop into a testis.

Testosterone A steroid hormone produced by the testis; the male sex hormone.

Anti-Müllerian hormone (AMH) A hormone produced by the developing testis that causes the breakdown of the Müllerian ducts in the embryo.

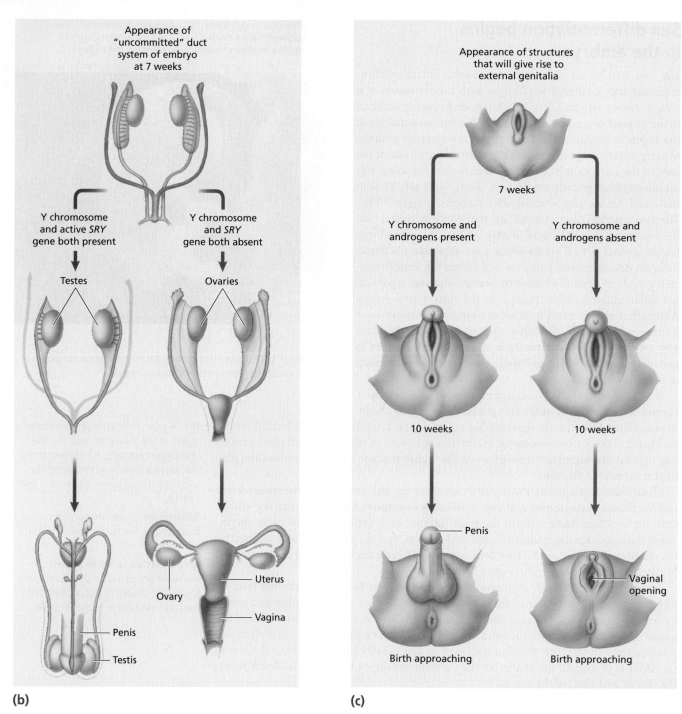

(b)

(c)

FIGURE 7.13 (continued) (b) Two duct systems (Wolffian and Müllerian) are present in the early embryo. They enter different developmental pathways in the presence and absence of a Y chromosome and the *SRY* gene. (c) Steps in the development of phenotypic sex from an undifferentiated stage to the male or female phenotype. The male pathway of development takes place in response to the presence of a Y chromosome and action of the *SRY* gene, followed by production of the hormones testosterone and dihydrotestosterone (DHT). Female development takes place in the absence of a Y chromosome and without those hormones.

(Figure 7.14). In males, this process begins with expression of the *SRY* gene on the Y chromosome, the presence of at least one X chromosome, and the expression of several autosomal genes. In females, the pathway involves the presence of two X chromosomes, the absence of the *SRY* gene, and the expression of a female-specific set of autosomal genes. These distinctions indicate that there may be important differences in the way genes in these pathways are activated, which may provide clues in the search for genes that regulate these pathways.

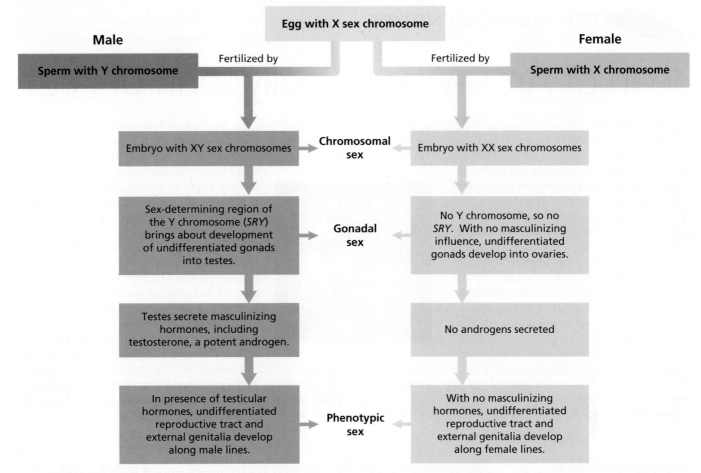

FIGURE 7.14 The major pathways of sexual differentiation and the stages at which genetic sex, gonadal sex, and phenotypic sex are established.

7.6 Mutations Can Uncouple Chromosomal Sex from Phenotypic Sex

Differentiation of the indifferent gonad can take a developmental pathway that results in a gonadal and/or sexual phenotype different than the one specified by the XX or XY chromosomal sex. These outcomes, which occur in about 1 in 2,000 births, may have several causes: chromosomal events that exchange segments of the X and Y chromosomes, mutations that affect the ability of cells to respond to the products of Y chromosome genes, or action of autosomal genes that control events on the X and/or Y chromosomes.

Androgen insensitivity can affect the sex phenotype.

The pathway from chromosomal sex to phenotypic sex can be disrupted at several stages. In one case, a mutation in a hormone receptor (*AR*, androgen receptor), causes XY males to become phenotypic females (Figure 7.15). This disorder is called **CAIS**, or **complete androgen insensitivity** (OMIM 313700).

In this case, the chromosomal sex is male (XY) and action of the *SRY* gene initiates testis formation and the production of AMH. Action of AMH causes degeneration of the Müllerian duct system, and no internal female reproductive tract is formed. However, because of the mutation, no receptors for male hormones are produced. Although DHT

Complete androgen insensitivity (CAIS) An X-linked genetic trait that causes XY individuals to develop into phenotypic females.

Joan of Arc—Was It Really John of Arc?

Joan of Arc, the national heroine of France, was born in a village in northeastern France in 1412, during the Hundred Years' War. At the age of 13 or 14, she began to have visions that directed her to help fight the English at Orleans. After victory, she helped orchestrate the crowning of the new king, Charles VII. During a siege of Paris, the English captured Joan, and in 1431 she was tried for heresy. Although her trial was technically a religious one conducted by the English-controlled church, it was clearly a political trial. Shortly after being sentenced to life imprisonment, she was declared a relapsed heretic, and on May 30, 1431, she was burned at the stake in the marketplace at Rouen.

In 1455, Pope Callistus formed a commission to investigate the circumstances of her trial, and a Trial of Rehabilitation took place over a period of 7 months in 1456. The second trial took testimony from over 100 individuals who had known Joan personally. Extensive documentation exists from the original trial and the Trial of Rehabilitation. This material has served as the source for the more than 100 plays and countless books written about her life. Although the story of her life is well known, perhaps more remains to be discovered. From an examination of the original evidence, R. B. Greenblatt proposed that Joan had phenotypic characteristics of complete androgen insensitivity syndrome (CAIS). By all accounts, Joan was a healthy female who had well-developed breasts. Those living with her at close quarters testified that she never menstruated, and physical examinations conducted during her imprisonment revealed a lack of pubic hair. Although such circumstantial evidence is not enough for a diagnosis, it provides more than enough material for speculation. This speculation also provides a new impetus for those medicogenetic detectives who prowl through history, seeking information about the genetic makeup of the famous, the infamous, the notorious, and the obscure.

Stockbyte/Getty Images

Pseudohermaphroditism An autosomal genetic condition that causes XY individuals to develop the phenotypic sex of females.

and testosterone are present, cells of the embryo cannot respond to their presence. As a result, development proceeds along the default pathway, which is female. The Wolffian duct system degenerates, and external genitalia develop as female structures. Affected individuals are chromosomal males but phenotypic females; they have well-developed breasts, little pubic hair, and do not menstruate (see Exploring Genetics: Joan of Arc—Was It Really John of Arc?).

Issel Kato/X90003/Reuters/Corbis

FIGURE 7.15 Santhi Soundarajan (green shorts), a phenotypic female who has an XY chromosomal constitution and androgen insensitivity.

Mutations can cause sex phenotypes to change at puberty.

Mutations in several different genes can produce a condition called **pseudohermaphroditism**. Affected individuals have both male and female structures, but at different times in their lives. At early stages of life, phenotypic sex does not match chromosomal sex, but later, the phenotypic sex changes do match chromosomal sex. One such mutation, an autosomal form of pseudohermaphroditism (OMIM 264300 and 605573) prevents conversion of hormone precursor molecules into testosterone and DHT. In these cases, the *SRY* gene initiates testes development, and the Wolffian ducts form the male duct system. AMH secretion prevents the development of female ducts. However, the failure to produce enough testosterone and DHT results in genitalia that are essentially female. The scrotum resembles the labia, a blind vaginal pouch is present, and the penis resembles a clitoris. Although chromosomally male, these individuals are often identified and raised as females.

At puberty, however, these females change into males. The testes move down into a developing scrotum, and what resembled a clitoris develops into a functional penis. The voice deepens, a beard grows, and muscle mass increases. In some cases, sperm production is normal. What causes these changes? The initial female phenotype is altered by the increased levels of testosterone secretion that accompany puberty. This condition is rare, but in a group of small villages in the Dominican Republic, more than 30 such cases are known. The high incidence is the result of common ancestry through intermarriage. In 12 of the 13 families in these villages, a line of descent can be traced from a single individual.

7.7 Equalizing the Expression of X Chromosome Genes in Males and Females

Dosage compensation A mechanism that regulates the expression of sex-linked genes.

Barr body A densely staining mass in the somatic nuclei of mammalian females; an inactivated X chromosome.

Females carry two X chromosomes and therefore have two copies of all the genes on that chromosome. Males are XY and have only one copy of all genes on the X chromosome. At first glance, it would seem that females should have higher levels of all products encoded by genes on the X chromosome. Is this true, or is there a mechanism that equalizes the expression of genes on the X chromosome so that males and females each have the same amounts of gene products encoded by genes on the X chromosome?

Dosage compensation makes XX equal XY.

Let's ask whether males and females have similar amounts of X-linked gene products. In Chapter 4, we discussed hemophilia A, an X-linked genetic disorder in which clotting factor VIII is missing. Because normal females have two copies of the clotting-factor gene and normal males have only one, does the blood of females contain twice as much of this clotting factor as males? The answer is straightforward: Careful measurements indicate that females have the same amount of this clotting factor as males. In fact, the same is true for all X chromosome gene products that have been tested: The amount of the encoded protein is the same in males and females.

How does this happen? A process called **dosage compensation** equalizes the amount of X chromosome gene products in both sexes. The explanation of how dosage compensation works in female mammals leads us from a physiologist working on cat nerves to a geneticist working on the inheritance of coat color in mice.

Mice, Barr bodies, and X inactivation can help explain dosage compensation.

In the late 1940s, Murray Barr and his colleagues were studying nerve conduction in cells from cats. Under the microscope, he saw a small, dense spot on the inside of the nuclear membrane in cells from female cats that did not appear in cells from male cats. A geneticist, Susumo Ohno, suggested that this spot—now called the **Barr body**—might actually be a genetically inactive X chromosome found in all female mammals (Figure 7.16).

About a decade later, Mary Lyon was studying the inheritance of coat color in mice. In female mice heterozygous for X-linked coat-color genes, Lyon found a unique phenotype that was different than either homozygous parent and was not a blend of parental coat colors. Instead, the female mice had patches of the parental colors in a random arrangement.

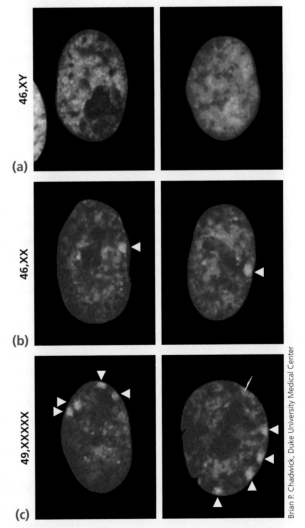

FIGURE 7.16 Relationship between X chromosome and Barr bodies. (a) XY males have no inactive X chromosomes and no Barr bodies. (b) XX females have one inactive X chromosome and one Barr body. (c) Females with five X chromosomes have four inactive X chromosomes and four Barr bodies. All X chromosomes except one are inactivated.

Brian P. Chadwick, Duke University Medical Center

Males, hemizygous for either gene, never showed such patches and had coats of uniform color. This genetic evidence suggested to Lyon that in heterozygous females, both alleles of the coat-color gene were active, but in different cells.

Mary Lyon put her genetic results together with Ohno's suggestion about Barr bodies in the cells of mammalian females and proposed her hypothesis (known as the **Lyon hypothesis**) about how dosage compensation works:

- Only one X chromosome is genetically active in the body cells of female mammals. The second X chromosome is inactivated and tightly coiled to form the Barr body.
- The inactivated chromosome can come from either parent.
- Inactivation takes place early in development. After four to five rounds of mitosis following fertilization, each cell of the embryo randomly inactivates one X chromosome.
- This inactivation is permanent (except in germ cells), and all descendants of a particular cell will have the same X chromosome inactivated.
- Because genes on only one X chromosome are expressed in females, this equalizes the amounts of products from X-linked genes in males and females.

ACTIVE FIGURE 7.17 The differently colored patches of orange and black fur on this calico cat result from X chromosome inactivation (white fur is controlled by a separate gene).

CENGAGENOW Learn more about X chromosome inactivation by viewing the animation by logging on to *login.cengage.com/sso* and visiting CengageNOW's Study Tools.

Lyon hypothesis The proposal that dosage compensation in mammalian females is accomplished by partially and randomly inactivating one of the two X chromosomes.

X inactivation center (Xic) A region on the X chromosome where inactivation begins.

Mammalian females can be mosaics for X chromosome gene expression.

The Lyon hypothesis means that female mammals are actually mosaics, constructed of two different cell types: Some cells express genes from the mother's X chromosome, and some cells express genes from the father's X chromosome. The pattern of coat color that Lyon observed in the heterozygous mice is a result of this inactivation. In females heterozygous for X-linked coat-color genes, patches of one color are interspersed with patches of another color. According to the Lyon hypothesis, each patch represents a group of cells descended from a single cell in which the inactivation event occurred.

The coat color of tortoiseshell cats and calico cats is a visible example of this mosaicism (Active Figure 7.17). In cats, an X-linked gene for coat color has two alleles: one that produces orange/yellow colored fur (*O*), and an alternative allele (*o*) that produces black fur. Heterozygous females (*O/o*) have patches of orange/yellow fur mixed with patches of black fur, called a tortoiseshell pattern. Cells expressing either the orange/yellow allele or the black allele cause this pattern. A cat with a tortoiseshell pattern on a white background is called a calico cat (white fur on the chest and abdomen in such cats is controlled by a different, autosomal gene). Therefore, tortoiseshell cats (and calico cats) are invariably female because males have only one X chromosome and would be either all orange/yellow or all black.

A mosaic pattern of gene expression also occurs in human females. An X-linked gene controls the formation of sweat glands. A rare recessive mutant allele blocks the formation of sweat glands. This condition is called anhidrotic ectodermal dysplasia (OMIM 305100). Heterozygous women have patches of skin (Figure 7.18) with sweat glands (cells in which the active X chromosome carries the normal allele) and patches of skin without sweat glands (cells in which the active X chromosome carries the mutant allele).

How and when are X chromosomes inactivated?

The process of X inactivation has presented researchers with several puzzling questions. How does the cell count the number of X chromosomes in the nucleus? If there are two X chromosomes in the nucleus, how is one X chromosome chosen to be turned off, but not the other? Finally, how is the chromosome inactivated?

Inactivation begins and is regulated from a region on the X chromosome called the **X inactivation center (Xic)**. One of the first steps is expression of the *XIST* gene, located in the *Xic*. When *XIST* is expressed, the X chromosome becomes coated with

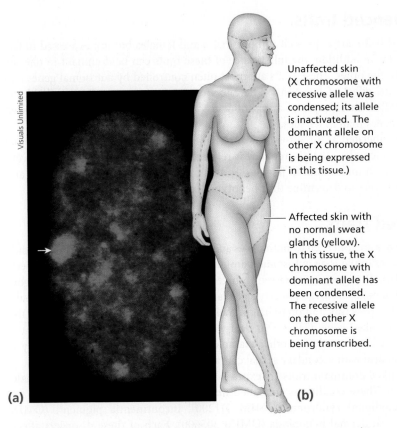

Unaffected skin (X chromosome with recessive allele was condensed; its allele is inactivated. The dominant allele on other X chromosome is being expressed in this tissue.)

Affected skin with no normal sweat glands (yellow). In this tissue, the X chromosome with dominant allele has been condensed. The recessive allele on the other X chromosome is being transcribed.

(a)

(b)

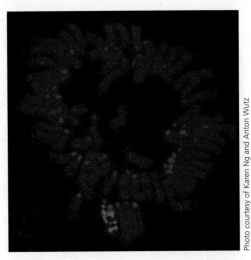

FIGURE 7.19 In the female mouse and other female mammals, expression of the *XIST* gene coats one X chromosome with *XIST* RNA (red), inactivating it. The active chromosomes in the set are stained blue.

FIGURE 7.18 (a) Photomicrograph of a Barr body (an inactive X chromosome) in a cell from a human female. (b) The mosaic pattern of sweat glands in a woman who is heterozygous for the X-linked recessive disorder anhidrotic ectodermal dysplasia.

KEEP IN MIND

One X chromosome is randomly inactivated in all the somatic cells of human females. This event equalizes the expression of X-linked genes in males and females.

XIST RNA (Figure 7.19). This causes almost all genes on the coated chromosome to become inactivated and form a Barr body. The small region not inactivated is called the pseudoautosomal region and contains genes homologous to those on the Y chromosome. Once an X chromosome is inactivated, all copies made in subsequent cell divisions are also inactivated.

In humans, both X chromosomes are genetically active in XX zygotes and all cells of early XX embryos. Random inactivation of one X chromosome usually occurs when the embryo has about 32 cells. Because there are only a small number of cells in the embryo at the time of inactivation and because inactivation occurs randomly in each cell, is it possible that all or almost all the mother's or father's X chromosome could be inactivated, just by chance? Yes. Figure 7.20 shows a pedigree in which one female monozygotic twin is colorblind, while the other is not. The females are heterozygotes for red-green color blindness through their color-blind father. The color-blind twin has three sons—two with normal vision and one who is color-blind—confirming that she carries the gene for color blindness (see pedigree).

Analysis of skin cells from the color-blind twin showed that almost all of the active X chromosomes came from her father and carry the allele for color blindness. In the twin with normal vision, the opposite situation is observed: Almost all of the active X chromosomes are maternal in origin and carry the normal allele for color vision.

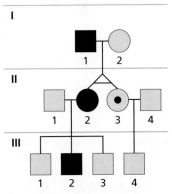

FIGURE 7.20 Pedigree showing monozygotic female twins (II-2 and II-3) discordant for color blindness. The twins inherited the allele for color blindness from their father. Almost all the active X chromosomes in the color-blind twin carry the mutant allele. Almost all the active X chromosomes in the twin who has normal vision carry the allele for normal vision.

7.8 Sex-Related Phenotypic Effects

In some cases, males and females differ in the phenotypic expression of a trait. This can arise in three situations: (1) sex-influenced traits, (2) sex-limited traits, and (3) imprinted genes.

FIGURE 7.21 Pattern baldness behaves as an autosomal dominant trait in males. It is an autosomal recessive trait in females. The degree of baldness in both males and females is related to hormone levels and other environmental influences.

Sex-influenced traits Traits controlled by autosomal genes that are usually dominant in one sex but recessive in the other sex.

Pattern baldness A sex-influenced trait that acts like an autosomal dominant trait in males and an autosomal recessive trait in females.

Sex-limited genes Loci that produce a phenotype in only one sex.

Imprinting A phenomenon in which expression of a gene depends on whether it is inherited from the mother or the father.

Sex-influenced traits.

Sex-influenced traits are expressed in both males and females but are expressed differently in each sex. In addition, the inheritance of these traits can be dominant in one sex but recessive in the other. These traits are most often controlled by autosomal genes and are related to the effect of hormonal differences on gene expression. **Pattern baldness** (OMIM 109200) is an example of a sex-influenced trait (Figure 7.21) that is expressed more often in males than in females.

Baldness is inherited as an autosomal dominant trait in males and as an autosomal recessive trait in females. The discrepancy is related to differences in testosterone levels, which are higher in males than in females. In this case, the hormonal environment and the genotype interact to determine the phenotypic expression of the gene.

Sex-limited traits.

Sex-limited genes are inherited by both males and females but are normally expressed only in one sex. One example is an autosomal dominant trait that controls precocious puberty (OMIM 176410). This gene is expressed in heterozygous males but not in heterozygous females. Affected males undergo puberty at 4 years of age or earlier. Heterozygous females are unaffected but pass this trait on to half of their sons, making it hard to distinguish this trait from a sex-linked gene. Genes that deal with traits such as breast development in females and facial hair in males are other examples of sex-limited genes, as are virtually all other genes that deal with secondary sexual characteristics.

Several X-linked dominant traits are expressed only in females because affected males die before birth. These conditions are called male-lethal X-linked dominant traits and include orofaciodigital syndrome (OMIM 311200), incontinentia pigmenti (OMIM 308300), and focal dermal hypoplasia (OMIM 305600). Each of these disorders affects multiple systems, including the skeleton, skin, teeth, and central nervous system. All reported cases are female. The exceptions are XXY males.

Duchenne muscular dystrophy (OMIM 310200) is an X-linked recessive disorder that for all practical purposes is a sex-limited trait. It affects 1 in 3,500 males and about 1 in 50,000,000 females. Because affected males die before reaching reproductive age, they cannot transmit the mutant gene to their daughters, and affected females are extremely rare. In most cases, affected females inherit an X chromosome with a mutant *DMD* gene from a carrier mother and a spontaneous mutation converts the normal *DMD* allele on the other X chromosome into a mutant allele. It has also been postulated that affected females with mild symptoms are heterozygotes but have undergone skewed X chromosome inactivation so that the active X chromosome in most body cells carries a mutant *DMD* allele.

Imprinted genes.

In humans, most cells of the body carry two copies of each gene, with one copy coming from each parent. Normally, either or both of these alleles can be expressed. However, in a small number of cases, only one of the two alleles is selectively expressed, depending on whether it was maternally or paternally inherited. This phenomenon is called **imprinting**. We will discuss imprinting briefly here; a more detailed discussion is presented in Chapter 11.

One example of imprinting involves *NOEY2* (OMIM 605193), a gene expressed in normal cells of the breast and ovary and some other cell types. Imprinting inactivates the maternal copy of the gene, and only the paternal copy is expressed. In breast cancer, both copies of *NOEY2* are imprinted and unexpressed, indicating that this gene may be important in controlling cell division and that silencing this gene may represent one of the steps in converting normal cells into cancer cells. We will discuss the role of gene mutation and the development of cancer in Chapter 12.

Genetics in Practice

Genetics in Practice case studies are critical-thinking exercises that allow you to apply your new knowledge of human genetics to real-life problems. You can find these case studies and links to relevant websites at *www.cengage.com/biology/cummings*

CASE 1

Melissa was referred for genetic counseling 16 weeks into her pregnancy because of a history of epileptic seizures. She was on medication (valproic acid) for her seizures and had not had a seizure in the last 3 years. Her obstetrician became concerned when he learned that she was still taking this medication, against his advice, during her pregnancy. He wanted her to speak to a genetic counselor about the possible effects of this medication on the developing fetus. The counselor took a detailed family history, which indicated that Melissa was the only family member with seizures and that no other genetic conditions were apparent in the family. The counselor asked Melissa why she continued to take valproic acid during her pregnancy. Melissa stated that she was "afraid my child would have epilepsy if she didn't take her medicine." Melissa went on to say that she was teased as a child when she had her "fits," and she wanted to prevent that from happening to her children.

With this in mind, the counselor reviewed the process of fetal development and why it is best that a physician carefully evaluate all medications that a woman takes while she is pregnant. Melissa's medication has been shown to cause spina bifida, which affected almost twice as many children who were exposed to it than children who were not exposed. Using illustrations, the counselor explained that spina bifida is a defect that occurs when the neural tube fails to close completely during embryonic development. The failure to fold exposes part of the spinal area when an infant is born. Valproic acid could also cause problems in the heart and the genitals. The counselor explained that prenatal diagnosis using ultrasound, and possibly amniocentesis, could help determine whether the baby's tube had closed properly.

Postscript: Melissa elected to have an ultrasound, which showed that the baby did not have a neural tube defect. However, she was offered an amniocentesis to rule out a possible false negative result of the ultrasound. She declined the amniocentesis and delivered a healthy baby boy.

1. As a genetic counselor, you have taken Melissa's family history. How can you address Melissa's fears that her child will develop epilepsy because she did?

2. From the perspectives of genetics, is Melissa at greater risk for having a child with epilepsy than is someone without epilepsy?

3. Women taking valproic acid have a 1% to 2% risk of having a child with a neural tube defect. Does the fact that Melissa had a normal child increase the risk that her next child will be affected? Why or why not?

4. The neural tube forms and closes during the first trimester of pregnancy. What does this suggest about Melissa's medication program in future pregnancies?

Summary

7.1 The Human Reproductive System

- The human reproductive system consists of gonads (testes in males, ovaries in females), ducts to transport gametes, and genital structures for intercourse and fertilization.

7.2 A Survey of Human Development from Fertilization to Birth

- Human development begins with fertilization and the formation of a zygote. Cell divisions in the zygote form a blastocyst. The embryo implants in the uterine wall, and a placenta develops to nourish the embryo.

7.3 Teratogens Are a Risk to the Developing Fetus

- The embryo and fetus are sensitive to chemical and physical agents, called teratogens, that can produce birth defects. Fetal alcohol syndrome is a preventable form of birth defect.

7.4 How Is Sex Determined?

- Mechanisms of sex determination vary from species to species. In humans, the presence of a Y chromosome is associated with male sexual development, and the absence of a Y chromosome is associated with female development.

7.5 Defining Sex in Stages: Chromosomes, Gonads, and Hormones

- Chromosomal sex is established at fertilization, but other aspects of sexual development depend on the interaction of genes and environmental factors, especially hormones.

7.6 Mutations Can Uncouple Chromosomal Sex from Phenotypic Sex

- Early in development, expression of the *SRY* gene on the Y chromosome signals the indifferent gonad to begin development as a testis. Hormones secreted by the testis control later stages of male sexual differentiation, including the development of phenotypic sex.

7.7 Equalizing the Expression of X Chromosome Genes in Males and Females

- Human females have one X chromosome inactivated in all somatic cells to balance the expression of X-linked genes in males and females.

7.8 Sex-Related Phenotypic Effects

- In sex-influenced and sex-limited inheritance, the sex of the individual affects whether and the degree to which the trait is expressed. This is true for both autosomal and sex-linked genes. Sex-hormone levels modify the expression of these genes, giving rise to altered phenotypic ratios.

Questions and Problems

CENGAGENOW Preparing for an exam? Assess your understanding of this chapter's topics with a pre-test, a personalized learning plan, and a post-test by logging on to *login.cengage.com/sso* and visiting CengageNOW's Study Tools.

The Human Reproductive System

1. How many chromosomes are present in a secondary oocyte as it leaves the ovary during ovulation?
2. Discuss and compare the products of meiosis in human females and males. How many functional gametes are produced from the daughter cells in each sex?
3. A human female is conceived on April 1, 1979, and is born on January 1, 1980. Onset of puberty occurs on January 1, 1992. She conceives a child on July 1, 2004. How long did it take for the ovum that was fertilized on July 1, 2004, to complete meiosis?

A Survey of Human Development from Fertilization to Birth

4. The gestation of a fetus occurs over 9 months and is divided into three trimesters. Describe the major events that occur in each trimester. Is there a point at which the fetus becomes more "human"?
5. FAS is caused by alcohol consumption during pregnancy. It can result in miscarriage, growth retardation, facial abnormalities, and mental retardation. How does FAS affect all of us—not just the unlucky children born with this syndrome? What steps need to be taken to prevent this syndrome?

How Is Sex Determined?

6. Describe, from fertilization, the major pathways of normal male sexual development; include the stages in which genetic sex, gonadal sex, and phenotypic sex are determined.
7. Which pathway of sexual differentiation (male or female) is regarded as the default pathway? Why?
8. The absence of a Y chromosome in an early embryo causes the:
   a. embryonic testis to become an ovary.
   b. Wolffian duct system to develop.
   c. Müllerian duct system to degenerate.
   d. indifferent gonad to become an ovary.
   e. indifferent gonad to become a testis.
9. Assume that humanlike creatures exist on Mars. As in the human population on Earth, there are two sexes and even sex-linked genes. The gene for eye color is an example of one such gene. It has two alleles. The purple allele is dominant to the yellow allele. A purple-eyed female alien mates with a purple-eyed male. All the male offspring are purple-eyed, whereas half the female offspring are purple-eyed and half are yellow-eyed. Which is the heterogametic sex? Why?

Mutations Can Uncouple Chromosomal Sex from Phenotypic Sex

10. Give an example of a situation in which genetic sex, gonadal sex, and phenotypic sex do not coincide. Explain why they do not coincide.
11. How can an individual who is XY be phenotypically female?
12. Discuss whether the following individuals (1) have male or female gonads, (2) are phenotypically male or female (discuss Wolffian/Müllerian ducts and external genitalia), and (3) are sterile or fertile.
    a. XY, homozygous for a recessive mutation in the testosterone biosynthetic pathway, producing no testosterone
    b. XX, heterozygous for a dominant mutation in the testosterone biosynthetic pathway, which causes continuous production of testosterone
    c. XY, heterozygous for a recessive mutation in the *MIH* gene
    d. XY, homozygous for a recessive mutation in the *SRY* gene that abolishes function

Sex-Influenced and Sex-Limited Traits

13. It has been shown that hormones interact with DNA to turn certain genes on and off. Use this fact to explain sex-limited and sex-influenced traits.
14. What method of sex testing did the International Olympic Committee previously use? What method did it use subsequently? Does either of these methods conclusively test for "femaleness"? Explain.
15. Explain why pattern baldness is more common in males than in females even though the gene resides on an autosome.

Equalizing the Expression of X Chromosome Genes in Males and Females

16. Calico cats are almost invariably female. Why? (Explain the genotype and phenotype of calico females and the theory of why calicos are female.)

17. How many Barr bodies would the following individuals have?
    a. normal male
    b. normal female
    c. Klinefelter male
    d. Turner female
18. Males have only one X chromosome and therefore only one copy of all genes on the X chromosome. Each gene is directly expressed, thus providing the basis of hemizygosity in males. Females have two X chromosomes, but one is always inactivated. Therefore, females, like males, have only one functional copy of all the genes on the X chromosome. Again, each gene must be directly expressed. Why, then, are females not considered hemizygous, and why are they not afflicted with sex-linked recessive diseases as often as males are?
19. Individuals with an XXY genotype are sterile males. If one X is inactivated early in embryogenesis, the genotype of the individual effectively becomes XY. Why will this individual not develop as a normal male?

Internet Activities

Internet Activities are critical-thinking exercises using the resources of the World Wide Web to enhance the principles and issues covered in this chapter. For a full set of links and questions investigating the topics described below, visit *www.cengage.com/biology/cummings*

1. *Embryological Development. The Visible Embryo* website provides free images and descriptions of human developmental stages from conception to stage 23. (Descriptions are available only for stages beyond 10 weeks.) Follow the stages and read about the development of the embryo.
   Further Exploration. Check out the *Morphing Embryos* video at Nova Online's *Odyssey of Life* website.
2. *Exploration of a Chromosomal Defect.* Fragile-X syndrome, which you may have researched as part of the Chapter 6 Internet Activities, affects males and females differently. Go to the *Your Genes, Your Health* website maintained by the Dolan DNA Learning Center at Cold Spring Harbor Laboratory and click on the "Fragile X Syndrome" link. At this page, choose the "How is it inherited?" link and explore how males and females inherit and display fragile-X syndrome.
   Further Exploration. To explore the complexities of the genetics of coat color in cats, including the genetics of X-linked characteristics such as tortoiseshell coat patterns, you can try "The Cat Color FAQ."

HOW WOULD YOU VOTE NOW?

The standard treatment for children born with genital abnormalities involves sex reassignment surgery, most often converting males into females. Now that you know more about how sex is determined and how sexual characteristics develop during pregnancy, what do you think? If you had a child with such a condition, would you consent to that kind of surgery for your child, or would you allow the child to make that decision upon reaching puberty? Visit the *Human Heredity* companion website at *www.cengage.com/biology/cummings* to find out more on the issue; then cast your vote online.

8 The Structure, Replication, and Chromosomal Organization of DNA

n February 2003, thousands of people in Asia became sick from a flulike disease with a high fever, headaches, and respiratory problems. In the next few months, the disease spread to more than two dozen countries across Asia, Europe, North America, and South America. Scientists around the globe mobilized to identify the cause of the illness and quickly isolated a virus, called a cornavirus, from infected individuals. The disease, named severe acute respiratory syndrome (SARS), is spread by droplets produced when an infected person sneezes or coughs. The SARS outbreak was eventually contained by isolating patients and screening travelers for infection. Despite those efforts, the World Health Organization (WHO) reported that just over 8,000 people were infected with SARS and about 10% of those died. In the United States, eight people developed SARS, but all had returned from parts of the world where SARS infections had been reported, and the disease did not spread in the United States.

By May 2003, scientists using genomic technology determined the DNA sequence of the viral chromosome, which carries 11 genes. To help fight future outbreaks, the development of a vaccine was a high priority. At that time, no vaccines against human cornaviruses had been developed, so researchers decided to try to develop a DNA vaccine.

To make a DNA vaccine, one of the virus' genes was isolated using recombinant DNA technology. The viral gene was injected into mice, where it directed the synthesis of a viral protein. The mouse's immune system responded by making antibodies against the protein, which protected the mouse against future infections with the SARS virus.

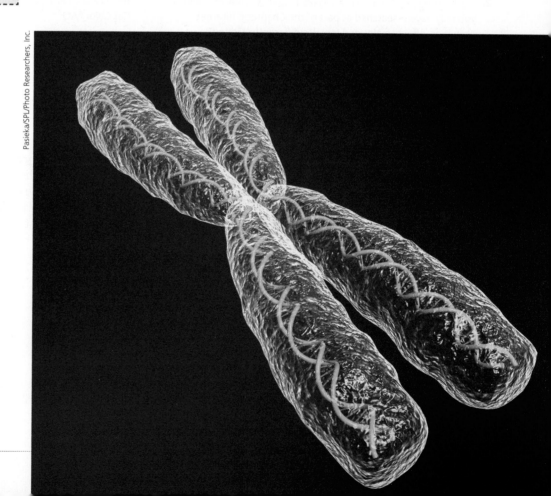

Pasieka/SPL/Photo Researchers, Inc.

A replicated chromosome consists of two sister chromatids, each of which contains one double-stranded DNA molecule.

The SARS DNA vaccine is now being tested on volunteers to see if it is effective and to resolve concerns about DNA vaccines. Will the viral DNA insert itself into a human chromosome and disrupt a gene, perhaps causing cancer? Will the immune system generate an autoimmune response in the volunteer instead of providing protection against the viral protein? Could injecting a SARS gene somehow increase susceptibility to SARS rather than prevent infection?

In this chapter, we describe the structure of DNA and the events that led to the confirmation of DNA as the molecule that carries genetic information. We also explore how DNA is replicated and what is known about the way DNA is organized into chromosomes. In Chapter 17, we will discuss the immune system and its genetic components.

8.1 DNA Is the Carrier of Genetic Information

Early in the 1860s, a chemist named Friedrich Miescher began to study the chemical composition of the nucleus. As a source of cells, he scraped pus from bandages supplied by a nearby surgical clinic. After collecting the cells, he broke them open by treating them with a protein-digesting enzyme called pepsin that he extracted from pig stomachs. After treatment, he noticed gray sediment at the bottom of the flask. In the microscope, he saw that the sediment was actually the nuclei from the pus cells. Miescher was therefore one of the first to isolate and purify a cellular organelle.

He chemically extracted the purified nuclei and obtained a substance he called nuclein. Chemical analysis revealed that it contained hydrogen, carbon, oxygen, nitrogen, and phosphorus. Nuclein was found in the nuclei of other cells, including kidney, liver, sperm, and yeast, and Miescher regarded it as an important component of cells. Many years later, it was shown that his nuclein contained deoxyribonucleic acid (DNA).

Research in the first few decades of the twentieth century established the fact that genes are carried on chromosomes. But which chemical component of a chromosome is the genetic material? Chromosomes contain DNA and proteins; which of these is the carrier of genetic information? As is often the case in science, the answer to this question came from an unexpected direction: in this case, the study of an infectious disease.

HOW WOULD YOU VOTE?

A DNA vaccine was developed quickly after the discovery of the SARS virus. Clinical trials are under way to assess its safety and effectiveness. Those trials will take several years to complete. Before the results of the clinical studies are in, if there is another outbreak of SARS or an epidemic caused by another virus like the H1N1 swine flu virus or a bioterrorist attack that releases anthrax or another potentially fatal disease-causing agent, would you agree to receive a DNA vaccine? Would you have members of your family treated with a DNA vaccine? Visit the *Human Heredity* companion website at *www.cengage.com/biology/cummings* to find out more on the issue; then cast your vote online.

DNA can transfer genetic traits between bacterial strains.

At the beginning of the twentieth century, pneumonia was a serious public health problem and was the leading cause of death in the United States. Medical researchers studied this infectious disease in an attempt to develop an effective treatment. The unexpected outcome of that research was the discovery that DNA carries genetic information.

By the 1920s, it was known that one form of pneumonia is caused by the bacterium *Streptococcus pneumoniae*. Two strains of *S. pneumoniae* were isolated. Strain S had a capsule that allowed the bacteria to evade the immune system and cause pneumonia. Strain R did not form a capsule, and did not cause pneumonia. Fredrick Griffith, an English microbiologist, showed that mice injected with live cells of strain R did not develop pneumonia and lived, but mice injected with living cells of strain S developed pneumonia and died soon thereafter (Active Figure 8.1).

In later experiments, Griffith killed strain S cells (by heating them) and injected the dead bacteria into mice. The mice did not develop pneumonia and survived. However, the most intriguing result came in an experiment where mice were injected with a mixture of heat-killed strain S and live strain R bacteria. Some of the mice developed pneumonia and died, and live strain S bacteria (with a capsule) were present in the dead mice. Griffith concluded that living strain R cells had somehow transformed into strain S cells in the injected mice.

He proposed that hereditary information had passed from the dead strain S cells into the living strain R cells. As a result, these R cells were able to make a capsule, transform into S cells, and cause pneumonia. He called this process **transformation** and called the unknown carrier of genetic information the **transforming factor**.

In 1944, Oswald Avery, Colin MacLeod, and Maclyn McCarty, working at the Rockefeller Institute in New York, discovered that the transforming factor is DNA. McCarty recounts the story of this discovery in a readable memoir: *The Transforming Principle: Discovering That Genes Are Made of DNA.*

In experiments that stretched over 10 years, Avery and his team purified the chemical components of S cells and separated them into carbohydrates, fats, proteins, and nucleic acids. Each component was mixed with live R cells and injected into mice. Mice got pneumonia and died only when injected with a mixture of S cell DNA and live R cells. Avery concluded that DNA from S cells was responsible for transforming the R cells into S cells. To confirm that DNA was responsible for transformation, they treated the DNA with enzymes to remove any residual protein or RNA before injection. Mice injected with live R cells and the highly purified DNA died from pneumonia, and living S cells were present in the dead mice. As a final test, the DNA preparation was treated

Transformation The process of transferring genetic information between cells by DNA molecules.

Transforming factor The molecular agent of transformation; DNA.

(a) Mice injected with live cells of harmless strain R.

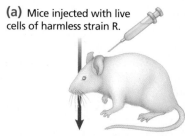

Mice do not die. No live R cells in their blood.

(b) Mice injected with live cells of killer strain S.

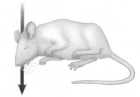

Mice die. Live S cells in their blood.

(c) Mice injected with heat-killed S cells.

Mice do not die. No live S cells in their blood.

(d) Mice injected with live R cells *plus* heat-killed S cells.

Mice die. Live S cells in their blood.

ACTIVE FIGURE 8.1 Griffith discovered that the ability to cause pneumonia is a genetic trait that can be passed from one strain of bacteria to another. (a) Mice injected with strain R do not develop pneumonia. (b) Mice injected with strain S develop pneumonia and die. (c) When the S strain cells are killed by heat treatment before injection, mice do not develop pneumonia. (d) When mice are injected with a mixture of heat-killed S cells and live R cells, they develop pneumonia and die. Griffith concluded that the live R cells acquired the ability to cause pneumonia from the dead S cells.

CENGAGENOW Learn more about the process of transformation by viewing the animation by logging on to *login.cengage.com/sso* and visiting CengageNOW's Study Tools.

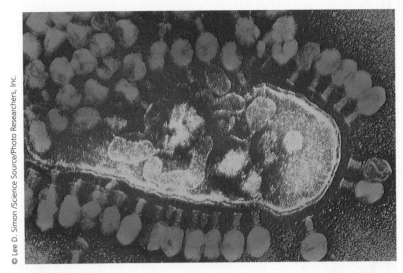

FIGURE 8.2 Bacteriophages are viruses that attack and kill bacteria. An electron micrograph shows bacteriophages attacking a bacterial cell.

© Lee D. Simon /Science Source/Photo Researchers, Inc.

with deoxyribonuclease, an enzyme that digests DNA, and the transforming activity was abolished.

The work of Avery and his colleagues produced two important conclusions:

- DNA carries genetic information. Only DNA transfers heritable information from one bacterial strain to another strain.
- DNA controls the synthesis of specific products. Transfer of DNA from strain S to strain R transfers the ability to synthesize a specific gene product (in the form of a capsule).

Although the evidence was strong, many in the scientific community were not persuaded that DNA was the carrier of genetic information. They remained convinced that proteins were the only molecule complex enough to perform that task. A few years later, Avery's conclusion that DNA carries genetic information was confirmed from studies using viruses. In spite of this groundbreaking work, neither Avery nor his colleagues ever received a Nobel Prize for their important discovery that genes are made of DNA.

DNA carries genetic information in viruses.

In the late 1940s and early 1950s, scientists began to study viruses that attack and kill bacterial cells (Figure 8.2). These viruses, known as bacteriophages (phages, for short), infect and copy themselves inside cells of *Escherichia coli*, a bacterium that inhabits the human intestinal tract.

Phages consist only of DNA and proteins, making them ideal candidates to help identify which of these molecules carry genetic information (Active Figure 8.3). In a classic series

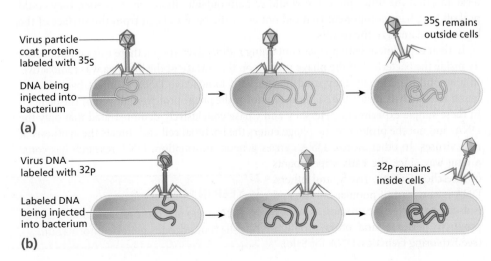

Virus particle coat proteins labeled with ^{35}S

DNA being injected into bacterium

^{35}S remains outside cells

(a)

Virus DNA labeled with ^{32}P

Labeled DNA being injected into bacterium

^{32}P remains inside cells

(b)

ACTIVE FIGURE 8.3 The Hershey-Chase experiments. Phages contain only DNA and protein; phage proteins contain sulfur but not phosphorus; and phage DNA contains phosphorus but not sulfur. Hershey and Chase designed two experiments to test whether DNA protein contained the genetic information needed to direct the replication of new phage particles. (a) In one experiment, they infected bacterial cells with virus particles whose protein was labeled with radioactive sulfur. (b) In a second experiment, they infected bacteria with viruses whose DNA was labeled with radioactive phosphorus. They found that only the radioactively labeled DNA entered the bacterial cell and directed the synthesis of new virus particles. This provided more evidence that DNA, not protein, is the genetic material.

CENGAGENOW Learn more about the Hershey-Chase experiment by viewing the animation by logging on to *login.cengage.com/sso* and visiting CengageNOW's Study Tools.

DNA for Sale

The magazine ad for the perfume reads: "Where does love originate? Is it in the mind? Is it in the heart? Or in our genes?" A perfume named DNA is marketed in a helix-shaped bottle. There is no actual DNA in the fragrance, but the image of the molecule is invoked to sell the idea that love emanates from the genes. Seem strange? Well, how about jewelry that actually contains DNA from your favorite celebrities? In this line of products, a process called the polymerase chain reaction (PCR) is used to amplify the DNA in a single hair or cheek cell. The resulting solution contains millions of copied DNA molecules and is added to small channels drilled into acrylic earrings, pendants, or bracelets. The liquid can be colored to contrast with the acrylic and be more visible. Just as people wear T-shirts with pictures of Elvis or Einstein, they can now wear jewelry containing DNA from their favorite entertainer, poet, composer, scientist, or athlete. For dead heroes, the DNA can come from a lock of hair; in fact, a single hair will do.

Or how about using DNA to protect licensed athletic jerseys and other clothing against unauthorized counterfeit copies? In the 2000 Summer Olympic Games in Sydney, DNA extracted from cheek cells swabbed from Australian athletes was amplified by PCR and mixed with the ink used to print souvenir shirts. More than 2,000 different types of items were created, and DNA testing of the labels was used to ensure that everything sold at the games was genuine.

DNA perfume created by Designer Bijan

Want music composed from the base sequence of DNA? Composers have translated the four bases of DNA (adenine, guanine, cytosine, and thymine) into musical notes. Long sequences of bases, retrieved from computer databases, are converted into notes, transferred to sheet music, and played by instruments or synthesizers as the music of the genes. DNA near chromosomal centromeres sounds much like the music of Bach or other Baroque composers, but music from other parts of the genome has a contemporary sound.

From a scientific standpoint, this fascination with DNA may be a little difficult to understand, but DNA has clearly captured the popular fancy and is being used to sell an ever-increasing array of products. DNA has name recognition. Over the last 40 years, DNA has moved from scientific journals and textbooks to the popular press and even to comic strips. The relationship between genes and DNA is well known—enough to be used in commercials and advertisements. In a few years, this fascination will probably fade and be replaced with another fad, but for now, if you want an item to sell, relate it to DNA.

---

of experiments, Alfred Hershey and Martha Chase used phages to determine whether protein or DNA was the genetic material. In preparation for their work, they first prepared two batches of phages: one with radioactive protein and one with radioactive DNA.

In one experiment, they added phages with radioactive protein to a tube of bacteria. After waiting a few minutes for the viruses to attach to and infect the bacterial cells, they put the mixture into a blender to separate the phages from the bacteria. Their idea was that only the molecule that carried phage genetic information would get inside the bacteria and that the other molecule would remain outside. By using a blender, they could pull off the phage component that did not enter the bacterial cell from the surface of the bacteria and analyze the results.

In their experiment with phages containing radioactive protein, there was no radioactivity inside the bacteria, but the phage protein on the surface of the bacteria was radioactive. In a second experiment, they used phages with radioactive DNA to infect bacteria and found radioactivity inside the bacteria, but not in the phage protein removed by the blender. In these simple experiments, Hershey and Chase conclusively demonstrated that only the DNA, and not the protein of the phage enters the bacterial cell and directs the synthesis of new viruses. In other words, DNA carries genetic information. DNA research has come a long way since the early experiments of Miescher, Griffith, Avery, and others. DNA is now part of popular culture and is used to advertise electronics, skin-care products, perfume, and other products (see Exploring Genetics: DNA for Sale).

> **KEEP IN MIND**
>
> DNA is the macromolecular component of cells that encodes genetic information.

8.2 The Chemistry of DNA

Recognition that DNA carries genetic information helped fuel efforts to understand the structure of DNA. From the mid-1940s through 1953, several laboratories made significant strides toward unraveling the structure of DNA, culminating in a model proposed in 1953 by James Watson and Francis Crick. Watson documented the scientific, intellectual, and personal intrigue that characterized the race to discover the structure of DNA in his book *The Double Helix*. The book provides a rare glimpse into the ambitions, jealousies, and rivalries that entangled scientists who were involved in the competitive and turbulent dash to the Nobel Prize.

Understanding the structure of DNA requires a review of some basic chemistry.

The structure of DNA and that of other molecules shown in this and later chapters are drawn using chemical terms and symbols. For this reason, we'll pause for a brief review of the terms and definitions used to describe molecules.

All matter is composed of atoms; different types of atoms are known as elements. In nature, atoms combine to form molecules, which are composed of two or more atoms chemically bonded together. Molecules are represented by formulas that indicate how many atoms of each type are present. Each atom has a symbol for the element it represents: H for hydrogen, N for nitrogen, C for carbon, O for oxygen, and so forth. For example, a water molecule, which is composed of two hydrogen atoms and one oxygen atom, has its chemical formula represented as H_2O:

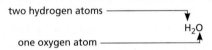

Many molecules in cells are large and have more complex formulas. A molecule of glucose contains 24 atoms and is written as

$$C_6H_{12}O_6$$

Atoms in molecules are held together by chemical links called **covalent bonds**. In its simplest form, a covalent bond is a pair of electrons shared between two atoms. Sharing two or more electrons can form more complex covalent bonds. Figure 8.4a shows how such bonds are written in chemical structures. A second type of atomic interaction involves a weak attraction known as a **hydrogen bond**. Hydrogen bonds are weak interactions between two atoms (one of which is always hydrogen) that carry partial but opposite electrical charges. Hydrogen bonds are usually represented in structural formulas as dotted or dashed lines that connect two atoms (Figure 8.4b).

In organisms, hydrogen bonds make an important contribution to the three-dimensional shape and functional capacity of biological molecules. Although individual hydrogen bonds are weak and can easily be broken, they hold molecules together by sheer force of numbers. As we see in a following section, hydrogen bonds hold together the two strands in a DNA molecule, and in Chapter 9 we will see that they are partly responsible for the three-dimensional structure of proteins.

Nucleotides are the building blocks of nucleic acids.

When Watson and Crick began their work, it was already known that organisms contain two types of nucleic acids: **deoxyribonucleic acid (DNA)** and **ribonucleic acid (RNA)**. Both are made up of subunits known as **nucleotides**. A nucleotide has three components (Figure 8.5):

1. A **nitrogen-containing base** (there are two types of bases: **purines** and **pyrimidines**). The purine bases **adenine** (A) and **guanine** (G) are found in both RNA and DNA. The pyrimidine bases are **thymine** (T), found only in DNA; **uracil** (U), found only in RNA; and **cytosine** (C), found in both RNA and DNA. RNA has four bases (A, G, U, C), and DNA has four bases (A, G, T, C).

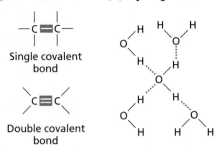

(a) Covalent bonds   **(b)** Hydrogen bonds

Single covalent bond

Double covalent bond

FIGURE 8.4 Representations of chemical bonds. (a) Covalent bonds are shown as solid lines that connect atoms. Depending on the degree of electron sharing, there can be one (*top*) or more (*bottom*) covalent bonds between atoms. Once formed, covalent bonds are stable and are broken only in chemical reactions. (b) Hydrogen bonds are usually represented as dotted lines that connect two or more atoms. As shown, water molecules form hydrogen bonds with adjacent water molecules. These hydrogen bonds are weak interactions that are easily broken by heat and molecular tumbling and can be re-formed with other water molecules.

Covalent bonds Chemical bonds that result from electron sharing between atoms. Covalent bonds are formed and broken during chemical reactions.

Hydrogen bond A weak chemical bonding force between hydrogen and another atom.

Deoxyribonucleic acid (DNA) A molecule consisting of antiparallel strands of polynucleotides that is the primary carrier of genetic information.

Ribonucleic acid (RNA) A nucleic acid molecule that contains the pyrimidine uracil and the sugar ribose. The several forms of RNA function in gene expression.

Nucleotide The basic building block of DNA and RNA. Each nucleotide consists of a base, a phosphate, and a sugar.

Nitrogen-containing base A purine or pyrimidine that is a component of nucleotides.

Purine A class of double-ringed organic bases found in nucleic acids.

Pyrimidine A class of single-ringed organic bases found in nucleic acids.

Adenine and guanine Nitrogen-containing purine bases found in nucleic acids.

Thymine, uracil, and cytosine Nitrogen-containing pyrimidine bases found in nucleic acids.

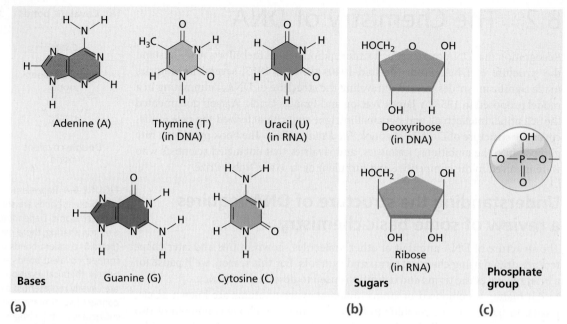

FIGURE 8.5 DNA is made up of subunits called nucleotides. Each nucleotide is composed of (a) a base, (b) a sugar, and (c) a phosphate group.

Sugar In nucleic acids, either ribose, found in RNA, or deoxyribose, found in DNA. The difference between the two sugars is an OH group present in ribose and absent in deoxyribose.

Phosphate group A compound containing phosphorus chemically bonded to four oxygen molecules.

2. A **sugar** (either ribose, found in RNA, or deoxyribose, found in DNA). The difference between the two sugars is an OH group present in ribose and absent in deoxyribose.
3. A **phosphate group**. The phosphate groups are strongly acidic and are the reason DNA and RNA are called acids.

The components of a nucleotide are assembled by covalently linking a base to a sugar, which in turn is covalently linked to a phosphate group (Figure 8.6). As shown in Figure 8.7, nucleotides can be linked together to form chains called polynucleotides.

Polynucleotides have a different structure at each end. At one end is a phosphate group; this is the 5′ (five prime) end. At the opposite end is an OH group attached to the

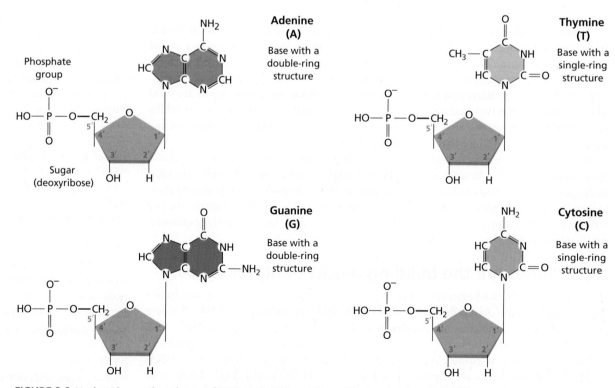

FIGURE 8.6 Nucleotides are the subunits of DNA. Nucleotides are formed by covalent bonding of the phosphate, base, and sugar. DNA contains four types of nucleotides.

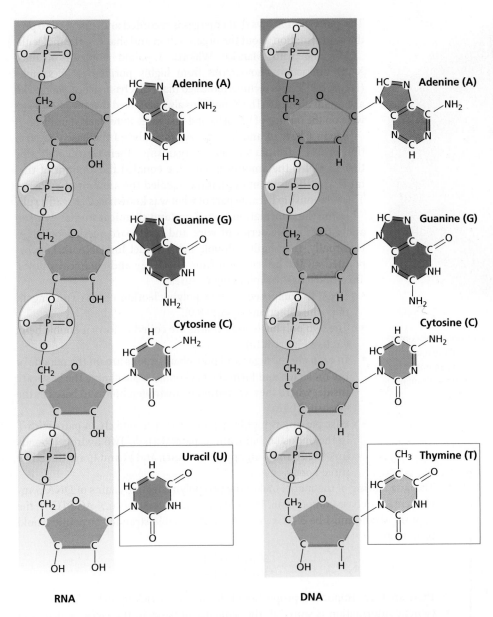

FIGURE 8.7 Nucleotides can be joined together to form chains called polynucleotides. Polynucleotides are polar molecules with a 5′ end (at the phosphate group) and a 3′ end (at the sugar group). An RNA polynucleotide is shown at the left, and a DNA polynucleotide is shown at the right.

RNA

DNA

sugar molecule; this is known as the 3′ (three prime) end of the chain (Figure 8.7). By convention, polynucleotide chains are written beginning with the 5′ end:

5′-CGATATGCGAT-3′

As we will see next, DNA contains two polynucleotide chains.

8.3 The Watson-Crick Model of DNA Structure

In the early 1950s, James Watson and Francis Crick began to work out the structure of DNA by sifting through and organizing the information about DNA that was already available. They focused on two sources of information:

- X-ray diffraction studies which provided clues to the helical structure of DNA
- chemical analysis of DNA base composition from many different organisms

In X-ray diffraction, molecules are placed in an X-ray beam, and as the beam passes through the molecule, some X-rays are deflected by atoms in the target molecule.

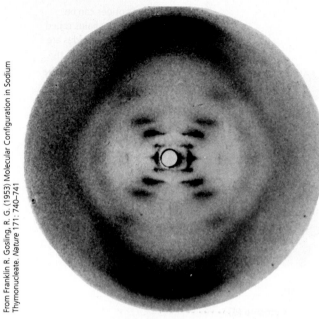

FIGURE 8.8 An X-ray diffraction photograph of DNA. The central X-shaped pattern is typical of helical structures, and the darker areas at the top and bottom indicate a regular arrangement of subunits in the molecule. Watson and Crick used this and similar photographs to construct their model of DNA.

The pattern of X-rays that emerges is recorded and analyzed to produce information about the organization and shape of the molecule.

Working with Maurice Wilkins, Rosalind Franklin obtained X-ray diffraction photographs from highly purified DNA samples (Figure 8.8). Those pictures indicated that DNA has a helical shape with a constant diameter. The X-ray films also suggested that the phosphates were on the outside of the helix and that bases were stacked inside.

Erwin Chargaff and his colleagues analyzed the base composition of DNA from a variety of organisms. Their results indicated that in DNA, the amount of adenine equaled the amount of thymine and the amount of guanine equaled the amount of cytosine. This relationship became part of what was known as Chargaff's rule.

Using the information from X-ray and chemical studies, Watson and Crick built a series of wire and cardboard models of DNA. Eventually, they built a double helix model for DNA that incorporated all the information from the X-ray and chemical studies. This model has the following features:

- DNA is composed of two polynucleotide chains running in opposite directions (Figure 8.9).
- The two polynucleotide chains are coiled to form a double helix (Active Figure 8.10).
- In each chain, sugar and phosphate groups are on the outside of the molecule and form the backbone of the chain. The bases are inside, where they are paired by hydrogen bonds to bases in the opposite chain.
- Base pairing is highly specific: The A in one chain pairs with T in the opposite chain, and C pairs with G. Two hydrogen bonds link the A and T in opposite chains, and G and C are linked by three bonds.
- The base pairing of the model makes the two polynucleotide chains of DNA complementary in base composition. If one strand has the sequence 5′-ACGTC-3′, the opposite strand must be 3′-TGCAG-5′, and the double-stranded structure would be written as

$$5'\text{-ACGTC-}3'$$
$$3'\text{-TGCAG-}5'$$

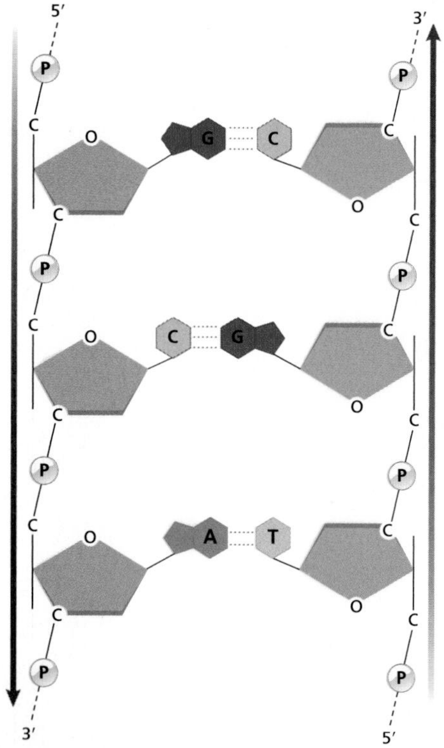

There are three important properties of the Watson-Crick model:

- Genetic information is stored in the sequence of bases in the DNA, which have a high coding capacity. A DNA molecule n base pairs long has 4^n combinations. That means that a sequence of 10 nucleotides has 4^{10}, or 1,048,576, possible combinations of nucleotides. The complete set of genetic information carried by an organism (its genome) can be expressed as base pairs of DNA (Table 8.1). Genome sizes vary from several thousand nucleotides that encode only a few genes (in viruses) to billions of nucleotides that encode 20,000 to 25,000 genes (in humans). The human genome consists of about 3.2×10^9, or about 3 billion, base pairs of DNA, distributed over 24 chromosomes (22 autosomes and two sex chromosomes).

FIGURE 8.9 The two polynucleotide chains in DNA run in opposite directions. The left strand runs 5′ to 3′, and the right strand runs 3′ to 5′. The base sequences in each strand are complementary. An A in one strand pairs with a T in the other strand, and a C in one strand is paired with a G in the opposite strand.

Table 8.1 Genome Sizes of Some Organisms

Organism	Species	Genome Size in Nucleotides
Bacterium	*E. coli*	4.6×10^6
Yeast	*S. cerevisiae*	1.2×10^7
Fruit fly	*D. melanogaster*	1.7×10^8
Tobacco plant	*N. tabacum*	4.8×10^9
Mouse	*M. musculus*	2.7×10^9
Human	*H. sapiens*	3.2×10^9

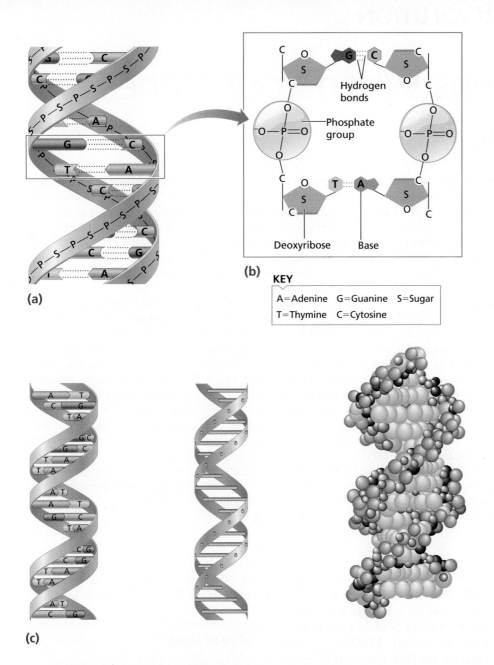

(b)

KEY

A=Adenine	G=Guanine	S=Sugar
T=Thymine	C=Cytosine	

(c)

- The model offers a molecular explanation for mutation. Because genetic information is stored as a linear sequence of bases in DNA, any change in the order or number of bases in a gene can result in a mutation that produces an altered phenotype. This topic is explored in more detail in Chapter 11.
- The complementary nature of the two polynucleotide DNA strands helps explain how DNA is copied; each strand can be used as a template to reconstruct the base sequence in the opposite strand.

In 1953, Watson and Crick described their model for DNA structure in a brief paper in the journal *Nature*. Although their work was based on the results of others, Watson and Crick correctly incorporated the physical and chemical data into a model that could also be used to explain the properties of genetic material. This discovery was of historic importance in biology. It paved the way for scientists to understand cellular processes in molecular terms and gave rise to the discipline of molecular biology. Present-day applications, including genetic engineering, genomics, and gene testing, can all be traced directly to that paper (see The Genetic Revolution: What Happens When Your Genes Are Patented?).

The 1962 Nobel Prize for Medicine or Physiology was awarded to Watson, Crick, and Wilkins for their work on the structure of DNA. Although Rosalind Franklin provided much of the X-ray data, she did not receive a share of the prize. There has been some

What Happens When Your Genes Are Patented?

Two genes associated with breast cancer, *BRCA1* and *BRCA2*, were discovered in 1994 and 1995, and shortly thereafter, were patented by Myriad Genetics, a company based in Utah. Under the patents, testing for mutations in these genes can only be performed by Myriad, at costs from $300 - $3000 dollars. Myriad also patented the process of analyzing the results of such tests, preventing anyone who obtains the sequence of their BRCA genes by other means (which itself would probably be patent infringement) from interpreting the information.

The idea that genes can be patented has been a contentious issue from the beginning. Patents are not granted for products of nature, meaning that genes inside the body are not patentable, but biotech companies successfully argued that by removing a gene from the human body, purifying it, and then obtaining its DNA sequence, they created a patentable invention. The U.S. Patent Office found the argument persuasive, but opponents argue that genes are parts of our bodies, and can be identified, but not invented. Biotech companies argue that without the protection offered by patents, they would have no incentive for research and development of diagnostic tests.

In Europe, patents for *BRCA1* and *BRCA2* were revoked in 2004 because they did not meet the standards for a patent. After more than a decade of legal disputes, the patents were partially restored in 2008 on a very restricted basis. In the United States, a lawsuit, focused on the patents for the BRCA genes, was filed in May of 2009. The suit challenges the basic idea that genes are patentable. In November of 2009, the judge ruled that the lawsuit can proceed, and the case is moving forward. In March 2010, the Federal court invalidated Myriad Genetics' patent on these genes. The decision has been appealed and may end up as a case for the U.S. Supreme Court.

What does this mean for you and your family? A new generation of genetic tests is now used to analyze hundreds of genes in a person's DNA to identify genetic disorders and susceptibility to complex diseases such as diabetes, cardiovascular diseases, and Alzheimer's disease. Tests now in development will scan someone's entire genome to identify risks for disorders. However, it may be difficult for you to receive the results of such tests if patented genes are involved, unless a license for each patented gene has been granted by the patent holder. At the moment, some 20% of all human genes have been patented, so obtaining the necessary licenses would be difficult and expensive. As things stand now, licenses are not even available for the BRCA genes and many other genes. The outcome of this lawsuit may have a big impact on your health care and that of your family.

---

controversy over this, but because only living individuals are eligible, she could not have shared in the prize. Franklin died of cancer in 1958 at the age of 37, 4 years before the Nobel Prize was awarded to Watson, Crick, and Wilkins. You can read about her life in science and her role in the discovery of the structure of DNA in a recent biography by Brenda Maddox: *Rosalind Franklin: The Dark Lady of DNA*. Often overlooked is the fact that although Erwin Chargaff made vital contributions to the Watson-Crick model, he did not receive a share of the Nobel Prize.

> **KEEP IN MIND**
>
> Watson and Crick built models of DNA structure using information from X-ray diffraction studies and chemical analyses of DNA from various organisms.

8.4 RNA Is a Single-Stranded Nucleic Acid

RNA (ribonucleic acid), a second type of nucleic acid, is found in both the nucleus and the cytoplasm. RNA has several roles: It transfers genetic information from the nucleus to the cytoplasm, it participates in the synthesis of proteins, and it is a component of ribosomes. The functions of RNA are considered in more detail in Chapter 9.

Nucleotides in RNA differ from those in DNA in two respects: The sugar in RNA nucleotides is **ribose** (**deoxyribose** in DNA), and the base uracil is used in place of the

Ribose and deoxyribose Pentose sugars found in nucleic acids. Deoxyribose is found in DNA, ribose in RNA.

Table 8.2 Differences Between DNA and RNA

	DNA	RNA
Sugar	Deoxyribose	Ribose
Bases	Adenine	Adenine
	Cytosine	Cytosine
	Guanine	Guanine
	Thymine	Uracil

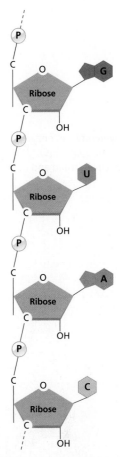

base thymine (Table 8.2). In most cases, RNA is single-stranded (Figure 8.11), but RNA molecules can fold back on themselves, forming short stretches that are double-stranded, as we will see in Chapter 9.

8.5 DNA Replication Depends on Complementary Base Pairing

DNA has two major functions in the cell: (1) it carries the cell's genetic information and (2) it replicates itself. Here we will explore how DNA replicates itself. In Chapter 9, we will investigate how the genetic information encoded in DNA is copied and transferred to the cytoplasm to make proteins.

Before a cell divides, its DNA must be copied so that each daughter cell will receive a complete set of genetic information. In their paper, Watson and Crick note: "It has not escaped our notice that the specific pairing we have postulated immediately suggests a possible copying mechanism for the genetic material." In a subsequent paper, they proposed a mechanism for DNA replication that depends on the complementary base pairing in the polynucleotide chains of DNA. If DNA is unwound, each strand can serve as a **template** for making a new, complementary strand (Figure 8.12). This process is known as **semiconservative replication**, because one old strand is conserved in each new molecule, and one new strand is synthesized. Later experiments by Matthew Meselson and Franklin Stahl confirmed that DNA is in fact replicated in a semiconservative fashion.

FIGURE 8.11 RNA is a single-stranded polynucleotide chain. RNA molecules contain a ribose sugar instead of a deoxyribose and have uracil (U) in place of thymine.

Template The single-stranded DNA that serves to specify the nucleotide sequence of a newly synthesized polynucleotide strand.

Semiconservative replication A model of DNA replication that provides each daughter molecule with one old strand and one newly synthesized strand. DNA replicates in this fashion.

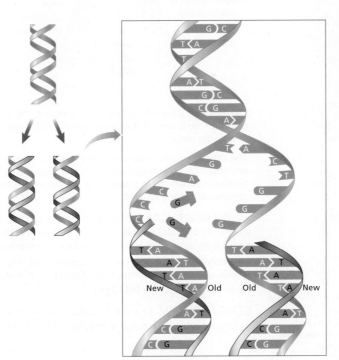

FIGURE 8.12 In DNA replication, the two polynucleotide strands uncoil, and each is a template for synthesizing a new strand. A replicated DNA molecule contains one new strand and one old strand. This mechanism is called semiconservative replication.

Stages of DNA replication.

DNA replication in all cells—from bacteria to humans—is a complex, multistep process requiring the action of many components. In humans, DNA replication begins in the S phase of the cell cycle (review the cell cycle in Chapter 2) at sites along the chromosome called origins of replication. At these origins, multiprotein complexes break the hydrogen bonds between strands and unwind the double helix for a short distance; these and other proteins also prevent the strands from rewinding. Once the strands are separated, the enzyme **DNA polymerase** reads the nucleotide sequence of the template strand and inserts complementary nucleotides in a 5′ to 3′ direction to form a newly synthesized strand (Active Figure 8.13a). As it moves along the template, DNA polymerase also "reads" the sequence, removing incorrect base pairs and synthesizing correct ones. As seen in the figure, the newly synthesized strand is made continuously on one template strand but is made in short stretches—called Okazaki fragments (named after the scientist who discovered them)—on the other strand. The gaps in these short strands are sealed by the action of an enzyme, DNA ligase to form a continuous strand (Active Figure 8.13b). After this step, proteins wind the template and the newly synthesized strands together to form a DNA double helix. The completed DNA molecule contains one old strand (the template strand) and one new strand (complementary to the old strand).

Recall that each chromosome contains one double-stranded DNA molecule running from end to end. When the S phase is over, chromosomes consist of two sister chromatids joined at a common centromere. Each chromatid contains a DNA molecule that

DNA polymerase An enzyme that catalyzes the synthesis of DNA using a template DNA strand and nucleotides.

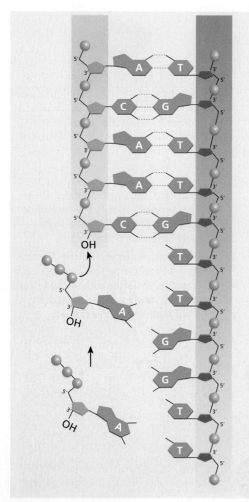

(a) Each DNA strand has two ends: one with a 5′ carbon, and one with a 3′ carbon. DNA polymerase can add nucleotides only at the 3′ carbon. In other words, DNA synthesis proceeds only in the 5′ to 3′ direction.

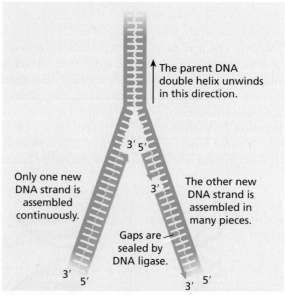

(b) Because DNA synthesis proceeds only in the 5′ to 3′ direction, only one of the DNA strands can be assembled in a single piece.

The other new DNA strand forms in short segments, which are called Okazaki fragments after the two scientists who discovered them. DNA ligase joins the fragments into a continuous strand of DNA.

ACTIVE FIGURE 8.13 A close-up look at the process of DNA replication. (a) As the strands uncoil, bases are added to the newly synthesized strand by complementary base pairing with bases in template strand. The new bases are linked together by DNA polymerase. (b) DNA synthesis can proceed only in the 5′ → 3′ direction; newly synthesized DNA on one template strand is made in short segments and linked together by the enzyme DNA ligase.

CENGAGENOW Learn more about DNA replication by viewing the animation by logging on to *login.cengage.com/sso* and visiting CengageNOW's Study Tools.

consists of one old strand and one new strand. When the centromeres divide at the beginning of anaphase, each chromatid becomes a separate chromosome that contains an accurate copy of the genetic information in the parental chromosome. In the next section, we will explore how DNA and proteins combine to form chromosomes.

8.6 The Organization of DNA in Chromosomes

Although an understanding of DNA structure was a landmark development in genetics, it doesn't tell us how a chromosome is organized or what regulates the uncoiling and re-coiling of chromosomes as cells move from interphase to mitosis and back. Knowledge of chromosome structure is important, because the spatial arrangement of DNA in the nucleus plays a key role in regulating the expression of genetic information. In addition, putting billions of nucleotides of DNA that make up the 46 human chromosomes into the cell nucleus requires packing a little more than 2 m (about 6.5 ft.) of DNA into a structure that measures about 5 μm in diameter. The length of DNA must be compacted thousands of times to fit into the nucleus.

Within this cramped environment, chromosomes unwind and become dispersed during interphase. While in the nucleus, they undergo replication, gene expression, homologous pairing during meiosis, and coiling to become visible during prophase. An understanding of chromosomal organization is necessary to understand these processes.

Chromosomes have a complex structure.

A combination of biochemical, molecular, and microscopic techniques has provided a great deal of information about the organization and structure of human chromosomes, although we still do not know all the details. In humans and other eukaryotes, each chromosome consists of one double-stranded DNA molecule combined with proteins to form **chromatin.** The major class of proteins in chromatin is **histones**, which play important roles in chromosomal structure and gene regulation.

At the lowest level of chromosome organization, DNA is wound around a core of histone molecules to form small spherical bodies called **nucleosomes** (Figure 8.14). Winding DNA around the histones compacts the DNA by a factor of 6 or 7. But because mitotic chromosomes are compacted by a factor of 5,000 to 10,000, there must be additional levels of organization between the nucleosome and the chromosome, each of which involves additional folding and/or compaction of DNA. Several models have been proposed to explain how nucleosomes are organized into more complex structures. Most of these are based on the idea that chromatin folds into loops and fibers that attach to and extend from a central protein scaffold. Studies using electron microscopy have partially clarified the organization of chromosomes. Figure 8.15 is a high-magnification scanning electron micrograph of a mitotic chromosome. At this magnification, individual fibers, produced by coiling and folding of chromatin, are visible. A model of chromosome organization incorporating these features is shown in Active Figure 8.16.

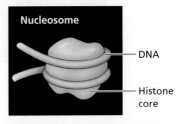

FIGURE 8.14 DNA coils around the outside of a histone cluster to form a nucleosome. This is the first level of DNA compaction in chromosome structure.

Chromatin The complex of DNA and proteins that makes up a chromosome.

Histones DNA-binding proteins that help compact and fold DNA into chromosomes.

Nucleosome A bead-like structure composed of histone wrapped with DNA.

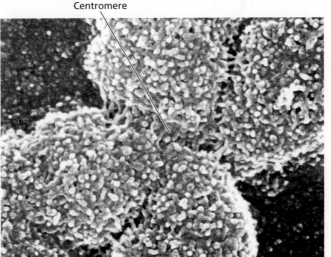

FIGURE 8.15 A scanning electron micrograph of the centromeric region of a replicated chromosome. Loops of chromatin can be seen as small fibers of about 30 nanometers in diameter in the sister chromatids. The centromere is where spindle fibers will attach in metaphase.

Reproduced with permission from Harrison, C. et al., 1982, High resolution scanning electron microscopy of human metaphase chromosomes, *Journal of Cell Science* 56: 409–422, Figure 3, The Company of Biologists, Limited.

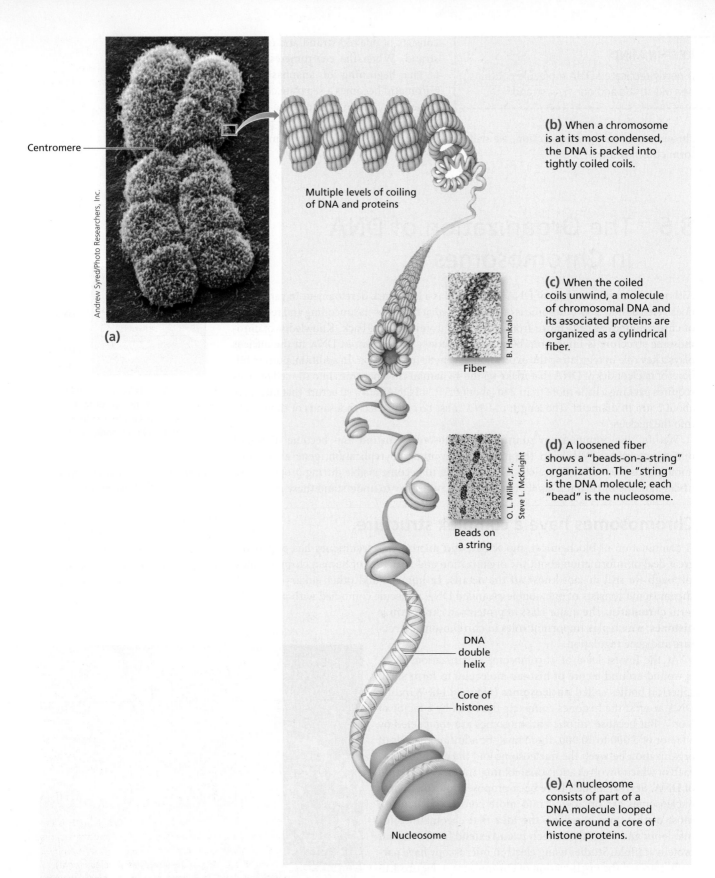

Centromere

(a)

Andrew Syred/Photo Researchers, Inc.

Multiple levels of coiling
of DNA and proteins

(b) When a chromosome
is at its most condensed,
the DNA is packed into
tightly coiled coils.

Fiber

B. Hamkalo

(c) When the coiled
coils unwind, a molecule
of chromosomal DNA and
its associated proteins are
organized as a cylindrical
fiber.

Beads on
a string

*O. L. Miller, Jr.,
Steve L. McKnight*

(d) A loosened fiber
shows a "beads-on-a-string"
organization. The "string"
is the DNA molecule; each
"bead" is the nucleosome.

DNA
double
helix

Core of
histones

Nucleosome

(e) A nucleosome
consists of part of a
DNA molecule looped
twice around a core of
histone proteins.

ACTIVE FIGURE 8.16 A model of chromosomal structure beginning with a double-stranded DNA molecule. The DNA is first coiled around histones to form nucleosomes. The nucleosomes are coiled again and again into fibers that form the body of the chromosome. Chromosomes undergo cycles of coiling and uncoiling in mitosis and interphase, so their structure is dynamic.

CENGAGENOW Learn more about chromosome structure by viewing the animation by logging on to *login.cengage.com/sso* and visiting CengageNOW's Study Tools.

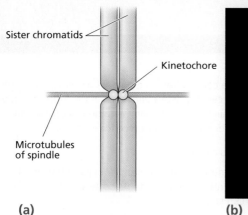

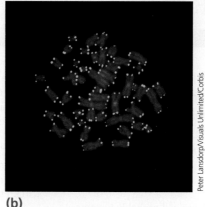

(a)   (b)

FIGURE 8.17 (a) Kinetochores are protein-containing structures that assemble at the centromeres of replicated chromosomes. Microtubules of the spindle apparatus attach to the kinetochore and are involved in moving chromosomes apart in anaphase. (b) The telomeres at the ends of chromosomes are highlighted.

Peter Lansdorp/Visuals Unlimited/Corbis

Centromeres and telomeres are specialized chromosomal regions.

Specific regions of chromosomes have important roles in chromosome function. Two of these are **centromeres** and **telomeres**. Each chromosome contains a centromere located at a constricted region where sister chromatids attach (you can see the centromere in Figure 8.15). Centromere locations differ for specific chromosomes (see Chapter 2). During cell division, the DNA in centromeres serves as a platform for the assembly of proteins that form a plate, called the kinetochore (Figure 8.17a). Spindle fibers attach to kinetochores during metaphase and move chromosomes apart during anaphase.

Telomeres are short DNA sequences repeated hundreds or thousands of times, located at each end of a chromosome (Figure 8.17b). The human telomere sequence is TTAGGG. Because of the way DNA replication takes place, some telomere repeats are not copied, making chromosomes a little shorter after each round of replication. In most cells of the body, telomeres shorten with each division, meaning that cells can divide only a certain number of times before they stop being able to divide. Telomere shortening may be linked to the aging process, but whether it is a cause or an effect is still debated.

Some cells, such as those forming sperm and eggs, contain an enzyme, **telomerase**, which makes extra telomere repeats, keeping the number of telomere repeats constant from one cell division to another. In addition, telomerase is active in more than 90% of cancers, allowing cancer cells to divide indefinitely. Because telomerase is not active in most cells of the body but is active in cancer cells, the development of a telomerase inhibitor may prove to be an effective treatment for cancer. If a drug can shut off telomerase, the chromosomes in cancer cells would shorten in each cell division, leading to a halt in tumor growth and the eventual death of the tumor cells.

Telomerase An enzyme that adds telomere repeats to the ends of chromosomes, keeping them the same length after each cell division.

The nucleus has a highly organized architecture.

The interphase nucleus is not a disorganized bag of chromosomes; instead, the nucleus has an internal structure in which each chromosome occupies a distinct region called a chromosome territory (Figure 8.18). This organization is closely linked with function. As a result, territories are not fixed; chromosomes move around in the nucleus at different times of the cell cycle, including the S phase, when DNA replication and chromosome duplication occur (review the cell cycle in Chapter 2). It has been proposed that DNA replication takes place at certain sites within the nucleus called "replication factories" and that chromosomes move to those locations during S phase. At other times, the chromosomes may be in territories where gene expression takes place. Much of what remains to be learned about how genes are turned on and off involves understanding the dynamics of chromosome organization in the nucleus.

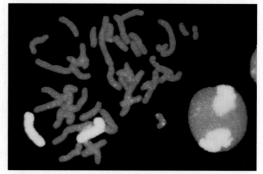

FIGURE 8.18 Chromosome painting highlights both copies of human chromosome 4 in a metaphase spread (*left*) and shows the chromosome territories occupied by chromosome 4 in the interphase nucleus (*right*). In the nucleus, each chromosome occupies a distinct territory, separated from other chromosomes by a region called the interchromosome domain.

Photo by Thomas Cremer. Courtesy of William C. Earnshaw from Lamond, A. I., and Earnshaw, W. C. Structure and Function in the Nucleus. Reprinted with permission from *Science*, 280: 547–553. © American Association for the Advancement of Science

KEEP IN MIND

DNA is packaged into chromosomes by several levels of coiling and compaction.

Genetics in Practice

Genetics in Practice case studies are critical-thinking exercises that allow you to apply your new knowledge of human genetics to real-life problems. You can find these case studies and links to relevant websites at *www.cengage.com/biology/cummings*

CASE 1

Tune in to news programs regularly and you will probably become aware of the considerable debate over the patenting of genes. This controversy is fueled largely by the work of the Human Genome Project and the biotechnology industry. Despite numerous meetings and publications on the subject, Congress has not used U.S. patent laws to shape a policy that allows maximum innovation from biotech inventions. The first gene patents, issued in the 1970s, were granted for genes whose full nucleotide sequence was known; the protein product was also known, and the protein's function was well understood.

Since that time, genome projects have produced new ways of finding genes. Short sequences, only 25 to 30 nucleotides in length, called expressed sequence tags (ESTs) can be used to identify genes but provide no information about the entire gene, the product, the function of the product, or its association with any genetic disorder. Using gene-hunting software, researchers can take a short sequence of DNA and use it to search gene databases, turning up theoretical information about the sequence. For example, the sequence may belong to a gene encoding a plasma membrane protein or may be similar to one in yeast that is involved in cell–cell signaling. At the present time, there are tens of thousands of ESTs and gene-hunting patent applications filed at the U.S. Patent Office.

The unresolved question at the moment is how much you need to know about a gene and its usefulness to file a patent application. How should utility be defined? The diagnosis of disease certainly meets the definition of utility. Many discoveries have identified disease-causing genes, such as those for cystic fibrosis, fragile-X syndrome, breast cancer, colon cancer, and obesity. Many of these discoveries have patents based on diagnostic utility. An increasing number of patent applications are being filed for discoveries of hereditary-disease-causing genes. These discoveries frequently lack immediate use for practical therapy, however, because gene discovery does not always include knowledge of gene function or a plan for developing a disease therapy.

The impact of a decision about gene patents is enormous. Pharmaceutical and biotech companies have invested hundreds of millions of dollars in identifying genes to be used in developing diagnostic tests and drugs. Without patents, it is unlikely that companies will invest in developing these drugs. However, patenting genes can lead to royalty-based gene testing accompanied by exorbitant fees and licensing arrangements requiring payment to companies that own the patent on a particular gene. A case now before the courts about the patents on two genes used to test for breast cancer may determine much of the future of patented genes. As the results of the Human Genome Project redefine health care, these issues take on importance for everyone.

1. What is a patent?

2. Is patenting a gene different from patenting another product or invention? Should patents be awarded for genes under any circumstances? Explain.

3. If patenting genes were not allowed, do you think it would slow gene research in a significant way?

CASE 2

A 34-year-old woman and her 1-month-old newborn were seen by a genetic counselor in the neonatal intensive care unit at a major medical center. The neonatologist was suspicious that the newborn boy had a genetic condition and requested a genetic evaluation. The newborn was very pale, was failing to thrive, had diarrhea, and had markedly increased serum cerebrospinal fluid lactate levels. In addition, he had severe muscle weakness with chart notes describing him as "floppy," and he had had two seizures since birth. The neonatologist reported that the infant had liver failure, which would probably result in his death in the next few days. The panel of tests performed on the infant led the neonatologist and the genetic counselor to the diagnosis of Pearson syndrome. The combination of marked metabolic acidosis and abnormalities in bone marrow cells is highly suggestive of Pearson syndrome.

Pearson syndrome is associated with a large deletion of the mitochondrial (mt) genome. The way the deletion-containing mtDNA molecules are distributed during mitosis is not known. However, it is assumed that during cell division daughter cells randomly receive mitochondria carrying wild type (WT) or mutant mtDNA. Mitochondrial DNA is, theoretically, transmitted only to offspring through the mother via the large cytoplasmic component of the oocyte. Nearly all cases of Pearson syndrome arise from new mutational events. Mitochondria have extremely poor DNA repair mechanisms, and mutations accumulate very rapidly. Most infants with Pearson syndrome die before age 3, often as a result of infection or liver failure.

A diagnosis of Pearson syndrome results in an extremely grave prognosis for the patient. Unfortunately, at this point, treatment can be directed only toward symptomatic relief.

1. How would a large deletion in the mitochondrial genome cause a disease?

2. Why doesn't the mother have the disease if she has mutant mitochondrial DNA?

3. How would you react to hearing this diagnosis? How would you counsel a couple through this kind of situation?

Summary

8.1 DNA Is the Carrier of Genetic Information

- At the turn of the twentieth century, scientists identified chromosomes as the cellular components that carry genes. This discovery focused efforts on identifying the molecular nature of the gene on the chromosomes and the nucleus. Biochemical analysis of the nucleus began around 1870 when Friedrich Miescher first separated nuclei from cytoplasm and described nuclein, a protein/nucleic acid complex now known as chromatin.

- Originally, proteins were regarded as the only molecular component of the cell with the complexity necessary to encode genetic information. This changed in 1944 when Avery and his colleagues demonstrated that DNA is the genetic material in bacteria.

8.2 The Chemistry of DNA

- DNA and RNA are nucleic acids composed of subunits called nucleotides. Each nucleotide has three components: a base, a sugar, and a phosphate group. DNA contains four bases (adenine, guanine, cytosine, and thymine), and RNA contains four bases (adenine, guanine, cytosine, and uracil). The sugar in DNA is deoxyribose; RNA contains the sugar ribose. Nucleotides can be linked together to form chains called polynucleotides. DNA contains two polynucleotide chains; RNA contains one chain.

8.3 The Watson-Crick Model of DNA Structure

- In 1953, Watson and Crick constructed a model of DNA structure that incorporated information from the chemical studies of Chargaff and the X-ray diffraction work of Wilkins and Franklin. Watson and Crick proposed that DNA is composed of two polynucleotide chains oriented in opposite directions and held together by hydrogen bonds between complementary bases in the opposite strand. The two strands are wound around to form a right-handed helix

8.4 RNA Is a Single-Stranded Nucleic Acid

- RNA is another type of nucleic acid. It contains a different sugar than DNA and uses the base uracil in place of thymine. RNA molecules are single stranded but can fold back on themselves to produce double-stranded regions. RNA has a variety of functions in the cell.

8.5 DNA Replication Depends on Complementary Base Pairing

- In DNA replication, proteins separate the strands of DNA, and DNA polymerase reads the base sequence in the template strand and inserts nucleotides into a complementary strand. The template strand and the newly synthesized complementary strand are wound together to form a double-stranded DNA molecule. This mechanism is called semiconservative replication.

8.6 The Organization of DNA in Chromosomes

- Within chromosomes, DNA is coiled around clusters of histones to form structures known as nucleosomes. Supercoiling of nucleosomes may form fibers that extend at right angles to the axis of the chromosome. The structure of chromosomes must be dynamic to allow the uncoiling and recoiling seen in successive phases of the cell cycle, but the details of this transition are not yet known.

Questions and Problems

CENGAGENOW Preparing for an exam? Assess your understanding of this chapter's topics with a pre-test, a personalized learning plan, and a post-test by logging on to *login.cengage.com/sso* and visiting CengageNOW's Study Tools.

DNA Is the Carrier of Genetic Information

1. Until 1944, which cellular component was thought to carry genetic information?
   a. carbohydrate
   b. nucleic acid
   c. protein
   d. chromatin
   e. lipid

2. Why do you think nucleic acids were originally not considered to be carriers of genetic information?

3. The experiments of Avery and his coworkers led to the conclusion that:
   a. bacterial transformation occurs only in the laboratory.
   b. capsule proteins can attach to uncoated cells.
   c. DNA is the transforming agent and is the genetic material.
   d. transformation is an isolated phenomenon in *E. coli*.
   e. DNA must be complexed with protein in bacterial chromosomes.

4. In the experiments of Avery, MacLeod, and McCarty, what was the purpose of treating the transforming extract with enzymes?

5. Read the following experiment and interpret the results to form your conclusion. Experimental data: S bacteria were heat killed and cell extracts were isolated. The extracts contained cellular components, including lipids, proteins, DNA, and RNA. The extracts were mixed with live R bacteria and then injected together into mice along with various enzymes (proteases, RNAses, and DNAses). Proteases degrade proteins, RNAses degrade RNA, and DNAses degrade DNA.

S extract + live R cells   mouse dies
S extract + live R cells + protease   mouse dies
S extract + live R cells + RNAase   mouse dies
S extract + live R cells + DNAase   mouse lives

Based on these results, what is the transforming principle?

6. Recently, scientists discovered that a rare disorder called polkadotism is caused by a bacterial strain, polkadotiae. Mice injected with this strain (P) develop polka dots on their skin. Heat-killed P bacteria and live D bacteria, a nonvirulent strain, do not produce polka dots when injected separately into mice. However, when a mixture of heat-killed P cells and live D cells were injected together, the mice developed polka dots. What process explains this result? Describe what is happening in the mouse to cause this outcome.

The Chemistry of DNA

7. List the pyrimidine bases, the purine bases, and the base-pairing rules for DNA.

8. In analyzing the base composition of a DNA sample, a student loses the information on pyrimidine content. The purine content is A = 27% and G = 23%. Using Chargaff's rule, reconstruct the missing data and list the base composition of the DNA sample.

9. The basic building blocks of nucleic acids are:
   a. phosphate groups
   b. nucleotides
   c. ribose sugars
   d. amino acids
   e. purine bases

10. Adenine is a:
   a. nucleoside
   b. purine
   c. pyrimidine
   d. nucleotide
   e. base

11. Polynucleotide chains have a 5′ and a 3′ end. Which groups are found at each of these ends?
   a. 5′ sugars, 3′ phosphates
   b. 3′ OH, 5′ phosphates
   c. 3′ base, 5′ phosphates
   d. 5′ base, 3′ OH
   e. 5′ phosphates, 3′ bases

12. DNA contains many hydrogen bonds. Are hydrogen bonds stronger or weaker than covalent bonds? What are the consequences of this difference in strength?

The Watson-Crick Model of DNA Structure

13. Watson and Crick received the Nobel Prize for:
   a. generating X-ray crystallographic data of DNA structure
   b. establishing that DNA replication is semiconservative
   c. solving the structure of DNA
   d. proving that DNA is the genetic material
   e. showing that the amount of A equals the amount of T

14. State the properties of the Watson-Crick model of DNA in the following categories:
   a. number of polynucleotide chains
   b. polarity (running in same direction or opposite directions)
   c. bases on interior or exterior of molecule
   d. sugar/phosphate on interior or exterior of molecule
   e. which bases pair with which
   f. right- or left-handed helix

15. Using Figure 8.7 as a guide, draw a dinucleotide composed of C and A. Next to this, draw the complementary dinucleotide in an antiparallel fashion. Connect the dinucleotides with the appropriate hydrogen bonds.

16. A beginning genetics student is attempting to complete an assignment to draw a base pair from a DNA molecule. The drawing is incomplete, and the student does not know how to finish. He asks for your advice. The assignment sheet shows that the drawing is to contain three hydrogen bonds, a purine, and a pyrimidine. From your knowledge of the pairing rules and the number of hydrogen bonds in A/T and G/C base pairs, what base pair do you help the student draw?

RNA Is a Single-Stranded Nucleic Acid

17. Chemical analysis shows that a nucleic acid sample contains A, U, C, and G. Is this DNA or RNA? Why?

18. How does DNA differ from RNA with respect to the following characteristics?
   a. number of chains
   b. bases used
   c. sugar used

19. RNA is ribonucleic acid, and DNA is *deoxy*ribonucleic acid. What exactly is deoxygenated about DNA?

DNA Replication Depends on Complementary Base Pairing

20. What is the function of DNA polymerase?
   a. It degrades DNA in cells
   b. It adds RNA nucleotides to a new strand
   c. It coils DNA around histones to form chromosomes
   d. It adds DNA nucleotides to a replicating strand
   e. none of the above

21. Which of the following statements is *not* true about DNA replication?
   a. It occurs during the M phase of the cell cycle
   b. It makes a sister chromatid
   c. It denatures DNA strands
   d. It occurs semiconservatively
   e. It follows base-pairing rules

22. Make the complementary strand for the following DNA template and label both strands as 5′ to 3′ or 3′ to 5′ (P = phosphate in the diagram). Draw an arrow showing the direction of synthesis of the new strand. How many hydrogen bonds are in this double strand of DNA?

    template: P—AGGCTCG—OH

    new strand:

23. How does DNA replication occur in a precise manner to ensure that identical genetic information is put into the new chromatid? See Figure 8.12.

The Organization of DNA in Chromosomes

24. Nucleosomes are complexes of:
    a. RNA and DNA
    b. RNA and histone
    c. histones and DNA
    d. DNA, RNA, and protein
    e. amino acids and DNA

25. Discuss the levels of chromosomal organization with reference to the following terms:
    a. nucleotide
    b. DNA double helix
    c. histones
    d. nucleosomes
    e. chromatin

Internet Activities

Internet Activities are critical-thinking exercises using the resources of the World Wide Web to enhance the principles and issues covered in this chapter. For a full set of links and questions investigating the topics described below, visit *www.cengage.com/biology/cummings*

1. *Experimenting with the Structure of DNA*. The Genetic Science Learning Center (a joint project of the University of Utah and the Utah Museum of Natural History) provides general genetics information to students and the community. Go to the link on "How to Extract DNA from Anything Living." This is an activity that you can do at home, but even if you choose not to try the experiment, you can still learn from the experimental design and discussion.
   Further Exploration. The Genetic Science Learning Center home page has a variety of review materials, interesting visuals, and fun activities.

2. *How Do Scientific Advances Occur? Access Excellence* is a website for "health and bioscience teachers and learners" run by the National Health Museum. Follow the "About Biotech" link to *Biotech Chronicles* and then click on "Pioneer Profiles." Read the profiles of Rosalind Franklin and James Watson. Then, from the home page, follow the "Activities Exchange" link to *Classic Collection* and read "A Visit with Dr. Francis Crick."
   Further Exploration. The "On-line Biology Book" has a good overview of DNA and molecular genetics that goes through the process of the discovery of DNA.

3. *Is the Pursuit of Science Always Objective and Unbiased? Access Excellence* was developed in 1993 by the pioneering biotechnology company Genentech. In 1999, the website was donated to the nonprofit National Health Museum but is still partially funded and intellectually supported by Genentech.

HOW WOULD YOU VOTE NOW?

No DNA vaccines have been approved for use in humans; however, clinical trials of such vaccines are under way to assess their safety and effectiveness. These trials are of DNA vaccines developed quickly after the discovery of the SARS virus. Because the trials will last several years, another outbreak of deadly SARS virus could occur before the results of the vaccine studies are in. There is also the threat of a bioterrorist attack releasing a potentially fatal-disease-causing organism before the studies are complete. Now that you know more about the structure and organization of DNA, what do you think? If another SARS outbreak or a bioterrorist attack occurred, would you agree to be treated with a DNA vaccine? Would you allow members of your family to be injected with a DNA vaccine? Visit the *Human Heredity* companion website at *www.cengage.com/biology/cummings* to find out more on the issue; then cast your vote online.

9 Gene Expression and Gene Regulation

Six months after earning a business degree from the University of Miami, Charlene Singh began to experience memory loss and behavioral changes. A short time later, she had difficulty walking. Within a year, she was diagnosed with the human form of mad-cow disease, known as variant Creutzfeldt-Jakob disease or vCJD. The disease takes many years to develop, and younger people are more likely to develop vCJD than are the elderly.

People can get vCJD by eating meat from cows infected with a brain-wasting disease called bovine spongiform encephalopathy (BSE). Charlene was born in England and moved with her family to the United States when she was a preteen. It is believed that Charlene became infected with vCJD while she was living in England. She died at age 25 in her home in Florida. Charlene was the first U.S. resident to die from vCJD, but in Great Britain, just over 160 people died from the disease from the 1980s through the early 1990s, and a second wave of 50 to 350 deaths is expected in the coming years.

BSE, vCJD, and several other diseases are called prion diseases. Prions are created when normal proteins in the body refold into a different and infectious three-dimensional shape that kills cells of the brain and nervous system. Normally, proteins perform tasks essential for life; they form most of the structures in a cell, transfer energy to drive all processes in a living cell, help copy chromosomes for cell division, control which genes are switched on and off, relay signals, fight infection, and repair damage caused by environmental agents. However, when a protein folds into a new shape, it can take on new functions, some of which can be deadly.

The information needed to make proteins is encoded in genes. In this chapter, we will examine the relationship between genes and proteins and the role of DNA as a carrier of genetic information. We will discuss the transfer of genetic information from the sequence of nucleotides in DNA to the sequence of amino acids in protein, the relationship between protein structure and function, and how the expression of genes is regulated.

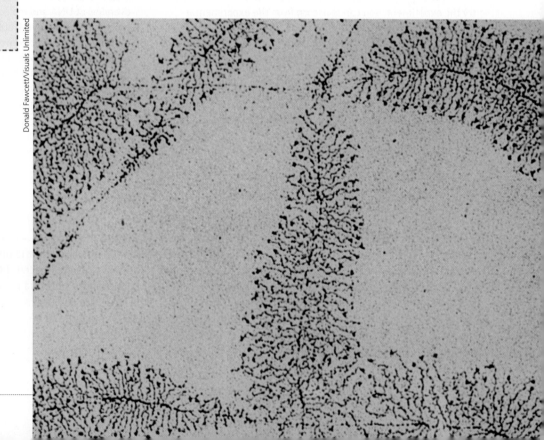

An electron micrograph showing transcription of a gene.

Donald Fawcett/Visuals Unlimited

9.1 The Link Between Genes and Proteins

At the turn of the twentieth century, just after Mendel's work received widespread recognition, Archibald Garrod proposed the existence of a relationship among genes, proteins, and phenotypes, based on his studies of infants with a condition called **alkaptonuria** (OMIM 203500). Newborns with this disorder have urine that turns black when exposed to air (Figure 9.1). Garrod discovered that the urine of affected infants contained large quantities of a compound he called alkapton (we call it homogentisic acid). He reasoned that because it is not present in the urine of most people, homogentisic acid is produced in the body but normally converted into other products. In infants with alkaptonuria, however, this conversion process must be blocked, causing the high concentration of homogentisic acid in the urine. He called this condition an "inborn error of metabolism" and proposed that some abnormal phenotypes were caused by a biochemical defect linked to a genetically inherited disorder.

HOW WOULD YOU VOTE?

Most cases of prion diseases caused by eating infected beef have been reported in the United Kingdom, not in the United States. Prions have also been transmitted by contaminated surgical and dental instruments and, in other cultures, by cannibalism. There is no cure for a prion infection, and prions cannot be destroyed by sterilization. Some countries are testing all beef used for human consumption, whereas others, such as the United States, are randomly testing only a small sample of cows. If you were traveling or living in a country with a history of infected cows, would you eat beef or allow your children to eat it? If not, what if infected beef was linked to human deaths in that country? Visit the *Human Heredity* companion website at *www.cengage.com/biology/cummings* to find out more on the issue; then cast your vote online.

How are genes and enzymes related?

Years later, George Beadle and Edward Tatum extended Garrod's work and established the connection among genes, enzymes, and phenotypes through their experimental work with *Neurospora*, a common bread mold. Beadle and Tatum showed that specific mutations in DNA caused loss of activity in a specific enzyme, which in turn produced a unique mutant phenotype. Establishing this connection was the key step in showing that genes produce phenotypes through the action of proteins. Beadle and Tatum received the Nobel Prize in 1958 for their work revealing the pathway from genes to proteins to phenotypes.

Genetic information is stored in DNA.

Proteins are the intermediary between a gene and a phenotype. The phenotype of a cell, tissue, and organism are all the result of protein function. When specific proteins are not functional or are not

Alkaptonuria An autosomal recessive trait with altered metabolism of homogentisic acid. Affected individuals do not produce the enzyme needed to metabolize this acid, and their urine turns black.

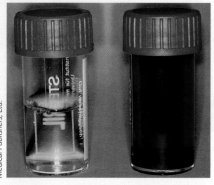

Figure 1 from Gubler Ch. Blaue Skleren und Thoraxschmerz. Schweiz Med Wochenschr 2000; 130 (17) 635. With permission from EMH Swiss Medical Publishers, Ltd.

FIGURE 9.1 (*left*) Urine sample from an unaffected person does not change color upon exposure to air. (*right*) Urine sample from someone with alkaptonuria turns black upon exposure to the air.

Amino acid One of the 20 subunits of proteins. Each contains an amino group, a carboxyl group, and an R group.

produced, the usual result is a mutant phenotype, which we call a genetic disorder. Because proteins are the products of genes and genes are composed of DNA, information encoded in DNA must control the kinds and amounts of proteins present in the cell. But how is genetic information carried in DNA? Based on their model, Watson and Crick proposed that genetic information is encoded in the sequence of nucleotides in DNA. The amount of genetic information that can be stored in any cell is a function of the number of DNA nucleotides carried by that cell. A molecule of DNA in a human chromosome can carry hundreds or thousands of genes. Each gene typically contains several hundred to several thousand nucleotides, with a beginning and an end marked by specific nucleotide sequences.

> **KEEP IN MIND**
>
> The information necessary to make proteins is encoded in the nucleotide sequence of DNA.

The relationship between genes and proteins.

How do genes (in the form of DNA) carry the information that specifies the structure of a protein? Proteins are linear molecules assembled from subunits called **amino acids**. Twenty different types of amino acids are used to make proteins. The diversity of protein structure and function results from how these 20 amino acids are combined in a protein. How much diversity is possible? As an example, let's consider a hypothetical protein composed of 5 amino acids. We can now rephrase the question as: How many different proteins of 5 amino acids can be made? Well, the first amino acid in the protein can be any one of the 20 different amino acids; the second amino acid can also be any of the 20 different amino acids, and so on. This means that the number of possible combinations of amino acids in a protein is 20^n, where 20 is the number of different amino acids found in proteins and n is the number of amino acids in a particular protein. In our example, the protein contains only 5 amino acids, so the number of possible combinations of amino acids in this protein is $20 \times 20 \times 20 \times 20 \times 20$, which is 20^5, or 3,200,000. Each of these 3.2 million proteins would have a different amino acid sequence and a potentially different function. When you consider that most proteins are composed of several hundred amino acids, it is easy to see that literally billions of different proteins are possible.

In Chapter 8, we learned that DNA contains 4 different nucleotides (A, T, C, and G). Because there are 20 different amino acids in proteins, the obvious question is: How can only 4 nucleotides encode the information for 20 amino acids?

9.2 The Genetic Code: The Key to Life

The linear sequence of nucleotides in a gene encodes the information that spells out the linear sequence of amino acids in a protein. Because DNA contains 4 different nucleotides, at first glance it seems difficult to envision how information for literally billions of different proteins can be carried in DNA. So the question is: How exactly does DNA encode genetic information?

To answer this question, let's start with some hypothetical cases:

1. If each nucleotide encoded the information for 1 amino acid, only 4 different amino acids could be inserted into proteins (4 nucleotides, taken 1 at a time, or 4^1).
2. If a sequence of 2 nucleotides encoded the information for 1 amino acid, only 16 combinations would be possible (4 nucleotides, taken 2 at a time, or 4^2).
3. However, a sequence of 3 nucleotides allows 64 combinations (4 nucleotides, taken 3 at a time, or 4^3), which at first doesn't seem right, because that leaves 44 more combinations than the 20 needed to encode amino acids.

So how many nucleotides are needed to encode one amino acid? Francis Crick and his colleagues answered this question in a series of experiments using a virus called T4. The results showed that 3 nucleotides are needed to carry the information for 1 amino acid. They also found that some amino acids could be specified by more than one combination of three nucleotides. This built-in redundancy uses up most of the other 44 combinations in the code. This important work established that the genetic code consists of:

- a linear series of 3 nucleotides (a triplet)
- these triplets each encode information that specifies an amino acid

After Crick's work, the question of which nucleotide combinations encode which amino acid was quickly solved, and the coding information of all 64 triplet combinations was established (Table 9.1). By convention, the genetic code is written as it appears in an RNA copy of the DNA template, and each group of 3 nucleotides is called a **codon**. Table 9.1 shows that 61 codons code for amino acids, but 3 (UAA, UAG, and UGA) do not; instead, each of these 3, called **stop codons**, signals the end of protein synthesis. One codon (AUG) has two functions: (1) it encodes the information for the amino acid methionine, and (2) it is always the first codon in a gene, called the **start codon**. The start codon marks the beginning of the coding sequence in a gene.

With a few exceptions, the same codons are used for the same amino acids in viruses and all living organisms, including bacteria, algae, fungi, and multicellular plants and animals. The nearly universal nature of the genetic code means that the code was established early in the evolution of life on this planet. The existence of such a code provides strong evidence that all living things are closely related and evolved from a common ancestor.

Codon Triplets of nucleotides in mRNA that encode the information for a specific amino acid in a protein.

Stop codon A codon in mRNA that signals the end of translation. UAA, UAG, and UGA are stop codons.

Start codon A codon present in mRNA that signals the location for translation to begin. The codon AUG functions as a start codon and codes for the amino acid methionine.

KEEP IN MIND

The three nucleotides in a codon are a universal language specifying the same amino acids in almost all organisms.

Table 9.1 The Genetic Code for Amino Acids

The messenger RNA codon for each amino acid.		
Amino Acid	Abbreviation	mRNA codons
Alanine	Ala	GCA GCC GCG GCU
Arginine	Arg	AGA AGG CGA CGC CGG CGU
Asparagine	Asn	AAC AAU
Aspartic acid	Asp	GAC GAU
Cysteine	Cys	UGC UGU
Glutamic acid	Glu	GAA GAC
Glutamine	Gln	CAA CAG
Glycine	Gly	GGA GGC GGG GGU
Histidine	His	CAC CAU
Isoleucine	Ile	AUA AUC AUU
Leucine	Leu	CUA CUC CUG CUU UUA UUG
Lysine	Lys	AAA AAG
Methionine	Met	*AUG
Phenylalanine	Phe	UUC UUU
Proline	Pro	CCA CCC CCG CCU
Serine	Ser	AGC AGU UCA UCC UCG UCU
Threonine	Thr	ACA ACC ACG ACU
Tryptophan	Trp	UGG
Tyrosine	Tyr	UAC UAU
Valine	Val	GUA GUC GUG GUU
Stop codons		UAA UAG UGA

Codon letters: A = adenine, C = cytosine, G = guanine, U = uracil.
*AUG encodes the amino acid methionine and signals "start" of translation when it occurs at the begining of a gene.

Now that we know how the information for proteins is encoded in DNA, let's turn our attention to another question: How is the linear sequence of DNA nucleotides in a gene converted into the linear sequence of amino acids in a protein? An answer to this question must take into account several things. First, almost all the cell's DNA is found in the nucleus (although some is in the mitochondria), but proteins are made in the cytoplasm. This means that the information that leads from gene to protein must originate in the nucleus and move to the cytoplasm. Second, how are genes turned on and off? If all genes in a cell are not active, what mechanisms are used to select specific genes for information transfer? We will answer each of these questions in turn.

9.3 Tracing the Flow of Genetic Information from Nucleus to Cytoplasm

pre-messenger RNA (pre-mRNA)
The transcript made from the DNA template that is processed and modified to form messenger RNA.

Transcription Transfer of genetic information from the base sequence of DNA to the base sequence of RNA, mediated by RNA synthesis.

messenger RNA (mRNA) A single-stranded complementary copy of the amino-acid coding nucleotide sequence of a gene.

Translation Conversion of information encoded in the nucleotide sequence of an mRNA molecule into the linear sequence of amino acids in a protein.

The transfer of genetic information from the linear sequence of nucleotides in a DNA molecule into the linear sequence of amino acids in a protein occurs in a series of steps (Figure 9.2). First, the information encoded in a gene is copied into an RNA molecule known as **pre-messenger RNA (pre-mRNA)**. This step, called **transcription**, takes place in the nucleus. The pre-mRNA is then processed into a finished **messenger RNA (mRNA)** that moves through a nuclear pore into the cytoplasm. Here, the information encoded in the mRNA is converted into the sequence of amino acids in a polypeptide chain, which is processed and folded to form a protein. This step is called **translation** (Figure 9.2).

The type and order of amino acids in the amino acid chain produced by translation, in turn, determines the structural and functional characteristics of the protein product and its role in phenotypic expression. In the next sections, we will examine transcription and translation in more detail.

> **KEEP IN MIND**
>
> Genetic information for proteins, in the form of mRNA, moves from the nucleus to the cytoplasm, where it is translated into the amino acid sequence of a polypeptide.

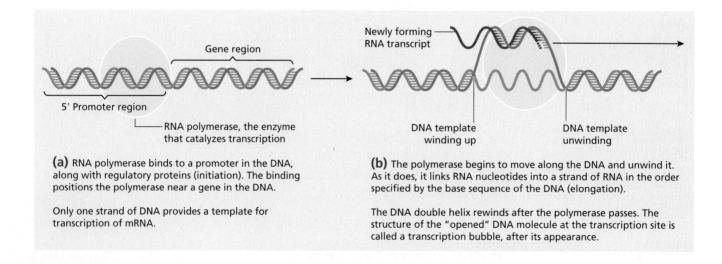

(a) RNA polymerase binds to a promoter in the DNA, along with regulatory proteins (initiation). The binding positions the polymerase near a gene in the DNA.

Only one strand of DNA provides a template for transcription of mRNA.

(b) The polymerase begins to move along the DNA and unwind it. As it does, it links RNA nucleotides into a strand of RNA in the order specified by the base sequence of the DNA (elongation).

The DNA double helix rewinds after the polymerase passes. The structure of the "opened" DNA molecule at the transcription site is called a transcription bubble, after its appearance.

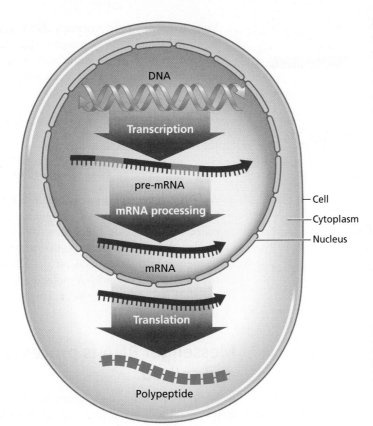

FIGURE 9.2 The flow of genetic information. One strand of DNA is transcribed into a strand of mRNA. The mRNA is processed and moves from the nucleus to the cytoplasm, where it is converted into the amino acid sequence of a polypeptide that folds to form a protein.

9.4 Transcription Produces Genetic Messages

Transcription begins when the DNA in a chromosome unwinds and one strand is used as a template to make a pre-mRNA molecule (Active Figure 9.3). This process involves several steps:

1. In a stage called initiation, RNA polymerase (an enzyme) and several regulatory proteins bind to a specific nucleotide sequence (called a promoter) that marks the beginning of a gene.
2. In a stage called elongation, the strands of DNA unwind, and RNA polymerase reads the nucleotide sequence of the template strand. As it moves along, it inserts and

Spotlight on...

Mutations in Splicing Sites and Genetic Disorders

Proper splicing of pre-mRNA is essential for normal gene function. Splicing defects cause several human genetic disorders. In one of these, a hemoglobin disorder called ß-thalassemia (OMIM 141900), mutations at the intron/exon border lower the efficiency of splicing and result in a deficiency in the amount of ß-globin mRNA and lower-than-normal amounts of the ß-globin protein, producing anemia as a phenotype.

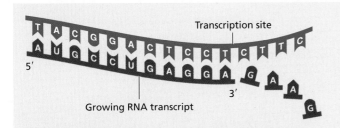

(c) What happened in the gene region? RNA polymerase catalyzed the covalent bonding of many nucleotides to one another to form an RNA strand. The base sequence of the new RNA strand is complementary to the base sequence of its DNA template—a copy of the gene.

(d) At the end of the gene region, the last stretch of the new transcript unwinds and detaches from the DNA template (termination).

ACTIVE FIGURE 9.3 Transcription of a gene. An enzyme, RNA polymerase, uses one strand of DNA as a template to synthesize a pre-mRNA molecule.

CENGAGE**NOW** Learn more about transcription by viewing the animation by logging on to *login.cengage.com/sso* and visiting CengageNOW's Study Tools.

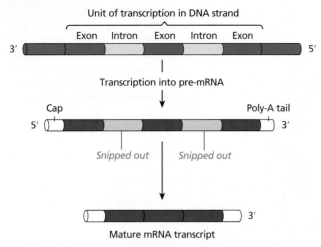

Unit of transcription in DNA strand

Exon | Intron | Exon | Intron | Exon

3'   5'

Transcription into pre-mRNA

Cap   Poly-A tail

5'   3'

Snipped out   *Snipped out*

3'

Mature mRNA transcript

ACTIVE FIGURE 9.4 Steps in the processing and splicing of mRNA. The template strand of DNA is transcribed into a pre-mRNA molecule. The ends of this molecule are modified, and the introns are spliced out to produce a mature mRNA molecule. The mRNA is then moved to the cytoplasm for translation.

CENGAGENOW Learn more about messenger RNA processing by viewing the animation by logging on to *login. cengage.com/sso* and visiting CengageNOW's Study Tools.

Introns DNA sequences present in some genes that are transcribed but are removed during processing and therefore are not present in mature mRNA.

Exons DNA sequences that are transcribed, joined to other exons during mRNA processing, and translated into the amino acid sequence of a protein.

Cap A modified base (guanine nucleotide) attached to the 5' end of eukaryotic mRNA molecules.

Poly-A tail A series of A nucleotides added to the 3' end of mRNA molecules.

links together complementary RNA nucleotides to form a pre-mRNA molecule. Remember that there is no T in RNA, so an A on the DNA template ends up as a U in the RNA transcript. For example, if the nucleotide sequence in the DNA template strand is

CGGATCAT

the mRNA will have the sequence

GCCUAGUA

3. As RNA polymerase moves along the DNA template, it eventually reaches the end of the gene, marked by nucleotides called a 3' termination sequence. When the RNA polymerase reaches the termination sequence, it stops adding nucleotides to the pre-mRNA and falls off the DNA template strand. In the process, the pre-mRNA molecule is released and the DNA strands re-form a double helix. This last stage of transcription is called termination.

Messenger RNA is processed and spliced.

In humans and other eukaryotes, transcription produces large pre-mRNA precursor molecules. These pre-mRNAs are processed in the nucleus to remove **introns**, which are nucleotide sequences present in genes that are not translated into the amino acid sequence of a protein. Introns occur between **exons**, the nucleotide sequences that remain in the mRNA. As introns are removed, the exons are spliced together to form mature mRNA molecules (Active Figure 9.4).

During processing, the ends of the mRNA are modified. A nucleotide **cap** is added at the 5′ end of the mRNA, and a tail of 30 to 100 A nucleotides called a **poly-A tail** is added at the 3′ end. The poly-A tail aids in export of the mRNA from the nucleus and plays a role in translation. The final product, a mature, processed mRNA, moves through a pore in the nuclear envelope into the cytoplasm, where translation takes place.

In a process called alternative splicing, exons can be retained or removed during splicing, allowing the processed mRNA to contain different combinations of exons (Figure 9.5). Alternative splicing allows one gene to encode information for several different forms of a protein. Overall, it is estimated that alternative splicing of pre-mRNA occurs in more than 95% of human genes.

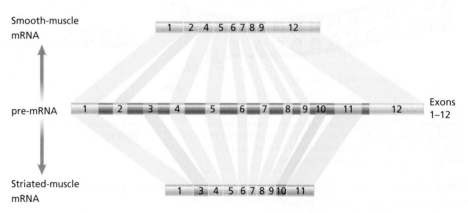

FIGURE 9.5 Alternative splicing of pre-mRNA for a muscle protein called α-tropomyosin. In smooth muscle (*top*), all the introns and exons 3, 10, and 11 are removed during processing. In striated muscle (*bottom*), all the introns and exons 2 and 12 are removed during processing. In this case, alternative splicing produces two different, tissue-specific forms of the same protein.

9.5 Translation Requires the Interaction of Several Components

Translation converts the nucleotide sequence of mRNA into the amino acid sequence of a protein, a process that requires several different components, each of which has a separate, specialized job. Before we examine the details of translation, let's look at the components.

Amino acids are subunits of proteins.

We have already explained that proteins are assembled from amino acids and that 20 different amino acids can be used to make proteins. The names and abbreviations for the 20 amino acids found in proteins are listed in Table 9.2. Each amino acid has three characteristic chemical groups: an **amino group** (NH_2), a **carboxyl group** (COOH), and an **R group** (Figure 9.6a). R groups are side chains that are different for each amino acid. During translation, amino acids are linked together by covalent chemical bonds formed between the amino group of one amino acid and the carboxyl group of a second amino acid (Figure 9.6b). Two linked amino acids form a dipeptide, 3 form a tripeptide, and 10 or more make a **polypeptide**. Each polypeptide (and protein) has a free amino group at one end, known as the **N-terminus**, and a free carboxyl group, called the **C-terminus**, at the other. By convention, the amino acid sequence of a protein is written beginning at the N-terminus and moving to the C-terminus.

Messenger RNA, ribosomal RNA, and transfer RNA interact during translation.

The nucleotide sequence of mRNA is converted into the amino acid sequence of a polypeptide by interactions with two other forms of RNA: ribosomal RNA (rRNA) and transfer RNA (tRNA).

Amino group A chemical group (NH_2) found in amino acids and at one end of a polypeptide chain.

Carboxyl group A chemical group (COOH) found in amino acids and at one end of a polypeptide chain.

R group Each amino acid has a different side chain, called an R group. An R group can be positively or negatively charged or neutral.

Polypeptide A molecule made of amino acids joined together by peptide bonds.

N-terminus The end of a polypeptide or protein that has a free amino group.

C-terminus The end of a polypeptide or protein that has a free carboxyl group.

Table 9.2 Amino Acids Commonly Found in Proteins

Amino Acid	3 Letter Abbreviation	1 Letter Abbreviation
Alanine	ala	A
Arginine	arg	R
Asparagine	asn	N
Aspartic acid	asp	D
Cysteine	cys	C
Glutamic acid	glu	E
Glutamine	gln	Q
Glycine	gly	G
Histidine	his	H
Isoleucine	ile	I
Leucine	leu	L
Lysine	lys	K
Methionine	met	M
Phenylalanine	phe	F
Proline	pro	P
Serine	ser	S
Threonine	thr	T
Tryptophan	trp	W
Tyrosine	tyr	Y
Valine	val	V

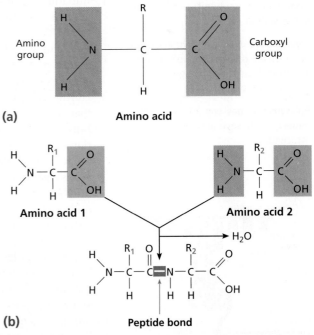

FIGURE 9.6 (a) An amino acid, showing the amino group, the carboxyl group, and the chemical side chain known as an R group. The R groups differ in each of the 20 amino acids used in protein synthesis. (b) Formation of a peptide bond between two amino acids.

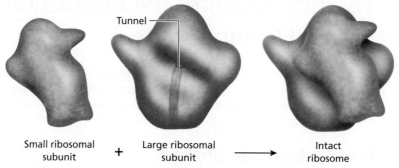

Tunnel

Small ribosomal
subunit

+

Large ribosomal
subunit

Intact
ribosome

FIGURE 9.7 Three-dimensional models of the small and large subunits of ribosomes.

Ribosomes Cytoplasmic particles composed of two subunits that are the site of protein synthesis.

ribosomal RNA (rRNA) RNA molecules that form part of the ribosome.

transfer RNA (tRNA) A small RNA molecule that contains a binding site for a specific type of amino acid and has a three-base segment known as an anticodon that recognizes a specific base sequence in messenger RNA.

Anticodon A group of three nucleotides in a tRNA molecule that pairs with a complementary sequence (known as a codon) in an mRNA molecule.

Initiation complex Formed by the combination of mRNA, tRNA, and the small ribosome subunit. The first step in translation.

Peptide bond A covalent chemical link between the carboxyl group of one amino acid and the amino group of another amino acid.

Polysomes A messenger RNA (mRNA) molecule with several ribosomes attached.

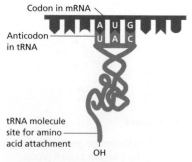

FIGURE 9.8 A transfer RNA (tRNA) molecule.

Ribosomes are cytoplasmic organelles with a large and a small subunit (Figure 9.7) and are the site of polypeptide synthesis. Ribosomes can float in the cytoplasm or attach to the outer membrane of the rough endoplasmic reticulum (RER) (review organelles in Chapter 2). Each ribosome subunit contains proteins and a type of RNA called **ribosomal RNA (rRNA)**. During translation, the rRNA in the large subunit acts as an enzyme, linking amino acids together to form a polypeptide.

Transfer RNA (tRNA) molecules (Figure 9.8) bring amino acids to the mRNA-ribosome complex during translation. Each tRNA molecule has two attachment sites: (1) a nucleotide sequence of three nucleotides called an **anticodon** that pairs with a complementary codon sequence in mRNA, and (2) a site for attachment of the amino acid specified by the mRNA codon. As we will see in the following section, tRNAs deliver their attached amino acids to the ribosome in the order specified by mRNA codons. Once at the ribosome, the amino acids are linked together by rRNA to form a polypeptide chain.

Translation produces polypeptides from information in mRNA.

Translation, like transcription, has three steps: initiation, elongation, and termination:

1. Initiation begins when mRNA binds to a small ribosomal subunit and the anticodon of the initiator tRNA carrying the amino acid methionine pairs with the AUG codon of mRNA (Active Figure 9.9a on page 206). Because AUG is the start codon and also encodes methionine, this amino acid is inserted first in all human proteins. Initiation is complete when a large ribosomal subunit binds to the complex (Active Figure 9.9b).

2. During elongation, amino acids are added to the growing polypeptide chain. Recall that during initiation, the initiator tRNA carrying methionine is added to the **initiation complex**, so each polypeptide chain begins with methionine. As elongation begins, a tRNA carrying the second amino acid (in this case, valine) pairs with the second mRNA codon, and the rRNA of the large subunit acts as an enzyme and forms a **peptide bond** between the two amino acids. As the ribosome moves along the mRNA, other tRNAs carrying amino acids pair with mRNA codons, adding amino acids to the growing polypeptide chain (Active Figure 9.9c–e).

3. Termination occurs when the ribosome reaches a stop codon. Recall that three codons (UAA, UAG, and UGA) do not code for amino acids, and there are no tRNA molecules with anticodons for stop codons. Proteins called release factors bind to stop codons, then the polypeptide, mRNA, and tRNA are released from the ribosome (Active Figure 9.9f).

Many antibiotics work by interfering with steps in protein synthesis, as described in Exploring Genetics: Antibiotics and Protein Synthesis.

Once a ribosome has started translation and moves away from the AUG start codon, new initiation complexes can form on an mRNA. mRNA molecules loaded with multiple ribosomes are called **polysomes**; this amplifies the number of protein molecules that can be made from a single mRNA (Figure 9.10).

Antibiotics and Protein Synthesis

Antibiotics are chemicals produced by microorganisms as defense mechanisms. The most effective antibiotics work by interfering with essential biochemical or reproductive processes. Many antibiotics block or disrupt one or more stages in protein synthesis. Some of these are mentioned here.

Tetracyclines are a family of chemically related compounds used to treat several types of bacterial infections. Tetracyclines interfere with the initiation of translation. The tetracycline molecule binds to the small ribosomal subunit and prevents binding of the tRNA anticodon during initiation. Both eukaryotic and prokaryotic ribosomes are sensitive to the action of tetracycline, but this antibiotic cannot pass through the plasma membrane of eukaryotic cells. Because tetracycline can enter bacterial cells to inhibit protein synthesis, it will stop bacterial growth, helping the immune system fight the infection.

Streptomycin is used in hospitals to treat serious bacterial infections. It binds to the small ribosomal subunit but does not prevent initiation or elongation; however, it does affect the efficiency of protein synthesis. Binding of streptomycin changes the way mRNA codons interact with the tRNA anticodons. As a result, incorrect amino acids are incorporated into the growing polypeptide chain. In addition, streptomycin causes the ribosome to randomly fall off the mRNA, preventing the synthesis of complete proteins.

Alex Cao/Digital Vision/ Getty Images

Puromycin is not used clinically but has played an important role in studying the mechanism of protein synthesis in the research laboratory. The puromycin molecule is the same size and shape as a tRNA-amino acid complex. When puromycin enters the ribosome, it can be incorporated into a growing polypeptide chain, stopping further synthesis because no peptide bond can be formed between puromycin and an amino acid, causing the shortened polypeptide to fall off the ribosome.

Chloramphenicol was one of the first broad-spectrum antibiotics introduced. Eukaryotic cells are resistant to its actions, and it was widely used to treat bacterial infections. However, its use is limited to external applications and serious infections. Chloramphenicol destroys cells in the bone marrow, the source of all blood cells. In bacteria, this antibiotic binds to the large ribosomal subunit and inhibits the formation of peptide bonds. Another antibiotic, erythromycin, also binds to the large ribosomal subunit and inhibits the movement of ribosomes along the mRNA.

Almost every step of protein synthesis can be inhibited by one antibiotic or another. Work on designing new synthetic antibiotics to fight infections is based on our knowledge of how the nucleotide sequence of mRNA is converted into the amino acid sequence of a protein.

9.6 Polypeptides Are Processed and Folded to Form Proteins

When released from the ribosome, most polypeptides are nonfunctional; only after they are processed, folded, and converted into their functional form are they called proteins. This processing, called post-translational modification, can include combinations of more than 200 different types of chemical reactions. These include the addition of lipids or sugars to specific amino acids, modifications to amino acids, and sometimes the removal of amino acids from the interior or the ends of the polypeptide chain. During processing, the polypeptide folds into a three-dimensional shape and becomes a functional protein. The three-dimensional shape of the processed protein is determined by its amino acid sequence and the results of chemical modifications and other processing events. Often, folding is guided by proteins called molecular chaperones.

How many proteins can human cells make?

Given the existence of alternative splicing and the many modifications that can occur during the processing of polypeptides, the larger question is: How many different proteins can human cells make? The answer appears deceptively simple. Humans carry

ACTIVE FIGURE 9.9 Steps in the process of translation.

CENGAGENOW Learn more about translation by viewing the animation by logging on to *login.cengage.com/sso* and visiting CengageNOW's Study Tools.

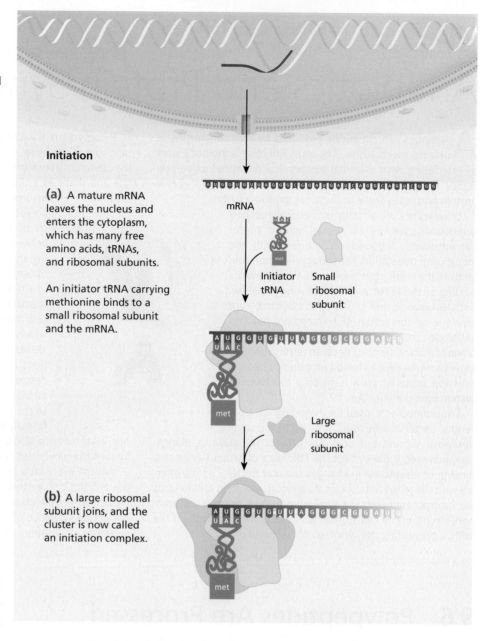

Initiation

(a) A mature mRNA leaves the nucleus and enters the cytoplasm, which has many free amino acids, tRNAs, and ribosomal subunits.

An initiator tRNA carrying methionine binds to a small ribosomal subunit and the mRNA.

mRNA

Initiator tRNA

Small ribosomal subunit

met

Large ribosomal subunit

(b) A large ribosomal subunit joins, and the cluster is now called an initiation complex.

met

Polysome

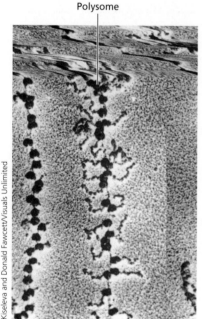

FIGURE 9.10 A polysome consists of an mRNA molecule loaded with many ribosomes, all engaged in translation.

between 20,000 and 25,000 protein-coding genes, and it would seem that there should be a direct correspondence between the number of genes and the number of encoded proteins. However, the set of proteins in a particular cell type, called its **proteome**, can be far greater than the number of genes in the genome. It is estimated that humans can make several hundred thousand different proteins from the 20,000 to 25,000 genes in the genome. Some of this diversity is produced by starting transcription at alternative sites, by differentially processing out exons during mRNA maturation, by modifications to the resulting polypeptide, and by other mechanisms we are only beginning to understand. These discoveries are one of the surprises of the Human Genome Project and are at the forefront of current research in human genetics.

KEEP IN MIND

Once polypeptides fold into a three-dimensional shape, are chemically modified, and become functional, they are called proteins. Mutations that prevent proper folding or cause misfolding can be the basis of disease.

Elongation

(c) An initiator tRNA carries the amino acid methionine, so the first amino acid of the new polypeptide chain will be methionine. A second tRNA binds the second codon of the mRNA (here, that codon is GUG, so the tRNA that binds carries the amino acid valine).

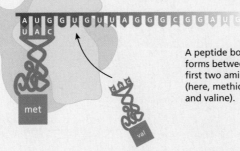

A peptide bond forms between the first two amino acids (here, methionine and valine).

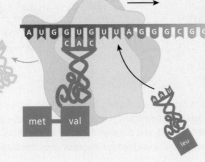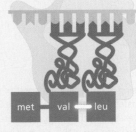

(d) The first tRNA is released and the ribosome moves to the next codon in the mRNA. A third tRNA binds to the third codon of the mRNA (here, that codon is UUA, so the tRNA carries the amino acid leucine).

A peptide bond forms between the second and third amino acids (here valine and leucine).

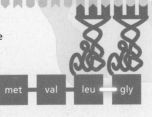

(e) The second tRNA is released and the ribosome moves to the next codon. A fourth tRNA binds the fourth mRNA codon (here, that codon is GGG, so the tRNA carries the amino acid glycine).

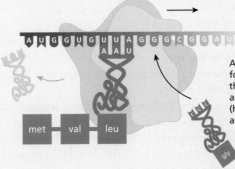

A peptide bond forms between the third and fourth amino acids (here, leucine and glycine).

Termination

(f) Steps d and e are repeated over and over until the ribosome encounters a stop codon in the mRNA. The mRNA transcript and the new poypeptide chain are released from the ribosome. The two ribosomal subunits separate from each other. Translation is now complete. Either the polypeptide chain will join the pool of proteins in the cytoplasm, the nucleus, or will enter the rough ER of the endomembrane system (Section 4.9).

Proteins are sorted and distributed to their cellular locations.

Proteins are made in the cytoplasm but are present in all parts of the cell: the nucleus, cytoplasm, internal membranes, and the plasma membrane. In addition, some proteins are secreted from the cell. How do proteins get to their destinations? Proteins made

Proteome The set of proteins present in a particular cell at a specific time under a particular set of environmental conditions.

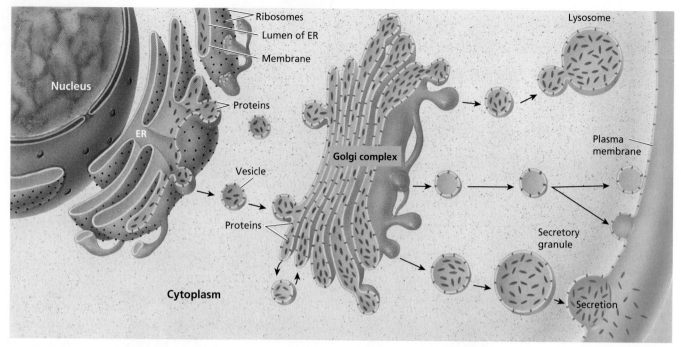

FIGURE 9.11 Processing, sorting, and transport of proteins synthesized in a human cell. Proteins made on ribosomes attached to the endoplasmic reticulum (ER) are transferred to the interior of the ER, where they are folded and chemically modified. Many of these proteins are transported to the Golgi complex in vesicles. In the Golgi, the proteins are further modified, sorted, and packaged into vesicles for delivery to other parts of the cell and are incorporated into organelles such as lysosomes or are transported to the surface for insertion into the plasma membrane. Proteins also can be packaged into vesicles for secretion.

on cytoplasmic ribosomes remain and function in the cytoplasm, but those made by ribosomes attached to the rough endoplasmic reticulum (RER) contain sorting signals that direct them to their proper locations within the cell. These signals are encoded in the DNA and are transcribed and translated.

Polypeptides made on the RER are secreted into the lumen, where the encoded signal is removed; there the polypeptide is processed, modified, and folded. These proteins are packaged into vesicles and moved to the Golgi complex for further processing, packaging, and transport to their destinations in the nucleus, other organelles, the plasma membrane, or for secretion from the cell (Figure 9.11).

9.7 Protein Structure and Function Are Related

Primary structure The amino acid sequence in a polypeptide chain.

Secondary structure The pleated or helical structure in a protein molecule generated by the formation of bonds between amino acids.

Tertiary structure The three-dimensional structure of a protein molecule brought about by folding on itself.

Quaternary structure The structure formed by the interaction of two or more polypeptide chains in a protein.

The three-dimensional shape and ultimate function of a protein comes from folding the polypeptide chain to produce up to four levels of protein structure (Active Figure 9.12):
- The first, called the **primary structure**, is the amino acid sequence of a polypeptide. The next two levels are determined mostly by interactions among these amino acids.
- The **secondary structure** is formed by the formation of hydrogen bonds among the amino acids in different parts of the protein. The secondary structure forms pleated sheets or coils, and most proteins contain both pleated and coiled regions.
- The third level, called the **tertiary structure**, is created when the helical or pleated sheet regions fold back on themselves.
- The fourth level, called the **quaternary structure**, is created when a functional protein is created by the physical association of two or more polypeptide chains.

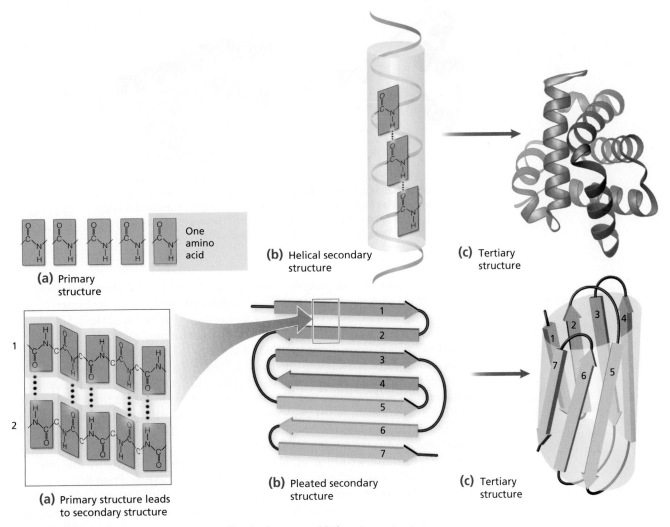

(a) Primary structure

(b) Helical secondary structure

(c) Tertiary structure

One amino acid

(a) Primary structure leads to secondary structure

(b) Pleated secondary structure

(c) Tertiary structure

ACTIVE FIGURE 9.12 Proteins can have several levels of structure. (a) The primary structure is the amino acid sequence, represented by the peptide units. (b) Hydrogen bonding between amino acids in the polypeptide chain can form a pleated sheet, an alpha helix, or a random coil. (c) Folding of the secondary structures into a functional three-dimensional shape creates the tertiary level of structure. Some functional proteins are made up of more than one polypeptide chain, and this level is the quaternary structure (not shown here).

CENGAGENOW Learn more about protein structure by viewing the animation by logging on to *login.cengage.com/sso* and visiting CengageNOW's Study Tools.

Improper protein folding can be a factor in disease.

Some mutations change the amino acid sequence (primary structure) of a polypeptide, causing misfolding of the mature protein. Misfolded proteins play a role in producing the abnormal phenotype seen in several genetic disorders, including cystic fibrosis (OMIM 219700), Alzheimer disease (OMIM 104300 and other numbers), and a metabolic disorder called MPS VI (OMIM 253200).

The cystic fibrosis gene encodes a protein (called CFTR) of 1,440 amino acids (review cystic fibrosis in Chapter 4) that is normally transported from the RER to the plasma membrane via the Golgi complex, where it acts as a channel to control the flow of chloride ions in and out of the cell. The most common mutation in CF is a deletion of the amino acid phenylalanine at position 508. This single amino acid change causes the polypeptide to fold improperly in the RER lumen. The misfolded CFTR is identified as defective and destroyed in the RER; as a result, it does not reach the plasma membrane. The phenotype of cystic fibrosis is produced because the CFTR protein is not present to regulate the flow of chloride ions through the plasma membrane.

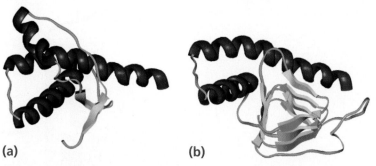

FIGURE 9.13 Misfolding or refolding of some proteins can result in disease. (a) The normal folding pattern for a prion protein. Most of the protein is in a helical configuration (red). (b) The protein has refolded to form a disease-causing prion. This refolding has altered the secondary and tertiary levels of protein structure. In this form, most of the protein is in pleated sheets (the yellow ribbon-like regions).

Prion A protein folded into an infectious conformation that is the cause of several disorders, including Creutzfeldt-Jakob disease and mad-cow disease.

Mad-cow disease A prion disease of cattle, also known as bovine spongiform encephalopathy, or BSE.

In other cases, certain proteins refold into a different three-dimensional shape and, as a result, become agents of disease. Protein refolding diseases are called **prion** diseases (Figure 9.13). In humans, Creutzfeldt-Jakob disease (CJD; OMIM 123400), Gerstmann-Straussler disease (OMIM 137440), and fatal familial insomnia (OMIM 600072) are all inherited forms of prion disease. In these disorders, a mutation changes one amino acid in the protein, predisposing it to refold into a different, disease-causing shape.

As described at the beginning of the chapter, Charlene Singh's death was caused by refolding of a protein in her body, causing degenerative changes in the nervous system leading to early death. In her case, she became infected with prions, presumably from eating beef from an infected cow. These prions cause other proteins of the same type to refold into the disease-causing conformation, producing the symptoms of the infectious form of the disease, known as variable Creutzfeldt-Jakob disease (vCJD). The process is slow, and the disease makes its appearance in about 5 to 15 years. Prion diseases such as **mad-cow disease** are infectious, and the disease is transmitted when refolded proteins are transferred from one individual to another. Case 2 at the end of the chapter deals with vCJD.

9.8 Several Mechanisms Regulate the Expression of Genes

Each human chromosome carries hundreds or thousands of the 20,000 to 25,000 genes that make us unique, but not all these genes are expressed in every cell. In fact, in a given cell, almost all genes are switched off most of the time. Liver cells, for example, do not express the genes for eye color, and brain cells do not make enzymes that help digest food. Only about 5% to 10% of the genes in most cells are active. The process of turning genes on and off is called gene regulation. As we saw earlier, gene expression involves a series of steps that take place in both the nucleus and the cytoplasm; each of these steps offers an opportunity to regulate the expression of genes. Thus, gene regulation can occur at the transcriptional, post-transcriptional, translational, and post-translational levels (Figure 9.14). Of these, the most important is regulation at the transcriptional level.

Chromatin remodeling and access to promoters.

Chromosomal DNA is tightly wound around histone proteins (review chromosome structure in Chapter 8) to form nucleosomes. The complex of DNA, histones, and other proteins that form chromosomes is called chromatin. The tight binding of DNA to histones in nucleosomes restricts access to promoters, located at the beginning or 5' end of

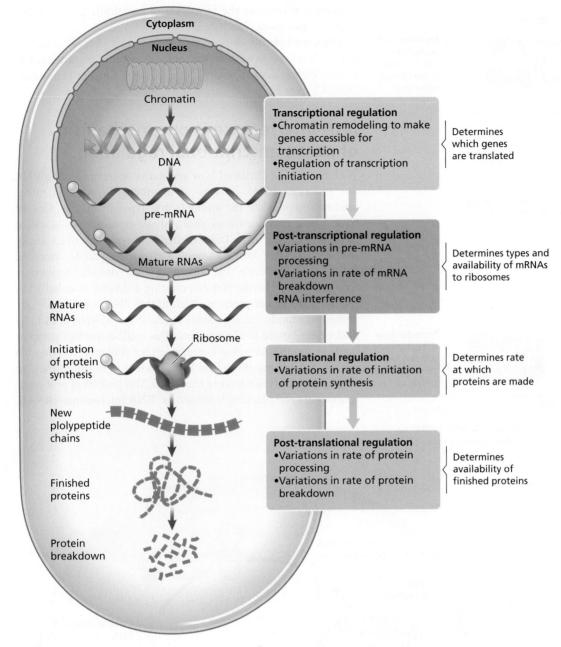

FIGURE 9.14 Stages of gene regulation in human cells.

Transcriptional regulation
•Chromatin remodeling to make genes accessible for transcription
•Regulation of transcription initiation

Determines which genes are translated

Post-transcriptional regulation
•Variations in pre-mRNA processing
•Variations in rate of mRNA breakdown
•RNA interference

Determines types and availability of mRNAs to ribosomes

Translational regulation
•Variations in rate of initiation of protein synthesis

Determines rate at which proteins are made

Post-translational regulation
•Variations in rate of protein processing
•Variations in rate of protein breakdown

Determines availability of finished proteins

Cytoplasm
Nucleus
Chromatin
DNA
pre-mRNA
Mature RNAs
Mature RNAs
Ribosome
Initiation of protein synthesis
New plolypeptide chains
Finished proteins
Protein breakdown

genes, keeping them inactive. Proteins called remodeling complexes chemically modify the histones and weaken the binding between DNA and histones, making promoters accessible for gene activation. Once the promoters are accessible, genes can be expressed when transcriptional proteins (including RNA polymerase) bind to the exposed promoters. Activated genes can be inactivated by reversing histone modification and removing the chemical groups added during activation. This reversible process, called **chromatin remodeling** (Figure 9.15), is one of the main mechanisms of gene regulation in humans.

Chromatin remodeling The process of chemical changes to the DNA and histones that activate and inactivate gene expression.

DNA methylation can silence genes.

Chromatin remodeling can also involve changes to chromosomal DNA instead of histones. The addition of methyl groups (CH_3) to cytosine bases in DNA turns genes off, a process called gene silencing. The presence of methyl groups may prevent the binding of transcriptional proteins to the gene; in addition, methylated DNA binds to non-histone

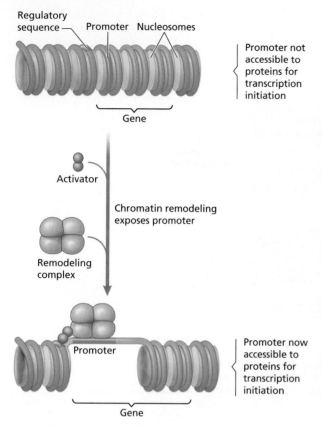

Regulatory
sequence — Promoter Nucleosomes

Gene

Promoter not
accessible to
proteins for
transcription
initiation

Activator

Chromatin remodeling
exposes promoter

Remodeling
complex

Promoter

Gene

Promoter now
accessible to
proteins for
transcription
initiation

FIGURE 9.15 Chromatin remodeling exposes the promoter region of a gene and makes it accessible to transcriptional protein complexes.

proteins that promote the formation of genetically inactive chromatin termed silent chromatin.

DNA methylation is associated with several processes, including the inactivation of X chromosomes, genomic imprinting, and cancer. We will discuss genomic imprinting in Chapter 11 and cancer in Chapter 12.

RNA interference is one mechanism of post-transcriptional regulation.

We have already discussed how alternative splicing of pre-mRNA can generate several versions of a protein. In addition, other mechanisms regulate the amounts and types of mRNA. One of these mechanisms involves short, single-stranded RNAs called micro-RNAs (miRNAs). These are derived from transcribed single-stranded RNA that folds back upon itself, forming double-stranded regions in the molecule. The remaining single-stranded regions are removed by an enzyme (Dicer), creating a double-stranded RNA (Figure 9.16). Other proteins attach to this RNA and degrade one of the two strands, leaving a small single-stranded RNA of 21 to 22 nucleotides bound to the proteins. This miRNA/protein complex binds to all mRNAs with a complementary sequence, cleaving the mRNA or blocking ribosomes from loading onto the mRNA. In either event, the mRNA is not translated. This process of post-transcriptional gene silencing is known as **RNA interference (RNAi)**.

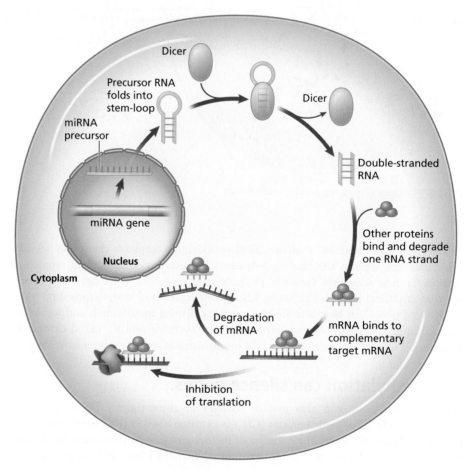

Dicer

Precursor RNA
folds into
stem-loop

Dicer

miRNA
precursor

Double-stranded
RNA

miRNA gene

Other proteins
bind and degrade
one RNA strand

Nucleus

Cytoplasm

Degradation
of mRNA

mRNA binds to
complementary
target mRNA

Inhibition
of translation

FIGURE 9.16 RNA interference (RNAi) regulates gene expression through the action of micro-RNAs (miRNAs).

Discovery of RNAi has led to the development of synthetic miRNAs, used as tools to block expression of specific genes as a way to dissect steps in development and differentiation. Researchers are also exploring the use of RNAi in cancer therapy by attempting to block expression of genes necessary for the uncontrolled growth of cancer cells.

RNA interference (RNAi)
A mechanism of gene regulation that controls the amounts of mRNA available for translation.

Translational and post-translational mechanisms regulate the production of proteins.

Regulation of protein synthesis occurs at several levels: mainly the rate at which proteins are translated from mRNA, and the ways in which polypeptides are processed into proteins, and selected proteins are broken down. Recall that during processing in the nucleus, a poly-A tail is added to many maturing mRNAs. In the cytoplasm, enzymes can lengthen or shorten the poly-A tail. When poly-A tails are made longer, there is increased translation, producing more protein molecules per mRNA. When poly-A tails are shortened, translation of proteins from those mRNAs is decreased. Exactly how the length of poly-A tails is adjusted and regulated is not yet known in detail, and this mechanism of gene regulation is an active area of research.

Once polypeptides have been produced, the rate at which they are processed and folded into functional proteins or reversibly modified are steps in gene regulation. For example, the reversible addition of phosphate groups to proteins can stimulate or repress their activity in the transduction of signals from the external environment through the cytoplasm into the nucleus.

The last point of regulation of gene expression is the control of protein breakdown. The life span of proteins ranges from a few minutes in some enzymes to almost a year for structural proteins such as collagen. By balancing protein translation, processing, and breakdown, cells can control the amount and activity of specific proteins needed for particular tasks or at certain points in development.

Genetics in Practice

Genetics in Practice case studies are critical-thinking exercises that allow you to apply your new knowledge of human genetics to real-life problems. You can find these case studies and links to relevant websites at *www.cengage.com/biology/cummings*

CASE 1

A genetic counselor was called to the pediatric ward to examine a 3-week-old infant who was diagnosed with a genetic disorder of sugar metabolism called galactosemia. The infant was admitted to the hospital because of failure to thrive and severe jaundice (yellowing of the skin resulting from liver problems). Upon examination, the physician determined that the infant had an enlarged liver, cataracts, and constant diarrhea and vomiting when fed milk. *Escherichia coli* infection is a common cause of death in infants who have galactosemia, and cultures were drawn from the infant. Laboratory results confirmed that the infant had a deficiency of the enzyme galactose-1-phosphate uridyltransferase and was infected with *E. coli*.

The counselor took a detailed family history and explained the condition to the parents. She indicated that the condition is due to the inheritance of a mutant gene from each parent (the trait is autosomal recessive) and that there is a 25%, or one in four, chance that each pregnancy they have together will produce a child with this condition. The counselor explained that there is wide variability in phenotype, ranging from very mild to severe. A blood test could determine which variant of the disease they carry.

1. What exists in blood that can be tested for a variant of a disease-causing gene?

2. What are possible treatments for this disease?

CASE 2

There have been recurring cases of mad-cow disease in the United Kingdom since the late 1990s. What started out as a topic of interest to a few cell biologists has become a huge public-interest story. Mad-cow disease is caused by a prion, an infectious particle that consists only of protein. In 1986, the media began reporting that cows were dying all over England from a mysterious disease. Initially, however, there was little interest in determining whether humans could be affected. For 10 years, the British government maintained that this unusual disease could not be transmitted to humans. However, in March 1996, the government did an about-face and announced that bovine spongiform encephalopathy (BSE), commonly known as mad-cow disease, can be transmitted to humans, where it is known as variant Creutzfeldt-Jakob disease (vCJD). As in cows, this disease eats away at the nervous system, destroying the brain and essentially turning it into a spongelike structure filled with holes. Victims experience dementia; confusion; loss of speech, sight, and hearing; convulsions; and coma and finally die. Prion diseases are always fatal, and there is no treatment. Precautionary measures taken in Britain to prevent this disease in humans may have begun too late.

Many of the victims today may have contracted it over a decade earlier, when the BSE epidemic began, and the incubation period is long (vCJD has an incubation period of 10 to 40 years).

1. How can a prion replicate itself without genetic material?

2. What measures have been taken to stop BSE?

3. If you were traveling in Europe, would you eat beef? Give sound reasons why or why not.

Summary

9.1 The Link Between Genes and Proteins

- At the beginning of the last century, Garrod proposed that genetic disorders result from biochemical alterations.

- Using *Neurospora*, Beadle and Tatum showed that mutations can produce a loss of enzyme activity and a mutant phenotype. They proposed that genes control the synthesis of proteins and that protein function is responsible for producing the phenotype.

9.2 The Genetic Code: The Key to Life

- The information transferred from DNA to mRNA is encoded in sets of three nucleotides, called codons. Of the 64 possible codons, 61 code for amino acids and 3 are stop codons.

9.3 Tracing the Flow of Genetic Information from Nucleus to Cytoplasm

- The processes of transcription and translation require the interaction of many components, including ribosomes, mRNA, tRNA, amino acids, enzymes, and energy sources. Ribosomes are the workbenches on which protein synthesis occurs. tRNA molecules are adapters that recognize amino acids and the complementary codon in mRNA, the gene transcript.

9.4 Transcription Produces Genetic Messages

- In transcription, one of the DNA strands is used as a template for making a complementary strand of RNA, called pre-mRNA. The pre-mRNA is processed and modified to produce mature messenger RNA (mRNA). Some pre-mRNAs can be processed in alternative ways to produce different versions of a protein.

9.5 Translation Requires the Interaction of Several Components

- Translation requires the interaction of tRNA molecules, amino acids, ribosomes, mRNA, and energy sources. Within the ribosome, tRNA anticodons bind to complementary codons in the mRNA. The ribosome moves along the mRNA, linking amino acids and producing a growing polypeptide chain. At termination, this polypeptide is released from the ribosome and undergoes a conformational change to produce a functional protein.

9.6 Polypeptides Are Processed and Folded to Form Proteins

- After synthesis, polypeptides fold into a three-dimensional shape, often assisted by proteins called chaperones. The importance of chaperones is underscored by the discovery that mutations in chaperone genes can cause genetic disorders. In addition, polypeptides can be chemically modified in many different ways, producing functionally different proteins from one polypeptide.

9.7 Protein Structure and Function Are Related

- Four levels of protein structure are recognized, three of which result from the primary sequence of amino acids in the backbone of the protein chain. Folding of the polypeptide into a three-dimensional structure produces a functional protein molecule.

9.8 Several Mechanisms Regulate the Expression of Genes

- Transcriptional regulation is the most basic mechanism of gene regulation. Other stages in the transcription and translation of mRNA offer opportunities to control the amount and availability of mRNA and the rate of protein synthesis, processing, and turnover.

Questions and Problems

The Link Between Genes and Proteins

1. The genetic material has to store information and be able to express it. What is the relationship among DNA, RNA, proteins, and phenotype?
2. Define replication, transcription, and translation. In what part of the cell does each process occur?

The Genetic Code: The Key to Life

3. If the genetic code used four bases at a time, how many amino acids could be encoded?
4. If the genetic code uses triplets, how many different amino acids can be coded by a repeating RNA polymer composed of UA and UC (UAUCUAUCUAUC . . .)?
   a. one   **b.** two   **c.** three   **d.** four   **e.** five
5. What is the start codon? What are the stop codons? Do any of them code for amino acids?

Transcription Produces Genetic Messages

6. The following segment of DNA codes for a protein. The uppercase letters represent exons. The lowercase letters represent introns. The lower strand is the template strand. Draw the primary transcript and the mRNA resulting from this DNA.

```
G C T A A A T G G C A a a a t t g c c g g a t g a c G C A C A T T G A C T C G G a a t c g a G G T C A G A T G C
C G A T T T A C C G T t t t a a c g g c c t a c t g C G T G T A A C T G A G C C t t a g c t C C A G T C T A C G
```

7. Is an entire chromosome made into an mRNA during transcription?
8. The promoter and terminator regions of genes are important in:
   a. coding for amino acids.
   b. gene regulation.
   c. structural support for the gene.
   d. intron removal.
   e. anticodon recognition.
9. What are the three modifications made to pre-mRNA molecules before they become mature mRNAs, are transported from the nucleus to the cytoplasm, and become ready to be used in protein synthesis? What is the function of each modification?
10. The pre-mRNA transcript and protein made by several mutant genes were examined. The results are given below. Determine where in the gene a likely mutation lies: the promoter region, exon, intron, cap on mRNA, or ribosome binding site.
    a. normal-length transcript, normal-length nonfunctional protein
    b. normal-length transcript, no protein made
    c. normal-length transcript, normal-length mRNA, short nonfunctional protein

d. normal-length transcript, longer mRNA, shorter nonfunctional protein
e. transcript never made

Translation Requires the Interaction of Several Components

11. Briefly describe the function of the following in protein synthesis.
    a. rRNA   **b.** tRNA   **c.** mRNA
12. What is the difference between codons and anticodons?
13. Determine the percent of the following gene that will code for the protein product. Gene length is measured in kilobases (kb) of DNA. Each kilobase is 1,000 bases long.

14. How many kilobases of the DNA strand below will code for the protein product?

15. Write the anticodon(s) for the following amino acids:
    a. met   **b.** trp   **c.** ser   **d.** leu
16. Given the following tRNA anticodon sequence, derive the mRNA and the DNA template strand. Also, write out the amino acid sequence of the protein encoded by this message.
    tRNA: UAC UCU CGA GGC
    mRNA:
    protein:
    How many hydrogen bonds would be present in the DNA segment?
17. Given the following mRNA, write the double-stranded DNA segment that served as the template. Indicate both the 5' and the 3' ends of both DNA strands. Also write out the tRNA anticodons and the amino acid sequence of the protein encoded by the mRNA message.
    DNA:
    mRNA: 5'-CCGCAUGUUCAGUGGGCGUAAACACUGA-3'
    protein:
    tRNA:
18. The following is a portion of a protein:
    met-trp-tyr-arg-gly-pro-thr-
    Various mutant forms of this protein have been recovered. Using the normal and mutant sequences, determine the

DNA and mRNA sequences that code for this portion of the protein, and explain each of the mutations.

a. met-trp-   b. met-cys-ile-val-val-leu-gln-

c. met-trp-tyr-arg-ser-pro-thr-

d. met-trp-tyr-arg-gly-ala-val-ile-ser-pro-thr-

19. Below is the structure of glycine. Draw a tripeptide composed exclusively of glycine. Label the N-terminus and C-terminus. Draw a box around the peptide bonds.

20. Indicate in which category, transcription or translation, each of the following functions belongs: RNA polymerase, ribosomes, nucleotides, tRNA, pre-mRNA, DNA, anticodon, amino acids.

Protein Structure and Function Are Related

21. In the most common form of cystic fibrosis, one amino acid is missing in the CF protein. Explain why none of the mutant protein product ends up in the cell's plasma membrane.

22. Polypeptide folding is often mediated by other proteins called chaperones. Describe how a mutant chaperone protein might be responsible for a genetic disorder involving an enzyme.

23. Do mutations in DNA alter proteins all the time?

24. a. Can a mutation change a protein's tertiary structure without changing its primary structure?

b. Can a mutation change a protein's primary structure without affecting its secondary structure?

Internet Activities

Internet Activities are critical-thinking exercises using the resources of the World Wide Web to enhance the principles and issues covered in this chapter. For a full set of links and questions investigating the topics described below, visit *www.cengage.com/biology/cummings*

1. *Review of Gene Expression.* At the *Cell Biology Topics 1: Ribosome* website, review the basics of translation after the mRNA leaves the nucleus.

2. *Quiz Yourself.* At University of Arizona's *The Biology Project: Molecular Biology* website, click on the "Nucleic Acids" link to access quizzes on DNA replication, transcription, and translation. Correct answers are rewarded with brief overviews; if you answer incorrectly, you will be linked to a short tutorial that will help you solve the problem.

3. *Control of Gene Expression.* At the *On-line Biology: Control of Gene Expression* website, read about the control of gene expression in bacteria, viruses, and eukaryotes.

a. How many different proteins and protein factors are involved in the various steps of gene expression? What would be the possible effects of a mutation that changed one of these proteins? Consequently, would you expect to see greater similarity or less similarity in the DNA sequences that code for these proteins in different organisms?

b. In some cases, the expression of multiple genes is controlled by a single protein factor, as in the operon model of transcriptional regulation proposed by Jacob and Monod. What might be the benefits of such a comparatively streamlined mechanism for the control of gene expression?

c. Compare the genome sizes for various eukaryotes. What percentage of the average eukaryotic genome actually codes for protein? What percentage of the human genome codes for protein? What function, if any, does the noncoding portion of the genome serve?

HOW WOULD YOU VOTE NOW?

Most cases of prion diseases caused by eating infected beef have been reported in the United Kingdom, not in the United States. Prions have also been transmitted by contaminated surgical and dental instruments and, in other cultures, by cannibalism. There is no cure for a prion infection, and prions cannot be destroyed by sterilization. Some countries are testing all beef used in human consumption, whereas others, such as the United States, are randomly testing only a small sample of cows. Now that you know more about proteins and the relationship between protein structure and function, what do you think? Would you eat beef or allow your children to eat it if you were traveling or living in a country with a history of infected cows? What if infected beef was linked to human deaths in that country? Visit the *Human Heredity* companion website at *www.cengage.com/biology/cummings* to find out more on the issue; then cast your vote online.

10 From Proteins to Phenotypes

The field of human biochemical genetics had its beginnings partly through the determination of a young Norwegian mother, Borgny Egeland, who had two mentally retarded children. Her daughter, Liv, did not walk until she was nearly 2 years of age and she spoke only a few words. Liv also had a musty odor that could not be washed away. Liv's younger brother, Dag, was also slow to develop and never learned to walk or talk. He had the same musty odor as his sister. Borgny was convinced that whatever was causing the odor was also causing her children's mental retardation. To learn why both of her children were retarded and had a musty odor, the mother went from doctor to doctor, but to no avail. Finally, in the spring of 1934, the persistent woman took the two children, then age 4 and 7 years, to Dr. Asbjorn Fölling, who was a physician and a biochemist.

Because the children's urine had a musty odor, Fölling tested the urine for signs of infection, but there was none. He discovered that the urine reacted with ferric chloride to produce a green color, indicating the presence of an unknown chemical. Beginning with 20 L (about 5 gallons) of urine collected from the children, he worked to isolate and identify the unknown substance. Over the next 3 months, he managed to purify the compound and work out its chemical structure. The chemical in the children's urine was a compound called phenylpyruvic acid. To confirm his finding, Fölling synthesized and purified phenylpyruvic acid from laboratory chemicals and showed that the compound he had isolated from the urine and his synthetic phenylpyruvic acid had the same physical and chemical properties.

Fölling proposed that the phenylpyruvic acid in the urine was produced by a metabolic disorder that prevented the breakdown of the amino acid phenylalanine. He further proposed that the accumulation of phenylpyruvic acid (the cause of the musty odor) in the bodies of the children was the cause of their mental retardation. To confirm this, he examined the urine of several hundred retarded patients and normal individuals. He found phenylpyruvic acid in the urine of eight retarded individuals but never in the urine of normal individuals. Less than 6 months after he began working on the problem, Fölling submitted a manuscript for publication that described the metabolic disorder he called phenylketonuria (PKU). His work helped establish the relationship among a gene, a gene product, and a phenotype. As a result of his pioneering work, PKU is now regarded as a prototype for metabolic genetic disorders.

As we discussed in the last chapter, DNA encodes information for the chemical structure of proteins. In this chapter, we will show how protein function is related to the phenotype and how mutations that alter or eliminate protein function can produce an abnormal phenotype we recognize as a genetic disorder.

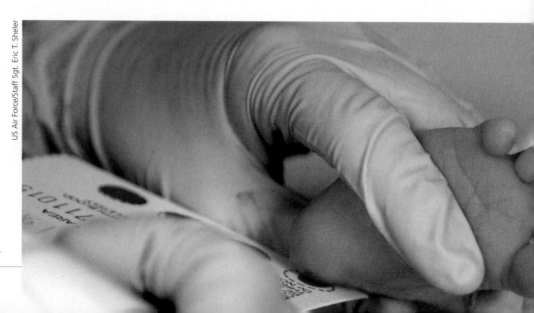

US Air Force/Staff Sgt. Eric T. Sheler

Blood samples being collected to test for phenylketonuria, a genetic disorder.

10.1 Proteins Are the Link Between Genes and the Phenotype

As outlined in Chapter 9, proteins are among the most important molecules in a cell. They are essential parts of all structures and biological processes carried out in cells. Proteins are essential parts of the cell membrane system and the internal skeleton. They are the glue that holds cells and tissues together. Proteins carry out biochemical reactions, destroy invading microorganisms, and act as hormones (Figure 10.1), receptors, and transport molecules. Even the replication of DNA and the expression of genes depend on the action of proteins. The many different functions of proteins are matched by their enormous diversity.

As we will see in this chapter, there is a direct link between a person's genotype, the proteins a person makes, and that person's phenotype. Mutations are heritable changes in a DNA sequence; they can occur in any cell of the body, including those that give rise to sperm and eggs. Mutated genes can produce either an abnormal, nonfunctional protein; a partially functional protein; or, in some cases, no protein at all. Other mutations can affect the timing and level of gene expression, while some may result in no phenotypic change and are essentially invisible. Mutations can also alter the amino acid sequence of a protein. Changes in proteins produced by mutations, in turn, produce changes in phenotype that range from insignificant to lethal. We will examine this link by using examples of mutations in proteins that act as enzymes and as transport

KEEP IN MIND AS YOU READ

- Phenotypes are the visible end product of a chain of events that starts with the gene, the mRNA, and the protein product.

- Phenylketonuria and several other metabolic disorders can be treated by dietary restrictions.

- Sickle cell anemia is caused by substitution of a single amino acid in beta globin.

- Small differences in proteins can have a large effect on our ability to taste, smell, and metabolize medicines.

Museo del Prado, Madrid, Spain/Giraudon/The Bridgeman Art Library

FIGURE 10.1 Portrait of a dwarf by Goya. Some genetic forms of dwarfism are caused by mutations in genes that encode proteins that act as growth hormones, receptors, and growth factors.

molecules. In addition, we will explore how variations in the proteins we make affect our reactions to drugs and environmental chemicals.

10.2 | Enzymes and Metabolic Pathways

Substrate The specific chemical compound that is acted on by an enzyme.

Product The specific chemical compound that is the result of enzymatic action. In biochemical pathways, a compound can serve as the product of one reaction and the substrate for the next reaction.

Metabolism The sum of all biochemical reactions by which cells convert and utilize energy.

Inborn error of metabolism The concept advanced by Archibald Garrod that many genetic traits result from alterations in biochemical pathways.

Many proteins in the body act as enzymes; these molecules facilitate biochemical reactions. They convert molecules known as **substrates** into **products** by catalyzing chemical reactions (Figure 10.2). In the cell, enzymatic reactions do not occur randomly; they are interconnected to form chains of reactions called metabolic pathways (Figure 10.3a); the sum of all the biochemical reactions going on in a cell is called **metabolism**.

In a metabolic pathway, the product of one reaction serves as the substrate for the next reaction (see Spotlight on Why Wrinkled Peas Are Wrinkled on page 222). If a mutation shuts down an enzyme that performs one step in a pathway, all the reactions beyond that point are also shut down, because there is no substrate available for reactions beyond the one that is blocked (Figure 10.3b). If one reaction in a pathway is shut down, it also results in the accumulation of products in the pathway leading up to the block.

In the early years of the twentieth century, Sir Archibald Garrod proposed that some human genetic disorders and abnormalities of metabolism are related. He studied a number of genetic disorders, including alkaptonuria (OMIM 203500), cystinuria (OMIM 220100), and pentosuria (OMIM 260800). He concluded that people with these disorders each carried a mutation that resulted in an enzyme defect that prevented them from carrying out a specific biochemical reaction. He called such disorders **inborn errors of metabolism**. From his work on families with these disorders, he concluded that those traits were inherited (Figure 10.4). His book, *Inborn Errors of Metabolism*, was a pioneering study in applying Mendelian genetics to humans and in understanding the relationship between genes and biochemical reactions.

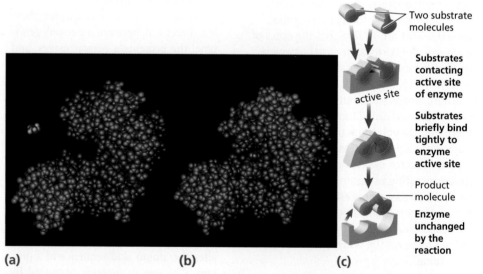

(a)   **(b)**   **(c)**

FIGURE 10.2 Each step in a metabolic pathway is a separate chemical reaction catalyzed by an enzyme in which a substrate is converted to a product. (a) The enzyme hexokinase (green) adds phosphate groups to the sugar glucose (gold). (b) When glucose enters the active site of the hexokinase molecule, the enzyme changes shape and closes around the glucose molecule, initiating the addition of a phosphate group to the glucose molecule. (c) A summary of an enzyme-catalyzed reaction in which two substrates (glucose and phosphate in the case of hexokinase) enter the active site of the enzyme are bound to the enzyme by a change in shape of the hexokinase, and undergo a chemical reaction that links them together. After the reaction is complete, the product (glucose phosphate) is released, and the enzyme resumes its previous shape and is ready to catalyze another reaction.

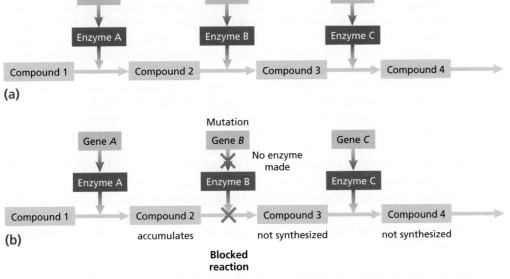

(a)

(b)

FIGURE 10.3 (a) The sequence of reactions in a metabolic pathway. In this pathway, compound 1 is present in the diet and is converted in the body to compound 2, which is then converted to compound 3. Finally, compound 3 is converted into compound 4. A specific enzyme catalyzes each of these reactions. Each enzyme is the product of a gene. (b) In this pathway, a mutation in gene B leads to the production of a defective protein that cannot function as an enzyme. As a result, compound 2 cannot be converted into compound 3. Because no compound 3 is made, compound 4 will not be produced, even though enzyme C is present. Compound 1 is supplied by the diet and is converted into compound 2, which accumulates because it cannot be metabolized. In this case, a genetic disorder might be associated with the accumulation of excess amounts of compound 2 or the lack of production of compound 4.

Mutations that eliminate or alter the activity of an enzyme can cause phenotypic effects in several ways. First, the substrate for the blocked reaction may build up and reach toxic levels, causing an abnormal phenotype. Second, the enzyme may control a reaction that produces a molecule needed for some cellular function. If this product is not made, a mutant phenotype may result. Mutations that affect enzymes can produce a wide range of phenotypes, ranging from inconsequential to lethal.

Essential amino acids Amino acids that cannot be synthesized in the body and must be supplied in the diet.

10.3 Phenylketonuria: A Mutation That Affects an Enzyme

To make the proteins required to maintain life, our cells need a supply of the 20 amino acids found in proteins. Our bodies can make just over half of those amino acids; however, the rest must be included in our diet. The amino acids we cannot synthesize are called **essential amino acids**. Humans require nine essential amino acids: histidine, isoleucine, leucine, lysine, methionine, phenylalanine, threonine, tryptophan, and valine. In other words, our diet has to be varied enough to provide 9 of the 20 amino acids from the foods we eat.

How is the metabolism of phenylalanine related to PKU?

Phenylalanine is an essential amino acid that is the starting point for a network of metabolic pathways. Here we will focus on what happens when the first step in this pathway is blocked by a mutation that prevents the conversion of phenylalanine into another amino acid, tyrosine. About two-thirds

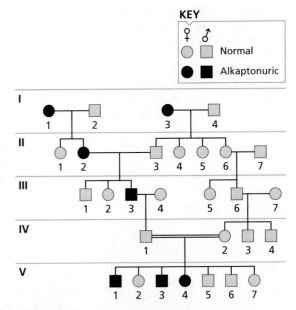

FIGURE 10.4 A pedigree of a family with alkaptonuria. Marriage between cousins (the double line) increases the chances that children will have an autosomal recessive disorder such as alkaptonuria.

Phenylketonuria (PKU) An autosomal recessive disorder of amino acid metabolism that results in mental retardation if untreated.

Spotlight on...

Why Wrinkled Peas Are Wrinkled

Wrinkled peas were one variety used by Mendel in his experiments. At the time, nothing was known about how peas became wrinkled or smooth. What Mendel knew is that a factor (which we now call a gene) controls seed shape and that it has two forms: a dominant one for smooth shape and a recessive one for wrinkled shape.

We now know how peas become wrinkled, providing a connection between a gene and its phenotype. In developing peas, starch is synthesized and stored as a food source. Starch is a large, branched molecule made up of sugar molecules linked together in long chains. The ability to form branches in these chains is controlled by an enzyme.

Normally, starch molecules are highly branched structures. In peas with the wrinkled genotype, the branching gene is inactive. Thus, the developing pea converts sugar into starch very slowly using other enzymes, and excess sugar accumulates. The excess sugar causes the pea to take up large amounts of water by osmosis, and the seed and its outer shell become swollen. In a final stage of development, water is lost from the seed. In homozygous wrinkled peas, more water is lost than in the smooth seeds, causing the outer shell of the pea to shrink and become wrinkled.

Mendel's contribution was to show that a specific gene controlled a trait and that a particular gene could have different forms. Now we know that genes exert their effect on phenotype through the production of a gene product.

of the phenylalanine we eat is converted to tyrosine by this reaction; the rest is incorporated into proteins. A mutation that prevents the conversion of phenylalanine to tyrosine results in a genetic disorder called **phenylketonuria (PKU)** (OMIM 261600). This is the disorder described at the beginning of the chapter that affected Liv and Dag Egeland. About 1 in every 12,000 newborns has PKU. In almost all cases, PKU is caused by a mutation in a gene for the enzyme phenylalanine hydroxylase (PAH), which normally converts phenylalanine to tyrosine.

In people with PKU, phenylalanine from protein-containing foods cannot be converted to tyrosine and builds up to high levels in the blood and tissues of the body (Figure 10.5). If untreated, newborns with high levels of phenylalanine become severely mentally retarded, have reflexes that cause their arms and legs to move in a jerky fashion, develop epileptic seizures, and never acquire language skills. Because the skin pigment melanin is also a product of the blocked metabolic pathway (Figure 10.5), most people with PKU usually have lighter hair and skin color than their siblings or other family members.

As you can see from Figure 10.5, there are two pathways leading from phenylalanine. Blockage of the pathway to tyrosine overloads the second pathway, producing high levels of other products, including phenylpyruvate, the molecule isolated from Liv and Dag's urine by Dr. Fölling.

How does the buildup of phenylalanine produce mental retardation?

Part of the answer is that high levels of phenylalanine, phenylpyruvic acid, and other by-products accumulate and cause damage to the developing brain in newborns. It's not

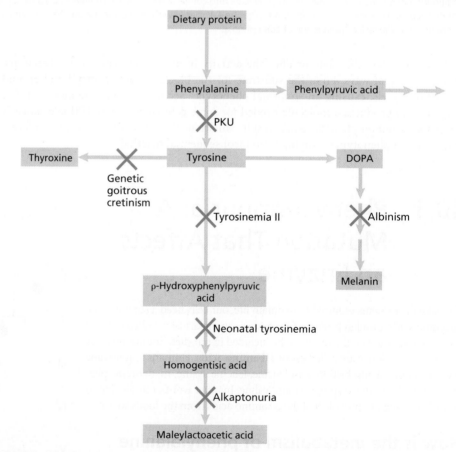

FIGURE 10.5 The metabolic pathway that leads from the essential amino acid phenylalanine. Normally, phenylalanine is converted to tyrosine, and from there into many other compounds. A metabolic block caused by a mutation in the gene that encodes the enzyme phenylalanine hydroxylase prevents the conversion of phenylalanine to tyrosine. People homozygous for this mutation have the phenotype of phenylketonuria (PKU). The diagram also shows other genetic disorders produced by mutations in genes that encode other enzymes in this pathway.

clear whether the brain damage results from too much phenylalanine or lowered levels of other amino acids. It may also be that breakdown products of phenylalanine accumulate in the nerve cells and cause the damage. The result, however, is brain damage, mental retardation, and the other neurological symptoms that result in the phenotype of PKU.

How effective is testing for PKU in newborns?

In newborns, the first sign of PKU is abnormally high levels of phenylalanine in the blood and urine. Since the 1960s, newborns in the United States have been routinely tested for PKU by analyzing blood or urine for phenylalanine levels (Figure 10.6). By the mid-1970s, many countries were testing newborns for PKU (see Chapter 16 for a discussion of genetic testing).

To date, more than 100 million infants have been screened in the United States, and over 10,000 cases of PKU have been detected and treated. All states require screening of newborns for PKU, so the number of untreated cases is very low. Screening and treatment allows PKU homozygotes to lead essentially normal lives.

PKU can be treated with a diet low in phenylalanine.

Most newborns with PKU have heterozygous parents. These infants developed normally before birth because the heterozygous mother has enough of the PAH enzyme in her body to break down the excess phenylalanine that accumulates in the fetus during prenatal development. After the children are born, this safeguard is no longer present, and newborns who are homozygous for PKU have neurological damage and become retarded when fed a normal diet. Why? Because protein sources in the normal diet such as meat, fish, and cheese contain phenylalanine, which PKU infants cannot metabolize. To avoid the consequences of PKU, dietary treatment must be started in the first month after birth. After 30 days, the brain is damaged beyond repair and treatment is less effective.

By carefully managing the amount of phenylalanine in the diet (see Exploring Genetics: Dietary Management and Metabolic Disorders on page 225), affected newborns who were diagnosed early have normal brain development and normal intelligence. However, managing PKU by controlling dietary intake is both difficult and expensive.

KEEP IN MIND

Phenylketonuria and several other metabolic disorders can be treated by dietary restrictions.

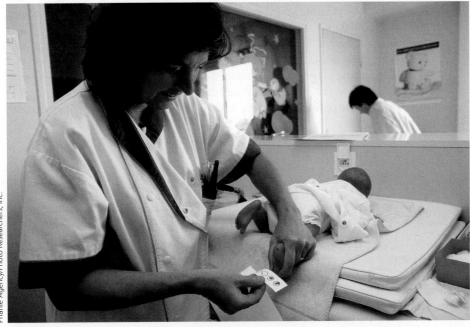

FIGURE 10.6 A drop of a newborn's blood collected from the heel will be used to test for phenylketonuria (PKU).

Phanie Agency/Photo Researchers, Inc.

One major problem is that phenylalanine is present in many protein sources, and it is impossible and in fact undesirable to eliminate all protein from the diet. Remember that phenylalanine is an essential amino acid, and there must be enough phenylalanine in the body to make proteins for normal development of the nervous system but phenylalanine levels must be low enough to prevent mental retardation.

Dietary treatment of PKU uses a two-pronged approach: restriction of protein intake and use of a phenylalanine-free amino acid supplement. The protein restriction means that in spite of being surrounded by normal foods and often a lot of peer pressure, children with PKU cannot eat hamburgers, chicken nuggets, pizza, ice cream, and many other favorite childhood foods. To get the amino acids needed to make proteins, they must drink a foul-tasting dietary supplement containing a synthetic mixture of amino acids (with very low levels of phenylalanine) along with vitamins and minerals.

How long must a PKU diet be maintained?

Up until the early 1980s, some clinicians felt that because brain development is about 95% complete by the age of 10, the diet could be discontinued after that age with little or no effects. Others proposed that the diet be continued until about the age of 14, when brain development is almost finished, and that higher levels of phenylalanine in the blood after this age posed little or no risk.

A study completed in 1983 and a follow-up study in 1998 each concluded that those who stayed on the diet as adults had lower rates of mental disorders, enhanced intellectual abilities, higher achievement test scores, and fewer behavioral problems than those who discontinued the diet. Based on these conclusions, it is now recommended that individuals with PKU stay on the protein-restricted diet for life and use amino acid supplements as part of their diet.

What happens when women with PKU have children?

As PKU children treated with diet therapy have matured and reached reproductive age, the question has arisen: Can a woman homozygous for the recessive PKU alleles have an unaffected child? The answer seems straightforward. If she has a child with a man who carries two normal (dominant) alleles, the child will be heterozygous and should be unaffected. Unfortunately, women homozygous for the mutant PKU allele who eat a normal diet during pregnancy will have mentally retarded children regardless of the child's genotype. Why? Because a pregnant PKU woman who eats a normal diet accumulates high levels of phenylalanine in her blood. This excess phenylalanine has a minimal effect on the woman because her nervous system is already developed. However, the high levels of phenylalanine cross the placenta and damage the developing nervous system of the fetus no matter what its genotype is.

To avoid this outcome, it is recommended that women with PKU stay on a low-phenylalanine diet throughout their reproductive years or return to the diet for several months before becoming pregnant and stay on it throughout pregnancy. In addition, if they do not wish to follow the protein-restricted diet, PKU women can select from a variety of other reproductive options, including *in vitro* fertilization and the use of surrogate mothers (see Chapter 16).

10.4   Other Metabolic Disorders in the Phenylalanine Pathway

The mutation that blocks the conversion of phenylalanine to tyrosine is not the only identified mutation in this metabolic pathway. Several other rare genetic disorders are caused by mutations that block enzymatic reactions leading from phenylalanine. For example, one pathway leads to synthesis of the thyroid hormones thyroxine and triiodothyronine.

Dietary Management and Metabolic Disorders

In several metabolic diseases, a restrictive diet is used to prevent full expression of the mutant phenotype. These diets replace metabolites that are not produced or prevent the buildup of toxic compounds caused by mutations. In addition to PKU, diet is used with varying degrees of success to treat several metabolic conditions, including galactosemia, tyrosinemia, homocystinuria, and maple syrup urine disease.

The diet for each disorder is usually available in two versions: one for infants with low levels of the restricted component and one for older children and adults that usually contains higher levels of the restricted compound and other nutrients. For management of PKU, the diet has two parts: dietary restriction of protein-containing food and the use of a phenylalanine-free amino acid supplement, which also contains fats, carbohydrates, vitamins, and mineral supplements. In one popular formula, casein (a protein extracted from milk) is enzymatically digested into individual amino acids. The amino acid mixture is treated to remove phenylalanine. Then sources of fat, carbohydrates, vitamins, and minerals are added, and the resulting powder is packaged. Individuals with PKU follow a protein-restricted diet that they supplement by mixing the powder with water at each meal as a source of amino acids. Protein restriction means that they cannot eat any meat (hamburgers, chicken, etc.) or any dairy products (milk, ice cream, etc.), or dairy-containing products (pizza, etc.). A typical menu for a school-aged child is shown below.

Breakfast
2 to 3 cups dry rice cereal
1 to 2 bananas
6 oz. formula

Lunch
1 to 2 cans vegetable soup
3 crackers
1 cup fruit cocktail
4 oz. formula

Dinner
2 cups low-protein noodles
1 to 2 cups meatless spaghetti sauce
1 cup of salad (lettuce)
French dressing
4 oz. formula

Snack
1 to 2 cups popcorn
1 tablespoon margarine

A mutation in this pathway causes the autosomal recessively inherited disorder called genetic goitrous cretinism (Figure 10.5). Newborn homozygotes are unaffected because, during prenatal development, thyroid hormones from the mother cross the placenta and promote normal growth. In the weeks after birth, physical development is slow, mental retardation begins, and the thyroid gland greatly enlarges. In this case, the phenotype is caused by the failure to synthesize an essential product (thyroid hormone), whereas in PKU, the problem is caused by accumulation of a dietary amino acid and its breakdown products. If diagnosed early, infants with goitrous cretinism can be treated with thyroid hormone and will develop normally.

In another pathway in this network, a mutation leads to the buildup of homogentisic acid and causes alkaptonuria, an autosomal recessive condition. This is the disorder first investigated by Garrod at the beginning of the twentieth century.

10.5 Genes and Enzymes of Carbohydrate Metabolism

Mutations in genes that encode enzymes are not limited to amino acid metabolic pathways. Other pathways, including those of lipid metabolism, nucleic acid metabolism, and carbohydrate metabolism, are also affected. We will briefly discuss some mutations in carbohydrate metabolism.

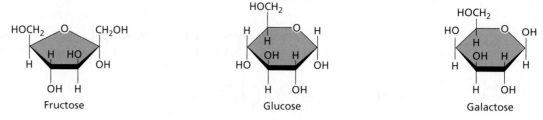

(a) Monosaccharides

Fructose

Glucose

Galactose

(b) Disaccharides

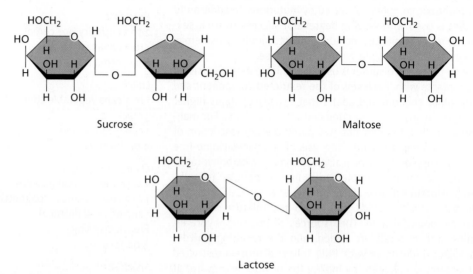

Sucrose

Maltose

Lactose

FIGURE 10.7 (a) The chemical structure of three common monosaccharides (simple sugars). (b) The structures of three disaccharides, molecules composed of two simple sugars.

Carbohydrates are organic molecules that include simple sugars and more complex molecules such as starches, glycogens, and celluloses. These complex carbohydrates are polymers made up of long chains of simple sugars. The simplest sugars are called monosaccharides (Figure 10.7a). Fructose, glucose, and galactose are all monosaccharides and are important energy sources for the cell. Two monosaccharides linked together form a disaccharide (Figure 10.7b). Some common disaccharides are sucrose (a glucose and a fructose molecule—the sugar you buy at the store), maltose (two glucose molecules—a sugar used in brewing beer), and lactose (a glucose and a galactose molecule—the sugar found in milk). Long chains of many sugars linked together form polysaccharides; these include the starches found in potatoes and wheat flour, as well as glycogen, found in animal muscles.

Mutations that cause metabolic blocks in the pathways that synthesize and break down sugars and complex carbohydrates can produce a wide range of phenotypic effects that range from normal to fatal. Some genetic disorders associated with the metabolism of the polysaccharide glycogen are listed in Table 10.1. We will examine two examples of how mutant enzymes of carbohydrate metabolism produce genetic disorders.

Galactosemia is caused by an enzyme deficiency.

Galactosemia A heritable trait associated with the inability to metabolize the sugar galactose. If it is left untreated, high levels of galactose-1-phosphate accumulate, causing cataracts and mental retardation.

Galactosemia (OMIM 230400) is an autosomal recessively inherited disorder caused by the inability to break down galactose, one of the two sugars found in lactose (Figure 10.8). Galactosemia occurs with a frequency of 1 in 57,000 births and is caused by mutations in the gene that encodes the enzyme galactose-1-phosphate uridyl transferase (GALT). When GALT is missing or inactive, galactose metabolism is disrupted

Table 10.1 Some Inherited Diseases of Glycogen Metabolism

Type	Disease	Metabolic Defect	Inheritance	Phenotype	OMIM Number
I	Glycogen storage disease, Von Gierke disease	Glucose-6-phosphatase deficiency	Autosomal recessive	Severe enlargement of liver, often recognized in second or third decade of life; may cause death due to renal disease	232200
II	Pompe disease	Lysosomal glucosidase deficiency	Autosomal recessive	Accumulation of membrane-bound glycogen deposits; first lysosomal disease known; childhood form leads to early death	232300
III	Forbes disease, Cori disease	Amylo-1,6-glucosidase deficiency	Autosomal recessive	Accumulation of glycogen in muscle, liver; mild enlargement of liver, with some kidney problems	232400
IV	Amylopectinosis, Andersen disease	Amylo-1,4-transglucosidase deficiency	Autosomal recessive	Cirrhosis of liver; eventual liver failure, death	232500

(step 3 in Figure 10.8), and a compound called galactose-1-phosphate accumulates and reaches toxic levels in the body.

As with PKU, homozygous recessive individuals usually have a heterozygous mother and are unaffected before birth but begin showing symptoms a few days later. Those symptoms include dehydration and loss of appetite; later, the infants develop jaundice, cataracts, and mental retardation. In some cases, the condition is progressive and fatal. Severely affected infants die within a few months, but mild cases may remain undiagnosed for many years. A galactose-free diet and the use of galactose- and lactose-free milk substitutes and foods lead to a reversal of symptoms. However, unless treatment is started within a few days after birth, mental retardation cannot be prevented.

Unlike PKU, a lifetime dietary treatment for galactosemia does not prevent long-term complications. Many affected individuals on a galactose-restricted diet develop problems in adulthood. Some have difficulties with balance or impaired motor skills, including problems with handwriting. It is not clear whether this is caused by low levels of damage to the nervous system that began during fetal development or whether dietary treatment is only partly effective.

Like the ABO blood type, galactosemia is an example of a multiple-allele system. In addition to the normal allele, G, and the recessive mutant allele, g, a third allele known as G^D (the Duarte allele—named after Duarte, California, the city in which it was discovered) has been found. The existence of three alleles produces six possible genotypic combinations and enzymatic activities that range from 100% to 0% (Table 10.2). This disease can be detected in newborns, and screening programs in many states test all newborns for galactosemia.

Lactose intolerance is a genetic variation.

Human milk is about 7% lactose, a major energy source for a nursing infant. The first step in breaking down lactose splits the molecule into glucose and galactose (step 1 in Figure 10.8). This step is controlled by the enzyme lactase. In most human populations, lactase levels are highest soon after birth and drop off in middle-to-late childhood. As a result, children and adults have less than 10% of the lactase activity found in infants. Those with low lactase levels cannot digest the lactose in milk and other dairy products. If they eat lactose-containing foods, they develop bloating, cramps, gas, and diarrhea, a condition called lactose intolerance. Most lactase-deficient adults learn to avoid dairy products. Lactose intolerance is not considered a genetic disorder but only a variation in gene expression. Most human populations have low adult lactase levels; however, in some populations, lactase activity persists. This trait, inherited in an autosomal dominant fashion, gives these individuals the ability to digest lactose throughout life. In populations around the world, the frequency of lactose intolerance varies from 0% to 100%. In Chapter 19, we will explore the role of natural selection in controlling the frequency of lactose intolerance in human populations.

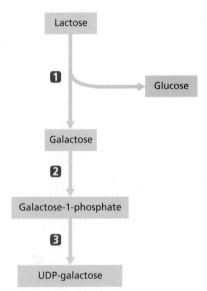

FIGURE 10.8 The metabolic pathway that begins with lactose, the main sugar in milk. Lactose (step 1) is broken down into glucose and galactose. In step 2, galactose is converted to galactose-1-phosphate. In the genetic disorder galactosemia, a mutation prevents the conversion of galactose-1-phosphate into UDP-galactose (step 3). As a result, the concentration of galactose-1-phosphate rises in the blood, causing the symptoms of galactosemia, which include mental retardation and blindness.

Table 10.2 Multiple Alleles of Galactosemia

Genotype	Enzyme Activity (%)	Phenotype
G^+/G^+	100	Normal
G^+/G^D	75	Normal
G^D/G^D	50	Normal
G^+/g	50	Normal
G^D/g	25	Borderline
g/g	0	Galactosemia

10.6 Defects in Transport Proteins: Hemoglobin

Hemoglobin, an iron-containing protein in red blood cells, transports oxygen from the lungs to the cells of the body. The hemoglobin molecule occupies a central position in human genetics. The study of hemoglobin variants led to an understanding of the molecular relationship among genes, proteins, and human disease in several ways:

- The discovery of variations in the amino acid composition of hemoglobin was the first example of inherited variations in protein structure.
- The altered hemoglobin in sickle cell anemia provided the first direct proof that mutations result in a change in the amino acid sequence of proteins.
- The mutation in sickle cell anemia provided evidence that a change in a single nucleotide is sufficient to cause a genetic disorder.
- The molecular organization of the globin gene clusters has helped scientists understand how genes evolve and how gene expression is regulated.

Heritable defects in globin structure or synthesis are well understood at the molecular level and are truly "molecular diseases," as Linus Pauling called them (See Exploring Genetics: The First Molecular Disease on page 233). In this section, we consider the structure of the hemoglobin molecule, the organization of the globin genes, and some genetic disorders related to globin structure and synthesis.

Hemoglobin is a composite molecule composed of four protein molecules called globins (Figure 10.9a). Each globin contains a heme group containing an iron atom (Figure 10.9b). In the lungs, oxygen enters red blood cells and binds to the iron for transport to cells throughout the body. Each red blood cell contains about 280 million molecules of hemoglobin, and there are between 4 and 6×10^{12} red blood cells in each liter of blood. About 2×10^6 red blood cells are made in the bone marrow each second, requiring the synthesis of about 5.6×10^{14} new hemoglobin molecules per second, making hemoglobin synthesis one of the body's major metabolic processes.

The hemoglobin shown in Figure 10.9a is the adult form (called HbA) and is composed of two alpha globins and two beta globins. Members of the alpha globin gene family are

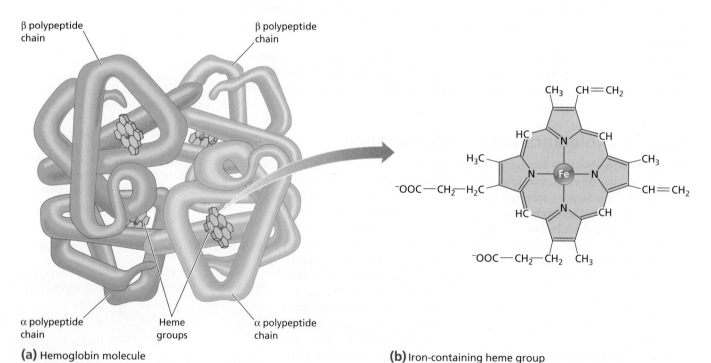

β polypeptide chain

β polypeptide chain

α polypeptide chain

Heme groups

α polypeptide chain

(a) Hemoglobin molecule

(b) Iron-containing heme group

FIGURE 10.9 (a) A hemoglobin molecule contains four proteins: two alpha globins and two beta globins. Within the folds of each globin molecule is an iron-containing heme group. (b) A heme group is a flat molecule that contains an iron atom, to which oxygen binds in the lungs for transport to the cells of the body.

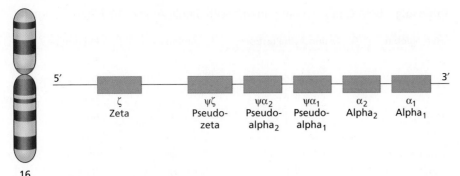

FIGURE 10.10 The organization and chromosomal location of the alpha globin gene cluster. Each copy of chromosome 16 contains two functional copies of the alpha globin gene (alpha$_1$ and alpha$_2$), two nonfunctional versions of the alpha globin gene (pseudo-alpha$_1$ and pseudo-alpha$_2$), and a zeta globin gene, which is expressed only in early embryonic development. A nonfunctional copy of the zeta gene (pseudo-zeta) is also present in the cluster.

grouped into a cluster on chromosome 16 (Figure 10.10). The beta globin gene family is located on chromosome 11 (Figure 10.11). We produce different types of hemoglobin throughout life; all contain different combinations of two alpha and two beta globins. Embryonic hemoglobin contains two zeta globins (alpha globins) and two epsilon globins (beta globins). Fetal hemoglobin consist of two alpha globins and two gamma (beta globins). After birth, HbA, with two alpha and two beta chains is synthesized (Figure 10.12).

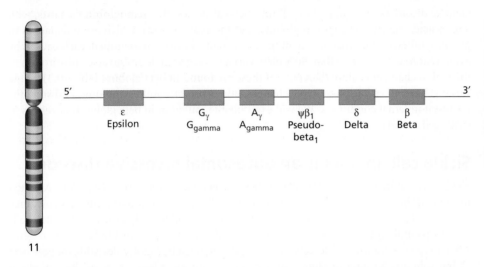

FIGURE 10.11 The organization and chromosomal location of the beta globin gene cluster. Each copy of chromosome 11 contains several versions of the beta globin gene, the epsilon globin gene (which is expressed only during embryogenesis), two gamma globin genes (G$_\gamma$ and A$_\gamma$) that are active during fetal development, and the delta globin and beta globin genes, which turn on at birth and are expressed throughout life.

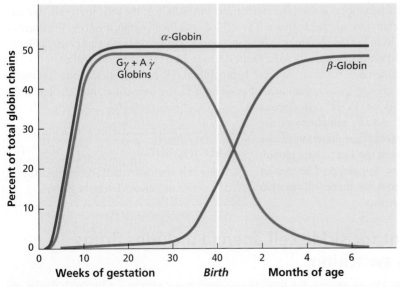

FIGURE 10.12 Patterns of globin gene expression at stages in the human life cycle. The alpha globin genes (blue line) are switched on early in embryonic development and continue to be expressed throughout life. The gamma globin genes (green line) are most active during fetal development and are turned off just before birth. The beta globin genes (red line) are expressed just before birth and are active throughout life.

Table 10.3 Beta Globin Chain Variants with Single Amino Acid Substitutions

Hemoglobin	Amino Acid Position	Amino Acid	Phenotype
HbA1	6	glu	Normal
S	6	val	Sickle cell anemia
C	6	lys	Hemoglobin C disease
HbA1	7	glu	Normal
Siriraj	7	lys	Normal
San Jose	7	gly	Normal
HbA1	58	tyr	Normal
Hb M Boston	58	his	Reduced O_2 affinity
HbA1	145	cys	Normal
Bethesda	145	his	Increased O_2 affinity
Fort Gordon	145	asp	Increased O_2 affinity

Hemoglobin variants Alpha and beta globins with variant amino acid sequences.

Thalassemias Disorders associated with an imbalance in the production of alpha or beta globin.

Hemoglobin disorders.

Genetic disorders of hemoglobin fall into two categories: the **hemoglobin variants**, with amino acid sequence changes in globins, and the **thalassemias**, which are imbalances in globin synthesis. More than 400 hemoglobin variants have been identified, each caused by a different mutation. More than 90% of all variants have a single amino acid substitution in the globin chain, and more than 60% of these are found in beta globin (Table 10.3). Some variants have no visible phenotypic changes, others produce mild symptoms, and still others result in lethal conditions. We will examine the mutation in beta globin that results in sickle cell anemia.

Sickle cell anemia is an autosomal recessive disorder.

Sickle cell anemia (OMIM 141900) is inherited as an autosomal recessive trait. Affected individuals have a wide range of symptoms, including weakness, abdominal pain, kidney failure, and heart failure (Active Figure 10.13), which lead to early death if left untreated.

This painful and disabling condition is caused by a mutation in the beta globin gene. After oxygen is transferred to cells in the body, hemoglobin molecules with mutant beta globin subunits become insoluble, stick together, and form long fibers in the cytoplasm (Figure 10.14). These fibers distort and harden the membrane of the red blood cell, twisting the cell into a characteristic sickle shape. The sickled cells break easily, lowering the number of red blood cells in the circulation, resulting in anemia. The sickled cells also clog capillaries and small blood vessels, producing pain, tissue damage, heart attacks, and strokes.

The only difference between normal hemoglobin (HbA) and sickle cell hemoglobin (HbS) is a change in an amino acid at position 6 in the beta chain; this change is the molecular basis of sickle cell anemia (Figure 10.15). All the symptoms of the disease derive from this alteration of one amino acid out of the 146 in beta globin. See Spotlight on Population Genetics of Sickle Cell Genes for more information on sickle cell anemia.

KEEP IN MIND

Sickle cell anemia is caused by substitution of a single amino acid in beta globin.

Treatment for sickle cell anemia includes drugs for gene switching.

If untreated, sickle cell anemia is a fatal disease, and many affected individuals die by age 2 years. Even with an understanding of the molecular basis of the disease, treatments are only partially successful in relieving the symptoms.

In the short term, the symptoms of sickle cell anemia can be treated with medications that include antibiotics and pain relievers, and in some cases, blood transfusions and

Spotlight on . . .

Population Genetics of Sickle Cell Genes

Sickle cell anemia is a genetic disorder caused by a mutation in the gene for beta globin, a component of hemoglobin. This mutation changes normal hemoglobin (HbA) to a mutant form called sickle cell hemoglobin (HbS). Individuals homozygous for sickle cell anemia who do not receive treatment often die prematurely.

The high frequency of heterozygotes in West Africa indicates that they have an advantage over homozygotes in certain environments. West Africa is an area where malaria is widespread, and this disease is a powerful force in changing genotype frequencies because resistance to malaria is about 25% greater in heterozygotes than in those with the normal homozygous genotype.

In the United States, the gene for hemoglobin S is decreasing as a result of early screening and testing of those at risk and because malaria is not present to enhance the survival of heterozygotes. However, if the Earth's atmosphere continues to warm, malaria may reemerge in the United States and become a force in changing genotype frequencies.

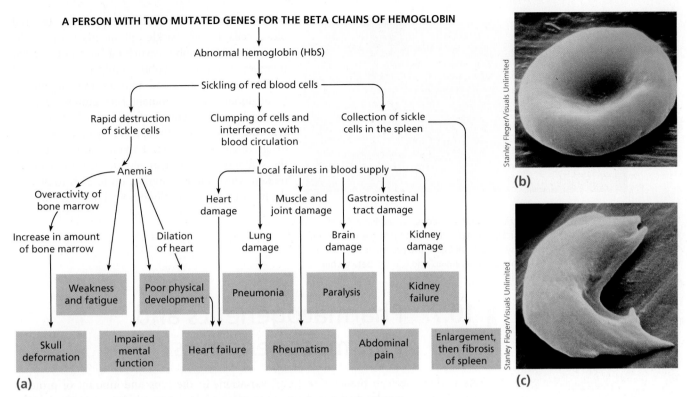

A PERSON WITH TWO MUTATED GENES FOR THE BETA CHAINS OF HEMOGLOBIN

Abnormal hemoglobin (HbS)

Sickling of red blood cells

Rapid destruction of sickle cells

Clumping of cells and interference with blood circulation

Collection of sickle cells in the spleen

Anemia

Local failures in blood supply

Overactivity of bone marrow

Heart damage

Muscle and joint damage

Gastrointestinal tract damage

Increase in amount of bone marrow

Dilation of heart

Lung damage

Brain damage

Kidney damage

Weakness and fatigue

Poor physical development

Pneumonia

Paralysis

Kidney failure

Skull deformation

Impaired mental function

Heart failure

Rheumatism

Abdominal pain

Enlargement, then fibrosis of spleen

(a)

(b)

Stanley Fleger/Visuals Unlimited

(c)

Stanley Fleger/Visuals Unlimited

ACTIVE FIGURE 10.13 (a) The cascade of phenotypic effects resulting from the mutation that causes sickle cell anemia. Homozygotes have effects at the molecular, cellular, and organ levels, all of which result from a mutation in the beta globin gene that substitutes one amino acid for another in the beta globin protein. (b) Normally shaped red blood cells. (c) Sickled red cells from people with sickle cell anemia.

CENGAGENOW Learn more about sickle cell anemia by viewing the animation by logging on to *login.cengage.com/sso* and visiting CengageNOW's Study Tools.

supplemental oxygen are used. The discovery that certain anticancer drugs change patterns of gene expression has created a new and effective treatment for sickle cell anemia.

The drug hydroxyurea shuts off cell division and is normally used to treat cancer patients. As a side effect, it was discovered that treatment with hydroxyurea leads to elevated levels of fetal hemoglobin. Treatment with hydroxyurea reactivates the gamma genes

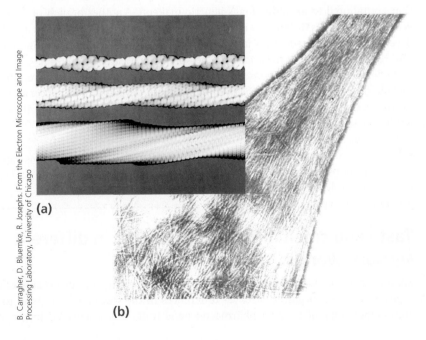

B. Carragher, D. Bluemke, R. Josephs. From the Electron Microscope and Image Processing Laboratory, University of Chicago

(a)

(b)

FIGURE 10.14 (a) A computer-generated image of the stages in the polymerization of hemoglobin containing the mutant form of beta globin. *Upper:* A pair of intertwined fibers formed from hemoglobin molecules. Each "bead" in the fiber represents a single hemoglobin molecule. *Middle:* Seven pairs of polymers form a larger fiber that distorts the membrane of red blood cells. *Lower:* A large fiber composed of many smaller fibers. (b) An electron micrograph of a ruptured sickled red blood cell, showing the internal fibers of polymerized hemoglobin.

Normal hemoglobin A (HbA)	1	2	3	4	5	6	7	8
DNA	CAC	GTG	GAC	TGA	GGA	CTC	CTC	TTC
mRNA	GUG	CAC	CUG	ACU	CCU	GAG	GAG	AAG
Amino acid	Val	His	Leu	Thr	Pro	Glu	Glu	Lys

Hemoglobin in sickle cell anemia (HbS)	1	2	3	4	5	6	7	8
DNA	CAC	GTG	GAC	TGA	GGA	CAC	CTC	TTC
mRNA	GUG	CAC	CUG	ACU	CCU	GUG	GAG	AAG
Amino acid	Val	His	Leu	Thr	Pro	Val	Glu	Lys

FIGURE 10.15 The first eight DNA triplets, mRNA codons, and amino acids of normal adult beta globin (HbA) and sickle cell beta globin (HbS). At the DNA level, the only difference is a T → A substitution in triplet 6. At the protein level, this causes substitution of valine for glutamic acid at amino acid 6 in the beta globin protein, which contains 146 amino acids.

and makes fetal hemoglobin reappear in the red blood cells. Because sickle cell anemia is caused by a defect in beta globin, switching on a non-mutant member of the beta globin family raises the level of fetal hemoglobin and reduces the amount of hemoglobin carrying mutant beta globins. This in turn reduces the number of sickled red blood cells, relieving many of the disorder's symptoms. Other drugs, including sodium butyrate, also switch on the synthesis of fetal hemoglobin. In some patients treated with sodium butyrate, up to 25% to 30% of the hemoglobin in the blood is fetal hemoglobin.

10.7 Pharmacogenetics and Pharmacogenomics

As we have seen in previous sections, variations in the type and amount of proteins produced by an individual can result in genetic disorders of metabolism. We are also discovering that protein variations affect the way individuals metabolize and react to foods, anesthetics, prescription drugs, and chemicals in the environment.

Observations from everyday life provide examples of this variation. For example, why is it that some people smoke cigarettes for years and never develop lung cancer? The answer may be in their genes. Alleles of genes for a family of enzymes called the P450 enzymes control the metabolism of carcinogens in cigarette smoke. Certain combinations of these alleles rapidly inactivate the carcinogens, offering protection against lung cancer.

Like certain metabolic disorders, phenotypic differences in drug metabolism or reaction to environmental chemicals appear only when an individual is exposed to the drug or chemical. These reactions are often the result of heritable variations in proteins and can be dominant or recessive traits. A branch of genetics known as **pharmacogenetics** studies the genetic variations that underlie differences in the body's response to drugs, while the related field of pharmacogenomics is focused on the development of drugs that are tailored to an individual's genetic makeup. Another branch of genetics called **ecogenetics** studies differences in individual reactions to environmental agents. We will describe some of the advances in pharmacogenetics and then discuss how ecogenetics is also revealing the extent to which each person is genetically unique.

Differences in drug responses can produce a range of phenotypic responses. These include: drug resistance, toxic sensitivity to low doses, development of cancer after prolonged exposure, or an unexpected reaction to a combination of drugs. Some of these responses are harmless, whereas others can be life threatening. In this section, we consider how exposure to drugs produces a wide range of phenotypic responses and describe the role of specific proteins in generating these phenotypes (if known).

Pharmacogenetics A branch of genetics concerned with the identification of protein variants that underlie differences in the response to drugs.

Ecogenetics A branch of genetics that studies genetic traits related to the response to environmental substances.

KEEP IN MIND

Small differences in proteins can have a large effect on our ability to taste, smell, and metabolize medicines.

Taste and smell differences: we live in different sensory worlds.

Shortly after Garrod proposed that we are all biochemically unique because of our genotypes, researchers began to discover differences in the way people respond to chemical compounds. One of the first pharmacogenetic traits was discovered in the 1930s as a

The First Molecular Disease

Linus Pauling, a two-time Nobel Prize winner, once recalled that when he first heard a description of how red blood cells change shape in sickle cell anemia, he had the idea that sickle cell anemia is really a molecular disease. He thought the disorder must involve an abnormality of the hemoglobin molecule caused by a mutated gene.

Early in 1949, Pauling and his student Harvey Itano began a series of experiments to determine whether there is a difference between normal hemoglobin and sickle cell hemoglobin.

They obtained blood samples from people with sickle cell anemia and from unaffected individuals. They extracted hemoglobin from the blood samples, placed it in a tube with an electrode at each end, and passed an electrical current through the tube. Hemoglobin from individuals with sickle cell anemia migrated toward the cathode, indicating that it had a positive electrical charge. Samples of normal hemoglobin migrated in the opposite direction (toward the anode), indicating that this hemoglobin had a net negative electrical charge. In the same year, James Neel, working with sickle cell patients in the Detroit area, demonstrated that sickle cell anemia is a genetic trait, inherited as an autosomal recessive condition.

Pauling and his colleagues published a paper on their results and incorporated Neel's findings into their discussion. They concluded that a mutant gene involved in the synthesis of hemoglobin causes sickle cell anemia. The idea that a genetic disorder can be caused by a defect in a single molecule was revolutionary. Pauling's idea about a molecular disease helped reignite interest in human biochemical genetics and played a key role in our understanding of the molecular nature of mutations.

After Watson and Crick worked out the structure of DNA, Crick was eager to show that mutant genes produce mutant proteins whose amino acid sequences differ from those of the normal protein. He persuaded Vernon Ingram to look for such differences. Because of Pauling's work, Ingram settled on hemoglobin as the protein he would analyze. He cut hemoglobin into pieces with the enzyme trypsin and separated the 30 resulting fragments. He noticed that normal hemoglobin and sickle cell hemoglobin differed in only one fragment—a peptide about 10 amino acids long. Ingram then worked out the amino acid sequence of that fragment in normal and sickle cell hemoglobin. In 1956, he reported that there is a difference of only one amino acid (glutamic acid in normal hemoglobin and valine in sickle cell hemoglobin) between the two proteins. This finding confirmed that there is a relationship between a mutant gene and a mutant gene product and established a way of thinking about mutations and disease that changed human genetics.

by-product of work on artificial sweeteners, when workers at DuPont discovered that some people cannot taste the chemical phenylthiocarbamide (PTC), but others find it very bitter tasting (OMIM 171200). Shortly thereafter, it was found that the ability to taste PTC depends on a single pair of alleles. The genotypes *TT* and *Tt* represent tasters, whereas those with the genotype *tt* are nontasters. The ability to taste PTC varies from population to population (Figure 10.16). In the United States, about 30% of adult whites are nontasters, whereas only about 3% of U.S. blacks are nontasters.

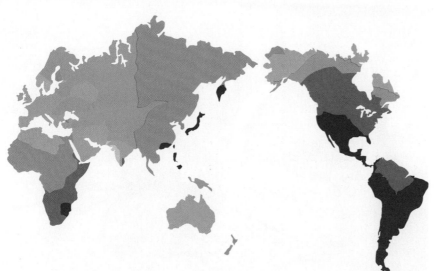

FIGURE 10.16 The distribution of PTC tasting ability in world populations. The lightest color represents populations with few tasters (about 5% of the population). The darkest color represents populations with the highest percentage of tasters (up to 85%).

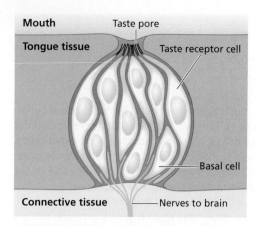

FIGURE 10.17 Taste buds are mainly located along the edges of moundlike bumps on the top surface of the tongue. Each receptor responds preferentially to one of the primary tastes: salty, sour, sweet, bitter, and umami. Another taste sensation, fatty, has been proposed as an addition to this list.

Mouth
Taste pore
Tongue tissue
Taste receptor cell
Basal cell
Connective tissue
Nerves to brain

In humans, taste is mediated by clusters of receptor cells in the taste buds (Figure 10.17). The ability to taste PTC, a related compound called PROP, and other bitter chemicals is controlled by proteins on the surface of the receptor cells. The gene for tasting PTC (OMIM 607751) and related compounds, located on chromosome 7, has two common alleles that encode variants of a receptor protein. Variations in the shape of the receptor protein determine how well PTC and related compounds bind to the receptor and trigger a nerve impulse that the brain interprets as taste.

Linda Bartoshuk at Yale University and other researchers found that tasters (*TT* and *Tt* individuals) can be divided into two groups: tasters and supertasters. The supertasters (*TT*) have intensely negative reactions to PTC and related compounds. Her surveys showed that about 25% of people are nontasters, 50% are tasters, and 25% are supertasters. Further work showed that supertasters can have up to 10 times as many tastebuds as nontasters and are very sensitive to tastes of all kinds.

How does such a discovery affect us? Some foods contain compounds similar to PTC and PROP. These plants, including kale, cabbage, broccoli, and Brussels sprouts (Figure 10.18), taste bitter to tasters and are intensely bitter to super-tasters. So, if you *really* don't like broccoli or Brussels sprouts, you may be a supertaster and be able to blame it on your genotype.

Other evidence indicates that PTC/PROP tasters and supertasters may live in a taste world different from that of nontasters. For example, capsaicin, the compound that makes hot peppers hot, has a more intense taste to PTC/PROP tasters; sucrose (table sugar) and artificial sweeteners are more intensely sweet to tasters. In addition, tasters have more food dislikes than nontasters do and usually do not like foods such as black coffee, dark beer, anchovies, and strong cheeses.

Are there relationships between our genotypes, our taste preferences, and our overall diets? For example, do tasters choose fruits and vegetables lower in cancer-*fighting* compounds, or do they choose foods that are lower in cancer-*causing* compounds? Or is there a relationship between genotype, diet preference, and obesity? Some research indicates that such relationships may exist.

In one study, Beverly Tepper and her colleagues at Rutgers University found that supertasters and tasters are more sensitive to the presence of fats and sugar in foods than nontasters. The study showed that nontasters prefer high-fat foods, while tasters do not. As a result, nontasters in this study weighed about 20% more than tasters. Other studies have shown a strong association between the nontaster genotype and increased body

FIGURE 10.18 Genetic differences in the number and location of taste receptors can make some vegetables—including broccoli, kale, and Brussels sprouts—taste very bitter to some people.

Grant Heilman Photography

FIGURE 10.19 Red and pink verbena flowers. Genetic differences in smell receptors allow many people to smell a fragrance from the pink flowers but not the red ones. Others can smell the red flowers but not the pink ones.

weight. If further work confirms this relationship, taster status may be a reliable indicator of weight-gain susceptibility.

The ability to smell is mediated by a family of 100 to 1,000 different proteins distributed on the surface of cells in the nose and sinuses. There are many combinations of alleles for these proteins, so each of us lives in a slightly different world of smell. In fact, our sensory worlds can be so different that some people cannot smell the odor released by skunks (OMIM 270350).

The garden flower *Verbena* comes in a variety of colors, including red and pink (Figure 10.19). Blakeslee discovered that about two-thirds of the people tested smelled a fragrance in the pink flowers but not in the red ones. The remaining one-third could detect a smell in the red flowers but not in the pink ones.

Although the genetics of taste and smell demonstrate that different genotypes may be responsible for our food preferences and the ability to smell flowers, the importance of pharmacogenetics lies in determining the genetic foundations for the wide range of reactions to therapeutic drugs.

Drug sensitivities are genetic traits.

In the last 50 years, tens of thousands of new drugs have been developed for use in medicine. In many cases, distinctive patterns of response to these drugs soon became apparent. Subsequent work has shown that many of the differences people experience in response to drugs are genetically controlled. Some patients metabolize drugs more slowly than others, causing higher drug levels than intended and sometimes leading to toxic or even fatal effects. Others metabolize drugs rapidly and require higher doses for effective treatment.

Sensitivity to anesthetics.

Succinylcholine is a muscle relaxant and a short-acting anesthetic. Soon after its introduction about 50 years ago, it became clear that some people took hours rather than minutes to recover from equal doses of the drug. Normally, the drug is rapidly inactivated by the enzyme serum cholinesterase. Those who take a long time to recover from the drug have a variant form of the enzyme that inactivates the drug very slowly, prolonging the effect of the anesthetic (OMIM 177400). Pedigree analysis indicates that this trait is inherited in an autosomal recessive manner. In a study of Canadians, the frequency of heterozygotes was 3% to 4%, and about 1 in 2,000 people were recessive homozygotes with a very long recovery time. The use of succinylcholine as an anesthetic in these homozygotes can lead to paralysis of the respiratory muscles and death.

Allele variations and breast cancer therapy.

More than 200,000 women in the United States are diagnosed with breast cancer each year. Almost 70% of all of these cancers are estrogen sensitive. This means that cells from

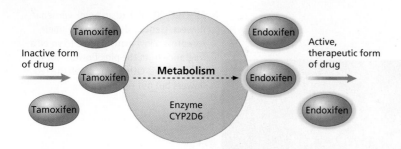

FIGURE 10.20 Tamoxifen, used to treat estrogen-sensitive breast cancer, is inactive until it is converted in the body into a powerful antiestrogen compound called endoxifen. This conversion is controlled by an enzyme called CYP2D6, which has several alleles. Depending on which allele combination a woman has, she may metabolize tamoxifen slowly, have intermediate rates of conversion, or rapidly produce endoxifen. Women who metabolize tamoxifen slowly have much higher rates of recurring breast cancer than those who are rapid metabolizers.

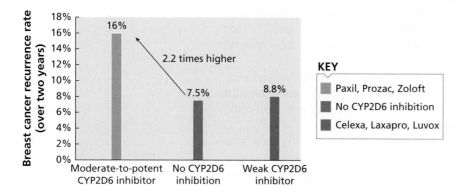

FIGURE 10.21 Some drugs, including antidepressants, interfere with tamoxifen conversion by competing for the CYP2D6 enzyme, thereby slowing tamoxifen conversion. Women taking tamoxifen and certain antidepressants have a 2.5 times higher risk of recurring breast cancer than those who are taking antidepressants that are not metabolized by the CYP2D6 enzyme.

these cancers have estrogen receptors on their surfaces and that estrogen in the body helps promote the growth of the tumors. One of the drugs widely used to treat estrogen-sensitive breast cancer is tamoxifen. Once in the body, tamoxifen is converted into endoxifen, a powerful anti-estrogen drug. Tamoxifen is converted to endoxifen by an enzyme called CYP2D6 (Figure 10.20). At least 46 alleles of the *CYP2D6* gene have been identified, and four distinct phenotypic groups are recognized. One group converts tamoxifen to endoxifen very slowly and has low levels of the active drug in the body. Another group has an intermediate rate of conversion of tamoxifen to endoxifen. Members of a third group are classed as extensive metabolizers of tamoxifen, and those in the fourth group are ultrarapid metabolizers. Alleles that abolish or decrease CYP2D6 activity lead to significantly reduced blood levels of endoxifen. Homozygotes for these alleles are poor or intermediate metabolizers of tamoxifen and may need higher doses of tamoxifen to benefit from the drug.

Clinical evaluation of breast cancer recurrence and *CYP2D6* genotypes indicates that women with the poorest conversion rate have a twofold to threefold higher risk of recurrence than women with higher metabolic rates. Results from these studies clearly show that genotype should be considered as an important factor in selecting drugs and dose rates for breast cancer treatment and that genotypic analysis to match drugs and dosage for each patient may improve the outcome.

The presence of other drugs in the body is another factor that influences the conversion of tamoxifen to endoxifen. For example, some antidepressants compete with tamoxifen for the CYP2D6 enzyme. If these drugs are present in the body, they lower the rate of tamoxifen conversion. Recent studies (Figure 10.21) show that women taking tamoxifen and one of these antidepressants had breast cancer recurrence rates more than 2.2 times higher than women on tamoxifen alone. This finding emphasizes the importance of understanding the interactions of the genotype with multiple therapeutic drugs for treating patients.

10.8 Ecogenetics

The scope of pharmacogenetics has expanded to study genetic differences in how people react to chemicals in food, in the workplace, and from exposure to industrial pollution. It is well known that health risks from environmental chemicals involve several factors, including the properties of the chemical itself, the dose received, and the length of exposure. It is now clear that the overall risks of environmental chemicals also depend on genetically determined variations in the proteins involved in transport, metabolism, and excretion of these chemicals.

What is ecogenetics?

Ecogenetics is the study of genetic variation that affects people's responses to environmental chemicals. Although more than 500,000 different chemicals are currently used in manufacturing and agriculture, only a few have been tested for their toxicity or ability to cause cancer. The recognition that some members of a population may be more sensitive or more resistant to the toxic effects of environmental chemicals has important consequences for research, medicine, and public health. In this section, we will focus on the ecogenetics of pesticides.

Sensitivity to pesticides varies widely in different populations.

Insects, weeds, fungi, and other pathogens destroy about 35% of the world's crops. After harvesting, another 10% to 20% is destroyed in storage by pests and fungi. Chemicals, including herbicides, insecticides, and fungicides, are used to control pests. In the United States, more than 600 million pounds of pesticides are used every year.

One class of agricultural insecticides is the organophosphates. Parathion is the most widely used member of this class of insecticides, and has been used in the United States for more than 50 years. Parathion is effective, but also highly toxic to humans (Figure 10.22). Its importation is banned in more than 50 countries worldwide. Exposure to parathion and other organophosphates can occur on the job (agricultural workers and forestry workers) or from eating contaminated food.

In the body, parathion is enzymatically converted to a more-toxic compound called paraoxon. This toxin disrupts the transmission of signals in the nervous system and can cause headaches, blurred vision, tremors, unconsciousness, and death by respiratory arrest. Paraoxon is inactivated by paraoxonase, an enzyme found in blood serum.

The gene for paraoxonase (*PON1*)(OMIM 168820) has two alleles (*Q* and *R*). The *R* allele encodes an enzyme with high levels of activity that detoxifies paraoxon and other organophosphate pesticides 10 times faster than the enzyme encoded by the *Q* allele. The proteins encoded by these two alleles differ only in a single amino acid at position 192 (the protein contains 355 amino acids). The *Q* allele has glutamic acid at position 192, and the *R* allele has arginine at position 192. People homozygous for the *R* allele (*R/R*) are more resistant to the toxic effects of parathion because they rapidly metabolize and inactivate the paraoxon produced from parathion. Conversely, those homozygous for low activity (*Q/Q*) are highly sensitive to parathion poisoning (OMIM 168820).

FIGURE 10.22 Parathion is a highly toxic organophosphate insecticide. Although it is banned in many countries, it has been used in the U.S. for over 50 years. To prevent poisoning, users must wear protective clothing, gloves, and a respirator. Genetic differences in the metabolism of parathion make many people more susceptible than others to this chemical.

PKU

Phenylketonuria (PKU), the recessively inherited genetic disorder of amino acid metabolism, is regarded as the gold-standard genetic disorder. Its discovery in 1935 laid the foundation for human biochemical genetics. Knowledge about PKU encompasses almost every area and most of the technology of human genetics. The three-dimensional structure of the enzyme (phenylalanine hydroxylase, or PAH) responsible for the disorder is known, and its regulatory, catalytic, and binding sites have been mapped. Additionally, the *PAH* gene has been cloned and its mutational sites identified. Biochemical and behavioral defects in PKU and the correlation between genotype and phenotype are well known, and the risks to fetuses carried by affected females are recognized. The disorder can be diagnosed prenatally, a treatment that prevents mental retardation is available, and a newborn-screening program is used throughout the developed world to detect and treat affected individuals. An animal model with a matching biochemical phenotype has been developed and used to study changes in brain structure, gene expression, and metabolism in this disorder.

PKU seems to be a straightforward disorder caused by a mutant allele encoding a defective enzyme, but the disorder has complexities still being revealed. In rare cases, the PAH gene is normal, but mutations in at least four genes encoding the enzyme's cofactor, BH4, cause PKU. Recent work using a mouse PKU model suggests that elevated blood levels of phenylalanine trigger changes in levels of gene expression and the expression of other genes, creating a unique metabolic environment in the brain that plays a role in the phenotype.

Dietary treatment with a nutritional supplement has been used for decades, but it is unpleasant in taste and has personal, social, and financial costs. Methods such as enzyme encapsulation have proven to be unsatisfactory, and gene therapy and liver transplantation have risks that far exceed the benefits of replacing a dietary supplement. Strategies for reducing blood levels of phenylalanine using yeast enzymes to metabolize the amino acid have not been successful. A new diet based on GMP, a protein by-product of cheese production with only trace amounts of phenylalanine, is showing promise. Whatever the outcome, the efforts in this area are a reminder that the social impact of genetics remains one of the most unexplored areas of this science.

If you were born in the United States or many other countries in the last 30 years, you were tested for PKU, and are a beneficiary of and participant in the genetic revolution that is part of human genetics.

In a study of pregnant women and their newborn children in an agricultural region in the western United States where organophosphate pesticides are used on a regular basis, researchers found that both levels *and* activity of the PON1 enzyme are important in determining sensitivity to these pesticides. On average, enzyme levels were fourfold lower in infants than in adults, but levels among the adults surveyed also varied widely.

Population studies also reveal significant differences in the frequency of the Q and R alleles and in genotype frequencies. The frequency of the Q allele in Latino populations is about 59%, in U.S. whites of northern European ancestry it is about 69%, and in U.S. blacks it is about 31%. This means that about 35% of the Latino population is homozygous Q/Q and extremely sensitive to parathion poisoning, compared with 47% of U.S. whites and 10% of U.S. blacks.

Setting standards for safe levels of exposure to organophosphate pesticides must take into account population differences in allele frequencies, genotype frequencies, and differences in the amount of the PON1 enzyme present in cells to ensure that the most sensitive members of the population, especially newborns and infants, are sufficiently protected.

The constellation of alleles for all the genes carried by each individual is the result of the random combination of parental genes and the sum of changes brought about by recombination and mutation. This genetic combination confers a distinctive phenotype upon each person. Garrod referred to this metabolic uniqueness as chemical individuality. Understanding the molecular basis for this individuality remains one of the great challenges of human biochemical genetics.

Genetics in Practice

Genetics in Practice case studies are critical-thinking exercises that allow you to apply your new knowledge of human genetics to real-life problems. You can find these case studies and links to relevant websites at *www.cengage.com/biology/cummings*

CASE 1

A couple was referred for genetic counseling because they wanted to know the chances of having a child with dwarfism. Both the man and the woman had achondroplasia, the most common form of short-limbed dwarfism. The couple knew that this condition is inherited as an autosomal dominant trait, but they were unsure what kind of physical manifestations a child would have if it inherited both mutant alleles. They were each heterozygous for the *FGFR3* gene that causes achondroplasia, and they wanted information on their chances of having a homozygous child. The counselor briefly reviewed the phenotypic features of individuals with achondroplasia. These include facial features (large head with prominent forehead; small, flat nasal bridge; and prominent jaw), very short stature, and shortening of the arms and legs. Physical examination and skeletal X-ray films are used to diagnose this condition. Final adult height is approximately 4 feet.

Because achondroplasia is an autosomal dominant condition, a heterozygote has a 1-in-2, or 50%, chance of having children with this condition. However, about 75% of those with achondroplasia have parents of average size. In these cases, achondroplasia is due to a new mutation. In the couple being counseled, each individual is heterozygous, and they are at risk for having a homozygous child with two copies of the mutated gene. Infants with homozygous achondroplasia are either stillborn or die shortly after birth. The counselor recommended prenatal diagnosis via serial ultrasound. In addition, a DNA test is available to detect the homozygous condition prenatally.

1. What is the chance that this couple will have a child with two copies of the dominant mutant gene? What is the chance that the child will have normal height?

2. Should the parents be concerned about the heterozygous condition as well as the homozygous mutant condition?

3. What if the couple wanted prenatal testing so that a normal fetus could be aborted?

CASE 2

Tina is 12 years old. Although symptomatic since infancy, she was not diagnosed with acid maltase deficiency (AMD; OMIM 232300) until she was 10 years old. The progression of her disease has been slow and insidious. She has difficulty walking and breathing because of severe muscle weakness. She relies on a respirator to assist her breathing. Tina has severe scoliosis (curvature of the spine), which further restricts her breathing and causes even greater difficulty in walking. She is extremely tired and experiences constant muscle pain. Although she is very bright and thinks like a normal teenager, her body won't let her function like one. She can no longer attend school. The future is bleak for Tina and other children like her. Death from the childhood form of AMD is frequently due to complications from respiratory infections, which are a constant threat. Life expectancy in this form of AMD is only to the second or third decade of life.

AMD, also called glycogen storage disease type II (or Pompe disease), is an autosomal recessive condition. Heterozygous parents have a 25% chance during each pregnancy that the child will have two abnormal genes and be affected.

Normally, glycogen is synthesized from sugars and is stored in the muscle cells for future use. The acid maltase enzyme breaks down the glycogen in the muscle cells when sugar is required as an energy source. Someone with AMD lacks this enzyme, and glycogen gradually builds up in the muscles, leading to progressive muscle weakness and degeneration.

There is no cure for AMD. Enzyme replacement is being used to successfully treat this disorder and offers new hope for children like Tina, but such treatments are expensive (often more than $100,000 per year) and must be continued for life.

1. What if Tina's parents cannot afford enzyme therapy? Should insurance be required to pay for this treatment? If not, who should pay?

2. As enzyme therapies for other disorders are developed, funding these therapies will become prohibitively expensive. Tina is 12 years old, and her life expectancy is 20 to 30 years. Should there be a lifetime limit on how much money should be spent on Tina's treatment?

Summary

10.1 Proteins Are the Link Between Genes and the Phenotype

- Proteins are the end product of the gene-expression pathway. Proteins are the link between genes and phenotype and as such are important components of cell structure, metabolic reactions, the immune system, hormonal responses, and cell-to-cell signaling systems.

10.2 Enzymes and Metabolic Pathways

- Biochemical reactions in the cell are linked together to form metabolic pathways. Mutations that block one reaction in a pathway can produce a mutant phenotype in several ways.

10.3 Phenylketonuria: A Mutation That Affects an Enzyme

- Phenylalanine is an essential amino acid and the starting point for a network of metabolic reactions. A mutation in a gene encoding the enzyme that controls the first step in this network causes phenylketonuria (PKU). The phenotype is caused by the buildup of phenylalanine and the products of secondary reactions.

10.4 Other Metabolic Disorders in the Phenylalanine Pathway

- The mutation that results in PKU is only one of several genetic disorders caused by the mutation of genes in the phenylalanine pathway. Others include defects of thyroid hormone, albinism, and alkaptonuria, the disease investigated by Garrod.

10.5 Genes and Enzymes of Carbohydrate Metabolism

- Mutations in genes encoding enzymes can affect the metabolic pathways of other biological molecules, including carbohydrates. Galactosemia is a genetic disorder caused by the lack of an enzyme in sugar metabolism. Lactose intolerance is not a genetic disorder but a genetic variation that affects millions of adults worldwide.

10.6 Defects in Transport Proteins: Hemoglobin

- In 1949, James Neel identified sickle cell anemia as a recessively inherited disease. This disorder is caused by a mutation in a gene encoding beta globin, a protein that transports oxygen from the lungs to cells and tissues of the body. Other mutations cause thalassemia, an imbalance in the production of globins that affects the transport of oxygen within the body.

10.7 Pharmacogenetics and Pharmacogenomics

- Individual differences in the reactions to therapeutic drugs represent a "hidden" set of phenotypes that are not revealed until exposure occurs. Understanding the genetic basis for these differences is the concern of pharmacogenetics and may lead to customized drug treatment for infections and other diseases.

10.8 Ecogenetics

- Ecogenetics is the study of genetic variation that affects responses to environmental chemicals. The fact that some members of a population may be sensitive or resistant to environmental chemicals, including pesticides, has important consequences for research, medicine, and public policy.

Questions and Problems

CENGAGENOW Preparing for an exam? Assess your understanding of this chapter's topics with a pre-test, a personalized learning plan, and a post-test by logging on to *login.cengage.com/sso* and visiting CengageNOW's Study Tools.

Enzymes and Metabolic Pathways

1. Many individuals with metabolic diseases are normal at birth but show symptoms shortly thereafter. Why?
2. List the ways in which a metabolic block can have phenotypic effects.
3. Enzymes have all the following characteristics except:
   a. They act as biological catalysts
   b. They are proteins
   c. They carry out random chemical reactions
   d. They convert substrates into products
   e. They can cause genetic disease

Questions 4 through 6 refer to the following hypothetical pathway in which substance A is converted to substance C by enzymes 1 and 2. Substance B is the intermediate produced in this pathway:

4. a. If an individual is homozygous for a null mutation in the gene that codes for enzyme 1, what will the result be?
   b. If an individual is homozygous for a null mutation in enzyme 2, what will the result be?
   c. What if an individual is heterozygous for a dominant mutation in which enzyme 1 is overactive?
   d. What if an individual is heterozygous for a mutation that abolishes the activity of enzyme 2 (a null mutation)?
5. a. If the first individual in Question 4 married the second individual, would their children be able to convert substance A into substance C?
   b. Suppose each of the adults mentioned in part *a* was heterozygous for an autosomal dominant mutation that prevents any enzyme function. List the phenotypes of their children with respect to compounds A, B, and C. (Would the compound be in excess, not present, and so on?)

6. An individual is heterozygous for a recessive mutation in enzyme 1 and heterozygous for a recessive mutation in enzyme 2. This individual marries an individual with the same genotype. List the possible genotypes of their children. For every genotype, determine the activity of enzymes 1 and 2, assuming that the mutant alleles have 0% activity and that each normal allele has 50% activity. For every genotype, determine if compound C will be made. If compound C is not made, list the compound that will be in excess.

Questions 7 to 11 refer to a hypothetical metabolic disease in which protein E is not produced. Lack of protein E causes mental retardation in humans. Protein E's function is not known, but it is found in all cells of the body. Skin cells from eight individuals who cannot produce protein E were taken and grown in culture. The defect in each of the individuals is the result of a single recessive mutation. Each individual is homozygous for her or his mutation. The cells from one individual were grown with the cells from another individual in all possible combinations of the two. After a few weeks of growth, the mixed cultures were assayed for the presence of protein E. The results are given in the following table. A plus sign means that the two cell types produced protein E when grown together (but not separately); a minus sign means that the two cell types still could not produce protein E.

	1	2	3	4	5	6	7	8
1	−	+	+	+	+	−	+	+
2		−	+	+	+	+	−	+
3			−	+	+	+	+	−
4				−	+	+	+	+
5					−	+	+	+
6						−	+	+
7							−	+
8								−

7. a. Which individuals seem to have the same defect in protein-E production?
   b. If individual 2 married individual 3, would their children be able to make protein E?
   c. If individual 1 married individual 6, would their children be able to make protein E?
8. a. Assuming that these individuals represent all possible mutants in the synthesis of protein E, how many steps are there in the pathway to protein-E production?
   b. Compounds A, B, C, and D are known to be intermediates in the pathway for production of protein E. To determine where the block in protein-E production occurred in each individual, the various intermediates were given to each individual's cells in culture. After a few weeks of growth with the intermediate, the cells were assayed for the production of protein E. The results for each individual's cells are given in the following table. A plus sign means

that protein E was produced after the cells were given the intermediate listed at the top of the column. A minus sign means that the cells still could not produce protein E even after being exposed to the intermediate at the top of the column.

		Compounds			
Cells	A	B	C	D	E
1	−	−	+	+	+
2	−	+	+	+	+
3	−	−	−	+	+
4	−	−	−	−	+
5	+	+	+	+	+
6	−	−	+	+	+
7	−	+	+	+	+
8	−	−	−	+	+

9. Draw the pathway leading to the production of protein E.
10. Denote the point in the pathway in which each individual is blocked.
11. a. If an individual who is homozygous for the mutation found in individual 2 and heterozygous for the mutation found in individual 4 marries an individual who is homozygous for the mutation found in individual 4 and heterozygous for the mutation found in individual 2, what will be the phenotype of their children?
    b. List the intermediate that would build up in each of the types of children who could not produce protein E.

Phenylketonuria: A Mutation That Affects an Enzyme
12. Essential amino acids are:
    a. amino acids the human body can synthesize
    b. amino acids humans need in their diet
    c. amino acids in a box of Frosted Flakes
    d. amino acids that include arginine and glutamic acid
    e. amino acids that cannot harm the body if not metabolized properly
13. Suppose that in the formation of phenylalanine hydroxylase mRNA, the exons of the pre-mRNA fail to splice together properly and the resulting enzyme is nonfunctional. This produces an accumulation of high levels of phenylalanine and other compounds, which causes neurological damage. What phenotype would be produced in the affected individual?
14. If phenylalanine was not an essential amino acid, would diet therapy (the elimination of phenylalanine from the diet) for PKU work?
15. Phenylketonuria and alkaptonuria are both autosomal recessive diseases. If a person with PKU marries a person with AKU, what will the phenotype of their children be?

Genes and Enzymes of Carbohydrate Metabolism

16. The normal enzyme required for converting sugars into glucose is present in cells, but the conversion never takes place and no glucose is produced. What could have occurred to cause this defect in a metabolic pathway?

17. Knowing that individuals who are homozygous for the G^D allele show no symptoms of galactosemia, is it surprising that galactosemia is a recessive disease? Why?

Defects in Transport Proteins: Hemoglobin

18. Describe the quaternary structure of the blood protein hemoglobin.

19. A person was found to have very low levels of functional beta globin mRNA and therefore very low levels of the beta globin protein. What problems would this cause for assembling functional hemoglobin molecules?

20. Look at Figure 10.15. If an extra nucleotide is inserted in the first exon of the beta globin gene, what effect will it have on the amino acid sequence of the globin polypeptides? Will the globin most likely be fully functional, partly functional, or nonfunctional? Why?

21. Transcriptional regulators are proteins that bind to promoters (the 5'-flanking regions of genes) to regulate their transcription. Assume that a particular transcription regulator normally promotes transcription of gene X, a transport protein. If a mutation makes this regulator gene nonfunctional, would the resulting phenotype be similar to a mutation in gene X itself? Why or why not?

Pharmacogenetics and Pharmacogenomics

22. Explain why there are variant responses to drugs and why these responses act as heritable traits.

Ecogenetics

23. Ecogenetics is a branch of genetics that deals with the genetic variation that underlies reactive differences to drugs, chemicals in food, occupational exposure, industrial pollution, and other substances. Cases have arisen in which workers claim that exposure to a certain agent has made them feel ill whereas other workers are unaffected. Although claims like these are not always justified, what are some concrete examples that prove that variation in reactions to certain substances exist in the human population?

Internet Activities

Internet Activities are critical-thinking exercises using the resources of the World Wide Web to enhance the principles and issues covered in this chapter. For a full set of links and questions investigating the topics described below, visit *www.cengage.com/biology/cummings*

1. *Sickle Cell Anemia.* At the *Sickle Cell Case Study* site, read about the genetics of sickle cell disease and the relationship between sickle cell disease and malaria. Read, too, about current research in sickle cell anemia.

2. *Enzyme Replacement Therapy and Pompe Disease.* At *Applied Biosystems's Biobeat* site, access and read the article on enzyme replacement therapy in the treatment of Pompe disease.

PKU and other metabolic disorders can be tested for in newborns, allowing for early treatment. All 50 states and the District of Columbia require testing for PKU and several other prevalent genetic disorders. However, the exact number of genetic diseases newborns are tested for varies from state to state, from as few as 6 genetic disorders to more than 40. One of the rationales given for testing for only a small number of disorders is that cost-benefit analysis shows that it is not cost efficient to test for a large number. That is, some diseases are so rare that the costs of testing all newborns appear to outweigh the health care costs for affected children, regardless of the severity of the problems caused by the disorder. Now that you know more about metabolic disorders and the relationships among genes, proteins, and phenotypes, what do you think? Should cost-benefit analysis be used as a determining factor in setting up and running newborn-testing programs? Visit the *Human Heredity* companion website at *www.cengage.com/biology/cummings* to find out more on the issue; then cast your vote online.

11 Mutation: The Source of Genetic Variation

During the past 40 years, research has demonstrated that radiation can help preserve food and kill contaminating microorganisms. Irradiation prevents sprouting of root crops such as potatoes; extends the shelf life of many fruits and vegetables; destroys bacteria and fungi in meat, fish, and grain; and kills insects and other pests in spices.

For irradiation, food is placed on a conveyor and moved to a sealed, heavily shielded chamber, where it is exposed to a radioactive source. An operator views the process on a video camera and delivers the dose. The food itself does not come in contact with the radioactive source, and the food is not made radioactive. Relatively low doses are used to inhibit the sprouting of potatoes and to kill parasites in pork. Intermediate doses are used to retard spoilage in meat, poultry, and fish. And high doses can be used to sterilize foods, including meats. The amount of food irradiated varies from country to country, ranging from a few tons of spices to hundreds of thousands of tons of grain.

NASA routinely has fed irradiated food to astronauts in space since 1972, and irradiated foods are sold in more than 40 countries, including the United States. The U.S. Food and Drug Administration (FDA) approved the first application for food irradiation in 1964, and approval has been granted for the irradiation of spices, herbs, fruits and vegetables, pork, beef, lamb, chicken, and eggs. All irradiated food sold in the United States must be labeled with an identifying logo (see inset).

Studies by U.S. government agencies have shown that there are benefits to irradiated foods. Irradiation kills all or most disease-causing organisms and does not change the nutritional value of the food. However, these studies also show that irradiation does not destroy toxins produced by bacteria before irradiation and does not prevent aging of fruits and vegetables.

Public concern about radiation has prevented the widespread sale of irradiated food in this country. Advocates point out that irradiation can eliminate the use of many chemical preservatives, lower food costs by preventing spoilage, and reduce the 76 million cases of food-borne illnesses and 5,000 deaths caused by contaminated food each year in the United States. For example, irradiation of lettuce has been shown to eliminate contamination with

A mutation produced the different eye colors in this dog.

Image copyright Bouvier Ben, 2010. Used under license from Shutterstock.com.

Escherichia coli O157:H7, a deadly strain responsible for over 70,000 cases of food-borne illness and over 60 deaths per year. Those opposed to food irradiation argue that irradiation produces mutation- and cancer-causing compounds in food and that the testing of irradiated food to detect cancer-causing effects is inadequate. Opponents also point out that treatment may select for radiation-resistant microorganisms.

In this chapter, we will consider the nature of mutations, how mutations are detected, the rate of mutation, and the role of radiation and chemicals in causing mutations.

KEEP IN MIND AS YOU READ

- Mutation can occur spontaneously as a result of errors in DNA replication or by exposure to environmental factors such as radiation or chemicals.

- Mutations in DNA can occur in several ways, including nucleotide substitution, deletion, and insertion.

- Damage to DNA can be repaired during DNA synthesis and by enzymes that repair damage caused by radiation or chemicals.

11.1 Mutations Are Heritable Changes in DNA

 HOW WOULD YOU VOTE?

E. coli contamination in meat causes 70,000 illnesses and about 60 deaths per year in the United States. Irradiation to kill *E. coli* in beef and *Salmonella* in poultry has been approved by the U.S. Food and Drug Administration and the U.S. Department of Agriculture. The World Health Organization and the American Medical Association have endorsed irradiation as an effective means of preventing disease and deaths. In spite of this approval, irradiated meat is not widely available. If such products were available in the supermarket, would you buy them? Why or why not? Visit the *Human Heredity* companion website at *www.cengage.com/ biology/cummings* to find out more on the issue; then cast your vote online.

Mutations are heritable changes in DNA and are the ultimate source of all genetic variation in humans and other organisms. There are two general categories of mutations: somatic mutations and germ-line mutations. Somatic mutations occur in the cells of the body that do not form gametes (liver cells, muscle cells, kidney cells, etc). If a mutation occurs in a somatic cell, it will be passed on to all daughter cells by mitosis, but the mutation cannot be transmitted to future generations. As we will see in Chapter 12, somatic mutations are the underlying cause of many types of cancer. Germ-line mutations occur in cells that will produce gametes, and these mutations are passed on to offspring. In this chapter, we will focus on germ-line mutations.

Mutations can also be classified as those that affect chromosomes and those that affect individual genes. Chromosomal mutations were discussed in Chapter 6; in this chapter, we focus on mutational changes in single genes—that is, heritable changes in the sequence or number of nucleotides in DNA. First, we'll consider how mutations can be detected and see how mutation rates are calculated, and finally, we'll examine how mutation works at the molecular level.

11.2 Mutations Can Be Detected in Several Ways

How do we know that a mutation has occurred? In humans, the sudden appearance of a dominant mutation in a family can be observed in a single generation. However, a mutation that changes a dominant allele into a recessive allele can be detected only in the homozygous condition, posing a challenge for human geneticists because the phenotype associated with this mutation may not appear for many generations.

If an affected individual appears in an otherwise unaffected family, the first question is whether the trait is caused by genetic or nongenetic factors. For example, if exposed to the rubella virus (which causes a form of measles called German measles) early in pregnancy, the fetus may have a phenotype similar to those seen in a number of genetic disorders. The phenotype caused by the rubella infection is not produced by mutation but by the effect of the virus on the developing fetus. To determine whether an abnormal phenotype is caused by a genetic disorder, geneticists depend on pedigree analysis and the study of births over several generations.

If a mutant allele is dominant, is fully expressed in everyone who carries the mutant allele (is fully penetrant), and appears in a family with no previous history of the condition, geneticists usually assume that a mutation has taken place. In the pedigree shown in Figure 11.1, a new trait—severe blistering of the feet—appeared in one of six children, although the parents were unaffected. The affected female transmitted this trait to six of her eight children, and the trait was passed to the next generation. This pattern is consistent with the inheritance of an autosomal dominant trait. A reasonable explanation for this pedigree is that a mutation to a dominant allele causing foot blisters first appeared in individual II-5. However, a number of uncertainties can affect this conclusion. For example, if the child's father is an affected male but is not the husband in the pedigree, it would only seem that a mutation had occurred. Uncertainty can be reduced by studying additional pedigrees with the same trait.

Mutations that produce recessive X-linked alleles can often be detected by examining males in the family line. However, it can be difficult to determine whether a heterozygous female who transmits a trait to her son is the source of the mutation or is only passing on a mutation that arose in one of her ancestors. The X-linked form of hemophilia that spread through the royal families of western Europe and Russia in the nineteenth and twentieth centuries probably originated with Queen Victoria (Figure 11.2; see Exploring Genetics: Hemophilia and History in

> **KEEP IN MIND**
>
> Mutation can occur spontaneously as a result of errors in DNA replication or by exposure to environmental factors such as radiation or chemicals.

FIGURE 11.1 A dominant trait, foot blistering, appeared in a member (II-5) of a family with no previous history of this condition. The trait is transmitted through subsequent generations in an autosomal dominant fashion. This pattern is consistent with the appearance of a mutation in this family.

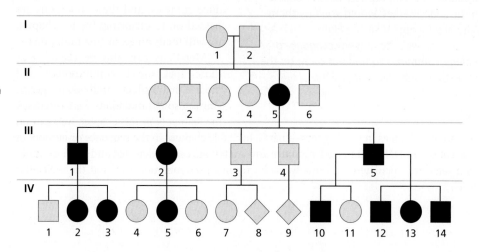

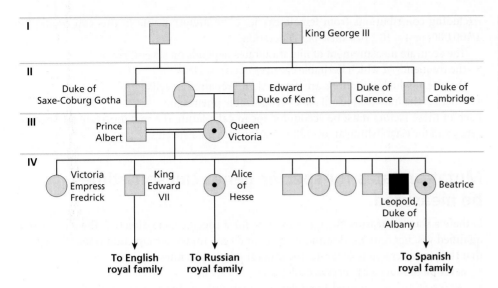

FIGURE 11.2 Pedigree of Queen Victoria of Great Britain, showing her immediate ancestors and children. Because she passed the mutant allele for hemophilia on to three of her children, she was probably a heterozygote rather than the source of the mutation.

Chapter 4). None of the males in previous generations of her family had hemophilia, but one of Victoria's sons was affected, and at least two of her daughters were heterozygous carriers.

Because there was no previous history of this disorder in her family and Victoria transmitted the mutant allele to a number of her children, it is reasonable to assume that she was a heterozygous carrier. Her father was not affected, and there is nothing in her mother's pedigree to indicate that hemophilia was present in her family. It is therefore likely that Victoria received a newly mutated allele in a gamete from one of her parents, but we can only speculate as to which parent. The role of hemophilia in the royal families of Europe and speculation about the origin of the mutant allele is explored in the book *Queen Victoria's Gene: Hemophilia and the Royal Family* by D. M. Potts and W. T. W. Potts.

If an autosomal recessive trait suddenly appears in a family, only homozygous individuals are affected, and it is usually difficult or impossible to trace the mutant allele through previous generations to identify the person or even the generation in which the mutation first occurred. A mutation of this type can remain undetected for generations as it is passed from heterozygote to heterozygote before being expressed in a homozygote.

11.3 Measuring Spontaneous Mutation Rates

Pedigree analysis shows that mutation does take place in the human genome. The available evidence suggests that it is a rare event, so how is it possible to accurately measure the rate of mutation?

Human geneticists define the **mutation rate** as the number of mutated alleles per gene per generation. Suppose that for a certain gene, 4 of 100,000 births have a phenotype derived from the mutation of a recessive to a dominant allele. Because each of these 100,000 individuals carries 2 copies of the gene, we have sampled 200,000 copies of the gene. The 4 births represent 4 mutated genes (we are assuming that the newborns are heterozygotes for a dominant mutation and carry only 1 mutant allele). In this case, the mutation rate is 4/200,000, or 2/100,000. In scientific notation, this would be written as 2×10^{-5} per allele per generation.

If the gene was X-linked and if 100,000 male births were examined and 4 individuals with mutant phenotypes were discovered, this would represent a sampling of 100,000 copies of the gene (because the males have only 1 copy of the X chromosome).

Mutation rate The number of events that produce mutated alleles per locus per generation.

Excluding contributions from female carriers, the mutation rate in this case would be 4/100,000, or 4×10^{-5} per allele per generation.

The accurate measurement of mutation rates depends on several factors:

- the frequency at which heritable changes occur in DNA
- the rate at which mutations are detected and repaired in cells
- whether the mutation results in a recognizable phenotype

Each of these factors must be accounted for in calculating the mutation rate for specific genes and for establishing an overall mutation rate for a species.

Mutation rates for specific genes can sometimes be measured.

Is there a way to measure the mutation rate for a specific gene directly? The answer is a qualified yes, but only for dominant alleles and only under certain conditions. To ensure that the measurement is accurate, the mutant phenotype must

- never be produced by recessive alleles
- always be fully expressed (completely penetrant) so that mutant individuals can be identified
- have clearly established paternity
- never be produced by nongenetic agents such as drugs or infection
- be produced by a dominantly inherited mutation of only one gene

One dominantly inherited trait, achondroplasia (OMIM 100800), is a form of dwarfism that results in short arms, short legs, and an enlarged skull (Figure 11.3). Several population surveys have studied mutations in this gene to estimate the overall mutation rate in humans. One survey found 7 achondroplastic children of unaffected parents in a total of 242,257 births. From those data, the mutation rate for achondroplasia has been calculated at 1.4×10^{-5}, or about one mutation in every 100,000 copies of the gene.

Although the mutation rate for a dominant trait such as achondroplasia can be measured directly, the next question is whether the mutation rate of this one gene is typical for all human genes. Perhaps this gene has an inherently high or low rate of mutation. Obviously, to get an accurate measure of the mutation rate in humans, it is important to measure mutation in a number of different genes before making any general statements. As it turns out, two other dominantly inherited mutations have widely different rates of mutation.

Neurofibromatosis I (OMIM 162200), an autosomal dominant condition, is characterized by pigmentation spots and tumors of the skin and nervous system (described in Chapter 4). Neurofibromatosis occurs in about 1 in 3,000 births, and about 50% of these births occur in families with no previous history of the disorder. This gene has a high mutation rate, calculated as 1 in 10,000 (1×10^{-4}) copies of the gene, which is one of the highest rates so far discovered in humans. At the other end of the spectrum, the mutation rate for Huntington disease (OMIM 143100) has been calculated as 1×10^{-6}, a rate 100-fold lower than that of neurofibromatosis and 10-fold lower than achondroplasia.

Measurements of the mutation rate in several human genes based on phenotypes are listed in Table 11.1. The average rate is approximately 1×10^{-5}. All the genes listed in the table are inherited as autosomal dominant or X-linked traits. It is almost impossible to measure directly the mutation rates in autosomal recessive alleles by pedigree analysis, but DNA sequencing and other genomic techniques are providing estimates of the rate and type of mutations found in many human genes, including those with autosomal recessive patterns of inheritance. These studies indicate that the overall mutation

FIGURE 11.3 The painting Las Meninas by Diego Velasquez shows Infanta Margarita of the seventeenth-century Spanish court accompanied by her maids and others, including an achondroplastic woman at the right. Achondroplasia (OMIM 100800) is a form of dwarfism caused by a dominant mutation.

Scala/Art Resource, NY

Table 11.1 Mutation Rates for Selected Genes

Trait	Mutants/Million Gametes	Mutation Rate	OMIM Number
Achondroplasia	10	1.4×10^{-5}	100800
Aniridia	2.6	2.6×10^{-6}	106200
Retinoblastoma	6	6×10^{-6}	180200
Osteogenesis imperfecta	10	1×10^{-5}	166200
Neurofibromatosis	50–100	$0.5–1 \times 10^{-4}$	162200
Polycystic kidney disease	60–120	$6–12 \times 10^{-4}$	173900
Marfan syndrome	4–6	$4–6 \times 10^{-6}$	154700
Von Hippel-Landau syndrome	<1	1.8×10^{-7}	193300
Duchenne muscular dystrophy	50–100	$0.5–1 \times 10^{-4}$	310200

rate in humans may be several times higher than previously thought, because many mutations detected by sequencing have no detectable phenotype. Still, many geneticists feel that to reduce any potential bias, a more conservative estimate of the mutation rate in humans should be used, and by convention 1×10^{-6}, or 1 mutation in every million copies of a gene is used as the average mutation rate for human genes.

Why do genes have different mutation rates?

Several factors influence the mutation rate:

- *Size of the gene.* Larger genes are bigger targets for mutation. The gene for neurofibromatosis (*NF-1*) is extremely large, spans more than 300,000 nucleotides, and has a high mutation rate. The gene for Duchenne and Becker muscular dystrophy (*DMD*; OMIM 310200), the largest gene identified to date in humans, contains more than 2 million base pairs. This gene also has a high mutation rate, and about one-third of all cases represent new mutations.

- *Nucleotide sequence.* Some genes contain short nucleotide sequences, called **trinucleotide repeats**. For example, in the gene for fragile-X syndrome (OMIM 309550), the sequence CGG is repeated some 6 to 50 times in normal individuals. Those with more than 230 copies experience symptoms of the disorder, and the symptoms become more severe as the number of CGG repeats increases. The presence of trinucleotide repeats may predispose a gene to mutate at a higher rate.

- *Spontaneous chemical changes.* Cytosine bases in DNA are especially susceptible to chemicals that can mutate this base, leading to a change in a gene's nucleotide sequence. These and other chemical changes are discussed in a later section of this chapter. Genes rich in G/C base pairs are more likely to undergo chemical changes than are those rich in A/T pairs, and genes with more G/C base pairs have a higher mutation rate than do A/T-rich genes.

Trinucleotide repeats A form of mutation associated with the expansion in copy number of a nucleotide triplet in or near a gene.

11.4 Environmental Factors Influence Mutation Rates

Now that we know what mutations are and how they are measured, we can discuss how they arise. In general, mutations result from mistakes that occur during normal cellular functions such as DNA replication or from the action of agents that attack DNA or disrupt cellular functions. These agents can originate from inside or outside the cell. We will begin by discussing two classes of environmental agents involved in mutation: radiation and chemicals.

Radiation is one source of mutations.

Radiation The process by which electromagnetic energy travels through space or a medium such as air.

Ionizing radiation Radiation that produces ions during interaction with other matter, including molecules in cells.

Background radiation Radiation in the environment that contributes to radiation exposure.

Simply defined, **radiation** is a process by which energy travels through space. For example, the heat from a fire in a fireplace travels through space and warms a room. There are two main types of radiation: waves (electromagnetic radiation) and particles (corpuscular radiation). Waves are electrical and magnetic energy, whereas particles are atoms or parts of atoms. Both forms of radiation can be **ionizing radiation** because they may form chemically reactive ions when they collide with molecules in cells.

Exposure to radiation is unavoidable. Everything in the physical world contains sources of radiation. This includes our bodies, the air we breathe, the food we eat, and the bricks in our houses. Some of this radiation is left over from the birth of the universe, and some has been created by the interaction of atoms on Earth with cosmic radiation. These natural sources of radiation are called **background radiation**. We are also exposed to radiation that results from human activity, including medical testing, nuclear testing, nuclear power, and consumer goods.

Radiation can cause biological damage at several levels. As ionizing radiation strikes the molecules in cells, it creates charged atoms. Such ionized molecules are highly reactive and can cause mutations in DNA. Because cells are about 80% water, radiation often splits water into hydrogen ions (H^+) and hydroxyl radicals ($\cdot OH$). These free radicals can produce mutations if they interact with DNA.

Cells can repair mutations; however, if too many mutations accumulate in a cell in a short time, the repair system can be overwhelmed. If mutations in somatic cells are not repaired, cell death or cancer can result. In germ cells, mutations that are not repaired are transmitted from generation to generation as newly mutated alleles.

How much radiation are we exposed to?

Rem The unit of radiation exposure used to measure radiation damage in humans. It is the amount of ionizing radiation that has the same effect as a standard amount of X-rays.

Millirem A rem is a measure of radiation dose equal to 1,000 millirems.

Radiation doses can be measured in several ways, but most often the dose is expressed in terms of damage. A **rem** (radiation equivalent in man, also called the Roentgen-equivalent man and named for the scientist who discovered X-rays) is the amount of radiation that causes the same damage as a standard amount of X-rays. Because people are usually exposed to very small amounts of radiation, the dose is expressed in **millirems** (1,000 mrem equals 1 rem). At a dose of approximately 100 rem (100,000 mrem), cells begin to die and radiation sickness results. At a dose of 400 rem, about 50% of people will die within 60 days if they are not treated. Case 1 at the end of this chapter discusses the impact of the Chernobyl reactor explosion in Ukraine on the surrounding populations.

In the United States, the average person is exposed to about 360 mrem/year, 81% of which is from background radiation (Figure 11.4). Because these doses are low, what are the major risks, if any, from this level of radiation exposure? Below 5,000 mrem, the major risk is mutations in somatic cells that increase susceptibility to cancer. Overall, however, this effect is very small, and risk analysis suggests that radiation is a small risk compared with others in our daily lives (Table 11.2).

Table 11.2   Various Sources and Doses of Radiation

Source	Dose (mrem)
Annual background exposure, Boston, MA	102
Annual background exposure, Denver, CO	180
Average annual dose, medical X-ray technicians	320
Average annual dose, airline crews	160
Average annual dose from nuclear power plants (U.S.)	0.002
Dose to bone marrow during dental X-rays	9–10
Dose to breast during mammogram	50–700
Dose from a full-body CT scan	1,200

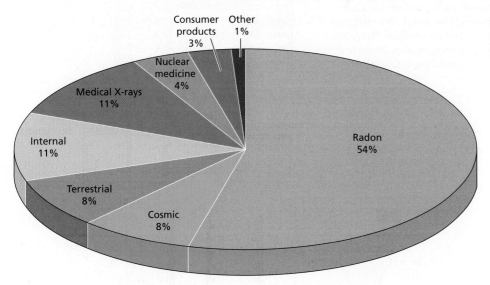

Sources of Radiation Exposure for the U.S. Population

Consumer products 3%
Other 1%
Nuclear medicine 4%
Medical X-rays 11%
Internal 11%
Terrestrial 8%
Cosmic 8%
Radon 54%

FIGURE 11.4 The sources of radiation that individuals are exposed to in the United States. The average dose is 360 mrem, just over 80% of which is from background radiation (radon, cosmic, terrestrial, and internal).

Chemicals can cause mutations.

We know of over 6 million chemical compounds, and almost 500,000 of those compounds are used in manufacturing processes and are part of everyday life in many ways: in packaging, food, building materials, and so forth. Unfortunately, we know little or nothing about whether most of those chemicals cause mutations (see Exploring Genetics: Flame Retardants: Are They Mutagens? on page 253). Chemicals cause mutations in several ways, and they are often classified by the type of damage they cause to DNA. Some chemicals cause nucleotide substitutions or change the number of nucleotides in DNA, whereas others structurally change the bases in DNA, causing a base-pair change after replication. Some of the ways chemicals act as mutagens are discussed here.

Base analogs.

Mutagenic chemicals that resemble nucleotides and are incorporated into DNA or RNA during synthesis are called **base analogs**. One of these, 5-bromouracil has a structure similar to that of thymine (Figure 11.5a) and is inserted into DNA in place of thymine (Figure 11.5b). However, once inserted into DNA, 5-bromouridine has the base-pairing properties of cytosine and in the next round of DNA synthesis will serve as a template for insertion of guanine in the newly synthesized strand, converting the original A/T base pair into a 5-Br/G base pair. After one more round of replication, the result is the creation of an A/T → G/C mutation (Figure 11.5b).

Base analogs A purine or pyrimidine that differs in chemical structure from those normally found in DNA or RNA.

Chemical modification of bases.

Chemical mutagens can modify the bases in a DNA molecule and alter their base-pairing properties. Some mutagens do this directly by attacking the bases in a DNA molecule, changing one base into another. For example, treatment of DNA with nitrous acid (HNO_2) changes cytosine into uracil (Figure 11.6a). What was a G/C base pair is converted into a G/U pair (Figure 11.6b). Uracil has the base-pairing properties of thymine (T). In the next round of DNA replication, the template DNA strand containing U will direct the insertion of A in the newly synthesized strand, forming an A/U pair in the new DNA molecule (Figure 11.6c). After another round of replication, a G/C → A/T mutation will be created. Nitrates and nitrites used in the preservation of meats (bacon and some lunch meats), fish, and cheese are converted into nitrous acid in the body. Although this has been studied extensively, the amount of mutation caused by these dietary chemicals has been difficult to assess.

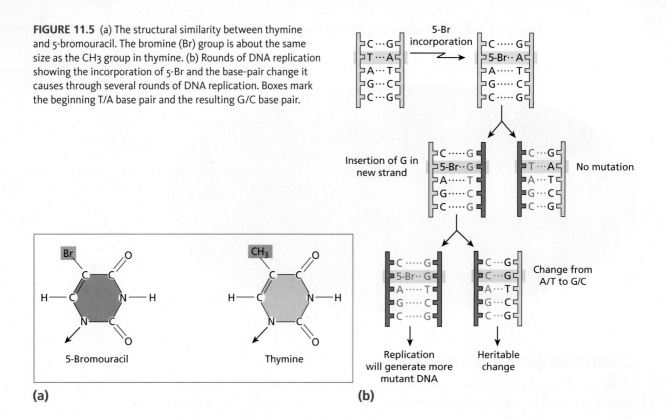

FIGURE 11.5 (a) The structural similarity between thymine and 5-bromouracil. The bromine (Br) group is about the same size as the CH3 group in thymine. (b) Rounds of DNA replication showing the incorporation of 5-Br and the base-pair change it causes through several rounds of DNA replication. Boxes mark the beginning T/A base pair and the resulting G/C base pair.

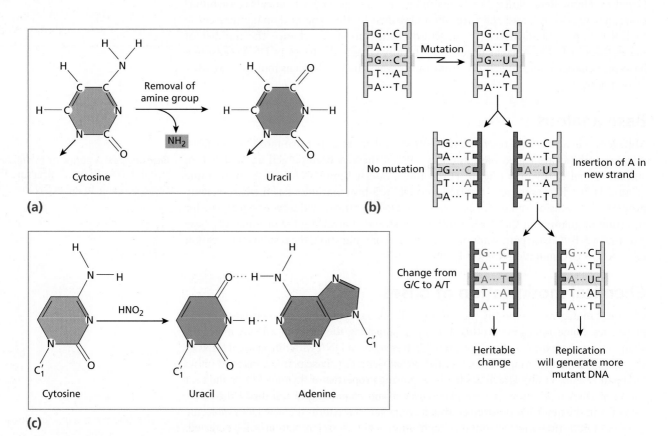

FIGURE 11.6 (a) The conversion of cytosine to uracil by the removal of an amine group, which can occur by the action of chemical mutagens. (b) Rounds of DNA replication after the conversion of cytosine to uracil in a DNA molecule. After two rounds of replication, the original G/C base pair (highlighted in the box) is converted into an A/T base pair. (c) The pairing between uracil and adenine.

Flame Retardants: Are They Mutagens?

In the 1970s, children's pajamas were treated with a flame retardant called tris-BP. It was banned after it was shown that this chemical caused cancer in test animals and was absorbed through the skin by those wearing treated clothing and converted into another powerful mutagen. However, other flame retardants chemically similar to tris-BP remain in wide use. In fact, these chemicals are almost a universal feature of consumer goods in the United States. They are found in polyurethane foam, carpets, mattresses, upholstery, televisions, computer monitors, printers, cell phones, and a long list of other items, often up to 10% to 20% by weight. One class of these retardants, the PBDEs (short for polybrominated diphenyl ethers), are chemically related to well-known and persistent environmental contaminants such as PCBs that cause cancer and other health problems in humans. PBDEs have not been linked directly to cancer in test animals, but recent studies of cancer patients who work in electronic recycling waste sites in China strongly suggest a link between cancer and high concentrations of PDBEs in the body. It is clear that PDBEs cause a number of

Image copyright EdwinAC, 2010. Used under license from Shutterstock.com.

developmental defects, including brain damage in mice. PBDEs are banned in many European countries but permitted in most states in the United States.

Measurements of PDBEs in human blood serum and milk in the United States show some disturbing trends. Analysis of serum samples collected between 1985 and 2002 showed a sixfold increase in levels of PBDEs. Samples of human milk collected from a milk bank and analyzed for PBDEs show levels 10 to 100 times greater than in European populations. PDBEs are also found in food, land sludge, house dust, freshwater fish, and even seals living above the Arctic Circle.

In the United States, several states have outlawed some PDBEs, but most states have taken no action. Other halogen-containing organic compounds such as PCBs persist in the environment for decades, and PDBEs will probably be around long after any global bans are put into effect. The link between PDBEs and cancer remains circumstantial, but as concentrations in human tissues continue to rise, the health consequences of exposure to these compounds need to be explored.

Chemicals that insert into DNA.

Chemicals that resemble base pairs, called intercalating agents, insert themselves into the DNA, distorting the double helix. The structure of one of these chemicals, acridine orange, is shown in Figure 11.7. This molecule is about the same size as a purine/pyrimidine base pair and wedges itself into DNA, distorting the shape of the double helix. When replication takes place in this distorted region, deletion or insertion of single base pairs can take place, resulting in a frameshift mutation. Some components and breakdown products of commonly used pesticides are intercalating agents.

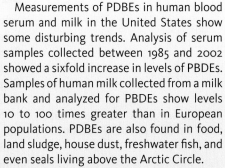

FIGURE 11.7 The molecular structure of acridine orange, an intercalating agent that inserts into the helical structure of DNA, distorting its shape. Replication in the distorted region can lead to the insertion or deletion of base pairs, producing a mutation.

11.5 Mutations at the Molecular Level: DNA as a Target

At the molecular level, mutations can involve substitutions, insertions, or deletions of one or more nucleotides in a DNA molecule. Mutations that alter the sequence but not the number of nucleotides in a gene are called **nucleotide substitutions**. Generally, these substitutions involve only one or a small number of nucleotides.

Frameshift mutations cause the *insertion* or *deletion* of bases in a DNA molecule. Because codons are composed of three bases, changing the number of bases in a gene can change the mRNA codon specified at the site of the insertion or deletion *and* the sequence of all following codons. These mutations cause large-scale changes in the amino acid sequence of the newly synthesized polypeptide chains and alter the structure

Nucleotide substitutions Mutations that involve replacement of one or more nucleotides in a DNA molecule with other nucleotides.

Frameshift mutations Mutational events in which a number of bases (other than multiples of three) are added to or removed from DNA, causing a shift in the codon reading frame.

and function of the derived protein. We will begin our discussion of mutation at this level by examining nucleotide substitutions and then consider frameshift mutations.

KEEP IN MIND

Mutations in DNA can occur in several ways, including nucleotide substitution, deletion, and insertion.

Many hemoglobin mutations are caused by nucleotide substitutions.

Scientists have identified several hundred single amino acid substitutions in the alpha and beta globin components of hemoglobin, the oxygen-carrying molecule in red blood cells. These variants provide examples of how a change in one nucleotide in a gene can affect both the structure and the function of a protein. In the following discussion, keep in mind that the term *DNA triplet* refers to the nucleotides that specify a codon and that the term *codon* refers to the sequence of three nucleotides in mRNA that specifies an amino acid.

Missense mutations Mutations that cause the substitution of one amino acid for another in a protein.

Missense mutations are single nucleotide changes in DNA that end up substituting one amino acid for another in a protein. These substitutions can have a variety of phenotypic effects, some of which cause genetic disorders. However, others do not affect protein function and often have no phenotypic consequences. To illustrate, let's consider three examples of the phenotypes associated with changes to amino acid number 6 in beta globin. Normal beta globin (HbA) mRNA has the codon GAG encoding glutamic acid as amino acid number 6 (Figure 11.8). In the first example, a single nucleotide substitution changes GAG (glu) to GUG, inserting valine (val) at position 6. When homozygous, this mutation causes sickle cell anemia (HbS), a condition with a potentially lethal phenotype (see Chapter 4). In the second example, a GAG → AAG change at codon number 6 changes amino acid number 6 from glutamic acid to lysine. In the homozygous condition, this change results in a disorder called HbC (hemoglobin C disease). HbC causes a mild set of clinical conditions that resembles but is less serious than sickle cell anemia. In the third example, a change in codon 6 from GAG (glu) → GCG (ala) produces a beta globin variant known as Hb Makassar. When heterozygous or homozygous, this amino acid substitution at position 6 causes no clinical symptoms and is regarded as harmless.

Sense mutations Mutations that change a termination codon into one that codes for an amino acid. Such mutations produce elongated proteins.

Normal HbA	1	2	3	4	5	6	7	8
DNA	CAC	GTG	GAC	TGA	GGA	CTC	CTC	TTC
mRNA	GUG	CAC	CUG	ACU	CCU	GAG	GAG	AAG
Amino acid	val	his	leu	thr	pro	glu	glu	lys

Sickle HbS	1	2	3	4	5	6	7	8
DNA	CAC	GTG	GAC	TGA	GGA	CAC	CTC	TTC
mRNA	GUG	CAC	CUG	ACU	CCU	GUG	GAG	AAG
Amino acid	val	his	leu	thr	pro	val	glu	lys

HbC	1	2	3	4	5	6	7	8
DNA	CAC	GTG	GAC	TGA	GGA	TTC	CTC	TTC
mRNA	GUG	CAC	CUG	ACU	CCU	AAG	GAG	AAG
Amino acid	val	his	leu	thr	pro	lys	glu	lys

Hb Makassar	1	2	3	4	5	6	7	8
DNA	CAC	GTG	GAC	TGA	GGA	CGC	CTC	TTC
mRNA	GUG	CAC	CUG	ACU	CCU	GCG	GAG	AAG
Amino acid	val	his	leu	thr	pro	ala	glu	lys

FIGURE 11.8 The first eight DNA triplets, mRNA codons, and amino acids of normal adult hemoglobin (HbA), sickle cell hemoglobin (HbS), hemoglobin C (HbC), and hemoglobin Makassar (HbMk). A single nucleotide substitution in codon 6 is responsible for the changes in these forms of hemoglobin.

In all three cases, the proteins differ only in amino acid number 6: Normal hemoglobin (HbA) has glutamic acid (glu), HbS has valine (val), HbC has lysine (lys), and Hb Makassar has alanine (ala). The other 145 amino acids in the protein are unchanged. In these examples, single nucleotide changes in the sixth codon of the beta globin gene result in phenotypes that range from harmless (Hb Makassar), to the mild clinical symptoms of HbC, to the serious and potentially life-threatening consequences of HbS (sickle cell anemia).

Other nucleotide substitutions produce proteins that are longer or shorter than normal. **Sense mutations** produce longer-than-normal proteins by changing a termination codon (UAA, UAG, or UGA) into one that codes for an amino acid. Several hemoglobin variants with longer-than-normal globin molecules are shown in Table 11.3. In each case, the extended polypeptide chain is produced by a single nucleotide substitution in the normal termination codon. In hemoglobin Constant Springs-1, the stop codon at position 142 in alpha globin mRNA is changed from UAA to CAA, replacing a stop codon with a glycine codon. In this case, the mRNA sequence following the CAA codon allows the insertion of 30 additional amino acids into the alpha globin molecule before another stop codon is reached.

Nonsense mutations change codons for amino acids into one of the three stop codons: UAA, UAG, or UGA (see Figure 9.3). This shortens the protein product. In the McKees Rock variant of beta globin, a change in codon 144 from UAU (tyr) to UAA (stop) results in a beta chain that is 143 amino acids long instead of 146. This change has little or no effect on the function of the beta globin molecule as a carrier of oxygen. However, other nucleotide substitutions can produce more drastic changes in polypeptide length and have more serious phenotypic effects. In one case, a single nucleotide substitution creates a UAG stop codon at position 39 in beta globin mRNA. This truncated mRNA is not functional, reducing the production of beta globin and causing a serious disorder called beta thalassemia.

Mutations can be caused by nucleotide deletions and insertions.

Nucleotide deletions and insertions within a gene can range from the deletion or insertion of one nucleotide to the duplication or deletion of an entire gene. Genomic analysis of human genes has revealed that deletions and insertions are a major cause of genetic disorders and account for 5% to 10% of all known mutations. As defined earlier, insertion or deletion of nucleotides within the coding sequence of a gene are frameshift mutations. Because codons consist of groups of three bases, adding or subtracting a base from one codon changes the coding sense of all the following codons in the gene. A frameshift mutation changes the amino acid sequence of the protein from the site of the mutation to the end of the protein. Suppose that a codon series reads as the following sentence:

THE FAT CAT ATE HIS HAT

A nucleotide (in this case, an A) inserted in the second codon destroys the sense of the remaining message:

insertion

THE FAA TCA TAT EHI SHA T

Similarly, a deletion in the second codon can also generate an altered message:

THE FTC ATA TEH ISH AT

A   deletion

In nucleotide substitutions, the number of nucleotides in the gene remains unchanged, and usually only one amino acid in the protein is altered (Figure 11.8). Frameshift mutations change the number of nucleotides in the gene and usually cause large-scale changes in the amino acid sequence of the protein (Active Figure 11.9). Most often, these changes result in a nonfunctional, shortened polypeptide. However, in some hemoglobin variants, the frameshift occurs near the end of the gene, producing a globin molecule with an extended chain that has a minimal impact on the function of the protein.

In normal alpha globin, the mRNA codons for the last few amino acids are as follows:

Position number	138	139	140	141	STOP
mRNA codon	UCC	AAA	UAC	CGU	UAA
Amino acid	ser	lys	tyr	arg	

In a variant of alpha hemoglobin called Hb Wayne, the last nucleotide (A) in codon 139 is deleted, producing a frameshift:

Position number	138	139	140	141	142	143	144	145	146	STOP
mRNA codon	UCC	AAU	ACC	GUU	AAG	CUG	GAG	CCU	CGG	UAG
Amino acid	ser	asn	thr	val	lys	leu	gln	pro	arg	

In this case, deletion of a single nucleotide (an A) shifts the codon reading frame one nucleotide to the left, so that the stop codon (UAA) at position 142 is split in two and changes from UAA to AAG. This destroys the stop codon and allows insertion of additional amino acids until another stop codon (created by the deletion) is reached. The result is an alpha chain variant with 146 amino acid residues instead of 141.

Table 11.3 Alpha Globins with Extended Chains Produced by Nucleotide Substitutions

Alpha Globins	Abnormal Chains
Constant Springs-1	gln (142) + 30 amino acids
Icaria	lys (142) + 30 amino acids
Seal Rock	glu (142) + 30 amino acids
Koya Dora	ser (142) + 30 amino acids

Nonsense mutations Mutations that change an amino acid specifying a codon to one of the three termination codons.

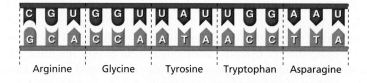

mRNA transcribed from the DNA

| C | G | U | G | G | U | U | A | U | U | G | G | A | A | U |

DNA TEMPLATE STRAND

| G | C | A | C | C | A | A | T | A | A | C | C | T | T | A |

Resulting amino acid sequence   Arginine   Glycine   Tyrosine   Tryptophan   Asparagine

Altered message in mRNA

| C | G | U | G | G | U | U | U | A | U | U | G | G | A | A | U |

A BASE INSERTION (RED) IN DNA

| G | C | A | C | C | A | A | A | T | A | A | C | C | T | T | A |

The altered amino acid sequence   Arginine   Glycine   Leucine   Leucine   Glutamic acid

ACTIVE FIGURE 11.9 Frameshift mutations. Insertions are mutations in which extra bases insert into genes. Insertions change the reading frame of the DNA code and mRNA codons, causing the wrong amino acids to be inserted into the protein product.

CENGAGENOW   Learn more about frameshift mutations by viewing the animation by logging on to *login.cengage.com/sso* and visiting CengageNOW's Study Tools.

Mutations can involve more than one nucleotide.

Up to this point, we have discussed mutations that involve single nucleotides. However, as outlined earlier, other mutations involve multiple nucleotide sequences called trinucleotide repeats. This class of mutations is associated with a number of genetic disorders. Trinucleotide repeats are sequences of three nucleotides repeated several times in consecutive order within or adjacent to a gene. Mutations expand the number of repeats within a gene, converting a normal allele into a mutant allele, a phenomenon called **allelic expansion**.

Allelic expansion was first observed in the X chromosome gene *FMR-1*, which is associated with fragile-X syndrome (OMIM 309550), the most common form of hereditary mental retardation. About 1% of all males institutionalized for mental retardation have this syndrome. The name of this condition derives from the observation that X chromosomes carrying the mutation have a stalk at the end of the long arm to which the chromosome tip is attached (Figure 11.10).

Normal alleles of the *FMR-1* gene contain from 6 to 52 CGG trinucleotide repeats. These alleles are stable and do not undergo expansion. Mutant *FMR-1* alleles have more than 230 CGG repeats, causing inactivation of *FMR-1* and producing mental retardation in all males and in 60% of heterozygous females who inherit this X chromosome (Figure 11.11).

More than a dozen genetic disorders are associated with expansion of trinucleotide repeats. Some of these are listed in Table 11.4. As with fragile-X syndrome, there is a correlation between the size of the expanded repeat, earlier age of onset, and the severity of symptoms.

Allelic expansion Increase in gene size caused by an increase in the number of trinucleotide sequences.

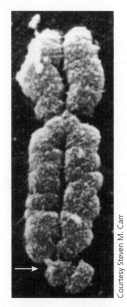

Courtesy Steven M. Carr

FIGURE 11.10 A human X chromosome showing a constriction (arrow) at the tip of the long arm, known as a fragile site. This site, caused by allelic expansion of a CGG repeat, is associated with fragile-X syndrome, one of the most common forms of mental retardation. About 1 in 1,250 male births have fragile-X syndrome.

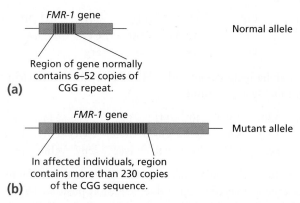

FMR-1 gene

Region of gene normally contains 6–52 copies of CGG repeat.

(a)

Normal allele

FMR-1 gene

In affected individuals, region contains more than 230 copies of the CGG sequence.

(b)

Mutant allele

FIGURE 11.11 Allelic expansion in the *FMR-1* gene at the fragile-X locus. (a) The gene normally contains 6 to 52 copies of a CGG trinucleotide repeat. (b) Affected individuals have more than 230 copies of the repeat.

Table 11.4 Some Mutations with Expanded Trinucleotide Repeats

Gene	Triplet Repeat	Normal Copy	Copy in Disease	OMIM Number
Spinal and bulbar muscular atrophy	CAG	12–34	40–62	313200
Spinocerebellar ataxia type 1	CAG	6–39	41–81	164400
Huntington disease	CAG	6–37	35–121	143100
Haw-River syndrome	CAG	7–34	54–70	140340
Machado-Joseph disease	CAG	13–36	68–79	109150
Fragile-X syndrome	CGG	6–52	230–72,000	309550
Myotonic dystrophy	CTG	5–37	50–72,000	160900
Friedreich ataxia	GAA	10–21	200–900	229300

Trinucleotide repeat expansion is related to anticipation.

The first five trinucleotide repeat disorders listed in Table 11.4 have some common characteristics. All are progressive, degenerative disorders of the nervous system inherited as autosomal dominant traits. All involve expansion of CAG repeats, and all show a correlation between the increasing size of the repeat and earlier age of onset in subsequent generations. In these disorders, mildly affected parents have more seriously affected offspring, who develop symptoms at an earlier age than their parents did.

The appearance of more severe symptoms at earlier ages in succeeding generations, which is called **anticipation**, was noted first in myotonic dystrophy (Figure 11.12). Although it was carefully documented by clinicians, geneticists originally discounted the phenomenon of anticipation because genes were regarded as highly stable entities that have only occasional mutations.

The discovery of anticipation means that the concept of mutation must take into account the existence of unstable genomic regions that undergo dynamic changes. In other words, when genes with trinucleotide repeats undergo allelic expansion, this event enhances the instability of the gene and the chance that there will be further expansions and the development of a disease phenotype. Forms of genome instability similar to trinucleotide expansion might explain genetic disorders that do not show simple Mendelian inheritance (for example, those showing incomplete penetrance or variable expression).

Anticipation Onset of a genetic disorder at earlier ages and with increasing severity in successive generations.

11.6 Mutations and DNA Damage Can Be Repaired

Not every mutation becomes a permanent genomic change. All cells have enzyme systems that repair mutations and damage to DNA. Table 11.5 estimates the rate of spontaneous DNA damage in a typical mammalian cell at 37°C (body temperature). Without repair systems, over time the accumulated damage would destroy much of the DNA in the cell. Cells that accumulate a lot of DNA damage can have several fates:

1. The cells can become dormant, a condition called senescence.
2. Control systems in the cell can induce cell suicide, a process called apoptosis.
3. The accumulated damage can cause the cell to escape the normal controls of the cell cycle and become cancerous.

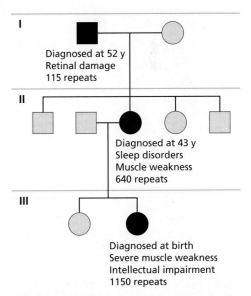

FIGURE 11.12 A pedigree showing the autosomal dominant inheritance of myotonic dystrophy. This disorder is associated with anticipation, in which the disorder increases in severity and is expressed at earlier and earlier ages over several generations. Myotonic dystrophy is associated with alleleic expansion of a CTG trinucleotide repeat.

In figure labels:
Diagnosed at 52 y
Retinal damage
115 repeats

Diagnosed at 43 y
Sleep disorders
Muscle weakness
640 repeats

Diagnosed at birth
Severe muscle weakness
Intellectual impairment
1150 repeats

Table 11.5 Rates of DNA Damage in a Mammalian Cell

Damage	Events/Hour
Depurination	580
Depyrimidation	29
Deamination of cytosine	8
Single-stranded breaks	2,300
Single-stranded breaks after depurination	580
Methylation of guanine	130
Pyrimidine (thymine) dimers in skin (noon Texas sun)	5×10^4
Single-stranded breaks from background ionizing radiation	10^{-4}

Fortunately, humans (and other organisms) have several highly efficient DNA repair systems (Table 11.6). However, because the rate of background damage is so high, it is easy to overload the repair systems.

KEEP IN MIND

Damage to DNA can be repaired during DNA synthesis and by enzymes that repair damage caused by radiation or chemicals.

Table 11.6 Maximum DNA Repair Rates in a Human Cell

Damage	Repairs/Hour
Single-stranded breaks	2×10^5
Pyrimidine dimers	5×10^4
Guanine methylation	10^4–10^5

Cells have several DNA repair systems.

To maintain the integrity of DNA, cells have a collection of enzyme systems that monitor and repair mutations and DNA damage. One of these systems corrects errors made during DNA replication. In humans, replication proceeds rapidly, with 10 to 20 nucleotides added each second to a DNA strand at each replication site. During replication, an incorrect nucleotide can be inserted into the newly synthesized strand, producing a potential mutation. However, DNA polymerase, the enzyme involved in replication, corrects many of these errors. In addition to directing DNA replication, the enzyme has a proofreading function. If an incorrect nucleotide is inserted, the enzyme can detect the error and reverse direction, move backwards and remove nucleotides until the incorrect nucleotide is eliminated (Active Figure 11.13). Then the enzyme inserts the correct nucleotide

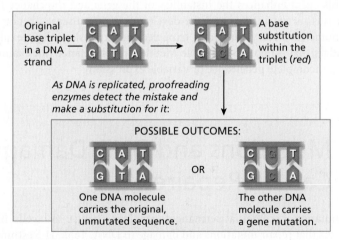

Example of a base pair substitution

ACTIVE FIGURE 11.13 Base-pair substitutions have two possible outcomes. The mutation can be detected by DNA polymerase proofreading and corrected (*bottom left*) or remain undetected and become a mutation.

CENGAGENOW Learn more about base pair substitutions by viewing the animation by logging on to *login.cengage.com/sso* and visiting CengageNOW's Study Tools.

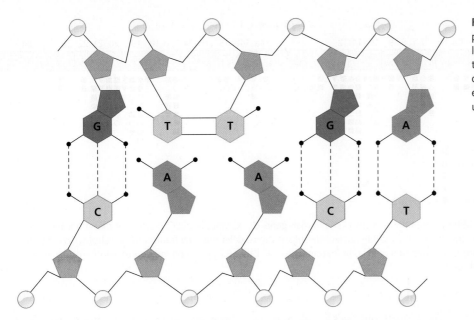

FIGURE 11.14 Thymine dimers are produced when ultraviolet light cross-links two adjacent thymine bases in the same strand of DNA. This structure causes a distortion in the DNA, and errors in replication are likely to occur unless corrected.

and moves forward, resuming replication. The few mistakes that elude the proofreading function of DNA polymerase remain as newly created mutations.

Other enzyme systems recognize and repair DNA damage in other phases of the cell cycle. These systems fall into several categories, each controlled by a number of system-specific genes. For example, exposure of DNA to UV light (from sunlight and tanning lamps) causes thymine bases adjacent to each other in the same DNA strand to pair with each other, forming **thymine dimers** (Figure 11.14). Thymine dimers distort the DNA molecule and can interfere with normal replication, producing mutations. These dimers are corrected by several different DNA repair mechanisms.

Thymine dimer A molecular lesion in which chemical bonds form between a pair of adjacent thymine bases in a DNA molecule.

Genetic disorders can affect DNA repair systems.

Because DNA repair is under genetic control, it also can undergo mutation. Several genetic disorders, including xeroderma pigmentosum (XP; OMIM 278700), are caused by mutations in genes that repair DNA. XP is an autosomal recessive disorder with a frequency of 1 in 250,000 individuals. Those affected with XP are extremely sensitive to sunlight (which contains UV light). Even short exposure to the sun causes dry, flaking skin and pigmented spots that can develop into skin cancer (Figure 11.15). Skin cancers are about 1,000 times more common in XP individuals. Early death from cancer is the usual fate of XP individuals who do not take extraordinary measures to protect themselves from UV light.

Several other genetic disorders of DNA repair are characterized by unusual sensitivity to sunlight and/or to other forms of radiation. These include Fanconi anemia (OMIM 227650), ataxia telangiectasia (OMIM 208900), and Bloom syndrome (OMIM 210900). The range of phenotypes seen in these disorders indicates that DNA repair is a complex process that involves many different genes.

11.7 Mutations, Genotypes, and Phenotypes

Sickle cell anemia was the first genetic disorder to be analyzed at the molecular level. Recall that a nucleotide substitution in codon 6 changes the amino acid in the beta globin polypeptide at position 6 and produces a distinctive set of clinical symptoms. All affected individuals and all heterozygotes have the same nucleotide substitution.

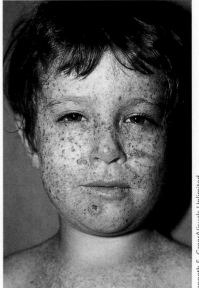

Kenneth E. Greer/Visuals Unlimited

FIGURE 11.15 Child affected with xeroderma pigmentosum. Affected individuals cannot repair damage to DNA caused by ultraviolet light from the sun and other sources.

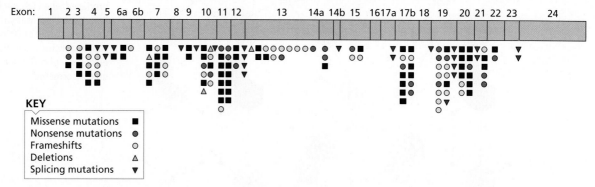

FIGURE 11.16 Distribution of mutations in the cystic fibrosis gene, *CFTR*. More than 1,600 different mutations have been discovered. The mutations shown here include nucleotide substitutions, deletions, and frameshift mutations. Any of these mutations in the homozygous condition or in combination with each other (i.e., a compound heterozygote) results in the phenotype of cystic fibrosis.

As it turns out, sickle cell anemia is probably an exception rather than the rule. Analysis of mutations in other genes—such as the gene for cystic fibrosis—reveals that more often than not a number of different mutations in a single gene can produce the phenotype associated with a genetic disorder (Figure 11.16).

Analysis of mutations in the cystic fibrosis gene provides a clear example of the types and numbers of mutations that can occur in a single gene, each of which can produce a disease phenotype. More than 1,600 different mutations have been identified in the cystic fibrosis gene (*CFTR*). These include single nucleotide substitutions and deletions and larger deletions that involve one or more regions of the gene, as well as frameshift mutations and mutations at intron-exon splice sites. Mutations are distributed in all regions of the *CFTR* gene (Figure 11.16), strengthening the idea that any mutational event that interferes with the expression of a gene produces an abnormal phenotype.

People with cystic fibrosis have a wide range of clinical symptoms. The relationship between the type and location of the *CFTR* mutation and the clinical phenotype has been investigated for a number of mutant alleles. In some mutations, such as the Δ508 deletion (present in 70% of all cases of CF), a mutant CFTR protein is synthesized in deletion homozygotes, but does not fold properly in the endoplasmic reticulum and is destroyed. As a result, no CFTR protein is inserted into the plasma membrane, chloride ion transport is disrupted, and clinical symptoms are severe. In other mutations, such as R117, R334, and R347 (Figure 11.17), mutant CFTR proteins are synthesized in homozygous individuals and inserted into the membrane, but these are only partially functional. Because these mutant CFTRs retain some function, the mutations are associated with milder forms of cystic fibrosis.

FIGURE 11.17 Mutations in the *CFTR* gene differ in their phenotypic effects. When homozygous, mutations R117, R334, or R347 allow between 5% and 30% of normal activity for the gene product and produce only mild symptoms. These nucleotide substitution mutations are not common: Together they account for about 2% of all cases of cystic fibrosis. The most common mutation in European populations, Δ508, causes an amino acid deletion in a cytoplasmic region of the protein and accounts for 70% of all mutations in the cystic fibrosis gene. This mutation inactivates the CFTR protein and is associated with severe symptoms.

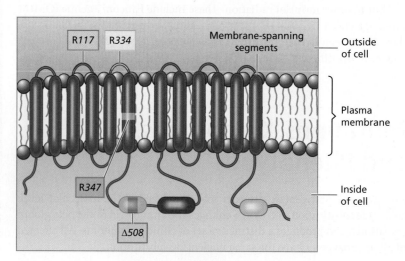

Because there are over 1,600 different mutations known in the *CFTR* gene, it is possible for someone with CF to carry two different mutant alleles and be a compound heterozygote (also known as a double heterozygote). This genetic variability further contributes to the phenotypic variability seen in cystic fibrosis.

11.8 Genomic Imprinting Is a Reversible Alteration of the Genome

We all carry two copies of each gene: one from our mothers and one from our fathers. As Mendel showed in his reciprocal crosses, normally there is no difference in the expression of the two copies. It is now known that parental origin often determines whether the maternal or paternal allele of certain genes will be expressed. This pattern of differential expression is called **genomic imprinting**.

The first evidence of imprinting in mammals was discovered in mice when two haploid nuclei were transplanted into eggs to produce zygotes carrying an all-female or all-male genome rather than the usual combination of one male and one female genome (Figure 11.18). Experimental embryos with a diploid male genome form abnormal embryos but have normal placentas. Embryos with only a diploid female genome develop normal embryos but have abnormal placentas. Both conditions are lethal, and we can conclude that both a maternal genome and a paternal genome are required for normal development.

Genomic imprinting Phenomenon in which the expression of a gene depends on whether it is inherited from the mother or the father; also known as genetic or parental imprinting.

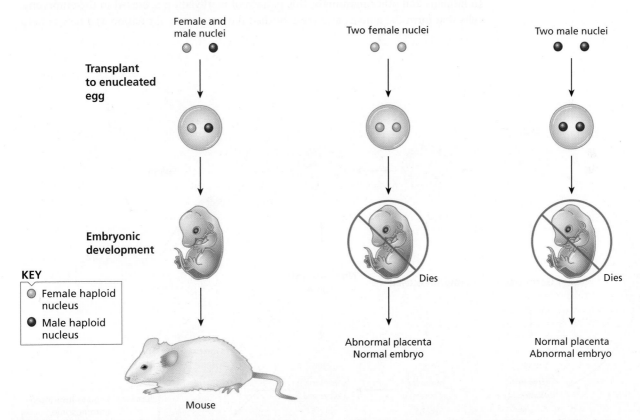

FIGURE 11.18 In mice and other mammals, including humans, embryonic development proceeds normally (*left*) when fertilized eggs contain a maternal and a paternal genome. However, for some genes, parent-specific expression is required for normal development. Fertilized mouse eggs containing two haploid female nuclei (*center*) develop into normal embryos but have abnormal placentas, and the embryos die. Fertilized eggs carrying two haploid male nuclei (*right*) develop into abnormal embryos with normal placentas. The abnormal embryos die. These experiments demonstrate that differential expression of maternal and paternal genes by imprinting is required for normal development.

Genomic imprinting plays a role in several human genetic disorders, including Prader-Willi syndrome (PWS; OMIM 176270) and Angelman syndrome (AS; OMIM 105830). Infants with PWS are small at birth, mentally retarded, and have trouble nursing. As children, they develop almost uncontrollable appetites and are often obese. About 80% of all cases of PWS carry a small deletion in one copy of chromosome 15. In PWS, the deleted chromosome is always inherited from the father. In other words, PWS is caused by a lack of paternal genes in a region of chromosome 15.

If the copy of chromosome 15 carrying the deletion is inherited from the mother, it does not result in PWS but instead produces a different disorder, called Angelman syndrome (AS). Children with AS have poor coordination, uncontrollable movements, seizures, and spontaneous laughter. These two disorders are caused by mutations in the same region of chromosome 15 but have very different symptoms. Deletion of paternal genes on chromosome 15 results in PWS, and deletion of maternal genes from the same region of chromosome 15 produces AS. The two disorders illustrate that for normal development, both parents must contribute genes from this region of chromosome 15.

Imprinting does not affect all genes. Only genes in certain regions of human chromosomes 4p, 8q, 15q, 17p, 18p, 18q, and 22q are imprinted. Imprinting is not a mutation or permanent change in a gene or a chromosome region. What is affected is the expression of a gene, not the gene itself; imprinting does not violate the Mendelian principles of segregation or independent assortment.

Remember that a chromosome received by a female from her father is transmitted as a maternal chromosome in the next generation. In each generation, the previous imprinting is erased, and a new pattern of imprinting defines the chromosome as either paternal or maternal (Figure 11.19). Imprinting events take place during gamete formation, and the effects are transmitted to all tissues of the offspring. The mechanism of imprinting involves methylation of DNA, a mechanism of gene silencing (see Chapter 9). In humans and other mammals, this pattern of methylation is erased in the embryonic cells that form the gonads and reestablished during gamete formation as a new pattern

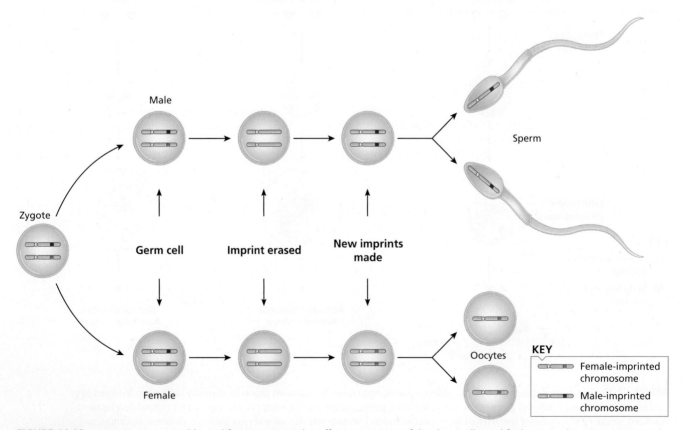

FIGURE 11.19 Imprinting is a reversible modification in DNA that affects expression of the chemically modified gene. A chromosome received by a female from her father has its imprint erased and is reprinted as a female chromosome before being transmitted to the next generation. Similarly, a chromosome received by a male from his mother is reprinted as a male chromosome before being passed to the next generation.

of imprinting. Although imprinting involves modification of the DNA, it is not strictly a mutational event. These modifications are one example of **epigenetics**. These and other epigenetic changes to the genome involve reversible chemical modifications of DNA (such as methylation of bases) that change the pattern of gene expression without affecting the nucleotide sequence of the DNA.

Epigenetics Reversible chemical modifications of chromosomal DNA (such as methylation of bases) and/or associated histone proteins that change the pattern of gene expression without affecting the nucleotide sequence of the DNA.

Genetics in Practice

Genetics in Practice case studies are critical-thinking exercises that allow you to apply your new knowledge of human genetics to real-life problems. You can find these case studies and links to relevant websites at *www.cengage.com/biology/cummings*

CASE 1

On April 26, 1986, one of the four reactors at the Chernobyl generating station in the Soviet Union melted down. It has been reported that the plant was running with disconnected safety measures. The result was fire, chaos, fear, a cloud of radioactive isotopes spreading across vast reaches of eastern Europe, and the radioactive contamination of thousands of people.

Unfortunately, this human tragedy is not being investigated as it should be, according to scientists who are trying to learn from it. Cancer-prevention specialists claim that there is a lack of resources available to look at cancer cases in the population exposed to radiation released from Chernobyl.

Scientists who want to study the aftermath of Chernobyl face many obstacles: (1) The dissolution of the Soviet Union split administrative, record-keeping, and medical responsibilities among Belarus, Ukraine, and Russia; (2) a general decline in living standards has reduced the level of medical care in the area; (3) individual doses cannot be reconstructed accurately and estimates of doses received by individuals have been complicated by the fact that some of the dose was external through exposure to radioactive dust and part was internal through eating contaminated food; and (4) little money is available for the types of large studies that are needed for extracting the best data.

Here are some of the facts—as best we know—about the Chernobyl meltdown. The accident released 1.85×10^{18} (1,850,000,000,000,000,000) international units of radioactive material. The releases contaminated an estimated 17 million people to some degree. The exact amount of exposure depended on location, wind direction, length of exposure, and eating habits, as well as whether a person was a "liquidator." These unfortunate heroes were pressed into service in a crude cleanup effort after the accident. One hundred thirty-four people showed signs of acute radiation sickness immediately after the accident. Many of the

28 people who died from acute radiation sickness had skin lesions covering 50% or more of their bodies. After the fire, 135,000 people evacuated the area around the reactor, and 800,000 liquidators moved in to try to decontaminate that area. Approximately 17% to 45% of the liquidators received doses between 10 and 25 rads (a rad is a unit of radiation dose and stands for *radiation absorbed dose*). For comparison, U.S. safety guidelines permit an annual dose to the general public of 0.1 rad; nuclear workers are permitted 5 rads.

Despite the early confusion, some medical information is available in Chernobyl's aftermath. The most compelling data involve radiation exposure and thyroid cancer, particularly in children. According to the International Chernobyl Conference (April 1996), radiation exposure caused "a substantial increase in reported cases of thyroid cancer in Belarus, Ukraine, and some parts of Russia, especially in young children." This is thought to be due to exposure to radioactive iodine during the early phases of the accident in 1986. By the end of 1995, approximately 800 cases of thyroid cancer had been reported in children who were under age 15 at the time of diagnosis. To date, three of these thyroid-cancer victims have died and several thousand more cases of thyroid cancer are expected. Ironically, most of the thyroid cancers could have been prevented if people in the contaminated areas had taken iodine tablets immediately after the accident. Most iodine in the body goes to the thyroid gland; if enough normal iodine is available, only a small amount of radioactive isotope irradiates the little gland, whose hormones help regulate growth. More information on the long-term effects of Chernobyl's explosion will be learned in the coming decades, as scientists and health care workers assess the full impact of this disaster.

1. How do you think radiation causes cancer?

2. The liquidators were exposed to large amounts of radiation. Are their families at risk even though they were not part of the cleanup effort?

3. What kind of compensation, if any, should liquidators receive from the government for their exposure to radiation?

4. An increase in thyroid cancer was reported after Chernobyl. Are these people at risk of passing a mutant cancer gene to their offspring?

Summary

11.1 Mutations Are Heritable Changes in DNA

- Without the phenotypic variations produced by mutations, it would be difficult to determine whether a trait is under genetic control and impossible to determine its mode of inheritance.

11.2 Mutations Can Be Detected in Several Ways

- Mutations can be classified in a variety of ways by using criteria such as pattern of inheritance, phenotype, biochemistry, and degrees of lethality.

- Dominant mutations are the easiest to detect because they are phenotypically expressed in the heterozygous condition. Accurate pedigree information can often be used to identify the individual in whom a mutation arose. It is more difficult to determine the origin of sex-linked recessive mutations, but an examination of the male progeny is often informative. If the mutation in question is autosomal recessive, it is almost impossible to identify the original mutant individual.

11.3 Measuring Spontaneous Mutation Rates

- Studies of mutation rates in a variety of dominant and sex-linked recessive traits indicate that mutations in the human genome are rare events, occurring about once in every 1 million copies of a gene.

11.4 Environmental Factors Influence Mutation Rates

- Environmental agents, including radiation and chemicals, can cause mutations.

11.5 Mutations at the Molecular Level: DNA as a Target

- Molecular analysis of mutations has shown a direct link between gene, protein, and phenotype. Mutations arise spontaneously as the result of errors in DNA replication or as the result of structural shifts in nucleotide bases. Environmental agents, including chemicals and radiation, also cause mutations. Frameshift mutations cause a change in the reading frame of codons, often producing dramatic alterations in the structure and function of proteins.

11.6 Mutations and DNA Damage Can Be Repaired

- Not all mutations cause genetic damage. Cells have a number of DNA repair systems that correct errors in replication and repair damage caused by environmental agents such as ultraviolet light, radiation, and chemicals.

11.7 Mutations, Genotypes, and Phenotypes

- In most genes associated with a genetic disorder, many different types of mutations can cause a mutant phenotype. In the cystic fibrosis gene, more than 1,600 different mutations have been identified, including deletions, nucleotide substitutions, and frameshift mutations.

11.8 Genomic Imprinting Is a Reversible Alteration of the Genome

- Genomic imprinting alters the expression of normal genes, depending on whether they are inherited maternally or paternally. Imprinting has been implicated in a number of disorders, including Prader-Willi and Angelman syndromes. Not all regions of the genome are affected, and only segments of chromosomes 4, 8, 17, 18, and 22 are imprinted. Genes are not altered permanently by imprinting but are reimprinted in each generation.

Questions and Problems

CENGAGENOW Preparing for an exam? Assess your understanding of this chapter's topics with a pre-test, a personalized learning plan, and a post-test by logging on to *login.cengage.com/sso* and visiting CengageNOW's Study Tools.

Measuring Spontaneous Mutation Rates

1. Define mutation rate.
2. Achondroplasia is an autosomal dominant form of dwarfism caused by a single gene mutation. Calculate the mutation rate of this gene given the following data: 10 achondroplastic births to unaffected parents in 245,000 births.
3. Why is it almost impossible to directly measure the mutation rates in autosomal recessive alleles?
4. What are the factors that influence the mutation rates of human genes?
5. Achondroplasia is a rare dominant autosomal defect resulting in dwarfism. The unaffected brother of an individual with achondroplasia is seeking counsel on the likelihood of his

being a carrier of the mutant allele. What is the probability that the unaffected client is carrying the achondroplasia allele?

Environmental Factors Influence Mutation Rates

6. Although it is well known that X-rays cause mutations, they are routinely used to diagnose medical problems, including potential tumors, broken bones, and dental cavities. Why is this done? What precautions need to be taken?

7. You are an expert witness called by the defense in a case in which a former employee is suing an industrial company because his son was born with muscular dystrophy, an X-linked recessive disorder. The employee claims that he was exposed to mutagenic chemicals in the workplace that caused his son's illness. His attorney argues that neither the employee, his wife, nor their parents have this genetic disorder, and, therefore, the disease in the employee's son represents a new mutation. How would you analyze this case? What would you say to the jury to refute this man's case?

8. Bruce Ames and his colleagues have pointed out that although detailed toxicological analysis has been conducted on synthetic chemicals, almost no information is available about the mutagenic or carcinogenic effects of the toxins produced by plants as a natural defense against fungi, insects, and animal predators. Tens of thousands of such compounds have been discovered, and he estimates that in the United States adults eat about 1.5 g of these compounds each day—levels that are approximately 10,000 times higher than those of the synthetic pesticides present in the diet. For example, cabbage contains 49 natural pesticides and metabolites, and only a few of these have been tested for their carcinogenic and mutagenic effects.
   a. With the introduction of new foods into the U.S. diet over the last 200 years (mangoes, kiwi fruit, tomatoes, and so forth), has there been enough time for humans to develop resistance to the mutagenic effects of the toxins present in those foods?
   b. The natural pesticides present in plants constitute more than 99% of the toxins we eat. Should diet planning, especially for vegetarians, take into account the doses of toxins present in the diet?

Mutations at the Molecular Level: DNA as a Target

9. Define and compare the following types of nucleotide substitutions. Which is likely to cause the most dramatic mutant effect?
   a. missense mutation
   b. nonsense mutation
   c. sense mutation

10. If the coding region of a gene (the exons) contains 2,100 base pairs of DNA, would a missense mutation cause a protein to be shorter, longer, or the same length as the normal 700 amino acid proteins? What would be the effect of a nonsense mutation? A sense mutation?

11. Two types of mutations discussed in this chapter are (1) nucleotide changes and (2) unstable genome regions that undergo dynamic changes. Describe each type of mutation.

12. What is a frameshift mutation?

13. A frameshift mutation is caused by:
   a. a nucleotide substitution
   b. a three-base insertion
   c. a premature stop codon
   d. a one-base insertion
   e. a two-base deletion

14. In the gene-coding sequence shown here, which of the following events will produce a frameshift after the last mutational site?

normal mRNA:	UCC	AAA	UAC	CGU	CGU	UAA
normal amino acids:	ser	lys	tyr	arg	arg	stop

   a. insertion of an A after the first codon
   b. deletion of the second codon (AAA)
   c. insertion of TA after the second codon and deletion of CG in the fourth codon
   d. deletion of AC in the third codon

15. Trinucleotide repeats cause serious neurodegenerative disorders such as Huntington disease, fragile-X syndrome, and myotonic dystrophy (DM). The process of anticipation causes the appearance of symptoms at earlier ages in succeeding generations. Describe the current theory of the way anticipation works.

16. Familial retinoblastoma, a rare autosomal dominant defect, arose in a large family that had no prior history of the disease. Consider the following pedigree (the darkly colored symbols represent affected individuals):

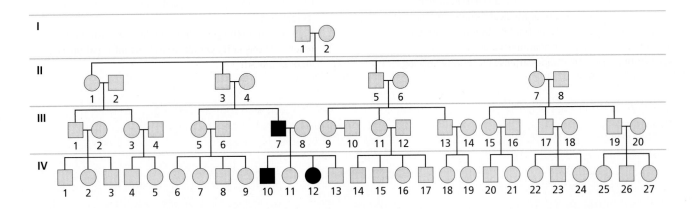

a. Circle the individual(s) in which the mutation most likely occurred.

b. Is the person who is the source of the mutation affected by retinoblastoma? Justify your answer.

c. Assuming that the mutant allele is fully penetrant, what is the chance that an affected individual will have an affected child?

17. Tay-Sachs disease is an autosomal recessive disease. Affected individuals do not often survive to reproductive age. Why has Tay-Sachs persisted in humans?

Mutations and DNA Damage Can Be Repaired

18. Replication involves a period of time during which DNA is particularly susceptible to the introduction of mutations. If nucleotides can be incorporated into DNA at a rate of 20 nucleotides/second and the human genome contains 3 billion nucleotides, how long will replication take? How is this time reduced so that replication can take place in a few hours?

19. Our bodies are not defenseless against mutagens that alter our genomic DNA sequences. What mechanisms are used to repair DNA?

Mutations, Genotypes, and Phenotypes

20. The cystic fibrosis gene encodes a chloride channel protein necessary for normal cellular functions. Let us assume that if at least 5% normal channels are present, the affected individual has mild symptoms of cystic fibrosis. Having less than 5% normal channels produces severe symptoms. At least 50% of the channels must be expressed for the individual to be phenotypically normal. This gene has various mutant recessive alleles:

Allele	Molecular Defect	Percentage of Functional Channels	Symptoms
CF100	Exon deletion	0%	Severe
CF1	Missense mutation in 5′ flanking region	25%	Mild
CF2	Missense mutation in exon	0%	Severe
CF3	Missense mutation	5%	Mild

Predict the percent of functional channels and severity of symptoms for the following genotypes:

a. heterozygous for CF100

b. homozygous for CF100

c. heterozygous, with one copy of CF100 and one of CF3

d. heterozygous, with one copy of CF1 and one copy of CF3

Internet Activities

Internet Activities are critical-thinking exercises using the resources of the World Wide Web to enhance the principles and issues covered in this chapter. For a full set of links and questions investigating the topics described below, visit *www.cengage.com/biology/cummings*

1. *Mutant Sequences.* Review the information on Pompe disease from the Internet Activities in Chapter 10. How many mutations have been found in humans for this one enzyme?

2. *Mutation Review.* Work through the Gene Action/Mutation worksheet at the *Access Excellence Activities Exch*ange to reinforce the concepts related to the various types of mutations and their consequences.

3. *Using a Mutation Database.* The *Human Gene Mutation Database* from the Institute of Medical Genetics in Cardiff, Wales, contains information about mutations identified in human genes. This information includes nucleotide substitutions, missense and nonsense mutations, splicing mutations, and small insertions and deletions. The data provided include the name and symbol for a gene, its chromosomal location, the mutant sequence codon number, and a reference to the paper that first identified the mutation.

a. At the website, do a search on "breast cancer." From the list of breast cancer genes, select "*BRCA1*" and scroll through the list of mutations, noting both the type of mutation and its location.

b. Do a search on the gene for cystic fibrosis. How do the results for this entry compare with the information available for *BRCA1*?

HOW WOULD YOU VOTE NOW?

E. coli contamination in meat causes 70,000 illnesses and 60 deaths per year in the United States. Irradiation to kill *E. coli* in beef and *Salmonella* in poultry has been approved by the U.S. Food and Drug Administration and the U.S. Department of Agriculture and endorsed by the World Health Organization and the American Medical Association as an effective means of preventing disease and deaths. In spite of this approval, there is public concern about possible mutation-causing compounds in irradiated meat, so it is not widely available. Now that you know more about mutations and the effects of radiation, what do you think? If such products were available in the supermarket, would you buy them? Visit the *Human Heredity* companion website at ***www.cengage.com/biology/cummings*** to find out more on the issue; then cast your vote online.

12 Genes and Cancer

ulie, a 23-year-old with a family history of breast cancer, sought counseling at the cancer risk assessment clinic. She had been visiting her local breast center to be treated for fibrocystic breast disease. After her last mammogram, she requested an operation to remove her breasts. Julie came to the counseling session with a maternal aunt who had undergone the surgery several years earlier. The genetic counselor explored the factors that had prompted Julie's request for surgery and reviewed her family history. Three of Julie's six maternal aunts had breast cancer in their late 30s or early 40s. One maternal first cousin had been diagnosed with breast cancer in her 30s. Julie's mother was in her 50s and had no history of cancer.

Julie told the counselor that women in her family believed that the only way to avoid the disease was to have their healthy breasts removed before cancer developed. Julie went on to say that her mother's breasts had been removed at age 34, and the aunt attending the session had had surgery the previous year, when she turned 40. The aunt told stories of caring for her dying sisters. She also stated, "It would be the same as a death sentence not to do this." This led the remaining aunts to seek surgery, and now Julie was being encouraged to take the same step.

The genetic counselor explained that the maternal aunts could take a genetic test to determine if a mutant gene was contributing to the development of breast cancer in the family. The counselor explained that if her aunts carried such a mutation, Julie could be tested for the same mutation. If Julie had the mutant gene, surgery could be a reasonable option. However, if Julie did not inherit the mutation, her risk for breast cancer would be the same as that of the general population (10%), and breast removal would be unnecessary. One of Julie's aunts with cancer was tested, and it was found that she carried a mutation for a predisposition to breast cancer. Julie and her mother were tested, and neither of them had the mutation. Julie continues to be followed by the local breast center for fibrocystic breast disease; however, she has decided not to have surgery.

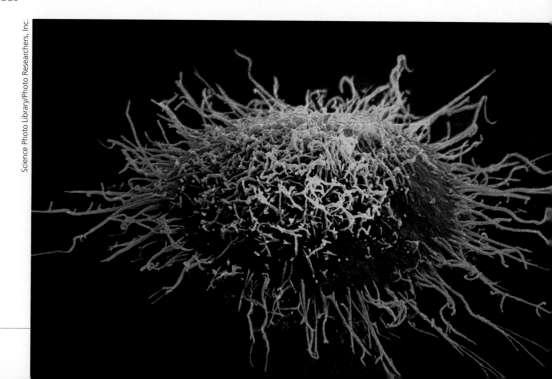

Science Photo Library/Photo Researchers, Inc.

A cervical cancer cell.

12.1 Cancer Is a Genetic Disorder of Somatic Cells

Cancer is a complex group of diseases that affects many different cells and tissues in the body. It is characterized by two properties: (1) uncontrolled cell division and (2) the ability of these cells to spread, or metastasize, to other sites in the body. If a cell begins to divide in an uncontrolled way, it will form a noncancerous growth (called a benign tumor). This type of tumor can be removed by routine surgery. If cells in the tumor acquire the ability to grow continuously and can break away and move to other locations, the tumor is malignant, or cancerous. Unchecked, the combination of uncontrolled growth and metastasis results in death, making cancer a devastating and feared disease. Improvements in medical care have reduced deaths from infectious disease and have led to increases in life span, but these benefits have also helped make cancer a major cause of illness and death in our society. The risk of many cancers is age related (Figure 12.1), and because more Americans are living longer, they are at greater risk of developing cancer. About one in three people will be diagnosed with cancer at some time in their life, and about one in four will die from cancer. Each year, more than a million new cases of cancer are identified in the United States (Table 12.1), and about 500,000 people die of cancer, making it one of the leading causes of death in developed countries. In the United States, over 10 million individuals are receiving medical treatment for cancer in hospitals and medical centers.

A link between cancer and genetics, originally proposed by Theodore Boveri in the nineteenth century, is supported by four lines of evidence:

1. A predisposition to more than 50 forms of cancer is inherited to one degree or another.
2. Most chemicals that are cancer causing are also mutagens.
3. Some viruses carry mutant genes that promote and maintain the growth of cancer in infected cells.
4. Specific chromosomal changes are found in certain forms of cancer, especially leukemia.

HOW WOULD YOU VOTE?

Women who have mutations in either of two cancer genes have a significantly higher risk of developing breast cancer and ovarian cancer. One of the mutant genes is present in members of Julie's family, but she and her mother do not carry that gene. On the basis of this information, Julie decided not to have surgery to remove her breasts or ovaries. If a female member of your family carried one of these cancer genes and asked your advice, would you recommend that she have her breasts and ovaries removed even though not everyone carrying the mutation will develop breast cancer? Visit the *Human Heredity* companion website at *www.cengage.com/biology/cummings* to find out more on the issue; then cast your vote online.

KEEP IN MIND AS YOU READ

- Cancer can be caused by an inherited susceptibility or a sporadic event.
- Cancer cells bypass cell-cycle checkpoints and divide continuously.
- Colon cancer is a multistep process that involves oncogenes and mutant tumor-suppressor genes.
- Several types of translocations associated with cancer create hybrid genes.
- Diet and behavior are two major factors in cancer prevention.

Table 12.1 Estimated New Cases of Selected Cancers in the United States, 2009

All Sites (except skin)	1,479,350
Breast	194,280
Brain, Nervous System	22,070
Colorectal	146,970
Lung	219,440
Lymphoma	74,490
Ovary	21,550
Prostate	192,280
Skin	>1,000,000
Urinary System	131,010>

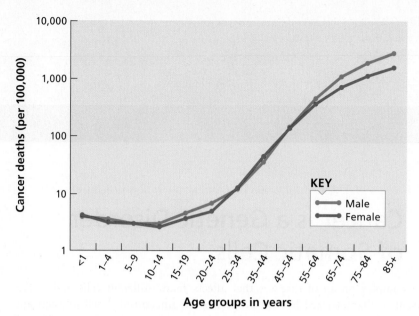

FIGURE 12.1 Age is the leading risk factor for cancer.

The link between cancer and genomic changes has been strengthened by completion of the Human Genome Project, which has helped identify the genes and the gene interactions important in transforming a normal cell into a cancerous cell. This knowledge is being used to develop individualized and often gene-specific methods of treating cancer.

Mutation is a universal feature of all cancers. In the vast majority of cases, these mutations take place in somatic cells, and the mutant alleles are not passed on to offspring. In about 1% of all cases, the cancer-related mutation occurs in germ cells and the mutant allele is passed on to succeeding generations as a predisposition to cancer. Mutations that cause cancer can include single nucleotide substitutions, insertions, deletions, and chromosome rearrangements. As we will see, cancer is a genetic disorder that begins in a single cell.

Mutation is the ultimate cause of cancer and because there is a constant background of spontaneous mutations, there will always be a baseline rate of cancer. The environment (ultraviolet light, chemicals, and viruses) and behavior (diet and smoking) also can play a significant role in cancer risks by increasing the rate of mutation.

12.2 Cancer Begins in a Single Cell

Cancers have several characteristics:

- Cancer begins in a single cell. All the cells in a cancerous tumor are clones directly descended from one cell.
- A cell becomes cancerous after it accumulates a number of specific mutations over a period of time; most cancers are age related.
- Once formed, cancer cells divide continuously. Mutations continue to accumulate, and the cancer may grow more aggressive over time.
- Cancer cells are invasive and can infiltrate surrounding tissues by breaking down the intercellular matrix. In addition, cancer cells can detach from the primary tumor and move to other sites in the body, forming new malignant tumors. This process is called **metastasis** (Active Figure 12.2). The ability to invade new tissues results from new mutations in cancer cells.

Metastasis A process by which cells detach from the primary tumor and move to other sites, forming new malignant tumors in the body.

In the following sections, we consider the genetic changes that take place within cells that lead to cancer.

12.3 Most Cancers Are Sporadic, but Some Have an Inherited Susceptibility

Families with high rates of cancer were identified hundreds of years ago, but in most cases no clear-cut pattern of inheritance can be identified, and the cancers are classified as familial. Why is it that some families have a rate of cancer that is much higher than average? Many explanations have been offered, including inheritance, environmental agents, and chance. A small fraction of these families do carry mutant alleles related to cancer, and studies using these families have helped identify a number of genes involved in cancer.

In these families, some members inherit a mutant allele that causes a predisposition to cancer (Table 12.2). The mutant allele is present in the germ cells and all the somatic cells of these individuals in a heterozygous condition. A mutation in the other, normal allele of the gene must occur before the cell can become cancerous. This mutational event is known as **loss of heterozygosity (LOH)**. In most cases, mutations in other genes in this cell are also needed to transform it from a normal cell into a cancerous one.

Only a small fraction (less than 5%) of all cancers is associated with an inherited predisposition; most cancers are sporadic and arise by the gradual accumulation of mutations in key genes in a single cell (Figure 12.3). The exact number of mutations needed to cause cancer is specific for different forms of cancer, but two mutations may be the minimum number needed. In most cases, cancer-causing mutations accumulate over a period of years; this explains why age is a leading risk factor for many cancers.

ACTIVE FIGURE 12.2 Cancer cells can move to new locations in the body, a process called metastasis.

1. Cancer cells break away from their original tissue.

2. The metastasizing cells become attached to the wall of a blood vessel or lymph vessel. They secrete digestive enzymes to create an opening. Then they cross the wall at the breach.

3. Cancer cells creep or tumble along inside blood vessels, then leave the bloodstream the same way they got in. They start new tumors in new tissues.

CENGAGENOW Learn more about metastasis by viewing the animation by logging on to *login.cengage.com/sso* and visiting CengageNOW's Study Tools.

Loss of heterozygosity (LOH) In a cell, the loss of normal function in one allele of a gene where the other allele is already inactivated by mutation.

KEEP IN MIND

Cancer can be caused by an inherited susceptibility or a sporadic event.

Table 12.2 Some Heritable Predispositions to Cancer

Disorder	Chromosome	OMIM Number
Early-onset familial breast cancer	17q	113705
Familial adenomatous polyposis	5q	175100
Hereditary nonpolyposis colorectal cancer	2p	120435
Li-Fraumeni syndrome	17p	151623
Multiple endocrine neoplasia type 1	11q	131100
Multiple endocrine neoplasia type 2	10q	171400
Neurofibromatous type 1	17q	162200
Neurofibromatous type 2	22q	101000
Retinoblastoma	13q	180200
Von Hippel-Lindau disease	3p	193300
Wilms tumor	11p	194070

FIGURE 12.3 Over time, cells acquire mutations. Two independently acquired mutations in the same cell may be sufficient to cause uncontrolled cell growth and cancer.

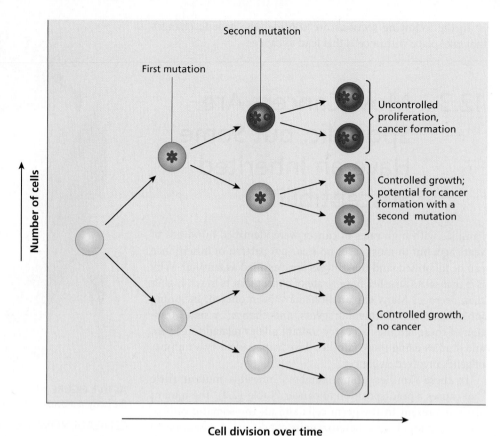

Second mutation

First mutation

Number of cells

Uncontrolled proliferation, cancer formation

Controlled growth; potential for cancer formation with a second mutation

Controlled growth, no cancer

Cell division over time

12.4 Mutations in Cancer Cells Disrupt Cell-Cycle Regulation

Most of the somatic cells in the body are structurally and functionally specialized and are nondividing. These differentiated cells perform highly specific tasks, such as muscle contraction, light reception, and nerve conduction. Many of these cell types (such as liver or kidney cells) can be stimulated to divide by external signals to replace dead and damaged cells. Other cells, such as those that line the surface of ducts and cavities inside the body (e.g., cells lining the intestines and ducts) or that cover the surface of the body (skin cells), divide more or less continuously throughout their lifetimes. These are called epithelial cells and are the source of about 80% of all cancers, including skin (Figure 12.4), breast, prostate, colon, and lung cancers.

Regulation of cell growth and division involves many genes, whose protein products respond to external signals and control progress through the cell cycle and the pathways leading to programmed cell death (apoptosis). Cancer cells carry mutant alleles of these regulatory genes and are characterized by uncontrolled cell division. These cells bypass checkpoints in the cell cycle that normally regulate cell division. As a result, studies of the cell cycle have become an important part of cancer research. In this section, we will outline the steps in the cell cycle, describe how the cycle is regulated, and discuss how mutations or aberrant expression of genes leads to cancer.

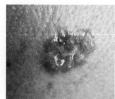 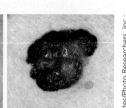

Basal-cell carcinoma

Squamous-cell carcinoma

Malignant melanoma

FIGURE 12.4 Skin cancers. Basal-cell carcinomas are slow growing and noninvasive. Squamous-cell carcinomas are fast growing and can be invasive. Melanomas are dark, fast growing, invasive, and are deadly if not caught early.

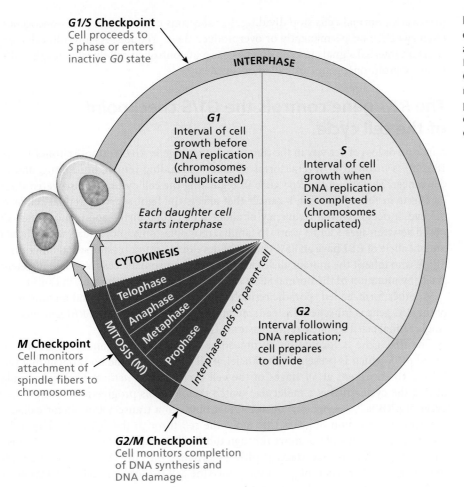

G1/S Checkpoint
Cell proceeds to
S phase or enters
inactive G0 state

INTERPHASE

G1
Interval of cell
growth before
DNA replication
(chromosomes
unduplicated)

*Each daughter cell
starts interphase*

S
Interval of cell
growth when
DNA replication
is completed
(chromosomes
duplicated)

CYTOKINESIS

Telophase
Anaphase
Metaphase
Prophase

MITOSIS (M)

Interphase ends for parent cell

G2
Interval following
DNA replication;
cell prepares
to divide

M Checkpoint
Cell monitors
attachment of
spindle fibers to
chromosomes

G2/M Checkpoint
Cell monitors completion
of DNA synthesis and
DNA damage

FIGURE 12.5 The eukaryotic cell cycle
consists of two parts: (1) interphase
and (2) mitosis followed by cytokinesis.
Interphase is divided into three stages—
G1, S, and G2—which make up the
major part of the cycle. Mitosis involves
partitioning replicated chromosomes to
daughter cells, followed by division of the
cytoplasm.

Recall from Chapter 2 that the cell cycle has two main parts: (1) interphase, the stage between divisions, and (2) mitosis, followed by cytokinesis (Figure 12.5). The cycle is regulated at three main checkpoints:

- in G1 just before cells enter S (the G1/S transition)
- at the transition between G2 and M (the G2/M transition)
- a point in late metaphase of mitosis called the M checkpoint

At the G1/S checkpoint, the cell either enters the next phase of the cycle or enters a nondividing state called G0 (G-zero). Many specialized cells, such as white blood cells, remain in G0 until stimulated by external signals to reenter the cycle and divide. These external signals are processed in the cell by a mechanism known as **signal transduction** to produce changes in gene expression, leading to cell division. Some cancers have abnormalities in signal-transduction systems. At each checkpoint, a combination of external signals and internal regulatory pathways determines whether the cell will proceed to the next stage of the cycle. The G2/M checkpoint ensures that the DNA has been replicated and that any damage to the DNA has been repaired. If not, progress through the cycle is stopped until these events have been completed. At the M checkpoint, attachment of spindle fibers to chromosomes is monitored; this ensures that anaphase separation of chromosomes occurs normally.

Mutations in the genes regulating these checkpoints are important in cancer development. Two classes of genes regulate the checkpoints: (1) genes that turn off or decrease the rate of cell division and (2) genes that turn on or increase the rate of cell division. The first class is known as **tumor-suppressor genes**. Products of these genes normally act at either the G1/S or the G2/M control point to inhibit cell division. If these genes are mutated or inactivated, cells pass through the checkpoints and divide in an uncontrolled manner.

The second class of checkpoint genes, called **proto-oncogenes**, encode proteins that start or maintain cell growth and division. When the products encoded by these genes are active, cells grow and divide. When proto-oncogenes and/or their proteins are

Signal transduction A cellular
molecular pathway by which an
external signal is converted into
a functional response.

Tumor-suppressor genes Genes
encoding proteins that suppress cell
division.

Proto-oncogenes Genes that initiate
or maintain cell division and that may
become cancer genes (oncogenes) by
mutation.

inactivated, normal cells stop dividing. In cancerous cells, mutant proto-oncogenes are often switched on permanently or overproduce their products. As a result, cells receive constant internal signals to keep dividing, and uncontrolled cell division results. Mutant forms of proto-oncogenes are called **oncogenes**.

The *RB1* gene controls the G1/S checkpoint of the cell cycle.

Mutation or loss of activity in the tumor-suppressor gene *RB1* (**retinoblastoma 1**; OMIM 180200) is involved in several forms of cancer, including retinal, bone, lung, and bladder cancer. The role of the *RB1* gene in regulating the cell cycle was discovered through its effects in retinoblastoma, a cancer that affects the light-sensitive retinal cell layer of the eye. It occurs with a frequency of about 1 in 15,000 births and is most often diagnosed between ages 1 and 3 years. In familial retinoblastoma, individuals who inherit one mutant allele of *RB1* have an 85% chance of developing retinoblastoma and other cancers. Those who inherit the mutant allele carry it in all their somatic cells, and loss of heterozygosity by mutation of the normal allele anywhere in the body can result in cancer.

Another form, called sporadic retinoblastoma, is extremely rare and no mutant allele of the *RB1* gene is inherited. Instead, mutations in both copies of the *RB1* gene occur in a single retinal cell and cause retinoblastoma (Figure 12.6).

The *RB1* gene is located on chromosome 13 (Figure 12.7) and encodes a protein called pRB, which is present in the nuclei of all cell types in the body, including retinal cells. pRB is present at all stages of the cell cycle, but its *activity* is closely regulated during the cycle. pRB is a molecular switch that controls progression through the cell cycle. If pRB is "switched on" during G1, it binds to a transcription factor called E2F, preventing expression of genes that move the cell through the cell cycle (Figure 12.8). As a result, the cell will not move through the G1/S checkpoint into S phase, but instead will enter the inactive G0 stage. If pRB is "switched off" in G1, it releases E2F, which then activates expression of genes that move the cell through the G1/S checkpoint into S phase, through G2, and on to mitosis. If both copies of the *RB1* gene are mutated or inactivated in a cell, the cell moves through the cell cycle in an unregulated way and forms a cancerous tumor.

> **KEEP IN MIND**
> Cancer cells bypass cell-cycle checkpoints and divide continuously.

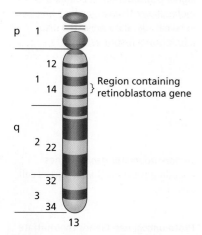

FIGURE 12.6 A child with retinoblastoma in one eye.

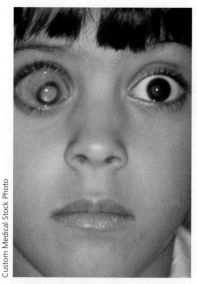

FIGURE 12.7 A diagram of chromosome 13, showing the retinoblastoma (*RB1*) locus.

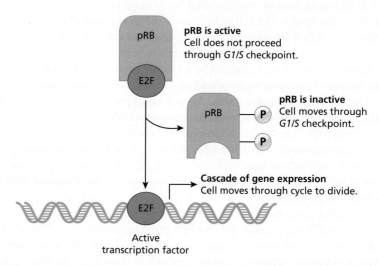

pRB

E2F

pRB is active
Cell does not proceed through *G1/S* checkpoint.

pRB — P

— P

pRB is inactive
Cell moves through *G1/S* checkpoint.

Cascade of gene expression
Cell moves through cycle to divide.

E2F

Active
transcription factor

FIGURE 12.8 When pRB is active in the nucleus, it binds E2F, a protein transcription factor that controls expression of a large gene set encoding genes that move the cell through the cycle. When bound to pRB, E2F is inactive, and the cell does not move from G1 into S. If pRB is inactivated by the addition of phosphate groups (P), E2F is released, binds to promoter regions, gene expression begins, and the cell moves into S and completes the cycle. If pRB is missing or mutated, E2F is always active, and the cell grows and divides continuously, forming a cancerous tumor.

The *ras* genes are proto-oncogenes that regulate cell growth and division.

Recall that proto-oncogenes switch on or maintain cell division. These genes are normally switched off or their protein products are inactivated when cell division stops. Mutant forms of these genes are oncogenes; they cause uncontrolled cell divisions and cancer.

What is the difference between a proto-oncogene in a normal cell and its mutant form as an oncogene in a cancer cell? Many differences are possible; these can include single base changes that produce an altered gene product, mutations that cause underproduction or overproduction of the normal gene product, and mutations that increase the number of copies of the normal gene. In fact, all these types of mutations have been identified in oncogenes or their adjacent regulatory regions. We'll examine an example from a proto-oncogene family that is mutated in more than 40% of all human cancers.

The *ras* proto-oncogene family (OMIM 190020) is a group of related genes that encodes signal-transduction proteins. These help transmit signals from outside the cell, through the plasma membrane and into the nucleus, initiating a cascade of gene expression that stimulates cells to divide. One member of the *ras* proto-oncogene family encodes a protein of 189 amino acids and is located in the cytoplasm at the border with the plasma membrane. The ras protein cycles between an active ("switched on") state and an inactive ("switched off") state. When a growth factor binds to the cell's plasma membrane, the ras protein is "switched on" and transfers growth-promoting signals from the plasma membrane to molecules in the cytoplasm and then to the nucleus, where changes in gene expression begin cell division. Once the signal has been transmitted, the ras protein returns to its inactive state and is "switched off." Analysis of mutant *ras* alleles from many different human tumors shows that single base changes at one of two sites in the gene is the only difference between the proto-oncogene in normal cells and the mutant oncogene in cancer cells. In all cases, the base change causes a change in a single amino acid in the ras protein at either amino acid 12 or amino acid 61 (Figure 12.9).

Changing glycine to valine at position 12 disrupts the structure of the ras protein and prevents it from folding into an inactive state. As a result, the mutant protein is locked into the active state and is permanently "switched on." It signals for cell growth at all times, even in the absence of an external signal. Cells carrying this mutation escape from growth control and become cancerous. An amino acid change at position 61 has a similar effect.

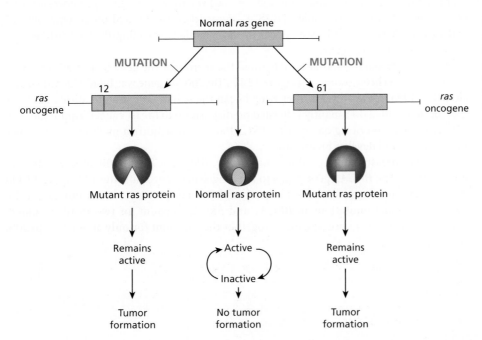

FIGURE 12.9 The *ras* proto-oncogene encodes a signal-transduction protein that processes signals for cell division and transmits them from the cytoplasm to the nucleus. Mutations at amino acid 12 or 61 cause the formation of an oncoprotein that permanently signals for cell division.

12.5 Mutant Cancer Genes Affect DNA Repair Systems and Genome Stability

All forms of cancer share several properties: (1) higher-than-normal rates of mutation, (2) abnormalities of chromosome structure and number, and (3) one or more forms of genomic instability. This instability is seen as progressive chromosomal changes as the cancer develops, including aneuploidy, loss of chromosomes, duplications, deletions, and other abnormalities. These changes are related to loss of the ability to repair DNA damage in cancer cells. Several forms of cancer associated with DNA repair defects have been identified, including breast cancer and a form of colon cancer.

Mutant DNA repair genes cause a predisposition to breast cancer.

Breast cancer is the most common form of cancer in women in the United States. Each year, more than 215,000 new cases are diagnosed (Figure 12.10). Although most cases of breast cancer are unrelated to heredity, geneticists struggled for years with the question, is there a genetic predisposition to breast cancer? After more than 20 years of work, the answer is clearly yes. Mutations in at least two different genes, *BRCA1* and *BRCA2*, can predispose women to breast cancer and ovarian cancer. Carriers of either of these mutations have a 10-to-30-fold higher risk of breast cancer than the general population.

The search for *BRCA1* began in the 1970s when Mary-Claire King and her colleagues analyzed pedigrees, searching for families with a clear history of breast cancer. Approximately 15% of the 1,500 families they studied had multiple cases of breast cancer. A statistical model predicted that about 5% (or 75/1,500) of these cases were genetic, but it was impossible to know which families had cases with a genetic predisposition and which had cases unrelated to heredity.

Instead of being discouraged, King decided on a brute-force approach. She and her team began testing as many families as possible, looking for a link between certain proteins and breast cancer. Later, the team switched to DNA markers and searched for linkage between markers and cases of breast cancer. Finally, in 1990, after testing hundreds of families and using hundreds of markers, they found linkage between breast cancer and a DNA marker called D17S74. The marker was located on chromosome 17 and was coinherited with breast cancer in some families, meaning that the marker was close to a breast cancer gene.

An international collaboration narrowed the search for the gene to a small region on the long arm of chromosome 17 (Figure 12.11). The *BRCA1* gene was finally identified and cloned in 1994, some 20 years after King began the search. The mutant allele of *BRCA1* is associated with a dominantly inherited predisposition to breast cancer. Approximately 85% of women inheriting one mutant *BRCA1* allele will acquire a mutation in the other *BRCA1* allele and develop breast cancer.

A second breast cancer predisposition gene, *BRCA2* (OMIM 600185), was discovered in 1995. The *BRCA2* gene maps to the long arm of chromosome 13 (Figure 12.11) and may be responsible for the majority of inherited predispositions not caused by *BRCA1*. Although mutations in *BRCA1* and *BRCA2* account for two-thirds of inherited predispositions to breast cancer, together they account for only about 15% to 20% of all cases.

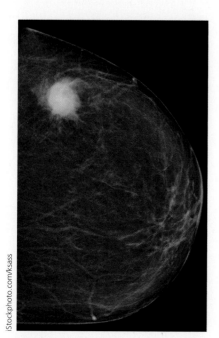

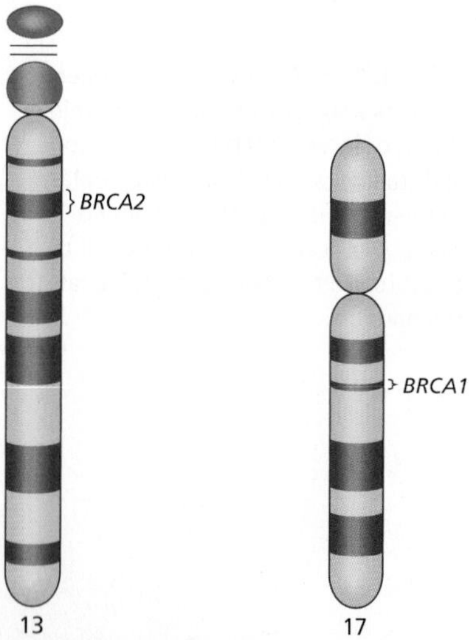

iStockphoto.com/ksass

FIGURE 12.10 Mammograms are used to detect breast cancer.

} BRCA2

} BRCA1

13   17

FIGURE 12.11 The chromosome locations for *BRCA1* and *BRCA2*. Together, these genes account for the majority of cases of breast cancer associated with a genetic predisposition.

BRCA1 and BRCA2 are DNA repair genes.

BRCA1 and BRCA2 are each involved in DNA repair. The normal alleles of both genes encode large proteins found only in the nucleus. In rapidly dividing cells, expression of BRCA1 and BRCA2 is highest at the G1/S border and into S phase.

The BRCA1 protein has several tumor-suppressing roles, all related to its ability to form complexes with other proteins in the nucleus. These BRCA1 complexes help maintain genome stability during DNA repair, regulation of cell-cycle checkpoints, and several other functions. We will focus on the role of BRCA1 in DNA repair.

BRCA1 is activated when breaks in DNA strands are detected. Unrepaired breaks increase the rate of cancer-causing mutations in genes located at or near breakpoints. During replication in S phase, DNA breaks are detected by a protein called Rap80, which then binds to BRCA1, recruiting it to the location of the breaks. Action of the activated BRCA1 protein stops DNA replication, and BRCA1 then participates in repairing the breaks. The mutant form of BRCA1 is unable to bind to Rap80, fails to move to the location of DNA breaks, and does not participate in DNA repair. As a result, mutations gradually accumulate within a cell and the cell becomes cancerous (remember that cancers begin with a single cell).

Breast cancer risks depend on genotype.

Together, mutations in BRCA1 and BRCA2 account for only 15% to 20% of all cases of breast cancer. Women who carry one mutant allele of BRCA1 or BRCA2 have up to an 85% risk of developing breast cancer by age 70. In contrast, women in the United States who carry two normal alleles of BRCA1 or BRCA2 have about a 12% risk of developing breast cancer by age 90.

Women carrying a mutant BRCA1 or BRCA2 allele also have an increased risk of developing ovarian cancer. For women with a BRCA1 mutation, the lifetime risk is about 55%, and for women with a BRCA2 mutation, the risk is about 25%. Women who carry normal alleles of BRCA1 or BRCA2 have roughly a 1.8% risk of ovarian cancer.

The risk for breast cancer associated with mutant alleles of BRCA1 and BRCA2 extends to men as well as women. Men who inherit a mutant BRCA1 or BRCA2 allele are also at risk of developing breast cancer (see Spotlight on Male Breast Cancer). In addition, men carrying these mutations are 3 to 7 times more likely to develop prostate cancer than men carrying normal alleles of these genes.

Spotlight on . . .

Male Breast Cancer

Although most people think of breast cancer as a disease of women, men also get breast cancer. In the United States, about 1% of breast cancers occur in males, and about 2,000 new cases are reported each year. Men who inherit a mutant BRCA1 or BRCA2 allele have about a 6% chance of developing breast cancer—a risk that is about 80 times higher than for men who carry normal alleles of these genes. In parts of Africa, however, the rates of male breast cancer are much higher. In Egypt, males account for 6% of all cases of breast cancer, and in Zambia, male breast cancer represents 15% of all cases. Risk factors for male breast cancer include age, a history of breast cancer in female family members, and occupational exposure to heat, gasoline, or estrogen-containing creams in the soap and perfume industry. Males of Eastern European Jewish ancestry and black males have relatively high rates of breast cancers. More advanced cases are found in men than in women, probably because of delayed detection.

12.6 Colon Cancer Is a Genetic Model for Cancer

As discussed earlier, the development of cancer requires the acquisition of a specific number of mutations in specific genes. In retinoblastoma, two mutational steps are required to convert a normal cell into a cancerous one. In other cases, a half-dozen-or-more mutations are required to initiate the formation of a cancer cell.

Colon cancer is one of the latter types. The number and order of genetic changes in colon-cancer development are a model for defining the transformation of a normal cell into a cancer cell. Colon/rectal cancer is one of the most common forms of cancer in the United States (Table 12.3), and worldwide, more than 1 million cases are diagnosed each year. Colon cancer involves genetic and environmental factors such as diet and lifestyle, as well as the interaction of these factors. Unraveling the genetic causes of colon cancer is a prerequisite to the development of improved screening, targeted therapies for treatment, and strategies for prevention.

Most cases of colorectal cancer are sporadic and occur in people with no family history of the disease. However, in about 5% of all cases, an inherited predisposition to colon cancer associated with mutations in specific genes has been identified.

Table 12.3 Colorectal Cancer in the United States

Estimated new cases, 2009	
Colon and rectal cancers	146,970
Mortality (estimated deaths, 2009)	
Colon and rectum	49,920 (10% of cancer deaths)
5-year survival rate (early detection)	
Colon	90%
Rectum	85%

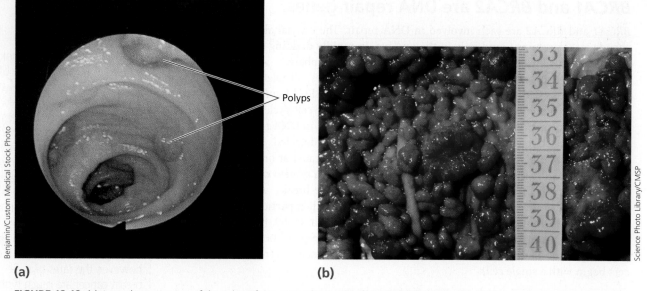

(a)

(b)

FIGURE 12.12 (a) An endoscopic view of the colon of someone who carries the normal alleles for the *APC* gene. The small number of polyps on the wall are a precursor to colon cancer. (b) Polyps in the colon of someone who carries a mutant APC allele. Hundreds of polyps are present. If a cell in one of these polyps acquires enough mutations, it will transform into a cancer cell, forming colon cancer.

Familial adenomatous polyposis (FAP) An autosomal dominant trait resulting in the development of polyps and benign growths in the colon. Polyps often develop into malignant growths and cause cancer of the colon and/or rectum.

Hereditary nonpolyposis colon cancer (HNPCC) An autosomal dominant trait associated with genomic instability of microsatellite DNA sequences and a form of colon cancer.

Inherited predispositions lead to colon cancer along one of two pathways, both of which are inherited as autosomal dominant traits. One form, called **familial adenomatous polyposis** (FAP; OMIM 175100), is coupled with chromosomal instability; the other, connected to failures in DNA repair, is called **hereditary nonpolyposis colon cancer** (HNPCC; OMIM 120435 and 120436). FAP accounts for only about 1% of all cases of colon cancer but has been useful in deriving the main features of the genetic model for colon cancer described below.

FAP causes chromosome instability and colon cancer.

The genetic pathway from the mutant FAP allele to colon cancer has two important features: (1) the presence of five to seven mutations in a single cell—if fewer mutations are present, benign growths or intermediate stages of malignant-tumor formation result; (2) a specific sequence of mutational events, indicating that *both* the number and the order of mutations are important in tumor formation.

In FAP-associated colon cancer, individuals inherit a mutation in the *APC* gene. In epithelial cells of the inner lining in the large intestine, cells are heterozygous for an *APC* mutation and partially escape control of the cell cycle. They divide to form small clusters of cells called polyps (Figure 12.12a). Heterozygotes carry a mutant copy of *APC* in all their cells and develop hundreds or thousands of polyps in the colon and rectum (Figure 12.12b). Each polyp is a clone, derived from a single cell that has partially escaped cell-cycle control. In these polyps, a single *APC* mutation is not enough to cause cancer—it is only the first step. Colon cancer develops only after mutations in several other tumor-suppressor genes cause the transition from polyp to colon cancer.

In the second step (Figure 12.13), mutation of one copy of the *K-RAS* proto-oncogene in a polyp cell transforms the polyp into an adenoma, a larger noncancerous tumor. These cells now carry two mutations—one in *APC* and one in *K-RAS*—but other mutations are still necessary to transform the polyp cells into a cancer. In subsequent steps, polyp cells must acquire mutations in *both* alleles of the downstream genes shown in Figure 12.13. These mutations usually occur through deletions in specific chromosome regions. The 18q region contains a number of genes involved in colon cancer, including *DCC*, *DPC4*, and *JV18-1*. Deletion of both alleles of any of these genes leads to the formation of late-stage adenomas with fingerlike outgrowths called villi. In the last step, mutations in both alleles of the *p53* gene cause the late-stage adenoma to become cancerous. Mutations in the *p53* gene are pivotal in the formation of other cancers, including lung, breast, and brain cancers.

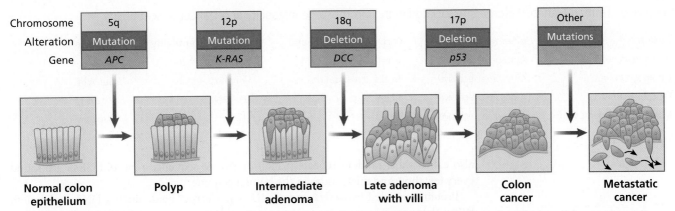

Chromosome	5q	12p	18q	17p	Other
Alteration	Mutation	Mutation	Deletion	Deletion	Mutations
Gene	APC	K-RAS	DCC	p53	

Normal colon epithelium → Polyp → Intermediate adenoma → Late adenoma with villi → Colon cancer → Metastatic cancer

FIGURE 12.13 A multistep model for colon cancer. The first mutation, in the *APC* gene, leads to the formation of polyps. Subsequent mutations in genes on chromosomes 12, 18, and 17 cause the transformation of a polyp into a malignant tumor. Subsequent mutations allow cells from the tumor to metastasize, move to other regions of the body, and form new tumors.

In summary, FAP-associated colon cancer requires a series of specific mutations that accumulate over time in a specific sequence within a single cell. Each mutation confers a slight growth advantage to the cell, allowing it to grow and divide to form a polyp, which enlarges and transforms in later stages as it gradually breaks away from cell-cycle controls. Eventually, one cell in the polyp accumulates enough mutations to escape completely from cell-cycle controls to become a malignant tumor. Later, additional mutations accumulate in some tumor cells, allowing them to become metastatic and break away to form tumors at remote sites. The FAP-associated form of colon cancer is linked to a sequence of chromosome deletions combined with a mutation in the *K-RAS* proto-oncogene.

Genomic techniques have been used to identify other cases in which cancer involves a number of mutations at specific chromosomal sites, often on different chromosomes (Table 12.4).

HNPCC is caused by DNA repair defects.

Most cancers are caused by mutations in two or more genes that accumulate over time. If one of these mutations is inherited, fewer mutations are required to cause cancer, resulting in a genetic predisposition to cancer. But if the overall mutation rate in cells is low (as we saw in Chapter 11), how do the multiple mutations needed for cancer formation accumulate in a single cell within the lifetime of an individual? Research on a second form of colon cancer has provided a partial answer to this question.

Mutations in several genes—including *MSH2* (OMIM 120435), *MLH1* (OMIM 120436), and at least five other genes—have been identified in hereditary nonpolyposis colon cancer (HNPCC). Mutations in *MSH2* and *MLH1* account for about 90% of all cases of HNPCC. Anyone who inherits mutant alleles of either *MSH2* or *MLH1* has a dominantly inherited predisposition to colon cancer that bypasses polyp formation. Mutations in either gene destabilize the genome, generating a cascade of mutations in short (2 to 9 nucleotides in length) DNA sequences called microsatellites. These repetitive DNA sequences are present thousands of times throughout the genome on many chromosomes. Clusters of microsatellite repeats (usually 10 to 100 copies of the sequence),

Table 12.4 Number of Mutations Associated with Specific Forms of Cancer

Cancer	Chromosomal Sites of Mutations	Minimal Number of Mutations Required
Retinoblastoma	13q14	2
Wilms tumor	11p13	2
Colon cancer	5q, 12p, 17p, 18q	5 to 7
Small-cell lung cancer	3p, 11p, 13q, 17p	10 to 15

Table 12.5 Human Genetic Disorders Associated with Chromosome Instability and Cancer Susceptibility

Disorder	Inheritance	Chromosome Damage	Cancer Susceptibility	Hypersensitivity
Ataxia telangiectasia	Autosomal recessive	Translocations on 7, 14	Lymphoid, others	X-rays
Bloom syndrome	Autosomal recessive	Breaks, translocations	Lymphoid, others	Sunlight
Fanconi anemia	Autosomal recessive	Breaks, translocations	Leukemia	X-rays
Xeroderma pigmentosum	Autosomal recessive	Breaks	Skin	Sunlight

also called short tandem repeats (STRs) or simple sequence repeats (SSRs), are found every few thousand nucleotides in the human genome.

Proteins encoded by *MSH2* and *MLH1* repair errors made during DNA replication. When these genes are inactivated by mutation, DNA repair is defective, and microsatellite mutation rates increase by at least 100-fold. These mutations include alterations in the number of microsatellite repeats as well as changes in repeat sequence. Cells from HNPCC tumors can carry more than 100,000 mutations in microsatellites scattered throughout the genome.

This genomic instability, spread through many different chromosomes, promotes mutations in nearby genes (including *APC* and other genes involved with growth control), eventually leading to colon cancer. HNPCC is also known as Lynch syndrome, partly to distinguish it from other cancers associated with microsatellite instability.

> **KEEP IN MIND**
>
> Colon cancer is a multistep process that involves oncogenes and mutant tumor-suppressor genes.

In summary, there are at least two paths to colon cancer: one (FAP) that begins with a mutation in the *APC* gene and the other (HNPCC) that begins with a mutation in a DNA repair gene. Mutation in *APC* causes formation of hundreds or thousands of polyps. These noncancerous growths progress slowly to cancer by accumulating mutations in other genes. Because there are thousands of polyps, there is a good chance that at least one of them will become cancerous. In HNPCC mutations, polyps are absent or accumulate slowly, forming only a small number of benign tumors. However, mutations in these polyps accumulate at an accelerated rate that is two to three times faster than that in normal cells, making it almost certain that at least one polyp will progress to colon cancer.

Mutations in DNA repair genes do not directly lead to cancer but do increase the mutation rate across the genome. Genetic disorders with DNA repair defects (Table 12.5) have chromosome instability and a susceptibility to cancer. The idea that DNA repair defects lead to genomic instability may help explain why many forms of cancer are characterized by genomic instability and aneuploidy.

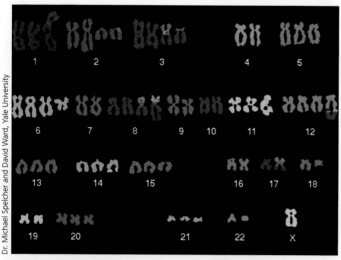

Dr. Michael Speicher and David Ward, Yale University

FIGURE 12.14 A karyotype from a cancer cell stained with chromosome-specific dyes. This cancer cell shows widespread genomic instability, including aneuploidy, translocations, and deletions.

12.7 Hybrid Genes, Epigenetics, and Cancer

Changes in the number and structure of chromosomes are a common feature of cancer cells (Figure 12.14). These changes include translocations, deletions, chromosome loss, aneuploidy, duplications, and amplification of certain genes. Cells from solid tumors grown in the laboratory continue to exhibit genomic instability by acquiring new chromosome aberrations as they divide. Although most of these chromosome changes are nonspecific, in other cases, specific chromosomal abnormalities are associated with certain forms of cancer.

Table 12.6   Chromosomal Translocations Associated with Human Cancers

Chromosomal Translocation	Cancer
t(9;22)	Chronic myelogenous leukemia (Philadelphia chromosome)
t(15;17)	Acute promyelocytic leukemia
t(11;19)	Acute monocytic leukemia, acute myelomonocytic leukemia
t(1;9)	Pre-B-cell leukemia
t(8;14),t(8;22),t(2;8)	Burkitt's lymphoma, acute lymphocytic leukemia of the B-cell type
t(8;21)	Acute myelogenous leukemia, acute myeloblastic leukemia
t(11;14)	Chronic lymphocytic leukemia, diffuse lymphoma, multiple myeloma
t(4;18)	Follicular lymphoma
t(4;11)	Acute lymphocytic leukemia
t(11;14)(p13;q13)	Acute lymphocytic leukemia

Some chromosome rearrangements cause leukemia.

The connection between chromosome aberrations is evident in certain leukemias. In these cancers (which involve the uncontrolled division of white blood cells), specific chromosome changes are well defined and diagnostic (Table 12.6).

One of the best-established links between cancer and a chromosomal aberration is a translocation between chromosome 9 and chromosome 22 in chronic myelogenous leukemia (CML) (Figure 12.15). Originally called the **Philadelphia chromosome** after the city in which it was discovered, this relationship was the first example of a chromosome translocation related to a human disease. Other cancers associated with specific chromosome translocations are listed in Table 12.6.

In the CML-associated translocation, the *C-ABL* gene on chromosome 9 is combined with part of the *BCR* gene on chromosome 22 to form a hybrid gene. The normal allele of the *C-ABL* gene encodes a protein involved in signal transduction, and the normal *BCR* allele encodes a protein that switches other proteins "on" or "off" by adding phosphate groups to them. These BCR target proteins control gene expression and cell growth. The translocation produces a hybrid gene with *BCR* sequences at its beginning and *C-ABL* sequences at the end (Figure 12.16). This hybrid gene encodes a unique protein that constantly signals the white blood cell to divide, even in the absence of external signals, resulting in CML.

Philadelphia chromosome An abnormal chromosome produced by translocation of parts of the long arms of chromosomes 9 and 22.

> **KEEP IN MIND**
>
> Several types of translocations associated with cancer create hybrid genes.

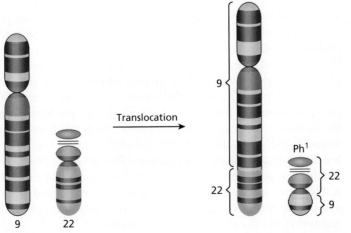

Translocation

FIGURE 12.15 A reciprocal translocation between chromosomes 9 and 22 results in the formation of a hybrid gene, whose expression leads to chronic myelogenous leukemia (CML).

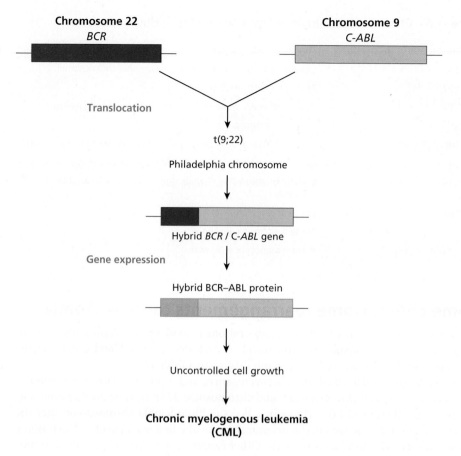

FIGURE 12.16 Gene fusion in the 9;22 translocation. Part of the *BCR* gene on chromosome 22 is moved to chromosome 9 and fuses with part of the *C-ABL* gene on that chromosome to form a hybrid gene. Expression of the hybrid gene produces a hybrid BCR-ABL protein that continuously signals for cell division in white blood cells. Overproduction of these cells results in CML.

Chromosome 22
BCR

Chromosome 9
C-ABL

Translocation

t(9;22)

Philadelphia chromosome

Hybrid *BCR* / *C-ABL* gene

Gene expression

Hybrid BCR–ABL protein

Uncontrolled cell growth

**Chronic myelogenous leukemia
(CML)**

12.8 Genomics and Cancer

Cancer can now be considered as a genomic disease generated by the accumulation of a number of specific mutations in tumor-suppressor genes and proto-oncogenes. Using strategies that include family studies, DNA sequencing, and genome-wide association studies, researchers are identifying the mutations and genetic modifications in cancer cells with the goal of cataloging all the mutations present in cancers. It is hoped that this knowledge will open the way to the development of new diagnostic tools and drugs for treatment.

Sequencing cancer genomes identifies cancer-associated genes.

With the sequence of the gene-coding regions of the human genome in hand and the development of newer sequencing methods, it is now feasible to sequence and analyze the protein-coding genes in cancer cells. The sequence of cancer genes can be compared with the sequence of genes in normal cells to systematically identify all the mutations present in a cancer cell. One such study sequenced over 13,000 genes in breast and colorectal cancers and found that individual tumors carried an average of 90 mutant genes. This unexpected finding revealed that the number of mutational events that occur during the transition from normal to malignant cells is much larger than previously thought. Secondly, the work showed significant differences in the types of mutations found in breast cancer and in colorectal cancers (Figure 12.17). Third, each tumor analyzed had a unique combination of mutations, and no tumor had more than 6 candidate cancer (CAN) genes in common with any other tumor analyzed. This emphasized the clonal and unique origins of cancer, underscoring the potential for the development of drug-resistant clones as a by-product of cancer therapy.

Genomic studies have greatly expanded the number of genes known to be associated with many forms of cancer. In breast cancer, for example, three classes of genes have been identified: (1) rare high-risk alleles (including *BRCA1* and *BRCA2*), (2) rare moderate-risk alleles, and (3) common but low-risk alleles (Table 12.7). In addition to cases of breast cancer caused by the action of a single high-risk gene such as *BRCA1*, it is now clear that other cases may be caused by mutations in a number of low-risk genes, in several possible combinations, with implications for both diagnosis and treatment.

The development of high-resolution methods for detecting genome-wide mutations in cancers has led to the creation of the Cancer Genome Atlas (see Exploring Genetics: The Cancer Genome Atlas on page 286), an international effort to use new technologies to catalog and understand the genetic changes associated with cancer and to improve the diagnosis and treatment of this disease.

Epigenetics and cancer.

As discussed in Chapter 11, epigenetics is a rapidly expanding research field that studies heritable alterations in gene expression caused by mechanisms that do not alter any DNA sequence. Epigenetic changes usually involve the chemical modification of DNA bases or the histone proteins associated with DNA (Figure 12.18). Some examples include the addition of methyl groups to DNA or the addition of methyl groups, acetyl groups, and phosphate groups to histone proteins. The genomic locations and patterns of these changes are heritable and can activate or inactivate genes.

Abnormal patterns of DNA methylation are associated with many types of cancer. Lower-than-normal levels of methylation (demethylation or hypopmethylation) are commonly observed in some types of cancer. Removal of methyl groups can activate genes involved with cell growth and can increase genomic instability. In other cancers— including retinoblastoma, breast cancer, and colon cancer—methylation inactivates key genes including *RB1*, *BRCA1*, and *APC*; these changes transform normal cells into cancer cells.

New methods of analyzing methylation patterns in cancer cells have improved cancer diagnosis. Analysis of methylation patterns is used to distinguish invasive from noninvasive forms of cancer and to diagnose subtypes of cancer. It is hoped that knowledge of methylation patterns in different cancers will lead to the development of new drugs that can reverse abnormal methylation of specific genes and normalize gene expression in cancer cells, reestablishing cell-cycle control and halting tumor growth.

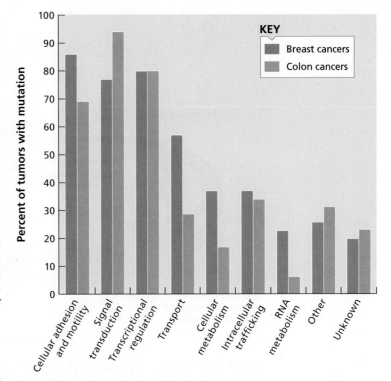

FIGURE 12.17 Genomic sequencing has identified hundreds of mutations in many classes of genes in both breast cancer and colon cancer, indicating widespread genomic instability in these cancers.

Table 12.7 Classes and Frequency of Breast Cancer Susceptibility Genes

Rare High-Risk Alleles (-0.01%)	Rare Moderate-Risk Alleles (-1.0%)	Common Low-Risk Alleles (-30%)
BRCA1 (OMIM 113705)	*BRIP1* (OMIM 605882)	*TOX3* (OMIM 611416)
BRCA2 (OMIM 600185)	*ATM* (OMIM 607585)	*FGFR2* (OMIM 176943)
PTEN (OMIM 601728)	*PALB2* (OMIM 610355)	*LSP1* (OMIM 153432)
CDH1 (OMIM 192090)	*CHEK2* (OMIM 604373)	*AKAP9* (OMIM 604001)
STK11 (OMIM 602216)		*CASP8* (OMIM 601763)
		MAP3K1 (OMIM 600982)

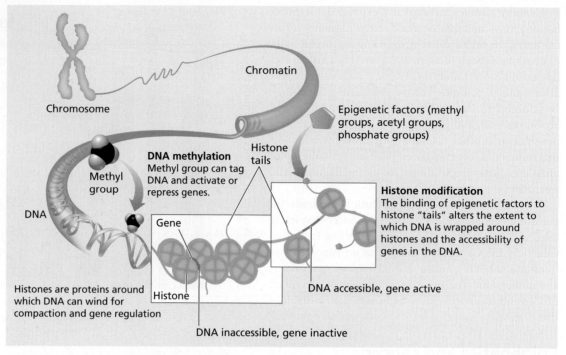

FIGURE 12.18 Epigenetic changes to DNA, such as methylation, change patterns of gene expression by activating or repressing transcription. Reversible modification of histones can also alter the pattern of gene expression.

Targeted therapy offers a new approach to treating cancer.

Cancer therapy has traditionally used radiation and chemicals to target and kill all rapidly dividing cells (see The Genetic Revolution: Cancer Stem Cells.). Although cancer cells are dividing rapidly, so are other cells in the body, including those in bone marrow (making red blood cells), the intestine (replacing worn-away cells), and many other tissues. All these cells are destroyed or damaged along with cancer cells during radiation treatment or chemotherapy, often with serious side effects for the patient.

Targeted therapy uses drugs that stop the growth of cancer cells by selectively blocking the action of oncogene proteins on the growth and division of malignant cells. This approach targets only cancer cells, rather than all dividing cells in the body. Targeted therapy currently uses two types of drugs: small molecules directed at specific cancers and monoclonal antibodies that bind to cancer cells, stop growth, and mark the cells for destruction by the body's immune system.

To illustrate how targeted therapy works, we will discuss two drugs: Gleevec (imatinib) for chronic myelogenous leukemia (CML) and Herceptin, a monoclonal antibody used to treat certain types of breast cancer.

As discussed above, CML cells contain a hybrid BCR-ABL protein that signals the cell to divide continuously. CML cells are the only cells in the body that contain the hybrid protein, offering an opportunity to develop a drug that targets only these cells. Researchers discovered that the hybrid protein folds to form a pocket for binding ATP, a molecular-energy source required for its signaling activity (Figure 12.19). Using that information, researchers designed a drug (called Gleevec) that fits into the ATP-binding pocket of the BCR-ABL protein and prevents ATP from entering the pocket. Without ATP to activate the hybrid protein (Figure 12.19), the BCR-ABL protein does not signal for division, and the cancer cells stop dividing. More than 90% of CML patients treated with this drug go into remission and show a dramatic reduction in the number of white blood cells carrying the Philadelphia chromosome. Gleevec has also proven effective in treating other forms of cancer, including some gastrointestinal cancers.

Cancer Stem Cells

For several decades, cancer therapy has used chemicals and radiation in an attempt to kill all cells in a tumor as the way to cure this disease. The underlying assumption in this approach is that all cells in a cancer can divide and metastasize, or move to other locations in the body and begin dividing, spreading the cancer. However, work on a number of cancers has led researchers to consider another model, called the cancer stem cell model. According to this model, cancers are organized much like normal adult tissue containing stem cells that can divide to form a limited number of cell types. Remember that stem cells have two properties: (1) self-renewal by division and (2) the ability to differentiate to form specific cell types, which divide only slowly or not at all. In the cancer stem cell model, the pathway of self-renewal has turned into one of continuous division, but leaving the other pathway still active.

If this is true, then most cancer cells in a tumor have followed the second pathway, and cannot divide fast enough (or at all) to contribute to tumor growth. Instead, attention should be focused on identifying and eliminating the small number of stem cells whose continuous division is driving the growth of the tumor.

Persuasive evidence for the stem cell model has shown that in breast cancer, brain cancer, and colon cancer, only a small number of cells are required to transfer the cancer into mice; most cells in these cancers are nontumorigenic, that is, they cannot transfer the cancer when transplanted into mice. Further work has shown that the stem cells from these and other cancers can be distinguished by biochemical markers that may prove to be the starting point for the development of drugs that target the stem cells.

Although some cancers may follow a stem cell model, this by no means shows that all cancers follow such a model. Other cancers may not contain two types of cells and all cells in these tumors may divide and have the potential to spread the cancer by metastasis. The stem cell model needs further study, but offers a new approach to therapy, that may be cancer-specific or patient-specific, changing the basic approach to cancer treatment.

Herceptin (tratuzumab) is a monoclonal antibody that binds to a receptor protein called HER2 on the surface of invasive breast cancer cells. The HER2 protein transfers signals from the external environment to the nucleus, where new programs of gene expression drive cell growth and division. In breast cancer, the HER2 protein is locked in the "on" position, continuously signaling cells to divide. When Herceptin binds to HER2, signaling is silenced and uncontrolled cell division stops. Binding also signals the immune system to attack and destroy the cancer cells.

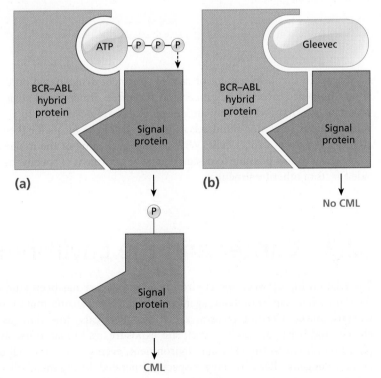

FIGURE 12.19 (a) In CML cells, ATP binds to the BCR-ABL protein and transfers a phosphate group to a signal molecule. The activated signal molecule moves to the nucleus and switches on gene expression, leading to cell growth and division. (b) The drug Gleevec competes for the ATP binding site, and when Gleevec is bound to the BCR-ABL protein, ATP cannot bind, keeping the signal molecule inactive and preventing the cell from completing the cell cycle and dividing.

The Cancer Genome Atlas (TCGA)

Based on the initial successes of cancer genome sequencing, The National Cancer Institute and the National Human Genome Research Institute began a pilot project in 2005 called The Cancer Genome Atlas (TCGA). The project will assess whether it is feasible to use genomic technology—especially large-scale genome sequencing of cancer cells—to catalog the genomic changes in cancer cells. The long-term goal is to provide the tools and information needed to improve the diagnosis, treatment, and prevention of cancer.

The project has several components. First, people with selected cancers are asked to donate a small part of tumor tissue as well as normal tissue. This material will become part of a bank of material that researchers can use for genomic analysis. Because the same cancer has different genomic alterations in different patients and cells within the tumors have variations in genomic changes, it will be necessary to analyze tissue from many different samples of each type of cancer to uncover the basic changes common to all cases. Genomic changes regarded as the most significant will be investigated by cancer genome sequencing centers. Information from these studies will be entered into a central public database: the TCGA Data Portal.

Initially, the project is focused on three types of cancer: a type of brain cancer called glioblastoma, one form of lung cancer (squamous carcinoma), and ovarian cancer. These cancers were selected for several reasons:

- Tissue samples for these cancers that meet the project's scientific and ethical standards and are already available in biorepositories.
- These cancers have a poor prognosis, and information gained may have immediate use in treatment.

- Together, these cancers account for more than 250,000 new cases per year.

To date, the Cancer Genome Characterization Centers have identified more than 6,000 candidate genes and types of RNA for sequencing. This is not a comprehensive list, but it does represent genes and RNA types that clearly have the potential for being associated with the development of cancer. From this list, genes and RNA types for each of the three types of cancer being analyzed are selected and sequenced in groups. Data analysis compares the sequence of selected genes in cancer cells and normal cells from the same individual, giving the results a resolution of one nucleotide. For example, 600 genes identified in glioblastoma were selected for sequencing in the first round. From the results in the first round of sequencing and input from experts in this form of cancer, an additional 700 genes were selected for sequencing.

This project has identified three previously unrecognized mutations in glioblastoma that occur with significant frequency in this cancer and discovered several disrupted cellular pathways. Perhaps most significant is the unearthing of a mechanism for the development of resistance to one of the drugs used to treat this form of cancer. It is hoped that this information can be used to develop a new generation of drugs for effective treatment of glioblastoma.

Work on lung cancer and ovarian cancer are also uncovering previously unknown mutations and networks of cellular disruptions that researchers are studying to use as starting points for new, targeted therapies.

Success in the development of cancer-treatment drugs based on detailed knowledge about the genes involved in cancer and the three-dimensional structure of their gene products has changed the way anticancer drugs are developed. In the past, such drugs were found by screening hundreds or thousands of chemicals for their ability to slow or stop the growth of cancer cells. With an understanding of the molecular events linked to cancer, it is now possible to design drugs for treatment of specific cancers, without the side effects of other treatments.

12.9 | Cancer and the Environment

The relationship between the environment and cancer has been studied for more than 50 years. Many environmental agents damage DNA, causing mutations, some of which lead to cancer. Certain viruses, radiation, chemicals, infection, as well as lifestyle choices and behaviors such as diet, sun exposure, and tobacco use are relevant examples. In addition to the external environment, events inside the cell can be mutagenic and carcinogenic. Reactive oxygen species generated during metabolism and unrepaired errors that are by-products of DNA replication can be cancer-causing events.

Some viral infections lead to cancer.

In humans, several viruses—including the papilloma viruses (HPV 16, HPV 18, HPV 31), herpes virus, and Epstein-Barr virus—are associated with cancer. It is estimated that about 15% of all cancers are associated with viruses, making viral infection a major cause of cancer.

Infections by HPV 16 and HPV 18 can cause cervical cancer. One of the proteins encoded by HPV (the E7 protein) binds to and inactivates the retinoblastoma protein pRB, which regulates the G1/S checkpoint. With pRB inactivated, the infected cell bypasses the checkpoint, completes the cell cycle, and divides. As long as E7 is present, the cell will continue to divide. Other HPV proteins inhibit different tumor-suppressor proteins, maintaining a cellular environment that allows continued division of the infected cells and the replication and dispersal of new virus particles.

Two vaccines, Gardasil and Cervarix were developed to prevent infection by HPV 16, HPV 18, and several other strains. The vaccines contain recombinant DNA–derived HPV coat proteins that stimulate the immune system to make antibodies to HPV. The vaccine is effective only if given before infection and protects against strains that cause about 70% of all cervical cancers.

> **KEEP IN MIND**
>
> Diet and behavior are two major factors in cancer prevention.

What other environmental factors are related to cancer?

The American Cancer Society estimates that 85% of lung cancer cases in men and 75% of cases in women are related to smoking. Smoking produces cancers of the oral cavity, larynx, esophagus, and lungs and is responsible for 30% of all cancer deaths. Most of these cancers have very low survival rates. Lung cancer, for example, has a 5-year survival rate of 13%. Cancer risks associated with tobacco are not limited to smoking; the use of snuff or chewing tobacco carries a 50-fold increased risk of oral cancer.

About 1 million new cases of skin cancer are reported in the United States every year, almost all related to ultraviolet light from the sun or tanning lamps. Skin cancer cases are increasing rapidly in the population (Figure 12.20), presumably as a result of exposure to ultraviolet light from tanning lamps or an increase in outdoor recreation. Epidemiological surveys show that lightly pigmented people are at much higher risk for skin cancer than heavily pigmented individuals. This supports the idea that genetic characteristics can affect the susceptibility of individuals or subpopulations to environmental agents that cause a specific form of cancer.

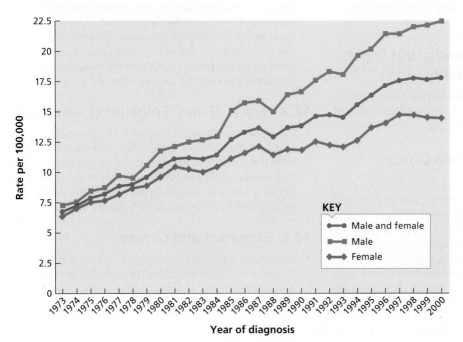

FIGURE 12.20 The age-adjusted rate of melanoma, a deadly form of skin cancer, from 1973 to 2000. Over that time period, the rate (expressed as cases per 100,000 individuals) rose from about 7 to about 17.5, an increase of 250%. Since 2000, the rate has continued to climb.

Genetics in Practice case studies are critical-thinking exercises that allow you to apply your new knowledge of human genetics to real-life problems. You can find these case studies and links to relevant websites at *www.cengage.com/biology/cummings*

CASE 1

Mike was referred for genetic counseling because he was concerned about his extensive family history of colon cancer. That family history was highly suggestive of hereditary nonpolyposis colon cancer (HNPCC). This predisposition is inherited as an autosomal dominant trait, and those who carry the mutant allele have a 75% chance of developing colon cancer by age 65. Mike was counseled about the inheritance of this condition, the associated cancers, and the possibility of genetic testing (on an affected family member). Mike's aunt elected to be tested for one of the genes that may be altered in this condition and discovered that she did have an altered *MSH2* gene. Other family members are in the process of being tested for this mutation.

1. Seventy-five percent of people who carry the mutant allele will get colon cancer by age 65. This is an example of incomplete penetrance. What could cause this?

2. Once a family member is tested for the mutant allele, is it hard for other family members to remain unaware of their own fate, even if they did not want this information? How could family dynamics help or hurt this situation?

3. Is colon cancer treatable? What are the common treatments, and how effective are they?

Summary

12.1 Cancer Is a Genetic Disorder of Somatic Cells

- Cancers are malignant tumors. The primary risk factor for cancer is age. Heritable predispositions to cancer usually show a dominant pattern of inheritance.

12.2 Cancer Begins in a Single Cell

- Cancer begins when a single cell acquires mutations over time that allow it to escape control of the cell cycle and begin uncontrolled division. Cancer cells are clonal descendants from a single mutant cell. Because mutations accumulate slowly, age is the primary risk factor for cancer.

12.3 Most Cancers Are Sporadic, but Some Have an Inherited Susceptibility

- Cancer can be caused by inheriting genes that cause a predisposition to cancer (inherited cancer) or by the accumulation of mutations in somatic cells (sporadic cancer).

12.4 Mutations in Cancer Cells Disrupt Cell-Cycle Regulation

- The study of two classes of genes—tumor-suppressor genes and oncogenes—has established the relationship between cancer, the regulation of cell growth and division, and the cell cycle. The discovery of tumor-suppressor genes that normally act to inhibit cell division has provided insight into the regulation of the cell cycle. These gene products act at control points in the cell cycle at G1/S or G2/M. Deletion or inactivation of these products causes cells to divide continuously.

12.5 Mutant Cancer Genes Affect DNA Repair Systems and Genome Stability

- Many of the basic properties of cancer—including high rates of mutation, chromosomal abnormalities, and genomic instability—result from the inability of cancer cells to repair damage to DNA. DNA repair genes are now recognized as a class of cancer-related genes along with tumor-suppressor genes and proto-oncogenes.

12.6 Colon Cancer Is a Genetic Model for Cancer

- Cancer is a multistep process that requires a number of specific mutations. Colon cancer has been studied to provide insight into the number and order of steps involved in transforming normal cells into cancer cells. Two pathways to colon cancer illustrate that some genes are gatekeepers, controlling the cell cycle, whereas others are caretakers, repairing DNA damage to prevent genomic instability.

12.7 Hybrid Genes, Epigenetics, and Cancer

- Other human disorders, including Down syndrome, are associated with high rates of cancer. This predisposition may result from the presence of an initial mutation or genetic imbalance that moves cells closer to a cancerous state. Other cancers, including leukemia, are caused by translocation events, some of which create hybrid genes that activate cell division.

12.8 Genomics and Cancer

- Sequencing the protein-coding genes in cancer cells has led to the identification of new classes of genes and genomic changes associated with cancer. These findings are being developed into new methods of diagnosis and treatment.

12.9 Cancer and the Environment

- It is now apparent that many cancers are environmentally induced. Occupational exposure to minerals and chemicals poses a cancer risk to workers in a number of industries. The widespread dissemination of these materials poses an undefined but potentially large risk to the general population. Social behavior contributes to approximately 50% of all cancer cases in the United States, most if not all of which are preventable.

Questions and Problems

CENGAGENOW Preparing for an exam? Assess your understanding of this chapter's topics with a pre-test, a personalized learning plan, and a post-test by logging on to *login.cengage.com/sso* and visiting CengageNOW's Study Tools.

Cancer Is a Genetic Disorder of Somatic Cells

1. Theodore Boveri predicted that malignancies would often be associated with chromosomal mutation. What lines of evidence substantiate this prediction?
2. Distinguish between a familial and a sporadic cancer.
3. Benign tumors:
   a. are noncancerous growths that do not spread to other tissues.
   b. do not contain mutations.
   c. are malignant and clonal in origin.
   d. metastasize to other tissues.
   e. none of the above.
4. What does it mean to have a malignant tumor?
5. Metastasis refers to the process in which:
   a. tumor cells die.
   b. tumor cells detach and move to secondary sites.
   c. cancer does not spread to other tissues.
   d. tumors become benign.
   e. cancer can be cured.

Most Cancers Are Sporadic, but Some Have an Inherited Susceptibility

6. Cancer is now viewed as a disease that develops in stages. What are the stages to which this statement refers?
   a. Malignant tumors become nonmalignant.
   b. Proto-oncogenes become tumor-suppressor genes.
   c. Younger people get cancer more than older people do.
   d. Small numbers of individual mutational events exist that can be separated by long periods of time.
   e. none of the above.
7. It is often the case that a predisposition to certain forms of cancer is inherited. An example is familial retinoblastoma. What does it mean to have inherited an increased probability of acquiring a certain form of cancer? What subsequent event(s) must occur?

Mutations in Cancer Cells Disrupt Cell-Cycle Regulation

8. A proto-oncogene is a gene that:
   a. normally causes cancer.
   b. normally suppresses tumor formation.
   c. normally functions to promote cell division.
   d. is involved in forming only benign tumors.
   e. is expressed only in blood cells.
9. What is the difference between a proto-oncogene and a tumor-suppressor gene?
10. Distinguish between dominant inheritance and recessive inheritance in retinoblastoma.
11. Describe the likelihood of developing bilateral (both eyes affected) retinoblastoma in the inherited versus the sporadic form of the disease.
12. The parents of a 1-year-old boy are concerned that their son may be susceptible to retinoblastoma because the father's brother had bilateral retinoblastoma. Both parents are normal (no retinoblastoma). They have their son tested for the *RB* gene and find that he has inherited a mutant allele. The father is tested and is found to carry a mutant allele.
    a. Why didn't the father develop retinoblastoma?
    b. What is the chance that the couple will have another child carrying the mutant allele?
    c. Is there a benefit to knowing that their son may develop retinoblastoma?
13. The search for the *BRCA1* breast cancer gene discussed in this chapter was widely publicized in the media (for example, *Newsweek*, December 6, 1993). Describe the steps taken by Mary-Claire King and her colleagues to clone this gene. How long did this process take?
14. What are the roles of cellular proto-oncogenes, and how are these roles consistent with their implication in oncogenesis?
15. Which of the following mutations will result in cancer?
    a. Homozygous recessive mutation in a tumor-suppressor gene coding for a nonfunctional protein.
    b. Dominant mutation in a tumor-suppressor gene in which the normal protein product is overexpressed.
    c. Homozygous recessive mutation in which there is a deletion in the coding region of a proto-oncogene, leaving it nonfunctional.
    d. Dominant mutation in a proto-oncogene in which the normal protein product is overexpressed.

16. In DNA repair, how does the normal allele of *BRCA1* work?

17. The following family has a history of inherited breast cancer. Betty (grandmother) does not carry the gene. Don, her husband, does. Don's mother and sister had breast cancer. One of Betty and Don's daughters (Sarah) has breast cancer; the other (Karen) does not. Sarah's daughters are in their 30s. Dawn, 33, has breast cancer; Debbie, 31, does not. Debbie is wondering if she will get the disease because she looks like her mother. Dawn is wondering if her 2-year-old daughter (Nicole) will get the disease.

  a. Draw a pedigree indicating affected individuals and identify all individuals.

  b. What is the most likely mode of inheritance of this trait?

  c. What are Don's genotype and phenotype?

  d. What is the genotype of the unaffected women (Betty and Karen)?

  e. A genetic marker has been found that maps very close to the gene. Given the following marker data for chromosomes 4 and 17, which chromosome does this gene map to?

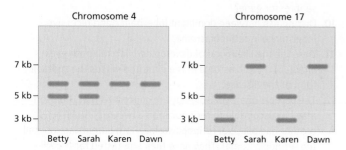

  f. Using the same genetic marker, Debbie and Nicole were tested. The results are shown in the following figure. Based on their genotypes, is either of them at increased risk for breast cancer?

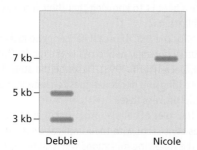

18. You are in charge of a new gene therapy clinic. Two cases have been referred to you for review and possible therapy.

  Case 1. A mutation in the promoter of a proto-oncogene causes the gene to make too much of its normal product, a receptor protein that promotes cell division. The uncontrolled cell division has caused cancer.

  Case 2. A mutation in an exon of a tumor-suppressor gene makes this gene nonfunctional. The product of this gene normally suppresses cell division. The mutant gene cannot suppress cell division, and this has led to cancer. What treatment options can you suggest for each case?

19. Explain how the *APC* gene starts the progress toward colon cancer.

Hybrid Genes, Epigenetics, and Cancer

20. Can you postulate a reason or reasons why children with Down syndrome are 20 times more likely to develop leukemia than children in the general population?

21. What mutational event is typically associated with Burkitt's lymphoma? Which chromosomes are involved in this tumor?

Genomics and Cancer

22. In the opening vignette, Julie is concerned that she may develop breast cancer, but testing shows that she does not carry the rare high-risk *BRCA1* and *BRCA2* alleles. What if further testing showed that some of her aunts, her mother, and she carried a common low-risk allele for breast cancer. What would you recommend to Julie if you were her genetic counselor?

23. One of the first drugs used for targeted cancer therapy is tamoxifin, used to treat estrogen-receptor breast cancer. In Chapter 10, review the use of this drug. What general conclusions can you draw about targeted drug therapy and genotype?

Cancer and the Environment

24. What are some factors that epidemiologists have associated with a relatively high risk of developing cancer?

25. Smoking cigarettes has been shown to be associated with the development of lung cancer. However, a direct correlation between how many cigarettes one smokes and the onset of lung cancer does not exist. A heavy smoker may not develop lung cancer, whereas a light smoker may develop the disease. Explain why this may be.

26. Research and discuss the relevance of epidemiologic and experimental evidence in recent governmental decisions to regulate exposure to asbestos in the environment.

27. Studies have shown that there are significant differences in cancer rates among different ethnic groups. For example, the Japanese have very high rates of colon cancer but very low rates of breast cancer. It has also been demonstrated that when members of low-risk ethnic groups move to high-risk areas, their cancer risks rise to those of the high-risk area. For example, Japanese who live in the United States, where the risk of breast cancer is high, have higher rates of breast cancer than do Japanese who live in Japan. What are some of the possible explanations for this phenomenon? What factors may explain why the Japanese have higher rates of colon cancer than do other ethnic groups?

Internet Activities

Internet Activities are critical-thinking exercises using the resources of the World Wide Web to enhance the principles and issues covered in this chapter. For a full set of links and questions investigating the topics described below, visit *www.cengage.com/biology/cummings*

1. *The "Cancer Gene" TP53 (Li-Fraumeni Syndrome).* The tumor-suppressor gene *TP53* (also known as *p53*) helps improve the rate of DNA repair; mutations in this gene are implicated in the onset of many cancers. At the *TP53 Cancer GeneWeb* website, you can explore links to many different sources of information on *TP53*. From this website, under the "Other Related Resources" heading, you can link to the Weizmann Institute's *p53* home page and read about the expression, cellular function, and involvement in disease of this critical gene.

2. *Family Histories of Breast Cancer.* Breast cancer screening followed by prophylactic (preventive) mastectomy has become an option for some women with strong family histories of breast cancer. At Lawrence Berkeley National Laboratory's *ELSI* website, link to the "Breast Cancer Screening" page. Within the main page, read about breast cancer in the introduction and then scroll down to "What Would You Do?" and consider how you would answer the questions posed in the various classroom scenarios.

3. *Cancer Case Studies.* The *Genetics of Cancer* website provides access to a number of case studies that include family history information and pedigrees and then poses personal and ethical questions. Select one of the cases, evaluate it, and answer the questions.

 HOW WOULD YOU VOTE NOW?

Women who carry mutations in either of two cancer genes, *BRCA1* and *BRCA2*, have a significantly higher risk of developing breast cancer and ovarian cancer. Julie, the woman discussed at the beginning of this chapter, has several female family members with one of these mutant genes. However, neither Julie nor her mother carries the mutant gene. On the basis of this information, Julie decided not to have surgery to remove her breasts. Now that you know more about cancer and the genetics it involves, what do you think? If a female member of your family carried one of the two breast cancer genes and asked your advice, would you recommend that she have her breasts removed even though not everyone carrying this mutation will develop breast cancer? Visit the *Human Heredity* companion website at *www.cengage.com/biology/cummings* to find out more on the issue; then cast your vote online.

13 An Introduction to Genetic Technology

On a farm in Wisconsin, Bob Schauf rises early every day to milk the cows in his herd and send out the collected milk for processing. After he is finished with the herd, he milks one last cow, Blackrose II. After collecting her milk, he saves some for his personal use and pours the rest down the drain. Why? Blackrose II is a clone, the offspring of Blackrose, a world champion Holstein. Blackrose II's DNA has not been genetically modified, only copied from her mother, and there are no regulations governing the sale of animal products from clones. The U.S. Food and Drug Administration (FDA) asked farmers and biotech companies to abstain temporarily from selling milk, meat, and other products from cloned cows. A 2002 National Academy of Sciences report concluded that meat and milk from cloned animals pose no risk to consumers. A 2005 study of the meat and milk from cloned cows concluded there are no significant differences in food products from cloned and noncloned animals. In 2006, the FDA invited public comment on animal cloning and food safety. After evaluating the comments, in January 2008, the FDA ruled that food products from cloned animals are safe and can be marketed. In September 2008, the FDA issued guidelines for industry and consumers about food products from cloned animals.

Why clone milk cows? The average cow produces about 3.9 gallons of milk a day. Blackrose II produces about 11.5 gallons of milk per day—triple the average. Herds of cloned cows or their offspring with high production would allow farmers to reduce herd size significantly without affecting milk production. There are other possible benefits for consumers as well. In recent years, dairy cows have been given growth hormones to stimulate milk production, a controversial practice that would be unnecessary now that milk from clones has been judged to be safe.

In this chapter, we will examine how plants, animals, and DNA molecules are cloned, and in the next chapter, we will discuss how this technology is being used.

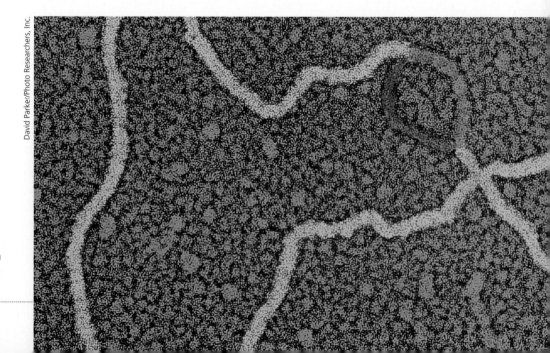

David Parker/Photo Researchers, Inc.

A colorized electron micrograph of a plasmid containing an inserted DNA fragment.

13.1   What Are Clones?

When a fertilized egg divides to form two cells, those cells usually stay attached to each other and become parts of one embryo. In a small number of cases, the cells separate from each other and form separate embryos that become identical twins. Because they are derived from a single ancestral cell (in this case a fertilized egg), identical twins are clones. **Clones** are defined as molecules, cells, or individuals derived from a single ancestor. Methods for producing clones are not new; the Greek philosopher Plato wrote about cloning fruit trees 3,000 years ago. However, cloning DNA molecules, cells, and animals became possible only in recent decades.

The ability to isolate and produce many copies of specific DNA molecules had a great impact on many areas of genetic research. That impact spread quickly beyond research laboratories, and the techniques are now used in many areas, including the criminal justice system, child support cases, archaeology, dog breeding, and environmental conservation. The commercial use of these methods gave rise to the multibillion-dollar biotechnology industry.

In this chapter, we will review the methods used to clone DNA molecules, identify them, and analyze them. In the next chapter, we will examine how cloning is used in research and everyday life, and in Chapter 15, we will discuss the Human Genome Project and the new field of genomics, direct outgrowths of recombinant DNA technology. Before we examine how to isolate and copy DNA molecules, let's look briefly at how whole organisms can be cloned.

HOW WOULD YOU VOTE?

Cloned cows do not contain any foreign genes and are not genetically modified. Cloning proponents argue that milk and meat from cloned cows is safe for human consumption. Watchdog groups that monitor genetically engineered foods are concerned about the safety of products from cloned animals and want more research to prove the safety of such products before they are allowed in the marketplace. Do you think that milk and meat from cloned animals is safe and should be sold without labeling? Visit the *Human Heredity* companion website at *www.cengage.com/biology/cummings* to find out more on the issue; then cast your vote online.

KEEP IN MIND AS YOU READ

- Cloned plants and animals are used in research, agriculture, and medicine.

- Cloning DNA uses techniques of biochemistry, genetics, and molecular biology.

- Finding a specific gene in a cloned library requires a molecular probe.

- The polymerase chain reaction (PCR) copies DNA without cloning.

Clones Genetically identical molecules, cells, or organisms, all derived from a single ancestor.

Animals can be cloned by several methods.

We have genetically manipulated plants and animals for thousands of years by means of selective breeding. Organisms with desirable characteristics are chosen for breeding, and the offspring with the best combination of characteristics are used as parents for breeding the next generation. Although this method, called artificial selection, seems slow and unreliable, Charles Darwin observed that after only a few generations, breeders produced many varieties of pigeons and minks.

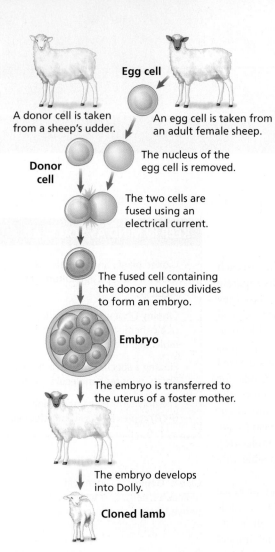

A donor cell is taken from a sheep's udder.

Egg cell

An egg cell is taken from an adult female sheep.

Donor cell

The nucleus of the egg cell is removed.

The two cells are fused using an electrical current.

The fused cell containing the donor nucleus divides to form an embryo.

Embryo

The embryo is transferred to the uterus of a foster mother.

The embryo develops into Dolly.

Cloned lamb

FIGURE 13.1 Cloning using an adult donor nucleus. Dolly the sheep was cloned by placing a nucleus from an udder cell into an egg cell that had its nucleus removed. The cells were fused by an electrical impulse and divided to form an embryo. The resulting embryo was implanted into a foster mother, resulting in the birth of Dolly.

On U.S. farms, artificial selection was supplemented more than 20 years ago by reproductive cloning. Two of the most widely used methods of cloning are embryo splitting and nuclear transfer. To clone an animal by embryo splitting, an unfertilized egg is collected from a donor mother and then fertilized in a dish through the use of a method called *in vitro* fertilization (IVF). The fertilized egg develops in the dish to form an embryo containing four to eight cells. After the cells are separated from one another, they divide to form genetically identical embryos. The embryos are then implanted into surrogate mothers for development. This method is a variation on nature's way of producing identical twins or triplets and can be used to clone any mammalian embryo, including human embryos.

The second method of animal cloning, nuclear transfer, is more technically difficult but can result in a larger number of cloned offspring. The first successful cloning of mammals by cell fusion was done in 1986. As the cell-fusion method was improved, it became possible to use cells from older and older donors to fuse with the enucleated eggs. The cloning of Dolly the sheep, reported in 1997, was the first time an adult cell (in this case from the udder) was successfully used as a donor to produce a cloned animal (Figure 13.1). Cloning Dolly was a significant event because it showed that even nuclei from highly specialized adult cells can direct all stages of development when transferred into eggs (Figure 13.2). However, the success rate in cloning animals by cell fusion is very low. A year after Dolly was cloned, a research team was able to clone over two dozen mice by directly injecting nuclei removed from adult cells into enucleated eggs (Figure 13.3). This method has a much higher success rate than cell fusion and opens the door to cloning many different animals.

Animal cloning has had a great impact on farming. Sheep, cattle, goats, and pigs have all been cloned. Farmers now can produce herds of genetically identical animals, all of which have valuable traits such as superior wool, milk, or meat production. The cloning of Blackrose described at the beginning of this chapter demonstrates one use of animal cloning. Case 2 at the end of the chapter discusses human cloning.

KEEP IN MIND

Cloned plants and animals are used in research, agriculture, and medicine.

FIGURE 13.2 Dolly the sheep, shown with her surrogate mother. Dolly was cloned by fusion of an adult cell and an enucleated egg.

Le Corre-Ribeiro/Liaison/Getty Images RF

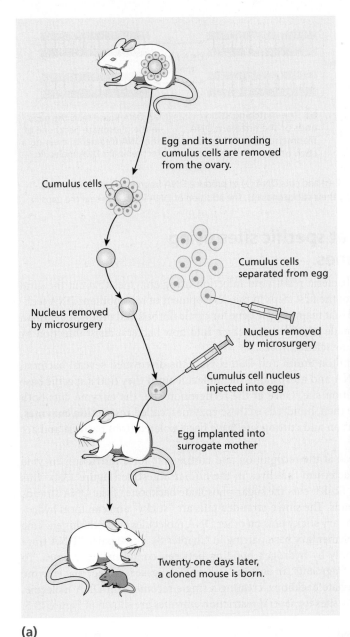

Egg and its surrounding cumulus cells are removed from the ovary.

Cumulus cells

Cumulus cells separated from egg

Nucleus removed by microsurgery

Nucleus removed by microsurgery

Cumulus cell nucleus injected into egg

Egg implanted into surrogate mother

Twenty-one days later, a cloned mouse is born.

(a)

(b)

May 27, 1999 by Dave Au. Courtesy of University of Hawaii.

FIGURE 13.3 (a) Cloning mice by nuclear injection. Eggs were removed from a mouse ovary along with the surrounding cumulus cells. A nucleus from a cumulus cell was injected into an enucleated mouse egg, and the resulting embryo was transferred to a foster mother. Nuclear injection is more effective than cell fusion as a method for cloning animals. (b) A cloned mouse (*on the bar*) and its parent (*lower left*).

13.2 Cloning Genes Is a Multistep Process

Although cloning of whole organisms has changed agriculture, **recombinant DNA technology** and the cloning of DNA molecules has revolutionized everything from laboratory research to health care to the food we eat. Using these methods, we can locate genes, map them, isolate them, and transfer them between species. Recombinant DNA technology is used to identify carriers of genetic disorders, perform gene therapy, and create disease-resistant food plants. The goal of cloning DNA (including genes) is to produce a large number of identical molecules, all of which are copies of one ancestral molecule. Once DNA clones are available, they can be used in research laboratories or in commercial applications.

When a source of DNA to be cloned has been selected (say, DNA from a human cell), cloning requires three things:

- a way to cut the DNA at specific sites to produce consistent, manageable pieces
- a carrier molecule to hold the DNA for transfer to a host cell for cloning
- a host cell in which many copies of the selected DNA molecule can be produced

Recombinant DNA technology A series of techniques in which DNA fragments from an organism are linked to self-replicating vectors to create recombinant DNA molecules, which are replicated or cloned in a host cell.

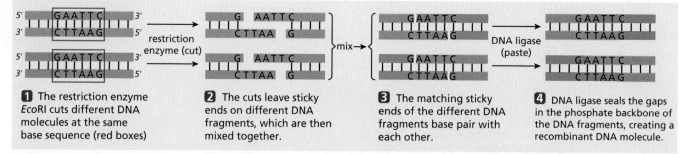

1 The restriction enzyme *Eco*RI cuts different DNA molecules at the same base sequence (red boxes)

2 The cuts leave sticky ends on different DNA fragments, which are then mixed together.

3 The matching sticky ends of the different DNA fragments base pair with each other.

4 DNA ligase seals the gaps in the phosphate backbone of the DNA fragments, creating a recombinant DNA molecule.

FIGURE 13.4 The restriction enzyme *Eco*RI recognizes the sequence GAATTC (1) and cuts DNA (2) to produce DNA fragments with sticky ends. When mixed together, DNA fragments from two different sources can pair by their sticky ends (3). The addition of DNA ligase can seal the gaps in the phosphate backbone to produce recombinant DNA molecules (4).

DNA can be cut at specific sites using restriction enzymes.

The discovery that some bacteria resist viral infection using enzymes that cut the viral DNA into pieces was one of the first steps in the development of recombinant DNA technology. It might seem odd that mapping the gene for cystic fibrosis and producing human insulin in bacteria were made possible by research into how bacteria resist infection by viruses, but this is often how science works.

In the mid-1970s, Hamilton Smith and Daniel Nathans discovered several bacterial enzymes that attach to DNA and move along the molecule until they find a specific base sequence called a recognition site. Once at the recognition site, the enzyme cuts both strands of the DNA. Since then, hundreds of these enzymes, called **restriction enzymes**, each with its own recognition and cutting site, have been isolated from bacteria and are used in DNA cloning.

As an example, let's look at the recognition and cutting site for a restriction enzyme from *Escherichia coli*, a bacterium that lives in the human intestine (Figure 13.4). This restriction enzyme, called *Eco*RI, cuts the sugar-phosphate backbone of the DNA strands to create single-stranded ends. The single-stranded tails are "sticky" and can form hydrogen bonds with complementary sticky tails on other DNA molecules cut with this enzyme (if necessary, review complementary base pairing in Chapter 8). These paired DNA fragments are held together only by hydrogen bonding between complementary bases. To close the gaps between the fragments, an enzyme called DNA ligase is added; that enzyme seals the gaps in the phosphate backbone, creating a single recombinant DNA molecule. The recognition and cutting sites for several restriction enzymes are shown in Figure 13.5.

Restriction enzyme A bacterial enzyme that cuts DNA at specific sites.

FIGURE 13.5 Several restriction enzymes and their cutting sites (*arrows*).

Enzyme	Recognition and cleavage sequence	Cleavage pattern	Source organism
*Eco*RI	GAATTC / CTTAAG	G AATTC / CTTAA G	*Escherichia coli*
*Hind*III	AAGCTT / TTCGAA	A AGCTT / TTCGA A	*Haemophilus influenzae*
*Bam*HI	GGATCC / CCTAGG	G GATCC / CCTAG G	*Bacillus amyloliquefaciens*
*Sau*3A	GATC / CTAG	GATC / CTAG	*Staphylococcus aureus*
*Hae*III	GGCC / CCGG	GG CC / CC GG	*Haemophilus aegypticus*

Vectors serve as carriers of DNA to be cloned.

DNA can be reproducibly cut into manageable pieces by restriction enzymes. Once a collection of DNA fragments has been produced by restriction-enzyme digestion, the next step is to link those pieces with carrier molecules for transfer to a host cell. Many species of bacteria carry small, circular DNA molecules called plasmids. During division, the bacterial cell copies its chromosome and any plasmids it carries, so that each daughter cell receives a chromosome and at least one copy of the cell's plasmids. By genetically engineering plasmids, researchers developed carrier molecules called **vectors** that can transfer foreign DNA sequences into bacterial host cells for cloning. A map of one such vector, a plasmid called pBR322, is shown in Figure 13.6. The middle section of the diagram shows the location of recognition sites that can be cut by restriction enzymes and used to insert DNA molecules from another organism.

Recombinant DNA molecules are inserted into host cells for cloning.

Now that we know that DNA can be cut with restriction enzymes and linked to vector DNA to produce recombinant DNA molecules, the next step is to transfer these molecules into a host cell for cloning. Let's look at an example using DNA from human cells (Active Figure 13.7). To create a recombinant molecule for cloning, DNA from human cells is cut with a restriction enzyme,

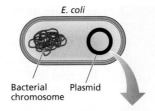

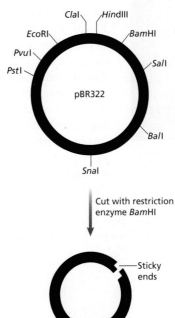

FIGURE 13.6 The plasmid pBR322 contains restriction sites used to carry DNA fragments for cloning. Cutting the plasmid with restriction enzymes produces molecules with sticky ends that can bind to DNA from other sources cut with the same enzyme.

Vectors Self-replicating DNA molecules that are used to transfer foreign DNA segments between host cells.

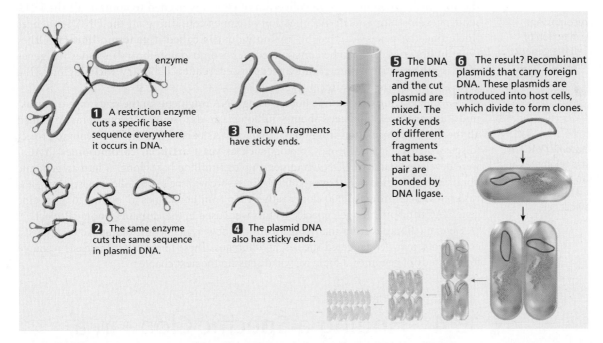

ACTIVE FIGURE 13.7 The steps in cloning DNA. (1) The DNA to be cloned is cut with a restriction enzyme. (2) A plasmid vector is cut with the same restriction enzyme. (3–4) Both types of DNA cut with this enzyme have complementary sticky ends. (5) When the DNA fragments and the cut plasmids are mixed, they can bind together by their sticky ends and link to form recombinant DNA molecules by DNA ligase. (6) The recombinant plasmids are transferred to bacterial host cells. At each bacterial cell division, the plasmid is replicated, producing many copies (or clones) of the foreign DNA.

CENGAGENOW Learn more about DNA cloning by viewing the animation by logging on to *login.cengage.com/sso* and visiting Cengage-NOW's Study Tools.

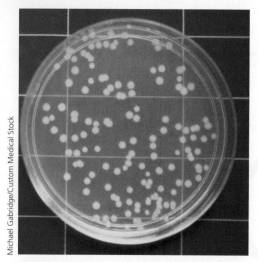

FIGURE 13.8 Colonies of bacteria on Petri plates. Each colony is descended from a single cell. Therefore, each colony is a clone.

producing sticky ends. The vector is cut with the same enzyme, also producing sticky ends. The human DNA fragments and the cut vector are mixed together, and in many cases human DNA fragments and vector molecules align to form recombinant plasmids containing human DNA and plasmid DNA. Once a DNA fragment and the vector have been linked together by DNA ligase, the resulting recombinant molecule is ready to be cloned. For cloning, the vector carrying a piece of human DNA is transferred into a bacterial host cell. The host cell is placed on a nutrient plate, where it grows and divides to form a colony (Figure 13.8). Because the cells in each colony are derived from a single ancestral cell, all the cells in the colony and the recombinant plasmids they contain are clones. Each time the bacterial cell divides (about every 20 minutes, under ideal conditions), each daughter cell receives one or more copies of the recombinant plasmid.

Scientists working in the field of recombinant DNA technology were among the first to realize that there might be unrecognized dangers in using and releasing recombinant organisms (see Exploring Genetics: Asilomar: Scientists Get Involved). After years of discussion and experimentation, there is now general agreement that such work poses little risk, but the episodes with *Escherichia coli* and and endangered species (see Spotlight on Can We Clone Endangered Species? on page 300) demonstrate that scientists are concerned about the risks and benefits of their work.

> **KEEP IN MIND**
>
> Cloning DNA uses techniques of biochemistry, genetics, and molecular biology.

13.3 Cloned Libraries

Genomic library In recombinant DNA terminology, a collection of clones that contains all the genetic information in an individual.

Because each cloned fragment of human DNA is relatively small compared with the size of the genome, a large collection of clones is needed to contain all the DNA sequences in a human cell. A collection of clones containing all the DNA sequences (including the genes) carried by an individual is called a **genomic library**. Other types of libraries can also be constructed. A chromosomal library includes the genes from a single chromosome, and an expressed sequence library includes only the genes expressed in a certain cell type.

If we made a human genomic library using DNA fragments in the size range that plasmids can carry, we would need about 8 million plasmids to make sure our library included all the genetic information from one human cell. As part of the Human Genome Project (discussed in Chapter 15), new vectors such as **yeast artificial chromosomes (YACs)** were developed that carry DNA fragments up to 1 million bases long. A human genomic library can be carried in just over 3,000 YACs. As we will see in the following sections, it is easier to search through 3,000 clones rather than 8 million clones in looking for a specific gene. YACs and other large-capacity vectors were used in the Human Genome Project.

Yeast artificial chromosome (YAC) A cloning vector that has telomeres and a centromere that can accommodate large DNA inserts and uses the eukaryote yeast as a host cell.

Genomic libraries from many organisms are now available, including bacteria, yeasts, crop plants, many endangered species, and humans. These libraries are the basic resource for the study of genomes, a topic we will discuss in the next chapter.

13.4 Finding a Specific Clone in a Library

If you want to study a specific gene, a genomic library must be searched to find the clone that carries that gene. Humans have around 20,000 to 25,000 genes. The question is how to search through the thousands or millions of clones in the library to find a gene. Most

Asilomar: Scientists Get Involved

The first steps in creating recombinant DNA molecules were taken in 1973 and 1974. Scientists immediately realized that modifying the genetic information in *Escherichia coli*, a bacterium from the human gut, could be potentially dangerous. A group of scientists asked the National Academy of Sciences to appoint a panel to assess the risks and the need to control recombinant DNA research. A second group published letters in two leading journals, *Science* and *Nature*, calling for a moratorium on certain kinds of experiments until the potential hazards could be assessed. Shortly afterward, an international conference was held at Asilomar, California, to consider whether recombinant DNA technology posed any dangers and whether this form of research should be regulated. In 1975 and 1976, guidelines resulting from this conference were published by the National Institutes of Health (NIH), a government agency that sponsors biomedical research in the United States. These guidelines prohibited certain kinds of experiments and listed other types of experiments that were to be conducted in containment to prevent release of bacterial cells carrying recombinant DNA molecules. In the meantime, legislation was proposed in Congress and in many states and local governments to regulate or prohibit the use of recombinant DNA technology.

After exhaustive testimony and reports, the federal legislation was withdrawn, and by 1978, research had demonstrated that the strain of *E. coli* was much safer for use as a host cell than was originally thought. Other work showed that recombinant DNA is produced in nature and that there are no detectable serious effects, and in 1982, NIH issued a new set of guidelines eliminating most of the constraints on recombinant DNA research. No experiments are currently prohibited.

The most important lesson from these events is that the scientists who developed the methods were the first to call attention to the possible dangers of recombinant DNA research, and they did so on the basis of its *potential* for harm. There were no known cases of the release of recombinant DNA–carrying host cells into the environment. Scientists voluntarily shut down their research work until the situation could be assessed properly and objectively. Only when they reached a consensus that there was no danger did their work resume. Contrary to how they are often portrayed in the popular media, scientists do care about the consequences of their work and do become involved in socially important issues.

---

often, a specific clone is identified by using a labeled, complementary single-stranded RNA or DNA molecule called a **probe**. Probes can identify genes in a library by their ability to base-pair with the complementary base sequence in the gene that is being searched for, which can be detected by the probe's fluorescent or radioactive label (Figure 13.9).

KEEP IN MIND

Finding a specific gene in a cloned library requires a molecular probe.

Probe A labeled nucleic acid used to identify a complementary region in a clone or genome.

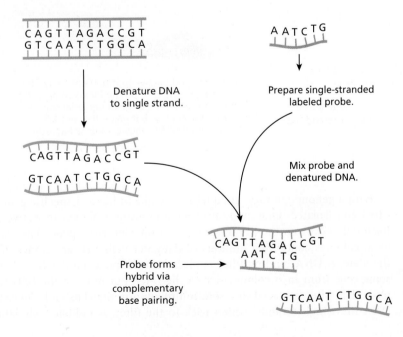

C A G T T A G A C C G T
G T C A A T C T G G C A

Denature DNA to single strand.

A A T C T G
Prepare single-stranded labeled probe.

CAGTTAGACCGT
GTCAATCTGGCA

Mix probe and denatured DNA.

Probe forms hybrid via complementary base pairing.

CAGTTAGACCGT
AATCTG

GTCAATCTGGCA

FIGURE 13.9 A DNA probe is a single-stranded molecule labeled for identification in some way. Both chemical and radioactive labels are used. The single-stranded probe forms a double-stranded hybrid molecule with regions complementary to the DNA being studied.

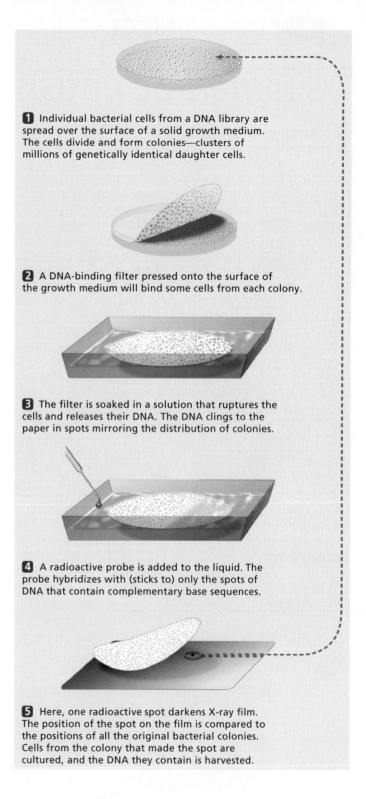

1 Individual bacterial cells from a DNA library are spread over the surface of a solid growth medium. The cells divide and form colonies—clusters of millions of genetically identical daughter cells.

2 A DNA-binding filter pressed onto the surface of the growth medium will bind some cells from each colony.

3 The filter is soaked in a solution that ruptures the cells and releases their DNA. The DNA clings to the paper in spots mirroring the distribution of colonies.

4 A radioactive probe is added to the liquid. The probe hybridizes with (sticks to) only the spots of DNA that contain complementary base sequences.

5 Here, one radioactive spot darkens X-ray film. The position of the spot on the film is compared to the positions of all the original bacterial colonies. Cells from the colony that made the spot are cultured, and the DNA they contain is harvested.

ACTIVE FIGURE 13.10 Using a probe to identify a colony in a genomic library carrying a specific DNA sequence or gene.

CENGAGENOW Learn more about finding specific genes by viewing the animation by logging on to *login.cengage.com/sso* and visiting CengageNOW's Study Tools.

With a genomic library and a labeled probe in hand, a specific gene can be identified and isolated. One method of gene hunting is shown in Active Figure 13.10. Bacterial cells from the library are grown on a nutrient plate. The cells divide and form colonies containing millions of daughter cells that are clones of each cell on the plate. A DNA-binding filter is placed on the plate, and as the filter is lifted off, some cells from each colony stick to the filter, but most of the cells remain on the plate. The filter is placed in a solution to rupture the cells and convert the released DNA into single strands, which stick to the filter. A radioactively-labeled probe is added to the solution and binds to any complementary sequences present in the DNA on the filter. The filter is washed to remove excess probe and laid down on a piece of X-ray film. Radioactivity from the probe exposes the film, making a spot on the film that corresponds to the location of the colony carrying the gene of interest. With this information, it is possible to return to the original plate, identify and isolate the colony, and grow large quantities of the bacteria carrying this clone for further study.

13.5 A Revolution in Cloning: The Polymerase Chain Reaction

Cloning with vectors and host cells is labor intensive and time consuming. Fortunately, it is not the only way to make many copies of a DNA molecule. A technique called the **polymerase chain reaction (PCR)**, invented in the 1980s, has revolutionized and in some cases replaced host-cell cloning.

PCR (Active Figure 13.11) is a form of DNA replication (see Chapter 8 for a discussion of DNA replication) in which sections of a DNA molecule are copied, rather than the whole molecule. PCR uses two components: single strands of DNA that serve as templates and a primer that binds to a region of the DNA to be copied. Primers serve as a starting point for the synthesis of the new DNA strand. Once made, the newly synthesized double-stranded molecule is separated into single strands, and each strand is once again used as a template in another round of PCR, so the amount of DNA doubles with each round of replication. In this way, even a single DNA molecule can be amplified to make millions or billions of copies in a few hours.

KEEP IN MIND

The polymerase chain reaction (PCR) copies DNA without cloning.

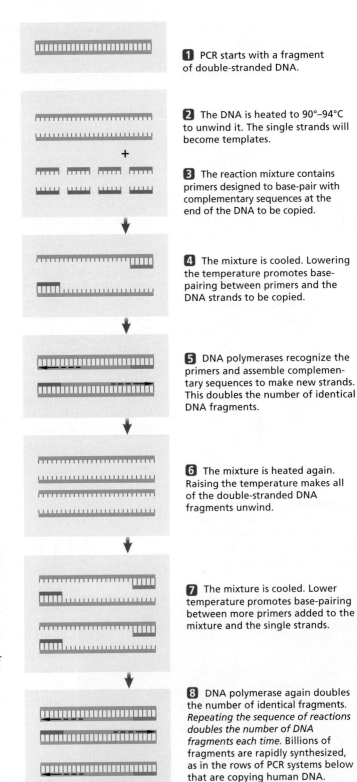

1 PCR starts with a fragment of double-stranded DNA.

2 The DNA is heated to 90°–94°C to unwind it. The single strands will become templates.

3 The reaction mixture contains primers designed to base-pair with complementary sequences at the end of the DNA to be copied.

4 The mixture is cooled. Lowering the temperature promotes base-pairing between primers and the DNA strands to be copied.

5 DNA polymerases recognize the primers and assemble complementary sequences to make new strands. This doubles the number of identical DNA fragments.

6 The mixture is heated again. Raising the temperature makes all of the double-stranded DNA fragments unwind.

7 The mixture is cooled. Lower temperature promotes base-pairing between more primers added to the mixture and the single strands.

8 DNA polymerase again doubles the number of identical fragments. *Repeating the sequence of reactions doubles the number of DNA fragments each time.* Billions of fragments are rapidly synthesized, as in the rows of PCR systems below that are copying human DNA.

ACTIVE FIGURE 13.11 The polymerase chain reaction. Repeated cycles can amplify the original DNA sequence by more than a million times.

CENGAGENOW Learn more about copying DNA with the polymerase chain reaction by viewing the animation by logging on to *login.cengage.com/sso* and visiting CengageNOW's Study Tools.

David Parker/SPL/Photo Researchers, Inc.

Table 13.1 DNA Sequence Amplification by PCR

Cycle	Number of Copies
0	1
1	2
5	32
10	1,024
15	32,768
20	1,048,576
25	33,544,432
30	1,073,741,820

Polymerase chain reaction (PCR) A method for amplifying DNA segments using cycles of denaturation, annealing to primers, and DNA polymerase-directed DNA synthesis.

Southern blot A method for transferring DNA fragments from a gel to a membrane filter, developed by Edwin Southern for use in hybridization experiments.

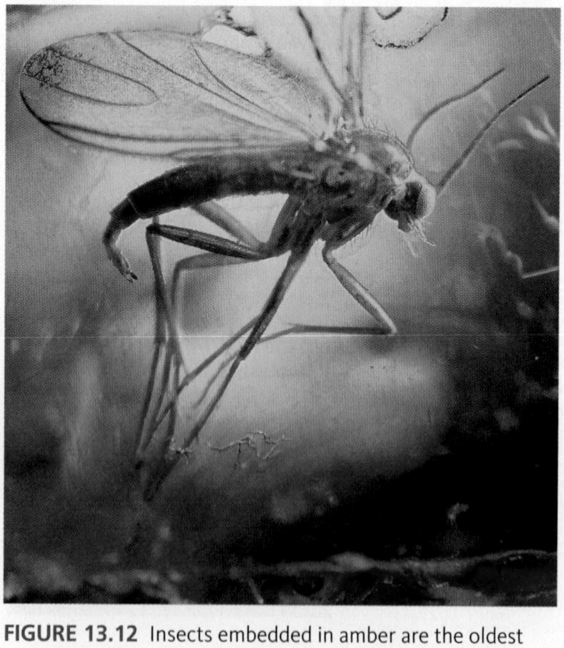

Alfred Pasieka/SPL/Photo Researchers, Inc.

FIGURE 13.12 Insects embedded in amber are the oldest organisms from which DNA has been extracted and cloned.

There are several steps in PCR:

1. A solution of DNA is heated to 90° to 100°C for a few minutes to break the hydrogen bonds between the two polynucleotide strands, producing two single-stranded molecules that will serve as templates.
2. The DNA solution is rapidly cooled to about 50°C, and primers for DNA replication bind to complementary regions on the single-stranded DNA fragments, marking the boundaries of the region to be copied.
3. After the primers are bound to the template strands, the enzyme Taq polymerase (a DNA polymerase from a bacterial species found in hot springs) synthesizes a complementary DNA strand beginning at the primers.

These three steps (Active Figure 13.11) make up one PCR cycle. The cycle is repeated (Steps 6–8) by heating the mixture and converting the double-stranded DNA into single strands, each of which will serve as a template in another round of synthesis. The amount of DNA present doubles with each PCR cycle. After n cycles, there is a 2^n increase in the amount of double-stranded DNA (Table 13.1). The power of PCR allows DNA to be copied in a test tube instead of a host cell, and millions or billions of copies can be made in hours rather than weeks.

PCR has several other advantages over host-cell cloning. The DNA to be amplified by PCR does not have to be purified and can be present in minute amounts; even a single DNA molecule can serve as a starting point. DNA from many sources has been used for PCR, including dried blood, hides from extinct animals such as the quagga (a zebra-like African animal that was hunted to extinction in the late nineteenth century), single hairs, mummified remains, and fossils. To date, the oldest DNA used in the PCR technique has been extracted from insects preserved in amber for about 30 million years (Figure 13.12). The DNA amplified from these insects is being used to study how specific genes have evolved over millions of years. PCR is also used in clinical diagnosis, forensics, and many other areas, including conservation. Some of these applications are discussed in the next chapter.

13.6 Analyzing Cloned Sequences

Once a cloned sequence containing all or part of a gene has been identified and selected from a library, it becomes a useful tool. It can be used as a probe to find and study regulatory sequences on adjacent chromosome regions or isolate the same gene from another species, investigate the internal organization of the gene, or study its expression in cells and tissues. For these studies, geneticists routinely use several methods, some of which are outlined in the following sections.

The Southern blot technique can be used to analyze cloned sequences.

Edwin Southern discovered a way to identify and isolate DNA fragments that have been separated by size. Once separated, the fragments are transferred to filters and screened with probes to identify a fragment of interest. This procedure, known as a **Southern blot**, has many applications. It is used to identify and analyze differences in normal and mutant alleles, identify related genes in other organisms, and study gene evolution.

To make a typical Southern blot, genomic DNA is extracted and cut into fragments with a restriction enzyme. The fragments are separated by gel electrophoresis. In this procedure, a solution containing the DNA restriction fragments is placed in a slot at the top of a porous gel made of agarose. The DNA fragments are separated by passing a small electrical current through the gel. Because DNA has a negative charge, the fragments migrate through the gel, moving from the negative pole at the top of the gel toward the positive pole at the bottom. Smaller fragments migrate through the agarose pores faster than larger fragments, and the DNA fragments become separated by size. After separation, the distribution of fragments is visualized by staining the DNA in the gel to show the number and location of the fragments (Figure 13.13). The fragments are converted into single strands and transferred (or blotted) onto a sheet of DNA-binding paper (Figure 13.14).

After transfer, fragments of interest are identified using a labeled probe that binds to complementary sequences in the fragments. After excess probe has been washed off, a piece of X-ray film is placed over the filter and the radioactive probe exposes the film, producing bands at the location of the fragments of interest. After the film is developed, the band pattern is analyzed and compared with patterns from other experiments.

DNA sequencing is one form of genome analysis.

The ability to determine the nucleotide sequence of individual genes and entire genomes has revolutionized genetics. The use of this technique has greatly enhanced our understanding of gene organization, the regulation of gene expression, and the evolutionary history of genes. The success of the Human Genome Project and hundreds of other genome projects is derived from advances in the technology of **DNA sequencing**.

Although newer DNA sequencing technologies that are faster and cheaper are becoming available, the most widely used method of DNA sequencing is one called the

DNA sequencing A technique for determining the nucleotide sequence of a DNA molecule.

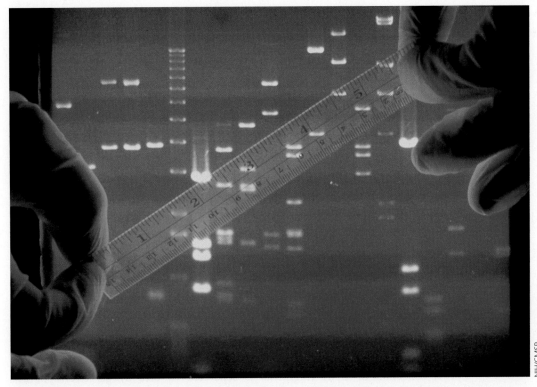

FIGURE 13.13 A gel stained to show the separation of restriction fragments by electrophoresis.

FIGURE 13.14 The Southern blotting technique. DNA is cut with a restriction enzyme, and the fragments are separated by gel electrophoresis. The DNA in the gel is denatured to single strands, and the gel is placed on a sponge partially immersed in buffer. The gel is covered with a DNA-binding membrane, layers of paper towels, and a weight. Capillary action draws the buffer up through the sponge, the gel, the DNA-binding membrane, and the paper towels. This movement of buffer transfers the pattern of DNA fragments from the gel to the membrane. The membrane is placed in a heat-sealed food bag with the probe and a small amount of buffer. After hybrids have been allowed to form, the excess probe is washed away, and the regions of hybrid formation are visualized by overlaying the membrane with a piece of X-ray film. After development, regions of hybrid formation appear as bands on the film.

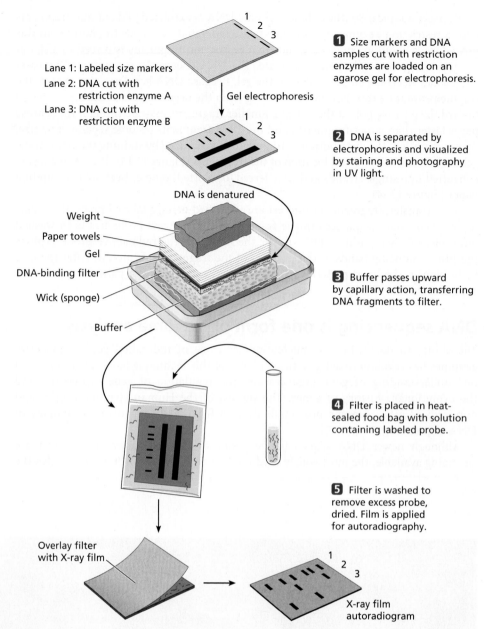

Lane 1: Labeled size markers
Lane 2: DNA cut with restriction enzyme A
Lane 3: DNA cut with restriction enzyme B

Gel electrophoresis

DNA is denatured

Weight
Paper towels
Gel
DNA-binding filter
Wick (sponge)
Buffer

Overlay filter with X-ray film

X-ray film autoradiogram

1 Size markers and DNA samples cut with restriction enzymes are loaded on an agarose gel for electrophoresis.

2 DNA is separated by electrophoresis and visualized by staining and photography in UV light.

3 Buffer passes upward by capillary action, transferring DNA fragments to filter.

4 Filter is placed in heat-sealed food bag with solution containing labeled probe.

5 Filter is washed to remove excess probe, dried. Film is applied for autoradiography.

Sanger method (Active Figure 13.15). In the Sanger method, DNA to be sequenced is first separated into single strands, each of which serves as a template for the synthesis of a complementary strand. The sequencing reaction uses the enzyme DNA polymerase, a primer, and the four deoxynucleotides (A, T, C, and G) found in DNA. The primer binds to the single-stranded DNA and serves as the starting point for DNA polymerase to synthesize a new strand. The key to this sequencing technique is the use of modified nucleotides called dideoxynucleotides (Figure 13.16). Instead of an –OH group needed to link nucleotides together, these modified nucleotides have an –H. Each of the four dideoxynucleotides is labeled with a different fluorescent dye. During synthesis, DNA polymerase randomly adds either a deoxynucleotide or a modified dideoxynucleotide to the growing new strand. If a deoxynucleotide is added, synthesis continues; however, if a dideoxynucleotide is added, replication stops because there is no –OH group to link another nucleotide to the new strand. Because this choice is made each time a nucleotide is added to the template, and because there are thousands of template strands in the reaction, new strands of every possible length accumulate in the reaction mixture. Each of these strands ends in a fluorescently labeled dideoxynucleotide.

This collection of DNA fragments is separated by size by gel electrophoresis (Active Figure 13.15). A laser beam scans the fragments on the gel to identify the end

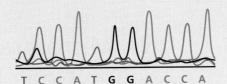

ddA ddC **ddG** ddT

1. The fragment of DNA to be sequenced is mixed with primers, DNA polymerase, and the four deoxynucleotides. Chemically modified nucleotides each labeled with a differently colored fluorescent dye are also added to the mixture.

5′ T C C A T G G A C C A 3′
3′ A G G T A C C T G G T 5′

T C C A T G G A C C A
T C C A T G G A C C
T C C A T G G A C
T C C A T G G A
T C C A T G G
T C C A T G
T C C A T
T C C A
T C C C
T C
T

2. The polymerase copies the DNA into new strands again and again. Synthesis at each new strand stops when a modified nucleotide gets added to it.

3. There are now many fragments of DNA in the mixture. Each is a truncated copy of the DNA template; each is tagged with a modified nucleotide.

4. Electrophoresis separates the fragments into bands according to their length. All fragments in each band are the same length, and all have the same modified nucleotide at their 3′ end. Thus, each band is a certain color.

— Electrophoresis gel

5. A computer detects and records the color of each band on the gel. The order of colors of the bands represents the sequence of the template DNA.

T C C A T G G A C C A

ACTIVE FIGURE 13.15 Automated DNA sequencing. Newly synthesized DNA fragments are labeled at their ends with a modified nucleotide that fluoresces in a specific color. The resulting fragments are separated by size, using gel electrophoresis. The gel is scanned with a laser and the data analyzed by software. Each peak in a printout shows the light fluoresced by a labeled nucleotide. The Human Genome Project used this method to determine the sequence of the 3.2 billion nucleotides in the human genome.

CENGAGENOW Learn more about DNA sequencing by viewing the animation by logging on to *login.cengage.com/sso* and visiting CengageNOW's Study Tools.

nucleotide by the color it fluoresces. Assembling the sequence of the nucleotides at the end of each fragment produces the DNA sequence.

DNA sequencing provides information about the size and organization of genes and the nature and function of the gene products they encode. Analysis of mutant genes provides information about how genes are altered by mutational events. Sequencing is also used to study the evolutionary history of genes and species-to-species variation.

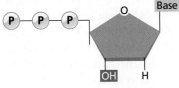

Deoxynucleotide

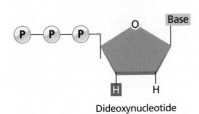

Dideoxynucleotide

FIGURE 13.16 The Sanger method of DNA sequencing uses DNA polymerase to add deoxynucleotides and modified nucleotides, called dideoxynucleotides to new DNA strands. Dideoxynucleotides do not have an –OH group on the nucleotide sugar, and after incorporation into a newly synthesized DNA strand, further synthesis stops.

DNA Sequencing

In 1977, Fred Sanger and his colleagues started the field of genomics (the study of genomes) by applying a newly developed method to sequence the 5,400 nucleotides in the genome of a virus called Φ X 174 (phi-X-174). The procedure was slow and laborious, and over the next few years only the very small genomes of other viruses were sequenced. Fast-forward 24 years to June 2001, to the announcement in the White House Rose Garden that a draft sequence of the 3.2 billion nucleotides in the human genome had been finished. The final human genome sequence, some 600,000 times larger than Φ X 174, was published in 2003. In the decades between those important milestones, the development of newer, automated sequencing methods made it possible to sequence the larger and more complex genomes of eukaryotes, including the 3.2 billion nucleotides that make up the human genome. The development of this technology was slow. The first genome of a free-living organism, a bacterium with 1,830,000 nucleotides in its genome, was not sequenced until 1995, 18 years after Sanger's work.

TEK IMAGE/Photo Researchers, Inc.

Progress in DNA sequencing was accelerated by the development of automated DNA sequencing machines and software to store, manipulate, and analyze DNA sequence information. Newer techniques now being introduced are again revolutionizing DNA sequencing, and lowering the cost of sequencing someone's genome to about $1,000. When this occurs in a few years, genome sequencing may become a routine part of medical care. This may enable physicians to determine whether someone has a genetic disorder or which predispositions to disease are present in their genome. This information will allow individuals to make lifestyle decisions to reduce the risk or impact of genetically determined disorders and to begin treatment as symptoms develop.

In addition to genome projects, DNA sequencing is used for applications that include drug discovery and development, identification of disease-causing organisms, analysis of environmental contaminants, and conservation. Along with cloning and PCR, DNA sequencing is one of the basic methods in recombinant DNA technology.

13.7 DNA Microarrays Are Used to Analyze Gene Expression

A new technology using devices called **DNA microarrays** is being used to analyze the expression of thousands of genes in a single experiment. In a typical application, microarrays are used to compare the expression of genes in two different tissues—one that is normal and one that represents a disease such as cancer.

The heart of a microarray is a small piece of glass (Figure 13.17) or quartz on which single-stranded fragments of DNA are attached as a spot called a field. Each field contains many copies of a specific DNA sequence, and the sequence in each field is different from the sequence in the thousands of other fields on the microarray. For an expression study (Figure 13.18), mRNA is isolated from normal cells and cancer cells derived from the same cell type as the normal cells. The isolated mRNAs are copied into single-stranded DNA molecules called cDNA (complementary DNA), using an enzyme called reverse transcriptase. cDNA molecules from the two cell types are labeled with different fluorescent dyes (green for normal, red for cancer). The labeled cDNA molecules are mixed together and applied to a microarray containing DNA fields representing thousands of genes. Unbound cDNA molecules are washed off, a laser scans the chip, and the image is sent to a computer for storage and analysis.

The pattern on the microarray appears as a series of dots; each dot represents one field. If a gene is expressed only in normal cells, the dot will be green, because only mRNA from normal cells binds to the DNA in that field. If the field is red, it means that the mRNA for that gene is expressed only in cancer cells. If the dot is yellow, it means that the gene is expressed equally in both the normal cells and the cancer cells. Intermediate colors mean that there are different amounts of cDNA in the normal and cancer cells and different levels of expression of these genes in the two cell types.

DNA microarrays have a wide range of applications, some of which are listed in Table 13.2.

Courtesy of Affymetrix, Inc., Santa Clara, CA

FIGURE 13.17 A commercial DNA microarray. Each chip can hold more than 50,000 fields.

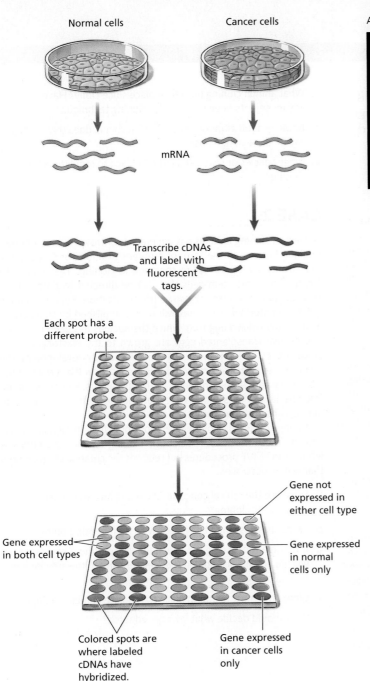

Normal cells   Cancer cells

mRNA

Transcribe cDNAs
and label with
fluorescent
tags.

Each spot has a
different probe.

Gene expressed
in both cell types

Colored spots are
where labeled
cDNAs have
hybridized.

Gene not
expressed in
either cell type

Gene expressed
in normal
cells only

Gene expressed
in cancer cells
only

Actual DNA microarray result

FIGURE 13.18 In a microarray assay for gene expression, DNA is isolated from normal cells and cancer cells. The mRNA is copied into single-stranded DNA molecules by the enzyme reverse transcriptase. The nucleotides incorporated into the cDNA from normal cells are labeled with green fluorescent dye and the cDNA from the cancer cells is labeled with a red fluorescent dye. The cDNA molecules are mixed together and applied to a microarray with single-stranded DNA molecules arranged in small spots called fields. Microarrays can contain DNA representing every gene in the human genome. The added cDNA molecules will bind to complementary sequences in the DNA bound to the microarray. Excess and unbound cDNA molecules are washed off the microarray, which is then scanned with a laser that quantifies the binding in each field. Red spots represent genes expressed only in cancer cells. Green spots represent genes expressed only in normal cells. Yellow spots represent genes expressed equally in both normal cells and cancer cells. Intermediate colors represent differential expression in normal and cancer cells. Gray fields represent genes that are not expressed in either cell type. Using this microarray, it is possible to determine the expression pattern for thousands of genes in a single experiment.

Table 13.2 Uses of Microarrays

Application	Description
Gene Expression	Assay expression levels of thousands of genes in a single experiment. Used to study development, disease versus normal states.
SNP Detection	Surveys alleles and/or populations to detect single nucleotide polymorphisms (SNPs). Used in genotyping, identifying predisposition to disease, forensic analysis, selection of patient-specific cancer therapy.
Alternative Splicing	Exon junction microarrays detect mRNAs for alternative splice forms of a gene in different tissues, stages of development.
Fusion Genes	Used to detect fusion genes for precise diagnosis of some leukemias and other cancers such as prostate cancer. Often used to help establish prognosis.
Organism Identification	Used to detect mislabeled food, often from endangered species, in the food supply. Also used to assay contaminating microorganisms in food.

Genetics in Practice

Genetics in Practice case studies are critical-thinking exercises that allow you to apply your new knowledge of human genetics to real-life problems. You can find these case studies and links to relevant websites at *www.cengage.com/biology/cummings*

CASE 1

Cloning an animal by means of nuclear transfer was first done in 1952, using frogs. This and later work showed that the first few cell divisions after fertilization produce cells with nuclei that form complete embryos when transferred into unfertilized eggs. As an embryo develops further, the cells lose this property, and the success of nuclear transfer rapidly declines. Nuclear transfer in mammals proved to be more difficult. In the 1980s, several groups developed transgenic livestock, and genetically modified pigs were considered as sources of organs for transplantation to human patients.

If animals can be derived from cells in culture, it should be possible to make genetic modifications, including the removal or addition of specific genes. This has been achieved in mice by using embryonic stem (ES) cells, but to date no one has been successful in obtaining ES cells from cattle, sheep, or pigs. Dr. Ian Wilmut at the Roslin Institute in Edinburgh, Scotland, thought nuclear transfer might provide an alternative to ES cells. A major breakthrough came in 1995 when Wilmut and his colleagues produced lambs by means of nuclear transfer from cells of early embryos.

So how was Dolly produced? Dolly was the first mammal cloned using a nucleus from an adult animal. Cells were collected from the udder of a 6-year-old ewe and cultured for several weeks in the laboratory. Individual cells were then fused with unfertilized eggs from which the nuclei had been removed. Two hundred seventy-seven of the reconstructed eggs—each now with a diploid nucleus from the adult animal—were cultured for 6 days in temporary recipients. Twenty-nine of the eggs appeared to have developed to the blastocyst stage and were implanted into surrogate Scottish Blackface ewes. One gave rise to a live lamb, Dolly, some 148 days later. Dolly was born on July 5, 1996.

1. Why did the birth of Dolly stun the scientific world, considering that cloning of animals was already taking place?

2. What animal cloning has taken place since Dolly? Have any advancements been made in the cloning technique?

3. Because adult cloning can be done in sheep, does that mean it can be done in humans?

4. What policies or laws need to be in place to regulate animal cloning?

CASE 2

Success in cloning animals has generated controversy over whether to clone humans in the form of embryos or adults or to treat disease using a method called therapeutic cloning. The goal is to create embryonic stem cells that can be directed to differentiate into functional tissues for use in transplantation. Reprogramming a differentiated adult nucleus will be accomplished by transferring it to an unfertilized egg from which the nucleus has been removed. The nuclear-transplanted eggs are grown to the blastocyst stage. However, instead of transferring the embryos for development, the embryo's stem cells are collected. The stem cells will be grown in the laboratory and, in the presence of growth factors, will differentiate into specialized cells such as liver cells, kidney cells, or insulin-producing cells of the pancreas.

Adding to the ongoing debate about human cloning and embryonic stem-cell research, therapeutic cloning will not use embryos left over from IVF procedures but requires the creation of an embryo that will be destroyed.

1. What are the ethical concerns in cloning human beings? In therapeutic cloning?

2. What diseases could be treated with therapeutic cloning?

3. If therapeutic cloning is allowed for the treatment of disease, will we begin a slide down the "slippery slope" toward cloning human beings?

4. What laws must be put in place to regulate human cloning?

5. Who should decide what laws are adopted?

Summary

13.1 What Are Clones?

- Cloning is the production of identical copies of molecules, cells, or organisms from a single ancestor. The development and refinement of methods for cloning higher plants and animals represents a significant advance in genetic technology that will speed up the process of improving crops and the production of domestic animals.

13.2 Cloning Genes Is a Multistep Process

- These developments have been paralleled by the discovery of methods for cloning segments of DNA molecules. This technology is based on the discovery that a series of enzymes known as restriction enzymes recognize and cut DNA at specific nucleotide sequences. Linking DNA segments produced by restriction-enzyme treatment with vectors such as plasmids or engineered yeast chromosomes produces recombinant DNA molecules.

- Recombinant DNA molecules are transferred into host cells, and cloned copies are produced as the host cells grow and divide. A variety of host cells can be used, but the most common is the bacterium *E. coli*. The cloned DNA molecules can be recovered from the host cells and purified for further use.

13.3 Cloned Libraries

- A collection of cloned DNA sequences from one source is a library. The clones in the library are a resource for work on specific genes.

13.4 Finding a Specific Clone in a Library

- Clones for specific genes can be recovered from a library by using probes to screen the library.

13.5 A Revolution in Cloning: The Polymerase Chain Reaction

- PCR is used to make many copies of a DNA molecule without using restriction enzymes, vectors, or host cells. It is faster and easier than conventional cloning.

13.6 Analyzing Cloned Sequences

- Cloned sequences are characterized in several ways, including Southern blotting and DNA sequencing.

13.7 DNA Microarrays Are Used to Analyze Gene Expression

- DNA microarrays can be used to analyze the expression of thousands of genes in a single experiment. This technology has a wide range of uses in genomics.

Questions and Problems

CENGAGENOW Preparing for an exam? Assess your understanding of this chapter's topics with a pre-test, a personalized learning plan, and a post-test by logging on to *login.cengage.com/sso* and visiting CengageNOW's Study Tools.

What Are Clones?

1. Cloning is a general term used for whole organisms and DNA sequences. Define what we mean when we say we have a clone.

2. Nuclear transfer to clone cattle is done by which of the following techniques?
   a. An 8-cell embryo is divided into two 4-cell embryos and implanted into a surrogate mother.
   b. A 16-cell embryo is divided into 16 separate cells, and those cells are allowed to form new 16-cell embryos that are implanted directly into surrogate mothers.
   c. A 2-cell embryo is divided into 2 separate cells and implanted into a surrogate mother.
   d. A 16-cell embryo is divided into 16 separate cells and fused with enucleated eggs. The fused eggs are then implanted into surrogate mothers.
   e. None of the above.

3. Dolly made headlines in 1997 when her birth was revealed. Why was cloning Dolly so important, considering that cattle and sheep were already being cloned through embryo splitting?

Cloning Genes Is a Multistep Process

4. What is meant by the term *recombinant DNA*?
   a. DNA from bacteria and viruses.
   b. DNA from different sources that normally are not found together.
   c. DNA from restriction-enzyme digestions.
   d. DNA that can make RNA and proteins.
   e. None of the above.

5. Restriction enzymes:
   a. recognize specific nucleotide sequences in DNA.
   b. cut both strands of DNA.
   c. often produce single-stranded tails.
   d. do all of the above.
   e. do none of the above.

6. Restriction enzymes are derived from bacteria. Why don't bacterial chromosomes get cut with the restriction enzymes present in the cell? Why do bacteria have these enzymes?

7. The following DNA sequence contains a six-base sequence that is a recognition and cutting site for a restriction enzyme. What is this sequence? Which enzyme will cut this sequence? (See Figure 13.5 for help.)
   5′ CCGAGGAAGCTTAC 3′
   3′ GGCTCCTTCGAATG 5′

8. Why is it an advantage that many restriction-enzyme recognition sites read the same on either strand of the DNA molecule when read from the 5′ end of the DNA? Remember: Restriction enzymes bind DNA and move along the molecule, looking for a recognition site.

9. Assume restriction-enzyme sites as follows: E = *Eco*RI, H = *Hin*dIII, and P = *Pst*I. What size bands (in kilobases) would be present when the DNA shown below is cut with
   a. *Eco*RI
   b. *Hin*dIII and *Pst*I
   c. all three enzymes

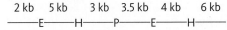

10. Which enzyme is responsible for covalently linking DNA strands together?
    a. DNA polymerase
    b. DNA ligase
    c. *Eco*RI
    d. restriction enzymes
    e. RNA polymerase

11. In cloning human DNA, why is it necessary to insert the DNA into a vector such as a bacterial plasmid?

12. Briefly describe how to clone a segment of DNA. Start with cutting the DNA of interest with a restriction enzyme.

Cloned Libraries

13. A cloned library of an entire genome contains
    a. the expressed genes in an organism.
    b. all the DNA sequences of an organism.
    c. only a representative selection of genes.
    d. a large number of alleles of each gene.
    e. none of the above.

14. You are given the task of preparing a cloned genomic library from a human tissue culture cell line. What type of vector would you select for this library and why?

15. The steps in the polymerase chain reaction (PCR) are:
    a. breaking hydrogen bonds, annealing primers, and synthesizing using DNA polymerase.
    b. restriction cutting, annealing primers, and synthesizing using DNA polymerase.
    c. ligating, restriction cutting, and transforming into bacteria.
    d. DNA sequencing and restriction cutting.
    e. transcription and translation.

16. You are running a PCR to generate copies of a fragment of the cystic fibrosis (CF) gene. Beginning with two copies at the start, how much of an amplification of this fragment will be present after six cycles in the PCR machine?

17. Why is PCR so revolutionary? Describe two applications of PCR.

Analyzing Cloned Sequences

18. Name three of the many applications for Southern blotting.

19. A new gene that causes an inherited form of retinal degeneration has been cloned from mice. This disease leads to blindness in affected individuals. As a researcher in a human vision laboratory, you want to see if a similar gene exists in humans (genes with similar sequences usually mean similar functions). With your knowledge of the Southern blot procedure, how would you go about doing this? Start with the isolation of human DNA.

20. A base change (A to T) is the mutational event that created the mutant sickle cell anemia allele of beta globin. This mutation destroys an *Mst*II restriction site normally present in the beta globin gene. This difference between the normal allele and the mutant allele can be detected with Southern blotting. Using a labeled beta globin gene as a probe, what differences would you expect to see for a Southern blot of the normal beta globin gene and the mutant sickle cell gene?

21. What exactly is a DNA sequence of a gene?

22. What kind of information can a DNA sequence provide to a researcher studying a disease-causing gene?

23. The cloning methods outlined in this chapter allow researchers to generate many copies of a gene they wish to study through the use of restriction enzymes, vectors, bacterial host cells, the creation of genetic libraries, and PCR. Once useful quantities of a disease-causing gene are available by cloning, what kinds of questions do you think should be asked about the gene?

Internet Activities

Internet Activities are critical-thinking exercises using the resources of the World Wide Web to enhance the principles and issues covered in this chapter. For a full set of links and questions investigating the topics described below, visit *www.cengage.com/biology/cummings*

1. *Biotechnology Review.* Cold Spring Harbor's *DNA Learning Center* provides interactive animations of PCR, Southern blotting, and cycle sequencing. Review these techniques by selecting "Resources" and then "Biology Animation Library."

2. *Recombinant DNA Problem Set.* At the University of Arizona's *Biology Project: Molecular Biology*, you can work through an interactive problem set on recombinant DNA technology. Correct answers will be summarized; if you answer incorrectly, you will be provided with a tutorial to increase your understanding.

3. *Cloning in the News: A Critical Analysis.* At the *Access Excellence About Biotech* website, Tom Zinnen presents an analysis of the accuracies and inaccuracies in news reporting about Dolly, the first cloned sheep.

4. *Conceiving a Clone.* At the student-created *Conceiving a Clone* website, you can access details on cloning and biotechnology, compare different cloning techniques, and view animations of cloning processes. Follow the "Interactions" link to create a clone on the Web, test your cloning knowledge, or participate in polls about cloning.

HOW WOULD YOU VOTE NOW?

Cloned animals are created through the manipulation of cells and genes; however, they do not contain any foreign genes and are not genetically modified. Livestock animals such as cows have been cloned, and proponents of cloning argue that milk from cloned cows is safe for human consumption. Watchdog groups that monitor genetically engineered foods are concerned about the safety of products from cloned animals and want more research to prove the safety of such products before they are allowed in the marketplace. Now that you know how cloned animals are made and have learned about recombinant DNA technology, what do you think? Is milk and meat from cloned animals safe, and should these products be sold without labeling? Visit the *Human Heredity* companion website at *www.cengage.com/biology/cummings* to find out more on the issue; then cast your vote online.

14 Biotechnology and Society

The story begins simply enough: A boy was born in England to parents who emigrated from Ghana. Because he was born in England, the boy was automatically a British citizen. As a youngster, he returned to Ghana to live with his father, leaving behind his mother, two sisters, and a brother. In 1985, he returned to England, intending to live with his mother and siblings. Immigration authorities suspected that the boy was an impostor and thought he was either an unrelated child or a nephew of the boy's mother. On the basis of their suspicions, the boy's application for residency was denied. The boy's family fought to establish his identity so that he could live in the country of his birth. A round of medical tests used blood types and genetic markers normally used to match organ donors and recipients. The results confirmed that the boy was closely related to the woman he claimed was his mother, but the tests could not tell whether she was his mother or an aunt.

The family turned to Alec Jeffreys, a scientist at the University of Leicester, for help. They asked if DNA fingerprinting, a technique developed in his research laboratory, could establish the boy's identity. The situation was somewhat complicated; the mother's sisters and the boy's father were not available for testing, and the mother was not sure about the boy's paternity. Despite these problems, Jeffreys agreed to take on the case. He took blood samples from the boy, the children he believed to be his brother and sisters, and the woman who claimed to be his mother. DNA was extracted from each sample and treated with restriction enzymes to cut the DNA at specific nucleotide sequences. The resulting fragments were separated by size and visualized by Southern blotting. The resulting pattern of bands, known as a DNA fingerprint, was analyzed to determine the boy's identity. The results showed that the boy had the same father as his brother and his sisters because they all shared DNA fragments contributed by the father. The most important question was whether the boy and his "mother" were related.

Jeffreys found that 25 fragments of the woman's DNA matched those of the boy, indicating that she was in fact the boy's mother. The chance that they were unrelated was calculated as 2×10^{-15}, or about one in a quadrillion. Faced with this evidence, immigration authorities reversed their position and allowed the boy to live in England with his family.

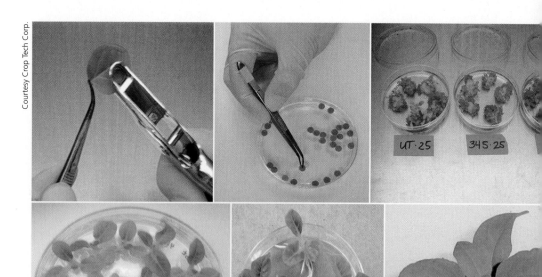

Courtesy Crop Tech Corp.

Plant cloning beginning with discs of cells punched from a leaf. All plants derived from this leaf are genetically identical to the parent plant and to each other.

DNA fingerprinting was one of the first of many techniques developed as part of the ongoing revolution in genetic technology, and this case was the first use of the method. As described in the last chapter, these developments began in the 1970s with the discovery that DNA from different organisms could be combined to create recombinant DNA molecules. In this chapter, we examine how the technology is being used in areas of human genetics, medicine, and agriculture.

Recombinant DNA technology was first used to transfer foreign genes into bacterial cells so that research scientists could clone large quantities of specific genes for their research. Later, these methods were used to produce human proteins in host cells and to transfer genes between species of plants and animals. As a result, the use of recombinant DNA has spread far beyond the research laboratory and now has a significant impact on our lives.

Biotechnology is the use of recombinant DNA and molecular biology to produce commodities or services. Biotechnology has changed the way we produce our food, diagnose and treat disease, trace our ancestry, gather evidence in criminal cases, identify human remains after disasters or war, and protect endangered species. It even helps establish whether our pets are really purebred. In this chapter, we will discuss a cross section of the ways biotechnology is being used to: (1) make pharmaceutical products in genetically altered plants and animals, (2) treat disease with stem cells, (3) produce new varieties of crop plants, (4) generate animal models of human diseases, and (5) prepare DNA profiles used in forensics and other fields. We will also consider the social, ethical, economic, and environmental controversies generated by the use of biotechnology.

Biotechnology The use of recombinant DNA technology to produce commercial goods and services.

14.1 Biopharming: Making Human Proteins in Animals

Mark Clarke/Photo Researchers, Inc.

FIGURE 14.1 Some diabetic patients must inject themselves with insulin several times a day.

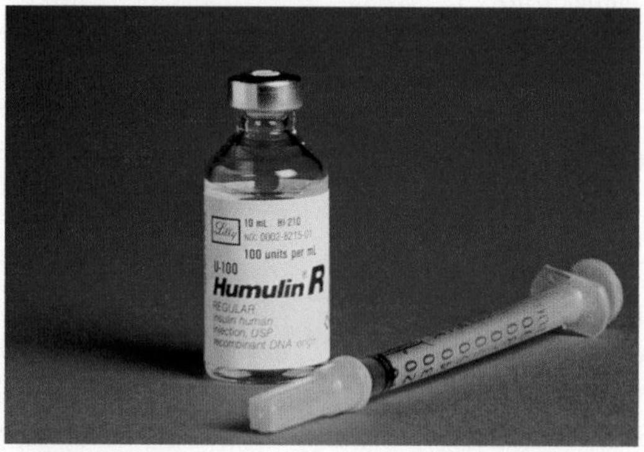

SIU/Visuals Unlimited, Inc.

FIGURE 14.2 Human insulin was the first human protein made using recombinant DNA technology.

One of the first uses of biotechnology was the production of human proteins for treating diseases. Insulin made by recombinant DNA technology was one of the first of these drugs. Children and adults with type I diabetes cannot make insulin, a protein that controls blood sugar levels. Left untreated, diabetes is a silent, crippling disease that kills its victims. To control blood sugar levels, diabetics must inject themselves with insulin every day (Figure 14.1). For decades, insulin was extracted from cow and pig organs, but in 1982, human insulin produced by recombinant DNA technology became available as a safe, pure, and reliable medication that is now widely used to treat diabetes (Figure 14.2).

Before the development of biotechnology, therapeutic proteins such as insulin, growth hormone, and blood-clotting factors were isolated from many sources, including animals in slaughterhouses, donated human blood, and even human cadavers. In some cases, contaminated proteins from these sources exposed people to serious and potentially fatal risks.

Hemophiliacs cannot make a protein necessary for forming blood clots. As a result, they can have episodes of uncontrollable bleeding, even after a minor injury. This clotting factor protein was first produced in the 1960s by concentrating blood serum from large quantities of donated blood. This serum-concentrated protein was given to hemophiliacs when they had bleeding episodes. When HIV (human immunodeficiency virus) infection and AIDS first became a serious public health problem in the early 1980s, donated blood was not tested for HIV. As a result, about 60% of all hemophiliacs became infected with HIV from contaminated clotting factor prepared from pooled blood donations. Many of those infected individuals went on to develop AIDS. Hemophiliacs who received the serum-concentrated clotting factor also faced a risk of hepatitis infection.

Production of clotting factors VIII (hemophilia A; OMIM 306700) and IX (hemophilia B; OMIM 306900) using recombinant DNA technology began in the early 1990s. The process does not use blood or blood cells, and since the introduction of these products, no new cases of infection with HIV or hepatitis from clotting factors have been reported. Clotting factor and insulin are only two of the dozens of human proteins made by biotechnology. Some others are listed in Table 14.1. In the next section, we will take a more detailed look at how human proteins are made by biotechnology and how they are used in treating genetic disorders.

Table 14.1 Some Products Made by Recombinant DNA Technology

Product	Use
Atrial natriuretic factor	Treatment for hypertension, heart failure
Bovine growth hormone	Improves milk production in dairy cows
Cellulase	Breaks down cellulose in animal feed
Colony stimulating factor	Treatment for leukemia
Epidermal growth factor	Treatment for burns, improves survival of skin grafts
Erythropoietin	Treatment for anemia
Hepatitis B vaccine	Prevents infection by hepatitis B virus
Human insulin	Treatment for diabetes
Human growth hormone	Treatment for some forms of dwarfism, other growth defects
Interferons (alpha, gamma)	Treatment for cancer, viral infections
Interleukin-2	Treatment for cancer
Superoxide dismutase	Improves survival of tissue transplants
Tissue plasminogen activator	Treatment for heart attacks

Human proteins can be made in animals.

Pompe disease (OMIM 232300) is an often-fatal autosomal recessive disorder that weakens and eventually disables heart and muscle function. This disease is caused by mutations in the gene for acid α-glucosidase (GAA), an enzyme that breaks down glycogen to make it available as an energy source for cells. In Pompe disease, mutations eliminate or reduce enzyme levels; as a result, glycogen accumulates in muscle cells, producing the symptoms of the disorder.

Pompe disease is treated using **enzyme replacement therapy**. Affected infants receive doses of GAA produced by biotechnology. The enzyme improves heart function, reduces glycogen accumulation, and improves survival of affected infants. GAA was originally manufactured by transferring the human GAA gene into fertilized rabbit eggs, producing transgenic adults. GAA was purified from milk of female transgenic rabbits. To increase production, GAA is now synthesized by transgenic hamster cells grown in large bioreactors (Figure 14.3a) and marketed as Myozyme® (Figure 14.3b).

Continued advancements in biotechnology may ultimately provide a safe and effective long-term treatment for Pompe disease. Several other genetic disorders are treated with enzyme therapy. In Chapter 16, we will discuss gene therapy, a treatment that transfers the normal alleles of disease genes into affected individuals.

Enzyme replacement therapy
Treatment of a genetic disorder by providing a missing enzyme encoded by the mutant allele responsible for the disorder.

Transgenic plants may replace animal hosts for making human proteins.

Although not widely used, transgenic plant cells have several benefits for producing human proteins. Costs are lower, and instead of using stainless steel bioreactors to grow bacteria or mammalian cells, plant cells can be grown in large disposable plastic bags. Plant cells also perform the posttranslational modifications often required to make

(a)

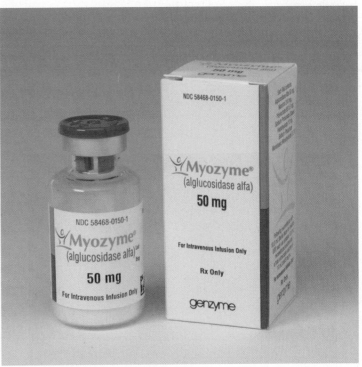

(b)

FIGURE 14.3 (a) Many therapeutic human proteins are synthesized in mammalian cells grown in large batches in bioreactors. (b) One of the proteins produced in bioreactors is Myozyme®, used to treat Pompe disease.

human proteins fully active. The main disadvantage to using plant cells has been the low yields of the transgenic protein. This problem has largely been solved, and human proteins made from transgenic plant cells are now being tested in clinical trials. Over 370 human therapeutic proteins made in plants are in various stages of development.

> **KEEP IN MIND**
>
> Transgenic organisms and cell lines are being used to make human proteins for treatment of disease.

Embryonic stem cells (ESC) Cells in the inner cell mass of early embryos that will form all the cells, tissues, and organs of the adult. Because of their ability to form so many different cell types, these cells are called pluripotent cells.

Pluripotent The ability of a stem cell to form any fetal or adult cell type.

Adult stem cells Stem cells recovered from bone marrow and other organs of adults. These cells can differentiate to form a limited number of adult cells, and are called multipotent cells.

Multipotent The restricted ability of a stem cell to form only one or a few different cell types.

Induced pluripotent stem cells (iPS) Adult cells that can be reprogrammed (induced) by gene transfer to form cells with most of the developmental potential of embryonic stem cells. Because of this developmental potential, such cells are pluripotent.

14.2 Using Stem Cells to Treat Disease

Stem cells (see Chapter 7) are unspecialized cells with two unique properties: They renew their numbers by cell division and can form many or most of the different cell types in the body. **Embryonic stem cells** form in the blastocyst stage of development in the inner cell mass (Figure 14.4a). During embryogenesis, they form all the different cell types, tissues, and organs of the body. Because they can form so many cell types, they are called **pluripotent** cells. Stem cells can be removed from an embryo and grown in the laboratory to establish an embryonic stem-cell line (Figure 14.4b). **Adult stem cells** can be recovered from certain adult tissues and from umbilical-cord blood. When needed, they divide to replace worn out or damaged cells in certain organs and tissues. Adult stem cells can only form a limited number of different cell types and are called **multipotent** cells.

Researchers have discovered that adult skin cells can be reprogrammed to form stem cells by transferring genes into their nuclei. These cells are called **induced pluripotent stem (iPS) cells** (Figure 14.5). To test whether iPS cells are pluripotent, a research team created iPS cells from mouse cells and used them to generate fertile adult mice

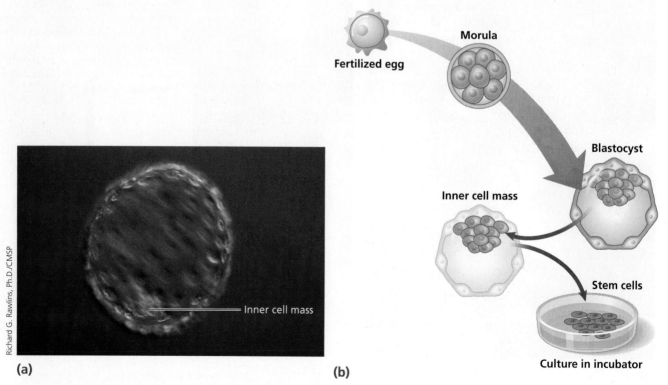

Richard G. Rawlins, Ph.D./CMSP

Inner cell mass

(a)

Fertilized egg

Morula

Blastocyst

Inner cell mass

Stem cells

Culture in incubator

(b)

FIGURE 14.4 (a) A human blastocyst with the inner cell mass. Cells from this region are pluripotent and will form all the 200 cell types in the body. (b) Cells from the inner cell mass can be removed from the embryo and grown to form embryonic stem-cell lines.

(Figure 14.6), establishing that these reprogrammed cells can form all the cells and tissues of an adult. However, other work shows that iPS cells may not have the same developmental potential as embryonic stem cells.

Stem cells provide insight into basic biological processes.

The ability to isolate and grow stem cells provides many research opportunities. Inducing stem cells to form adult cell types in the laboratory will add to our knowledge of how patterns of gene expression lead to the formation of tissues and organs during human development. This, in turn, may lead to an understanding of what errors produce birth defects such as cleft lip, clubfoot, and spina bifida.

The ability to produce iPS cells from individuals with genetic disorders offers a way to study the development and expression of both normal and abnormal phenotypes. For adult-onset disorders such as Huntington disease, iPS cell technology offers researchers an opportunity to study the early events in the disease process that occur before changes in the phenotype are seen. iPS cell lines representing several single-gene disorders and complex disorders are now available for research. These include severe-combined immunodeficiency (SCID), Gaucher disease, Huntington disease, and type I juvenile diabetes. These reprogrammed adult cells are quickly becoming a widely used research tool in human genetics. To facilitate their use, a stem-cell center at Harvard University is establishing a collection of iPS cell lines from a large number of genetic disorders and making them available to researchers.

Stem-cell-based therapies may treat many diseases.

The greatest promise for the use of stem cells is their potential to treat a wide range of diseases. Stem cells can be induced to form specific cell types with specialized functions. Several diseases, including type I diabetes, Huntington disease, and Parkinson's disease, are caused by the death of one or a small number of cell types and are good candidates for treatment with pluripotent stem cells. Because embryonic stem cells have to form all of the 200-plus different cell types that make up the tissues in our body, most proposed therapies use these cells.

Several clinical trials using human embryonic stem cells are poised to begin in the near future. Some of these are listed in Table 14.2. However, in the United States, many individuals and organizations strongly oppose the use of embryonic stem cells because they believe that the zygote and its derived embryo is a human life and should not be destroyed to recover stem cells. Others strongly advocate the use of embryonic stem-cell therapy as a lifesaving form of treatment. Induced pluripotent cell lines may provide an alternative way to develop lifesaving therapies without the destruction of embryos, but much more research on the use of iPS cells is needed before they can be used in treatment.

Although cell therapy with embryonic stem cells is controversial, adult stem cells have been used to treat diseases for several decades (Figure 14.7). Bone marrow transplants, stem cells isolated from blood, or umbilical cord blood are all familiar and widely used

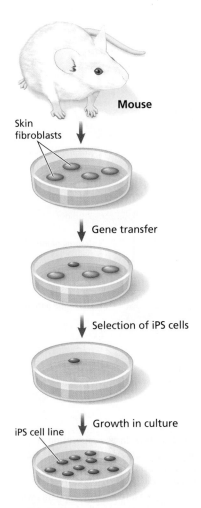

FIGURE 14.5 Pluripotent stem cells can be generated from adult skin cells by transferring a small number of master control genes into their nuclei. Cells that express all the transferred genes are selected and used to form cell lines. Cells produced in this way are called induced pluripotent stem (iPS) cells.

Table 14.2 Pending Embryonic Stem-Cell Therapy Trials

Cell Type	Condition Treated	Organization
Neural cells	Spinal cord injuries	Geron
Retinal pigment cells	Stargardt's macular dystrophy	Advanced Cell Technology
Keratinocytes	Burns	Institute for Stem Cell Therapy
Beta cells	Diabetes	Novocell Geron
Cardiac progenitor cells	Heart damage	George Pompidou Hospital

Boland, M. J., et al., 2009. "Adult mice generated from induced pluripotent stem cells," *Nature, 461*, pp. 91–96.

(a) iPS mouse: 4 weeks **(b) iPS mouse progeny**

FIGURE 14.6 (a) To test whether iPS cells are truly pluripotent, these cells were used to create mice, indicating that the cells can form all the cells of the embryo and the adult. (b) Mice produced from iPS cells are fertile.

forms of treatment for leukemias, other cancers, and blood disorders. Newer therapies in use and under development are expanding the use of adult stem cells in treating disease. In one application, multipotent adult stem cells are recovered from a patient's bone marrow and reprogrammed to become heart muscle cells. These stem cells are infused or injected into the heart to repair dead and damaged cells that follow a heart attack. In another application, an innovative method for treating second-degree burns starts with a biopsy of normal skin. A suspension of skin stem cells is prepared from the biopsied cells. This suspension is then sprayed onto the burned area (Figure 14.8), and the stem cells, called basal cells, divide to form a layer of skin over the burn. Clinical results are comparable to those obtained with skin grafts. Altogether, more than 70 conditions have been successfully treated using adult stem cells.

These approaches have several advantages. Adult stem-cell therapy is not controversial and does not involve embryo destruction. Because the stem cells originate from the patient, there is no rejection of the transplanted cells by the immune system. The disadvantages of adult stem-cell therapy are that these cells are very rare and difficult to isolate from the body. In addition, because they are adult stem cells, they have only limited developmental potential and can be used to treat a relatively small number of diseases. Finally, adult stem cells are more difficult to grow and reprogram than embryonic cells. In sum, stem cells offer great promise for the development of effective therapies to treat a wide range of diseases, but technical, social, and political issues remain to be resolved.

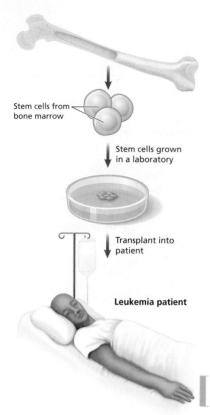

FIGURE 14.7 In one application of adult stem-cell therapy, stem cells are removed from the bone marrow of a donor, grown in the laboratory, and then infused into leukemia patients who have had their stem cells destroyed by radiation. Leukemia is a disease caused by overproduction of white blood cells by stem cells in the bone marrow. The transplanted stem cells will migrate to the bone marrow and begin production of normal blood cells.

Genetically modified organisms (GMOs) A general term used to refer to transgenic plants or animals created by recombinant DNA techniques.

FIGURE 14.8 Basal stem cells removed from a small region of skin cells are sprayed over burn areas, where the stem cells re-form skin, bypassing the need for painful skin grafts.

14.3 Genetically Modified Foods

Humans have been genetically modifying crop plants for thousands of years by artificial selection, which has dramatically increased the yield and nutritional value of crop plants. In the last 50 to 60 years alone, corn crop yields have increased about fourfold.

There is now an alternative to artificial selection; genes can be transferred into crop plants using biotechnology. These transgenic plants are produced faster and more precisely than by plant breeding and carry specific new characteristics such as resistance to insects or herbicides (Active Figure 14.9). The transferred gene can originate from another plant, an animal, or even from fungi or bacteria. In contrast to selective breeding, which transfers hundreds or thousands of genes in each cross, transgenic plants receive only one or a few specific genes. In the media, transgenic plants are often called **genetically modified organisms (GMOs)** or genetically modified (GM) plants. However, in the larger picture, all crop plants have been genetically modified over thousands of years by selection and crossbreeding, and only recently by biotechnology.

Transgenic crop plants can be made resistant to herbicides and disease.

Weeds destroy more than 10% of all crops worldwide, and pesticide runoff from treated fields is a major cause of pollution in local water supplies. The U.S. Geological Survey reports that more than 90% of all U.S. streams contain pesticides (see Spotlight on Bioremediation: Using Bugs to Clean Up Waste Sites). Herbicide-tolerant (HT) crops carry genes for resistance to broad-spectrum herbicides such as glyphosate and glufosinate, which kill all plants in the field except the crop plant. These herbicides break down quickly in the soil, reducing the number and amount of herbicides needed and also reducing runoff and environmental impact.

Other transgenic crops, including corn, cotton, and soybeans, are resistant to insect pests (Figure 14.10). These crops (known as Bt crops) carry a bacterial gene that produces a toxin that is released in the insect's gut, causing death.

Globally, transgenic crops are planted in over 25 countries, and the land devoted to these crops is increasing every year. Use of these crops has reduced the number and amount of pesticides needed to control weeds and insects. More importantly, transgenic crops have increased yields and reduced production costs.

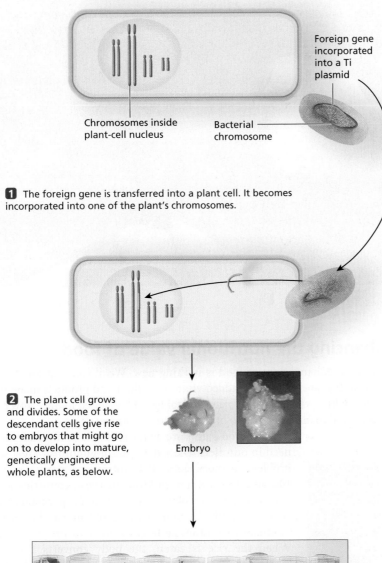

Chromosomes inside plant-cell nucleus

Bacterial chromosome

Foreign gene incorporated into a Ti plasmid

1 The foreign gene is transferred into a plant cell. It becomes incorporated into one of the plant's chromosomes.

ACTIVE FIGURE 14.9 A gene incorporated into a bacterial plasmid is used to transfer a genetic trait into a plant cell. As the plant cell divides, some cells form embryos that develop into transgenic plants. Such plants are sometimes called genetically modified organisms (GMOs). Genes for herbicide resistance and resistance to insect pests were transferred to crop plants in this way.

CENGAGE**NOW** Learn more about making transgenic plants by viewing the animation by logging on to *login.cengage.com/sso* and visiting CengageNOW's Study Tools.

2 The plant cell grows and divides. Some of the descendant cells give rise to embryos that might go on to develop into mature, genetically engineered whole plants, as below.

Embryo

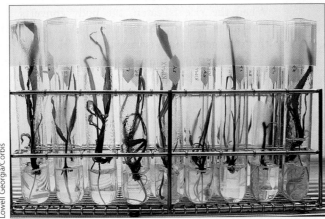

Lowell Georgia/Corbis

About 60% to 70% of the processed foods in your supermarket contain at least a small amount of one or more of the transgenic plants approved for commercial production in the United States (Table 14.3). Products made from corn, soybeans, and cottonseed and canola oils account for almost all the foods that contain transgenic ingredients. As more transgenic plants—including nutritionally enhanced grains and functional foods—are approved, this overall percentage will probably increase.

KEEP IN MIND

Many crop plants have been genetically modified.

Spotlight on. . .

Bioremediation: Using Bugs to Clean Up Waste Sites

Bacteria are being genetically engineered to break down dangerous chemicals such as chlorinated solvents, which are used in industrial processes and then discarded as waste products at factory sites, in landfills, and in other areas. The bacteria have been genetically modified to break down and use the solvents in essential cellular processes, rendering the solvents harmless.

FIGURE 14.10 Transgenic cotton (*left*) has been modified to resist infestation, which destroys normal cotton bolls (*right*).

Agricultural Research Service/USDA.

Enhancing the nutritional value of foods.

Nutritional enhancement of food is nothing new. We already eat fortified foods every day. Read the labels: Milk has added vitamin D, flour and products made from flour are enriched with B vitamins and folate, and table salt has added iodine. Now biotechnology is being used to modify crops to increase their nutritional value.

Most crop plants are deficient in one or more of the nutrients we need in our diets. Diets deficient in vitamin A are a serious health problem in more than 70 countries, primarily in Asia. Between 100 and 150 million children in these countries are vitamin A deficient, and 500,000 of them become permanently blind each year as a result. To combat this problem, rice has been genetically modified to synthesize beta-carotene, a precursor to vitamin A. When the rice is eaten, the body converts beta-carotene into vitamin A. Because the beta-carotene turns the rice a light yellow, this transgenic grain is called golden rice (Figure 14.11).

Locally adapted varieties of golden rice are being developed for free distribution to low-income farmers in many Southeast Asian countries. This use of biotechnology, in combination with other programs, is being used to reduce vitamin A deficiency in the diet of millions of people. Other strains of golden rice enriched with other micronutrients and increased protein content are now under development.

Professor Ingo Potrykus, Chair of the Golden Rice Humanitarian Board.

FIGURE 14.11 Normal rice grains (*left*) are white. Grains of golden rice (*right*), produced by genetic engineering, contain a precursor to vitamin A. This transgenic rice is one tool in fighting vitamin A deficiency, which causes blindness and health problems in many parts of the world.

Table 14.3 Genetically Modified Crops Approved for Commercial Use in the United States

Crop Plants	New Trait	Source of Gene
Corn, cotton, potato, tomato	Insect resistance (Bt crops)	Soil bacterium
Canola, corn, cotton, flax, rice, soybeans, sugar beets	Herbicide resistance	Several bacteria
Tomato	Delayed ripening	Tomato
Canola, soybeans	Altered oil content	Soybean
Corn, radicchio (chicory)	Prevention of cross-pollination with wild plants	Soil bacterium

Functional foods and health.

The first agricultural applications of biotechnology focused on weed and insect control, and these products were rapidly adopted by farmers in the United States and around the world. The golden-rice project marked a turning point in agricultural biotechnology by emphasizing benefits to the consumer rather than the farmer. In addition to enhancing the nutritional value of food, several research centers are developing functional foods, defined as foods that provide health benefits beyond basic nutrition. For example, plants are being modified by gene transfer to increase their levels of fatty acids, antioxidants, and minerals. Fatty acids known as omega fatty acids are important in eye and brain development and the prevention of cardiovascular disease. Fish are the usual source of omega fatty acids, but overfishing has depleted wild populations, and fish and fish oils are often contaminated with mercury, pesticides, and flame retardants.

Transgenic plants that produce these fatty acids—including rape (from which canola oil is produced), sunflowers, and soybeans—provide an enhanced dietary source of omega fatty acids. Overall, more than 40% of the world's population has a diet deficient in one or more nutrients, and the use of transgenic functional foods is one way to combat this problem.

What are some concerns about genetically modified organisms?

Transgenic crops have generated controversy in many developed nations, including the United States and countries of western Europe, as well as in developing nations in Africa. Transgenic crops are strongly opposed by some activists. Early in their use, questions about safety and environmental impact were raised: Is it safe to eat food carrying part of a viral gene that switches on transgenes? Will Bt-resistant insects develop? Will disease-causing bacteria acquire the antibiotic-resistant genes used as markers in transgenic crops?

Most food from transgenic crops contains one or more proteins encoded by transferred genes. Are foods containing the new proteins safe to eat? In general, if the proteins are not toxic or allergenic and do not have any negative physiological effects, they are not considered a significant health hazard by regulatory agencies. In the case of herbicide-tolerant food plants, the enzyme produced by the transgene is readily degraded in digestive fluids, is nontoxic to mice at doses thousands of times higher than any potential human exposure, and has no similarity to known toxins or allergens. After 15 years of widespread use, no human health risks from transgenic plants have been demonstrated.

Several potential environmental risks associated with transgenic crops have been identified, including transfer of herbicide resistance or insect resistance from crop plants to weeds and wild plants, loss of biodiversity caused by transgenic plants hybridizing with wild varieties of the same species, and possible deleterious impacts on ecosystems. However, most of these problems are identical to those we face in using conventional crop plants, and there is no current scientific evidence that transgenic crops are inherently different from crops developed by artificial selection. Transgenic crops have a different evolutionary history and are genetically different from conventional crops, but there are no demonstrated ecological or environmental risks unique to transgenic crops. However, as the modification of crop plants becomes more complex, some combinations of traits may be generated that may require specific crop-management procedures.

Humans have been genetically modifying plants and animals by artificial selection for 8,000 to 10,000 years. The genetic makeup of corn, sheep, cattle, tobacco, and even dogs and cats has been shaped by choosing individuals with desired characteristics to be the parents of the next generation. The development of biotechnology has changed the ways these modifications are made and the range of species that can donate genes. To continue improving crops, farm animals, pets, and garden plants by conventional breeding techniques and biotechnology, relevant scientific questions, economic issues, and public perceptions must be addressed by research, testing, and public education.

14.4 Transgenic Animals as Models of Human Diseases

Genomics has revealed that many human genes have identical counterparts in other organisms, including the mouse and even the fruit fly, *Drosophila*. More than 90% of the genes known to be involved in human diseases are present in the mouse. This makes the mouse an excellent model to explore the genetic, biochemical, and cellular aspects of human diseases. Mouse models of inherited human diseases have been created by transferring mutant human disease alleles into mice and using the resulting transgenic mice to achieve several goals: (1) to produce an animal with symptoms that mirror those in humans, (2) to use the model to study the early stages of development and progress of diseases, and (3) to test drugs to treat symptoms or, hopefully, cure the diseases.

Mice have been extremely useful as models of human neurodegenerative diseases. As an example, let's look at how a transgenic mouse model of Huntington disease (HD; OMIM 143100) was constructed. As discussed in Chapter 4, HD is a neurodegenerative disorder that develops in adulthood and is inherited as an autosomal dominant trait. Affected individuals gradually lose control of their arms and legs in middle age, undergo personality changes and dementia, and die 10 to 15 years after symptoms begin. The mutant allele causes the accumulation of huntingtin, a protein that disrupts many cellular functions and kills cells in the nervous system. Mouse models of HD are explored in more detail in Chapter 18; here we will focus on how a model is constructed.

To make transgenic HD mice, copies of the mutant human *HD* allele were cloned into vectors and microinjected into the nuclei of fertilized mouse eggs (Figure 14.12). The injected eggs were implanted into foster mothers, which gave birth to transgenic mice. DNA was isolated from tail tips of the newborn mice and analyzed for the presence and expression of the mutant *HD* allele. Adult transgenic mice expressing the human *HD* allele develop symptoms that parallel those seen in humans, including the formation of protein aggregates and cell death in the brain and cells of the nervous system.

Scientists use animal models to study human diseases.

HD mice are used to study the progressive destruction of brain structures that occurs in the earliest stages of the disease process—something that is impossible to do in humans. In addition, HD mice are used to link specific changes in brain structure with behavioral changes.

Animal models have identified several molecular mechanisms that play important roles in the early stages of HD. These mice are also being used to screen drugs to identify

Transgenic Refers to the transfer of genes between species by recombinant DNA technology; transgenic organisms have received such a gene.

FIGURE 14.12 Microinjection of fertilized eggs is one way to transfer genes into mammals.

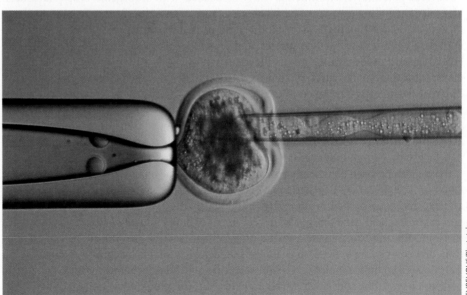

EURELIOUS/Phototake.

those that improve symptoms and/or reverse brain damage. Several candidate drugs have been identified and are being used in human clinical trials as experimental treatments for HD. Similar methods have been used to construct animal models of other human genetic disorders and infectious diseases (Table 14.4).

14.5   DNA Profiles as Tools for Identification

The use of DNA to identify individuals was developed in the 1980s and first used variations in the length of repeated DNA sequences called **minisatellites**, located at many different chromosomal sites. **DNA fingerprints** using minisatellites were used to identify the relationships among people in the story at the beginning of this chapter.

Today, law-enforcement agencies use other, shorter nucleotide sequences called **short tandem repeats (STRs)** instead of minisatellites, and the term DNA fingerprint has been replaced by the term **DNA profile**. DNA profiles are used in criminal cases and many other ways, including: paternity cases, studying human evolution, tracing ancient migrations, and monitoring contamination of food by microorganisms.

> **KEEP IN MIND**
>
> DNA profiles are based on variations in the copy number of DNA sequences.

Minisatellite Nucleotide sequences 14 to 100 base pairs long, organized into clusters of varying lengths, on many different chromosomes; used in the construction of DNA fingerprints.

DNA fingerprint Detection of variations in minisatellites used to identify individuals.

Short tandem repeat (STR) Short nucleotide sequences 2 to 9 base pairs long found throughout the genome that are organized into clusters of varying lengths; used in the construction of DNA profiles.

DNA profile The pattern of STR allele frequencies used to identify individuals.

Making DNA profiles.

STRs are short, repeated sequences from 2 to 9 base pairs long, distributed across the genome. For example, the repeat

$$\text{CCTTCCCTTCCCTTCCCTTCCCTTCCCTTC}$$

contains six copies of the five-nucleotide sequence CCTTC. There are thousands of different STRs on different chromosomes, and because each one of these can have dozens of alleles of different lengths, heterozygosity is common. The number of repeats in any particular STR varies so much from one person to another that by analyzing a number of different STRs in a DNA sample, a DNA profile that is unique to an individual (except, of course, in the case of identical twins) can be produced (Figure 14.13).

To make a profile, specific STRs are amplified from a DNA sample by polymerase chain reaction (PCR) before Southern blotting and visualization. Because PCR is used, profiles can be prepared from very small DNA samples from many possible sources, including single hairs, licked envelope flaps, toothbrushes, cigarette butts, and dried saliva on the back of a postage stamp. DNA profiles can also be obtained from very old samples of DNA, increasing their usefulness in legal cases. In fact, DNA profiles have been prepared from mummies more than 2,400 years old.

In the United States, a standardized set of 13 STRs called the CODIS panel (Combined DNA Index System) is used by law enforcement and other government agencies in preparing DNA profiles.

DNA profiles are used in forensics.

Forensics is the use of scientific knowledge in civil and criminal law. DNA was first used as scientific evidence in a criminal case in 1986. In the small English village of Narborough, two girls were raped and murdered within a 3-year period. Faced with few leads, the police asked Alec Jeffreys at the nearby University of Leicester to help. He analyzed DNA samples recovered from

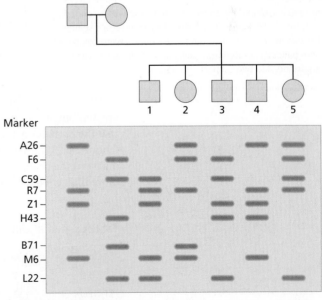

FIGURE 14.13 A DNA profile from a family, showing that each of the bands in the children's (1–5) DNA are represented in the parents and that each child has a unique band pattern.

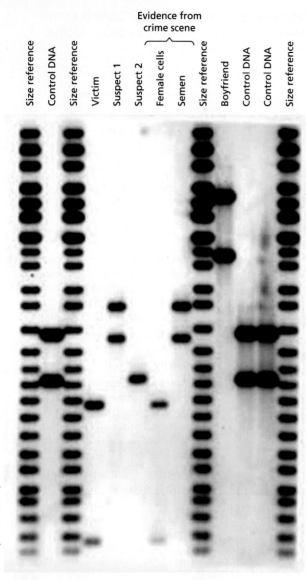

Evidence from crime scene

Size reference · Control DNA · Size reference · Victim · Suspect 1 · Suspect 2 · Female cells · Semen · Size reference · Boyfriend · Control DNA · Control DNA · Size reference

Courtesy of Genelex Corp

FIGURE 14.14 A DNA profile prepared for a criminal case of sexual assault. One short tandem repeat was amplified by PCR from evidence that included the perpetrator's semen and the victim's cells. Bands from these samples were compared with bands generated from DNA samples from the victim, her boyfriend, and two suspects (1 and 2).

the victims and concluded that the same man had committed both murders; he then prepared a DNA fingerprint of the killer. In court, this DNA fingerprint was admitted as evidence to free an innocent man who had been jailed for one of the murders. To find the murderer, police asked over 4,000 men from the surrounding area to provide a DNA sample for analysis. The killer, a man named Colin Pitchfork, paid someone else to give a DNA sample in his place. Pitchfork would have escaped detection but was caught when his substitute mentioned the deception in a pub. This historic case was described in the book *The Blooding* by Joseph Wambaugh.

Each year in the United States, DNA profiles are used in about 10,000 criminal cases and about 200,000 civil cases (mostly for paternity testing). DNA profiles are created and analyzed in state and local police crime labs, private labs, and the Federal Bureau of Investigation (FBI) lab in Washington, D.C. In criminal cases, DNA is often extracted from biological material left at a crime scene. This can include blood, tissue, hair, skin fragments, and semen. DNA profiles are prepared from evidence and compared with those of the victim and any suspects in the case (Figure 14.14). Usually, a profile is constructed using at least four different STRs from the CODIS panel. If a suspect's profile does not match the evidence, he or she can be excluded as the criminal. About 30% of DNA profile results clear innocent people by exclusion.

Analysis of DNA profiles combines probability theory, statistics, and population genetics to estimate how frequently a particular profile is found in an individual and in the general population. This is done in a series of steps. First, the sample is analyzed to establish which alleles of the CODIS panel are present. Once the STR alleles are known, the population frequencies for all the STR alleles are multiplied together to produce an estimate of the probability that anyone carries that combination of alleles. The frequencies are multiplied because STRs on different chromosomes are inherited independently.

Table 14.5 shows how STR alleles provide a profile with a high degree of reliability. In the example, an individual carries alleles of the following CODIS markers: allele 1 of CSF1PO, allele 2 of TPOX, allele 3 of vWA, and allele 4 of D5S818. The CSF1PO allele 1 is found in 1 of 25 individuals in the population, and allele 2 of TPOX is found in 1 of every 100 people in the population. On the basis of these frequencies, the probability of someone carrying both the CSF1PO allele 1 *and* the TPOX allele 2 is 1 in 2,500 (1/25 × 1/100), meaning that this combination of alleles is found in 1 in every 2,500 individuals.

When the population frequencies of the alleles for CSF1PO, TPOX, vWA, and D5S818 are combined, the frequency of this combination is 1 in about 60 million. This means that the probability that someone has this combination of alleles is 1 person in 60 million (about 4 or 5 individuals in the United States). If an analysis includes all 13 CODIS STR markers, the chance that anyone has a particular combination is about 1 in 100 trillion—more than enough to identify someone in Earth's population of 6 billion.

Table 14.5 Calculating Probabilities in DNA Profiles

Locus	Allele	Frequency in Population	Combined Frequencies
CSF1PO	1	1 in 25 (0.040)	—
STRTPOX	2	1 in 100 (0.010)	Alleles 1 × 2 = 1 in 2,500
STRvWA	3	1 in 320 (0.0031)	Alleles 1 × 2 × 3 = 1 in 800,000
STRD5S818	4	1 in 75 (0.0133)	Alleles 1 × 2 × 3 × 4 = 1 in 60,000,000

Death of a Czar

Czar Nicholas Romanov II of Russia was overthrown in the Bolshevik Revolution that began in 1917. He and Empress Alexandra (granddaughter of Queen Victoria); their daughters Olga, Tatiana, Marie, and Anastasia; and their son Alexei (who had hemophilia) were taken prisoner. In July 1918, revolutionists executed the czar, but for many years the fate of his family was unknown. In the 1920s, a Russian investigator, Nikolai Solokof, reported that the czar, his wife and children, and four others were executed at Ekaterinburg, Russia, on July 16, 1918, and that their bodies were buried in a grave in the woods near the city. Other accounts indicated that at least one family member, Anastasia, escaped to live in western Europe or the United States. Over the years, the mystery surrounding the family generated several books and movies.

In the late 1970s, two Russian amateur historians began investigating Solokof's accounts, and, after a painstaking search, nine skeletons were dug from a shallow grave at a site 20 miles from Ekaterinburg in July 1991. All the skeletons bore marks and bullet wounds indicating violent death. Forensic experts examined the remains and, using computer-assisted facial reconstructions and other evidence, concluded that the remains included those of the czar, the czarina, and three of their five children. The remains of two children were missing: the son, Alexei, and one daughter.

DNA was extracted from bone fragments and used for sex testing, for DNA typing to establish family relationships, and for mitochondrial DNA testing to trace maternal relationships. The results indicated that the skeletons were of four males and five females, confirming the results of physical analysis. Five STRs were used to test family relationships. The results showed that five skeletons were a family group and that two were parents and three were children.

To determine whether the remains belonged to the Romanovs, investigators conducted mitochondrial DNA (mtDNA) testing. Because mtDNA is maternally inherited, living relatives of the czarina—including Prince Philip, the husband of Queen Elizabeth of England—were included in the tests. This analysis showed an exact match between the remains of the czarina, the three children, and living relatives, and the mtDNA from the czar matched that of two living maternal relatives, confirming that the remains are those of the czar, his wife, and three of his children. The remains of the czar and his family were reburied in the cathedral of St. Peter and Paul in St. Petersburg in 1998.

In August, 2007, excavations at a nearby site recovered bones thought to be those of the young prince Alexei and his elder sister Maria. In 2009, the publication of forensic DNA test results confirmed that the remains belonged to the two children of the czar, ending the mystery about the fate of the royal family.

This study, stretching over several years and two continents, overcame several technical challenges but clearly established the identity of the skeletons as those of the Romanovs. The results show the power that genetic technology can bring to solving a historical mystery. Robert Massie describes the intrigues, the mysteries, and the science surrounding the search for the Romanovs in his book *The Romanovs: The Final Chapter*.

Bettmann/Corbis

The FBI began cataloging DNA profiles of convicted felons in 1998, and the database now contains more than 3,600,000 profiles. Many states are collecting DNA samples from anyone arrested for a felony and entering those DNA profiles in a database. As they accumulate more and more profiles, DNA databases are becoming important tools in solving crimes. In Virginia, which has one of the oldest and largest DNA databases in the United States, more than 3,600 crimes without suspects were solved over a 6-year period by matching DNA profiles from the crime scene with profiles already in the state database.

DNA profiles have many other uses.

DNA profiles have many uses outside the courtroom. Biohistorians use DNA analysis to identify bodies and body parts of famous and infamous people whose graves have been moved several times or whose graves have been newly discovered. The identification of Czar Nicholas II and many of his family members in unmarked graves in Russia is one example of how DNA profiles are used in biohistory (see Exploring Genetics: Death of a Czar on page 325). DNA analysis has been completed or is planned for the remains of Christopher Columbus, Juan Pizarro, Jesse James, and several Egyptian pharaohs.

14.6 Social and Ethical Questions About Biotechnology

The development of methods for cloning plants and animals by recombinant DNA technology is the foundation of the biotechnology industry, a growing and dynamic multibillion-dollar component of our economy. These methods are revolutionizing biomedical research, the diagnosis and treatment of human diseases, agriculture, the justice system, and even the production of chemicals. Like many other technologies, biotechnology raises numerous social and ethical issues that need to be identified, discussed, and resolved.

> **KEEP IN MIND**
>
> The uses of biotechnology have produced unresolved ethical issues.

Unfortunately, the technology and its applications are developing much faster than public policy, legislation, and social norms. The fact that we *can* do some things does not mean we *should* do them without a consensus from society. For example, should very short children receive recombinant DNA–produced growth hormone to make them average in height? The U.S. Food and Drug Administration approved the use of growth hormone for this purpose in 2003. If we accept that we can treat short children with growth hormone to increase their adult height, is it ethical to treat children of normal or above-average height to enhance their chances of becoming professional basketball or volleyball players?

The current guidelines of the Food and Drug Administration do not require identifying labels on food produced by recombinant DNA technology. Should such food be labeled, and if so, why? It can be argued that food products have been genetically manipulated for thousands of years and that gene transfer is simply an extension of past practices. It can also be claimed that consumers have a right to know that gene transfer has altered the food products they buy and use.

With genome scanning, it may be possible to map out someone's lifetime risks for all the major diseases, including those that may not develop for decades. This technology raises many ethical questions. Testing often reveals information about disease genes carried by the parents, siblings, and children of the one who is tested. What should be done with that information? How can we protect access to the results of genetic testing? Should we test for serious or fatal disorders for which there is no treatment? As citizens, it is our responsibility to become informed about these issues and to participate in formulating policy about the use of this technology.

Genetics in Practice

Genetics in Practice case studies are critical-thinking exercises that allow you to apply your new knowledge of human genetics to real-life problems. You can find these case studies and links to relevant websites at *www.cengage.com/biology/cummings*

CASE 1

Todd and Shelly Z. were referred for genetic counseling because of advanced maternal age (Shelley was over 40 years old) in their current pregnancy. While obtaining the family history, the counselor learned that during their first pregnancy, the couple had elected to have an amniocentesis for prenatal diagnosis of cytogenetic abnormalities because Shelly was 36 years old at the time. During that pregnancy, Shelly and Todd reported that they were also concerned that Todd's family history of cystic fibrosis (CF) increased their risk for having an affected child. Todd's only sister had CF, and she had severe respiratory complications.

The genetic counseling and testing was performed at an outside institution, and the couple had not brought copies of the report with them. They did state that they had completed studies to determine their CF carrier status and that Todd had been found to be a CF carrier but Shelly's results were negative. The couple was no longer concerned about their risk of having a child with CF on the basis of those results. To support their belief, they had a healthy 5-year-old son who had a negative sweat test at age 4 months. The counselor explained the need to review the records and scheduled a follow-up appointment.

The test report from their first pregnancy showed that they had been tested only for the delta F508 mutation, the most common mutation in CF. The report confirmed that Shelly was not a carrier of this mutation and that Todd was. This report reduced Shelly's risk status from 1 in 25 to about 1 in 300. More information has been learned about the different mutations in the CF gene since the last time they received genetic counseling. The counselor conveyed the information about recent advances in CF testing to the couple, and Shelly decided to have her blood drawn for CF mutational analysis with an expanded panel of mutations. Her results showed that she was a carrier for the W1282X CF mutation. The family was given a 25% risk for CF for each of their pregnancies on the basis of their combined molecular test results. They proceeded with the amniocentesis because of the risks associated with advanced maternal age and requested fetal DNA analysis for CF mutations. The fetal analysis was positive for both parental mutations, indicating that the fetus had a greater than 99% chance of being affected with CF.

1. How is it that the fetus has a greater than 99% chance of being affected with CF if each parent carries a different mutation? Is the fetus homozygous or heterozygous for these mutations?

2. If this child has CF, what are the chances that any future child will have this disease? Does the fact that they have a healthy 5-year-old son affect the chances of having future children with CF?

CASE 2

Can DNA profiling identify the source of a sample with absolute certainty? Because any two human genomes differ at about 3 million sites, no two persons (except identical twins) have the same DNA sequence. Unique identification with DNA profiling is therefore possible if enough sites of variation are examined. However, the systems used today examine only a few sites of variation and have only limited resolution for measuring the variability at each site. There is a chance that two persons might have DNA patterns (i.e., genetic types) that match at the small number of sites examined. Nonetheless, even with today's technology, which uses three to five loci, a match between two DNA patterns can be considered strong evidence that the two samples came from the same source. How is DNA profiling currently being used?

- Paternity and maternity testing: Because a person inherits his or her STRs from his or her parents, STR patterns can be used to establish paternity and maternity. The patterns are so specific that a parental STR pattern can be reconstructed even if only the children's STR patterns are known (the more children produced, the more reliable the reconstruction). Parent-child STR pattern analysis has been used to solve standard father-identification cases and more complicated cases confirming nationality and, in instances of adoption, biological parenthood.

- Criminal identification and forensics: DNA isolated from blood, hair, skin cells, or other genetic evidence left at the scene of a crime can be compared, through STR patterns, with the DNA of a criminal suspect to determine guilt or innocence. STR patterns are also useful in establishing the identity of a homicide victim. However, DNA profiles have limitations. In cases where one identical twin has committed a crime, DNA evidence may not be helpful.

- Personal identification: The notion of using DNA fingerprints as a sort of genetic bar code to identify individuals has been discussed, but this is not likely to happen any time in the foreseeable future. The technology required to isolate, keep on file, and analyze millions of very specific STR patterns is both expensive and impractical. Social security numbers, picture IDs, and other more mundane methods are much more likely to remain ways of establishing personal identification.

However, the FBI and other police agencies and the military are enthusiastic proponents of DNA data banks. The practice has civil-liberties implications because police can go on "fishing expeditions" with DNA readouts.

1. A police chief in a large U.S. city has proposed that everyone who is arrested be DNA fingerprinted; this information would be stored in a national database even if those arrested are later found to be innocent. The chief is actively seeking funding from his state legislature to begin such a program. Do you support such a program? Why or why not? Could there be abuses of the program?

2. Several companies advertise DNA-fingerprinting services on the Internet for genealogy research. In light of the fact that most ancestors are dead and not available for such testing, how can this service be of value in genealogy? Are there privacy concerns about what the companies do with the results of such testing?

Summary

14.1 Biopharming: Making Human Proteins in Animals

- Genetic engineering is used to manufacture proteins used in treating human diseases. This provides a constant supply that is uncontaminated by disease-causing agents. These proteins are made in bacteria, cell lines from higher organisms, animals, and even plants.

14.2 Using Stem Cells to Treat Disease

- Stem cells have the potential to provide an unlimited source of cells that can be used to treat a number of diseases. Stem cells can be isolated from embryos and adult tissues. Researchers have also reprogrammed adult skin cells to form stem cells.

14.3 Genetically Modified Foods

- Gene transfer into crop plants has conferred resistance to herbicides, insect pests, and plant diseases. In addition, gene transfer is being used to increase the nutritional value of foods as a way to combat diseases such as vitamin A deficiency.

14.4 Transgenic Animals as Models of Human Diseases

- The transfer of disease-causing human genes creates transgenic organisms that are used to study the development of human diseases and the effects of drugs and other therapies as methods for treating these disorders.

14.5 DNA Profiles as Tools for Identification

- DNA profiles use variations in the length of short repetitive DNA sequences to identify individuals with a high degree of accuracy and reliability. This method is used in many areas, including law enforcement, biohistory, conservation, and the study of human populations.

14.6 Social and Ethical Questions About Biotechnology

- Recombinant DNA technology and its many applications have developed faster than societal consensus, public policy, and laws governing its use. In addition, efforts to inform legislators, members of the legal and medical professions, and the public at large about the applications of this technology have often lagged behind its development and commercial use. An effort to educate and debate the risks and benefits of the ways in which we use genetic engineering will ensure a balanced approach to this technology.

Questions and Problems

CENGAGENOW Preparing for an exam? Assess your understanding of this chapter's topics with a pre-test, a personalized learning plan, and a post-test by logging on to *login.cengage.com/sso* and visiting CengageNOW's Study Tools.

Biopharming: Making Human Proteins in Animals

1. Making human proteins to treat diseases in animal and plant hosts offers several advantages over extracting those proteins from human and animal sources. What are some of those advantages?

Using Stem Cells to Treat Disease

2. What is the difference between pluripotent and multipotent stem cells?
3. How was it established that induced pluripotent stem cells derived from adult skin cells are really pluripotent?

Genetically Modified Foods

4. The creation of transgenic crop plants using recombinant DNA methods involves the transfer of just one gene or a small number of genes to the plants, in contrast to classical breeding methods in which hundreds or even thousands of genes are transferred at once. Explain why this is true. If fewer genes are transferred during the creation of transgenic crops, why are some people afraid that they are dangerous?
5. What are "Bt crops"? What potential risks are associated with this technology?

6. "Golden" rice strains are transgenic crops that have been genetically engineered to produce elevated levels of beta-carotene. These strains were developed to help prevent:
   a. tuberculosis
   b. malaria
   c. vitamin A deficiency
   d. protein deficiency
   e. hypertension

Transgenic Animals as Models of Human Diseases

7. You are a neuroscientist working on a particular neurodegenerative disease and discover that the disease is caused by the expression of a mutant gene in the brain. A colleague suggests that you use this new information to develop a transgenic mouse model of the disease. How might you do this? How would you use the animal to study the disease or look for new treatments?

DNA Profiles as Tools for Identification

8. STRs are
   a. used for DNA profiles.
   b. repeated sequences present in the human genome.
   c. highly variable in copy number.
   d. all of the above
   e. none of the above

9. A crime is committed, and the only piece of evidence the police are able to gather is a small bloodstain. The forensic scientist at the crime lab is able to
   - extract DNA from the blood,
   - cut the DNA with a restriction enzyme,
   - separate the fragments by electrophoresis, and
   - transfer the DNA from a gel to a membrane and probe with radioactive DNA.

Probe 1 is used to visualize the pattern of bands. The forensic scientist compares the band pattern in the evidence (E) with the patterns from the suspects (S1, S2). The first probe is removed, the membrane is hybridized using another probe (Probe 2), and the band patterns are compared. This process is repeated for Probe 3 and Probe 4.
   a. On the basis of the results of this testing, can either of the suspects be excluded as the one who committed the crime?
   b. If so, which one? Why?
   c. Is the pattern from the evidence consistent with the band pattern of one of the suspects? Which one?

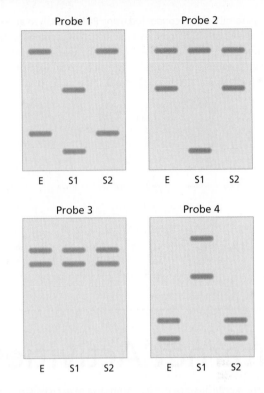

10. You are serving on a jury in a murder case in a large city. The prosecutor has just stated that DNA fingerprinting shows that the suspect must have committed the crime. He says that for cost reasons, two STR probes were used instead of four, but that the results are just as accurate. The first probe detected a locus that has a population frequency of 1 in 100 (1%), and the second probe has a population frequency of 1 in 500 (0.2%).
   a. What is the combined frequency of these alleles in the population?
   b. Does this point to the suspect as the perpetrator of the crime?
   c. Is there any other evidence you would like from the DNA lab?
   d. Do you think that most people on juries understand basic probabilities, or do you think that DNA evidence can be used to mislead jurors into reaching false conclusions?

11. DNA profiles have been used in identifying criminals, establishing bloodlines of purebred dogs, identifying endangered species, and studying extinct animals. Can you think of other uses for DNA profiles?

12. A paternity test is conducted using PCR to analyze an RFLP that consistently produces a unique DNA fragment pattern from a single chromosome. Examining the results of the following Southern blot, which male(s) can be excluded as the father of the child? Which male(s) could be the father of the child?

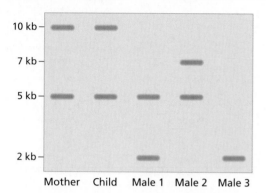

Social and Ethical Questions About Biotechnology

13. Answer the following questions with a yes or a no, and support each view with a coherent reason.
   a. Should very short children receive recombinant human growth hormone (HGH) to reach average height?
   b. Should genetically modified foods be labeled in grocery stores?
   c. Is it acceptable for physicians to treat children of normal height with HGH to make them taller if they and their parents want this?

Internet Activities

Internet Activities are critical-thinking exercises using the resources of the World Wide Web to enhance the principles and issues covered in this chapter. For a full set of links and questions investigating the topics described below, visit *www.cengage.com/biology/cummings*

1. *DNA Analysis.* This technique can be used to identify and compare human DNA samples on the basis of DNA profiles. Conclusions from these comparisons can be used in paternity cases, criminal trials, and many other applications. The University of Arizona's *Biology Project: Human Biology* Web page provides several problem sets and activities to enhance your understanding of DNA analysis. The two DNA Forensics problem sets allow you to try your hand at answering questions about DNA data and also provide tutorials to explain difficult concepts. The two Blackett Family DNA activities provide opportunities to work with DNA fingerprints and the newer STR (short tandem repeat) technique.

2. *Discussing Biotechnology and Genetically Engineered Products.* The *Union of Concerned Scientists* website provides information and a forum for discussing scientific aspects of biotechnology, as well as the marketing of genetically engineered products. At the website, click on the "Genetically Engineered Food" link to find the answers to frequently asked questions about biotechnology and fact sheets on current biotechnology issues. Read the position statement on the "Genetically Engineered Food" Web page. What are the positions—and potential biases—of the scientists in this organization? Do you share them, or do you disagree? Don't forget to check out the "Transgenic Cafe"!

HOW WOULD YOU VOTE NOW?

DNA identification cards offered by several companies contain a photo; a thumbprint or retinal scan; personal information such as height, weight, and eye color; and a CODIS DNA profile that makes personal identification a certainty. Law enforcement officials in Britain want to expand the national DNA database to include all citizens, not just criminals, to provide a means of personal identification. Advocates point out that such ID cards can protect children who are lost or kidnapped, help locate missing persons, prevent fraud, identify actual and potential terrorists, and be a deterrent to crime. Critics argue that DNA databases were set up to solve crimes, not to put everyone's profile in a database to deter crime, and that this technology can be used by governments to track the normal day-to-day activities of citizens instead of fighting crime. Now that you know more about the power of DNA identification, what do you think? Would you get a government DNA identification card if one was offered, and would you support making such cards mandatory? Visit the *Human Heredity* companion website at *www.cengage.com/biology/cummings* to find out more on the issue; then cast your vote online.

15 Genomes and Genomics

When Baby T was brought to the hospital, the five-month-old infant was suffering from severe dehydration and other health problems. A family history revealed that Baby T's parents were first cousins and that the mother had previously had two miscarriages and a premature baby who had died. The physicians suspected that Baby T might have a genetic disorder and made a preliminary diagnosis of Bartter syndrome—a rare, life-threatening mutation that causes a defect in kidney function. To confirm this diagnosis, they sent a blood sample from Baby T to a lab at Yale University for testing. This lab was pioneering a genomic method called whole exome sequencing. Instead of sequencing the entire genome, the method focuses on the 1% of the genome that encodes proteins, called the exome. Using this method, researchers identified a mutation in a gene called *SLC26A3*, responsible for an autosomal recessive condition called congenital chloride diarrhea that prevents proper ion transport in the intestines. The physicians at Baby T's hospital confirmed this diagnosis. Baby T's case is the first in which whole exome sequencing (or genome sequencing of any kind) has been used to identify a genetic disorder. Following Baby T's case, a team at the University of Washington used whole exome sequencing to identify the gene responsible for Miller syndrome, a rare disorder that causes multiple physical malformations. This study showed that whole exome sequencing of a small number of individuals is a powerful method for identifying mutant genes that underlie genetic disorders. It also points to a future when exome sequencing and similar methods will be used as a routine part of medical care.

The availability of this technology raises many questions. If whole exome sequencing reveals that you are at risk for developing heart disease, breast cancer, or diabetes, it may influence you to adopt lifestyle strategies that can lower those risks. On the other hand, if sequencing shows that you will develop a devastating and fatal disease in middle age, should you inform your siblings, who may also be at risk for the disease? Should you have your children tested to see if they will develop this disorder as adults?

This chapter examines the methods, results, and impact of sequencing the human genome, including the ethical, legal, and social implications of genomics—the study of genomes.

A bank of DNA sequencing machines used in genome projects.

Celera, Inc.

15.1 Genome Sequencing Is an Extension of Genetic Mapping

Mutations play a central role in genetics. Historically, transmission of mutant alleles from generation to generation was studied to discover the basic principles of genetics. In addition, geneticists used induced and spontaneous mutations in laboratory organisms to dissect biological processes such as metabolism, cell structure, and development. As information about the genetics of experimental organisms grew, catalogs of mutations were established, and research using these mutations provided insight into genome organization (how many chromosomes, what genes are on which chromosomes, etc.). Gradually, mutational analysis became the method of choice for establishing how many genes are contained in an organism's **genome**.

While this approach worked well for experimental organisms, human genetics grew very slowly because of its reliance on indirect and inferential methods for collecting and analyzing information about the inheritance of genes in our species. Although analysis of mutants is a useful tool in genetics, the other side of mutations is that they are responsible for many inherited disorders in humans.

If we hope to learn how many genetic disorders humans can have and how to develop treatments for those diseases, we need to know how many genes are in our genome, know where they are located on our chromosomes, and have ways of establishing their functions. Genetic mapping is one step to accomplish these goals. In human genetics, this approach began in the 1930s, when Julia Bell and J. B. S. Haldane used pedigree analysis to show that the genes for hemophilia (OMIM 306700) and color blindness (OMIM 303800) are both on the X chromosome. Genes on the same chromosome tend to be inherited together and are said to show **linkage**.

Although X-linked genes can be identified by their unique pattern of inheritance, it is more difficult to map genes to individual autosomes. Linkage studies of autosomal genes

KEEP IN MIND AS YOU READ

- The Human Genome Project grew out of methods originally developed for basic research: recombinant DNA technology and DNA sequencing.

- Genomics relies on interconnected databases and software to analyze sequenced genomes and identify genes.

- The human genome has a surprisingly small number of genes and produces a surprisingly large number of proteins, using a number of different mechanisms.

- Genomics is affecting basic research in biology and generating new methods of diagnosis and treatment of disease.

Genome The set of DNA sequences carried by an individual.

Linkage A condition in which two or more genes do not show independent assortment. Rather, they tend to be inherited together. Such genes are located on the same chromosome. When the degree of recombination between linked genes is measured, the distance between them can be determined.

333

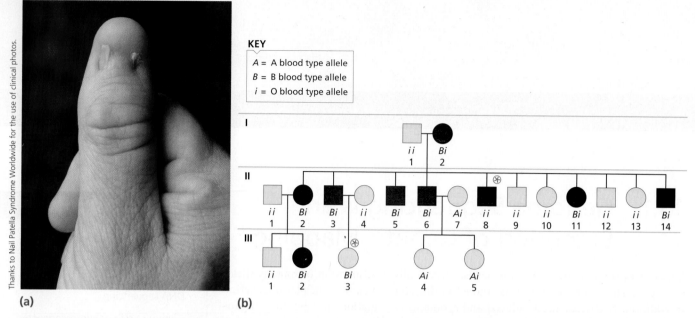

KEY

A = A blood type allele
B = B blood type allele
i = O blood type allele

(a)

(b)

FIGURE 15.1 (a) The phenotype of nail-patella syndrome. (b) Linkage between nail-patella syndrome and the ABO blood-type locus. The darker-shaded symbols in this pedigree represent members with nail-patella syndrome, an autosomal dominant trait. Genotypes for the ABO locus are shown below each symbol. Individual I-2 has both the nail-patella syndrome allele and the *B* allele; these two alleles tend to be inherited together in this family, and they are identified as linked genes. Individuals marked with an asterisk (II-8 and III-3) inherited the nail-patella allele or the *B* allele alone. This separation of the two alleles occurred by recombination.

require multigenerational pedigrees of large families carrying two genetic disorders. Because most genetic disorders are rare, it is difficult to find such families. To get around this problem, geneticists decided to use a common trait such as blood types and search for a rare genetic disorder linked to a specific blood-type allele. Because everyone has a blood type, it was thought that this would make the search for linkage easier. This search also began in the 1930s, but it took until 1955 to discover linkage between an allele of the ABO (the *I* locus; OMIM 110300) blood type and an autosomal dominant condition called nail-patella syndrome (OMIM 161200). Nail-patella syndrome causes deformities in the nails and kneecaps (Figure 15.1a).

In a pedigree showing this linkage (Figure 15.1b), type B blood and nail-patella syndrome occur together in individual I-2. Over the next two generations, the two traits are inherited together (linked) in most cases. However, the pedigree also shows that two individuals inherited only *one* of the two alleles—either the nail-patella allele (II-8) or the B blood-type allele (III-3), but not both. The separation of the two alleles is the result of crossing over between the two genes (Figure 15.2). The frequency of crossing over between two linked genes (discussed in Chapter 2) can be used to construct a genetic map showing the order and the distance between genes on a chromosome.

Recombination frequencies are used to make genetic maps.

A genetic map is made in two steps: (1) Finding linkage between two genes establishes that they are on the same chromosome, and (2) measuring how frequently crossing over takes place between them establishes the distance between them. The units of distance are expressed as a percentage of crossing over (Figure 15.3), where 1 map unit is equal to a 1% crossover frequency. (This unit is also known as a **centimorgan**, or **cM**.)

We can calculate the distance between alleles by measuring how often they are separated by crossing over. The allele for nail-patella and the type B allele underwent crossing over in 2 of the 16 family members (individuals II-8 and III-3 in Figure 15.1). From the frequency of recombination (2/16 = 0.125, or 12.5% recombination), we can calculate the distance between the ABO locus and the nail-patella locus as 12.5 map units.

Centimorgan (cM) A unit of distance between genes on chromosomes. One centimorgan equals a frequency of 1% crossing over between two genes.

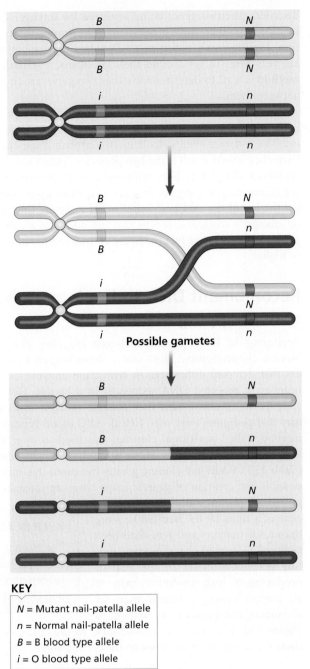

FIGURE 15.2 Crossing over between homologous chromosomes during meiosis involves the exchange of chromosomal parts. In this case, crossing over between the genes for blood type (alleles *B* and *i* of gene *I*) and nail-patella syndrome (*N*) produces new combinations of alleles in gametes that are passed on to offspring and recorded in pedigrees. The frequency of crossing over is proportional to the distance between the genes, allowing construction of a genetic map for this chromosomal region.

Possible gametes

KEY

N = Mutant nail-patella allele
n = Normal nail-patella allele
B = B blood type allele
i = O blood type allele

For accuracy, either a much larger pedigree or many more pedigrees need to be examined to determine the extent of recombination between these two genes. When a large series of families is combined in an analysis, the map distance between the ABO locus and the nail-patella locus is about 10 map units. It took about 20 years to find the first example of linkage between autosomal genes, and by 1969, only five cases of linkage had been discovered. It was apparent that the effort to map all human genes by linkage analysis using pedigrees was going nowhere.

Linkage and recombination can be measured by lod scores.

Although pedigree analysis is a basic method in human genetics, recall that linkage studies using pedigrees is difficult because they require large multigenerational pedigrees with two genetic

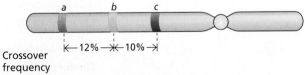

Crossover frequency

FIGURE 15.3 Linked genes are carried on the same chromosome. Crossover frequencies are used to construct genetic maps, giving the order and distance between genes on a chromosome.

disorders (the two genes being analyzed for linkage). In most cases, pedigrees do not include more than three generations—the grandparents, parents, and children. In addition, families with two genetic disorders that might map to the same chromosome are very rare. To get around these problems, a statistical technique known as the **lod method** is used to determine whether two genes are linked and to measure the distance between them.

Lod scores can be calculated by using software programs such as LINKMAP. To begin, the frequency of crossing over between two genes is derived from pedigree studies. The software then calculates two probabilities: the probability that the observed frequency would result if the two genes are linked and the probability that the observed frequency would have been obtained even if the two genes were not linked. The results are expressed as the $\log_{10}$ of the ratio of the two probabilities, or the **lod score** (*lod* stands for the log of the odds). A lod score of 3 means that the odds are 1,000 to 1 in favor of linkage; a score of 4 means that the odds are 10,000 to 1 in favor of linkage. Most geneticists agree that two genes are considered to show linkage when the lod score is 3 or higher.

Lod method A probability technique used to determine whether two genes are linked.

Lod score The ratio of probabilities that two genes are linked to the probability that they are not linked, expressed as a $\log_{10}$. Scores of 3.0 or higher are taken as establishing linkage.

Recombinant DNA technology radically changed gene-mapping efforts.

Beginning in 1980, geneticists began mapping cloned DNA sequences to specific human chromosomes. Most often, those sequences weren't genes but simply markers that detected differences in restriction-enzyme cutting sites or differences in the number of repeated DNA sequences in a cluster. However, once these markers were assigned to chromosomes, they became valuable tools in linkage studies in the same way that pedigrees were used to link ABO blood types and nail-patella syndrome. This method, called **positional cloning**, was used to map and isolate the genes for cystic fibrosis, neurofibromatosis, Huntington disease, and dozens of other genetic conditions (Table 15.1). Positional cloning greatly increased the number of mapped genes and made possible the creation of genetic maps covering most human chromosomes. Although positional cloning only identified one gene at a time, by the late 1980s, more than 3,500 markers and genes had been assigned to human chromosomes, putting geneticists closer to the goal of producing a high-resolution map of all human genes. A genetic map for a human chromosome is shown in Figure 15.4. The map is drawn adjacent to the chromosome, and the connecting lines show the locations of the genes on the chromosome.

Positional cloning A recombinant DNA–based method of mapping and cloning genes with no prior information about the gene product or its function.

> ### KEEP IN MIND
>
> The Human Genome Project grew out of methods originally developed for basic research: recombinant DNA technology and DNA sequencing.

Table 15.1 Some of the Genes Identified by Positional Cloning

	OMIM		OMIM
Chromosome 4		**Chromosome 17**	
Huntington disease	143100	Breast cancer (*BRCA1*)	113705
		Neurofibromatosis (*NF1*)	162200
Chromosome 5			
Familial polyposis (APC)	175100	**Chromosome 19**	
		Myotonic dystrophy	160900
Chromosome 7			
Cystic fibrosis	219700	**Chromosome 21**	
		Amyotrophic lateral sclerosis	105400
Chromosome 11			
Wilms tumor	194070	**X chromosome**	
Ataxia-telangiectasia	208900	Duchenne muscular dystrophy	310200
		Fragile-X syndrome	309550
Chromosome 13		Adrenoleukodystrophy	300100
Retinoblastoma	180200		
Chromosome 16			
Polycystic kidney disease	173900		

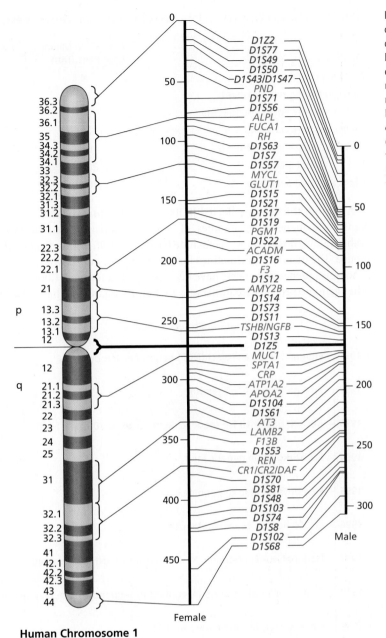

FIGURE 15.4 Genetic map of human chromosome 1. At the left is a drawing of the chromosome. The two vertical lines at the right represent genetic maps derived from studies of recombination in males and females. Between the genetic maps are the order and location of 58 loci, some of which are genes (*red*) and others of which are genetic markers (*blue*) detected using recombinant DNA techniques. The map in females is about 500 cM long, and in males it is just over 300 cM. This is a result of differences in the frequency of crossing over in males and females. This map provides a framework for locating genes on the chromosome as part of the Human Genome Project.

15.2 Genome Projects Are an Outgrowth of Recombinant DNA Technology

The development of recombinant DNA technology and its use in positional cloning and gene mapping provided a springboard for discussions on how to map all the genes in the human genome (recall that a genome is the set of all DNA sequences carried by an individual). After several years of planning, in 1990, the Human Genome Project (HGP) began in the United States.

Instead of finding and mapping markers and disease genes one at a time, the HGP set out to sequence all the DNA in the human genome and to use this information to identify and map the thousands of genes we carry on the 24 chromosomes (22 autosomes and the X and Y chromosomes) and to establish their functions.

Table 15.2 Model Organisms Included in the Human Genome Project

Organism	Genome Size Million Base Pairs (Mb)	Estimated Number of Genes
Escherichia coli (bacterium)	4.6	4,300
Saccharomyces cerevisiae (yeast)	12	6,000
Caenorhabditis elegans (roundworm)	97	20,000
Arabidopsis thaliana (plant)	120	25,000
Drosophila melanogaster (fruit fly)	165	13,600
Mus musculus (mouse)	3,000	30,000
Homo sapiens (human)	3,200	20,000–25,000

Although this is not reflected in its name, the Human Genome Project also included plans to sequence the genomes of model organisms used in experimental genetics (Table 15.2). Those organisms include bacteria, yeast, a roundworm, the fruit fly, and the mouse. The history and timelines for these projects are shown in Figure 15.5. Since then, genome-sequencing projects have been completed for over 1,000 species, and over 5,000 other genome projects are underway, including organisms that cause human diseases.

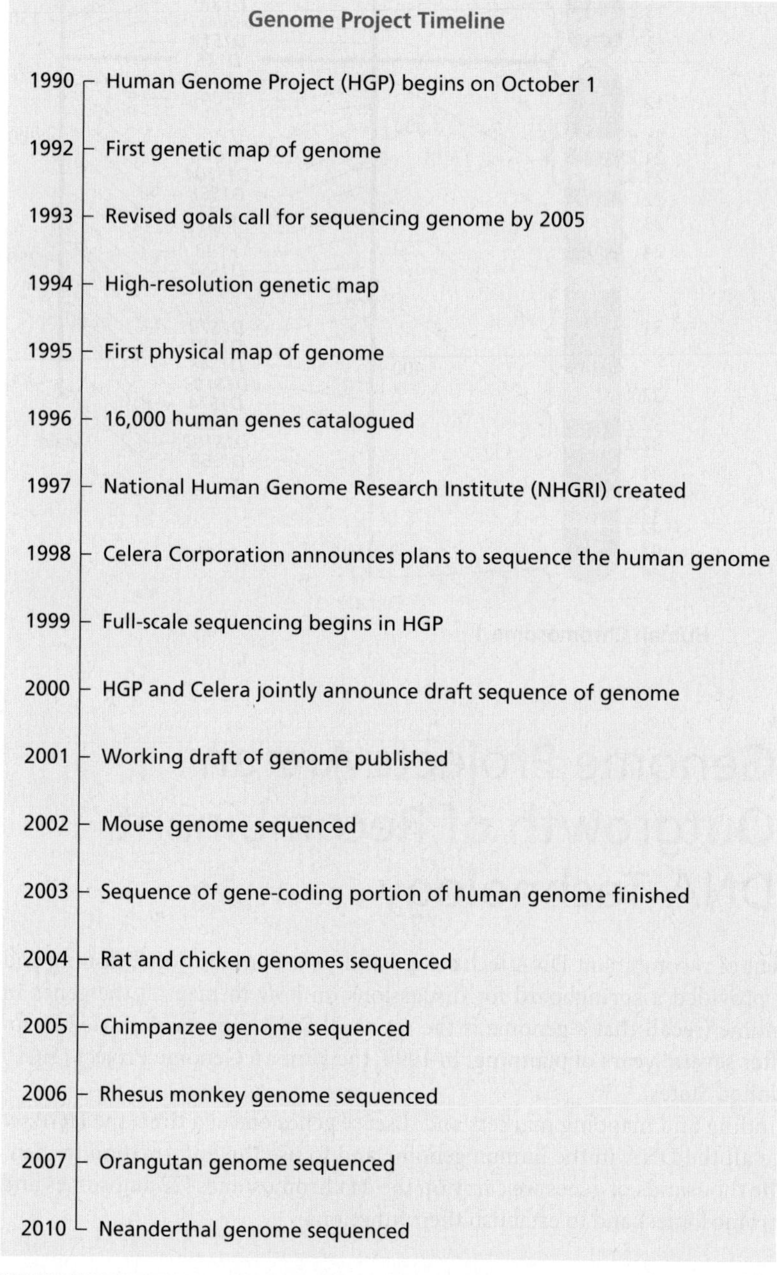

FIGURE 15.5 The history and timeline for a number of genome projects.

Genome Project Timeline

1990 — Human Genome Project (HGP) begins on October 1

1992 — First genetic map of genome

1993 — Revised goals call for sequencing genome by 2005

1994 — High-resolution genetic map

1995 — First physical map of genome

1996 — 16,000 human genes catalogued

1997 — National Human Genome Research Institute (NHGRI) created

1998 — Celera Corporation announces plans to sequence the human genome

1999 — Full-scale sequencing begins in HGP

2000 — HGP and Celera jointly announce draft sequence of genome

2001 — Working draft of genome published

2002 — Mouse genome sequenced

2003 — Sequence of gene-coding portion of human genome finished

2004 — Rat and chicken genomes sequenced

2005 — Chimpanzee genome sequenced

2006 — Rhesus monkey genome sequenced

2007 — Orangutan genome sequenced

2010 — Neanderthal genome sequenced

15.3 Genome Projects Have Created New Scientific Fields

The human genome contains about 3.2 billion nucleotides of DNA, but even that number is dwarfed by the size of some other genomes. For example, the marbled lungfish genome contains over 132.8 billion nucleotides. The sheer size of most eukaryotic genomes required the development of new technologies, including automated methods of DNA sequencing; advances in software to collect, analyze, and store the information derived from sequencing; and the creation of Web-based databases and research tools to access genome sequence information. The study of genomes through the use of these methods is called **genomics**. The goals of genomics are outlined in Table 15.3.

Methods for sequencing DNA were developed in the 1970s (see Chapter 13). The original Sanger method used four separate chemical reactions with radioactive-labeled nucleotides (one for each base—A, T, C, and G). The fragments generated by these reactions are separated by size in four adjacent lanes in a gel (Figure 15.6). The sequence can be read from the bottom up. In Figure 15.6, the lowest fragment is in the A lane, so that is the first base in the sequence. Reading from this base upward, the beginning of the sequence is AATCGGCCGCT. This method is useful for sequencing small amounts of DNA but is somewhat labor intensive, requiring several days to produce a few hundred bases of sequence.

To sequence whole genomes, new technology was needed to speed and automate the process. Genome projects refined DNA sequencing so that only a single reaction is needed. Each base is labeled with a different fluorescent dye, and the resulting sequence can be read from a single lane on a gel by a scanner. Machines called DNA sequencers were developed to automatically sequence several hundred thousand bases per day. The HGP and Celera Corporation used banks of hundreds of DNA sequencing machines, linked to computers that controlled the machines and stored the resulting data (Figure 15.7). New, faster, and less expensive methods of DNA sequencing are now becoming available, and as we saw in the story of Baby T, these methods will have a significant impact on medicine.

Once a genome sequence has been deciphered, it must be stored in a database so that it can be assembled and analyzed to identify genes. A new field called **bioinformatics** was developed to use computer hardware and software to store, analyze,

Table 15.3 Goals of Genomics

- Create genetic and physical maps of genomes.
- Catalog all non-gene sequence families in the genome and determine their copy number.
- Find the chromosomal location of all genes in a genome and annotate each gene.
- Compile lists of expressed genes and nonexpressed sequences.
- Elucidate gene function and gene regulation.
- Identify all proteins encoded by a genome and their functions.
- Compare genes and proteins between species.
- Characterize DNA variations with and between genomes.
- Implement and manage Web-based databases.

Genomics The study of the organization, function, and evolution of genomes.

Bioinformatics The use of computers and software to acquire, store, analyze, and visualize the information from genomics.

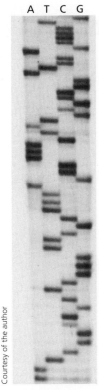

FIGURE 15.6 A DNA sequencing gel showing the separation of fragments in the four sequencing reactions. The sequence is read from the bottom, starting with the lowest band in any lane, then the next lowest, and so on. In this gel, the sequence begins with AATCGGCCGCT.

Courtesy of the author

FIGURE 15.7 Gene sequencing computers used in human genome research at Celera Corporation in Maryland.

Volker Steger/SPL/Photo Researchers, Inc.

Comparative genomics Compares the genomes of different species to look for clues to the evolutionary history of genes or a species.

Structural genomics A branch of genomics that generates three dimensional structure of proteins from their amino acid sequences.

Pharmacogenomics A branch of genetics that analyzes genes and proteins to identify targets for therapeutic drugs.

Map-based sequencing A method of genome sequencing that begins with genetic and physical maps; clones are sequenced after they have been placed in order.

Whole genome sequencing A method of genome sequencing that selects clones at random from a genomic library and, after sequencing them, assembles the genome sequence by using software analysis.

visualize, and access genomic sequence data. Bioinformatics is a rapidly growing field that encompasses more than just genome sequence data. It plays important roles in other fields, including the following:

- **Comparative genomics**—compares genomes of different species, looking for similarities and differences in genes and clues to the evolutionary history of a species.
- **Structural genomics**—derives three-dimensional structures for proteins.
- **Pharmacogenomics**—analyzes genes and proteins to identify targets for therapeutic drugs.

In the following sections, we will see how genomics and bioinformatics are used to collect, analyze, and store information about the human genome.

15.4 Genomics: Sequencing, Identifying, and Mapping Genes

Genomics begins by sequencing an organism's genome. Because most genomes are large, they are cut into fragments of manageable size before sequencing. The task of sequencing each of these fragments is straightforward, but correctly assembling the sequence from thousands or millions of fragments is a difficult task. To manage this process, two sequencing strategies are used. The first is called **map-based sequencing** (also called the clone-by-clone method), and the second is called **whole genome sequencing** (also called the shotgun method). Both methods were used in the Human Genome Project.

In map-based sequencing (Figure 15.8a), detailed genetic maps and physical maps are prepared using genetic markers and identified genes spaced at regular intervals along the

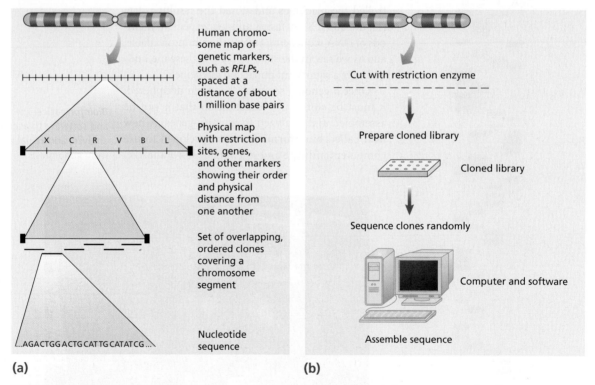

(a)   **(b)**

FIGURE 15.8 Two cloning strategies. (a) The map-based or clone-by-clone method of genome sequencing. First, genetic maps and physical maps are constructed for each chromosome. Clones from each chromosome library are organized into overlapping sets, and each clone is sequenced. The genome sequence is assembled as the clones are sequenced, using the markers from the genetic and physical maps as guides. (b) In the whole genome or shotgun method, genomic libraries are prepared using several restriction enzymes, and clones are chosen randomly for sequencing. Software programs assemble the sequence. The Human Genome Project, sponsored by the National Institutes of Health and the Department of Energy, started its work using the map-based method, and Celera Corporation's project used the whole genome, or shotgun, method.

length of each chromosome. Distances on **genetic maps** are measured in centimorgans, and on **physical maps**, distances are measured in base pairs.

Next, individual chromosomes are isolated and cloned chromosomal libraries containing large cloned DNA fragments are prepared. Several libraries are prepared, each using a different restriction enzyme, creating libraries with overlapping fragments that differ in the location of restriction-enzyme cutting sites. Using the markers and maps prepared earlier, these clones are sorted and arranged in overlapping order to cover the entire chromosome or a chromosome segment. In a third step, these large clones are cut with restriction enzymes into smaller fragments that are cloned and sequenced. The genome sequence is assembled from the overlapping clones using the genetic and physical map markers.

In whole genome sequencing (Figure 15.8b), the need for genetic and physical maps is bypassed. Instead, the genome is cut into small fragments and cloned into genomic libraries. Several libraries are constructed, each using a different restriction enzyme; clones are selected at random from each library and sequenced. Software called assembly programs are then used to organize the information into a sequenced genome. This method lends itself to automation, and large banks of sequencing machines can feed information into banks of computers, where software assembles the genomic sequence.

Initially, the government-sponsored genome project used the clone-by-clone method and by 1998 had assembled most of the maps and clones needed for sequencing. In 1999, a privately funded human genome project coordinated by Celera Corporation was announced. Celera proposed using whole genome sequencing to sequence the human genome in 18 months. This announcement set off an intense competition between the two projects to be first with the sequence. At a White House press conference in the summer of 2000, the public and private projects jointly announced that they had completed a draft of the human genome sequence, and in February 2001, they each published their results. In 2003, more sequence from the gene-coding portion of the human genome was published, mostly ending this phase of the project.

Genetic map A diagram of a chromosome showing the order of genes and the distance between them based on recombination frequencies (centimorgans).

Physical map A diagram of a chromosome showing the order of genes and the distance between them measured in base pairs.

Scientists can analyze genomic information with bioinformatics.

Scientists are now using large-scale databases to store, analyze, visualize, and share the results of their work on genomes. After a genome is sequenced, the information must be organized (called compiling the sequence) and checked for accuracy. To ensure that the sequence is error free, genomes are sequenced more than once. It is estimated that the finished human genome sequence is 99.99% accurate, with less than one error per 100,000 base pairs—a level of accuracy much higher than originally anticipated.

KEEP IN MIND

Genomics relies on interconnected databases and software to analyze sequenced genomes and identify genes.

Annotation is used to find where the genes are.

Once a genome sequence has been compiled, it is analyzed to find all the genes that encode proteins and RNA. This process is called **annotation**. Looking at a DNA sequence (Figure 15.9), it is not obvious whether or not it contains genes. If the sequence does contain one or more genes, it isn't clear where they begin and end. How, then, are genes identified in a DNA sequence?

Genes leave certain identifiable footprints that clue us in to their location in a DNA sequence. If a DNA sequence encodes a protein, its nucleotide sequence is an **open reading frame**, or **ORF**, that encodes amino acids. Promoter regions at the beginning of genes are marked by identifiable sequences (CAAT or CCAAT). Splice sites between exons and introns have a predictable sequence, as do the sites at the ends of genes where a poly-A tail is added. Computer software scans sequence data, searching for ORFs and other features of genes. Analysis of the sequence in Figure 15.9 shows a control region and three ORFs (exons) of a gene. The two unshaded regions between the exons are introns. Searching genome databases from other species, we discover that this is the sequence of the human beta globin gene.

Annotation The analysis of genomic nucleotide sequence data to identify the protein-coding genes, the nonprotein-coding genes, their regulatory sequences, and their function(s).

Open reading frame (ORF) The codons in a gene that encode the amino acids of the gene product.

FIGURE 15.9 A stretch of DNA sequence generated in a sequencing project. Analysis using gene-searching software shows several ORFs (exons, shown in *blue*), bordered by the splice junctions between introns and exons. A sequence just before the coding region marks the site at which transcription begins (*green*). Database searches show that this sequence encodes the human beta globin gene. This protein is part of hemoglobin, the oxygen-carrying protein found in red blood cells.

```
gagccacacc  ctagggttgg  ccaatctact  cccaggagca  gggagggcag  gagccagggc
tgggcataaa  agtcagggca  gagccatcta  ttgcttacat  ttgcttctga  cacaactgtg
ttcactagca  acctcaaaca  gacaccatgg  tgcacctgac  tcctgaggag  aagtctgccg
ttactgccct  gtggggcaag  gtgaacgtgg  atgaagttgg  tggtgaggcc  ctgggcaggt
tggtatcaag  gttacaagac  aggtttaagg  agaccaatag  aaactgggca  tgtggagaca
gagaagactc  ttgggtttct  gataggcact  gactctctct  gcctattggt  ctattttccc
acccttaggc  tgctggtggt  ctacccttgg  acccagaggt  tctttgagtc  ctttgggggat
ctgtccactc  ctgatgctgt  tatgggcaac  cctaaggtga  aggctcatgg  caagaaagtg
ctcggtgcct  ttagtgatgg  cctggctcac  ctggacaacc  tcaagggcac  ctttgccaca
ctgagtgagc  tgcactgtga  caagctgcac  gtggatcctg  agaacttcag  ggtgagtcta
tgggaccctt  gatgttttct  ttccccttct  tttctatggt  taagttcatg  tcataggaag
gggagaagta  acagggtaca  gtttagaatg  ggaaacagac  gaatgattgc  atcagtgtgg
aagtctcagg  atcgttttag  tttcttttat  ttgctgttca  taacaattgt  tttcttttgt
ttaattcttg  ctttcttttt  ttttcttctc  cgcaattttt  actattatac  ttaatgcctt
aacattgtgt  ataacaaaag  gaaatatctc  tgagatacat  taagtaactt  aaaaaaaaac
tttacacagt  ctgcctagta  cattactatt  tggaatatat  gtgtgcttat  ttgcatattc
ataatctccc  tactttattt  tcttttattt  ttaattgata  cataatcatt  atacatattt
atgggttaaa  gtgtaatgtt  ttaatatgtg  tacacatatt  gaccaaatca  gggtaatttt
gcatttgtaa  ttttaaaaaa  tgctttcttc  ttttaatata  cttttttgtt  tatcttattt
ctaatacttt  ccctaatctc  tttctttcag  ggcaataatg  atacaatgta  tcatgcctct
ttgcaccatt  ctaaagaata  acagtgataa  tttctggggtt  aaggcaatag  caatatttct
gcatataaat  atttctgcat  ataaattgta  actgatgtaa  gaggtttcat  attgctaata
gcagctacaa  tccagctacc  attctgcttt  tatttatgg   ttgggataag  gctggattat
tctgagtcca  agctaggccc  ttttgctaat  catgttcata  cctcttatct  tcctcccaca
gctcctgggc  aacgtgctgg  tctgtgtgct  ggcccatcac  tttggcaaag  aattcacccc
accagtgcag  gctgcctatc  agaaagtggt  ggctggtgtg  gctaatgccc  tggcccacaa
gtatcactaa  gctcgctttc  ttgctgtcca  atttctatta  aaggttcctt  tgttccctaa
gtccaactac  taaactgggg  gatattatga  agggccttga  gcatctggat  tctgcctaat
aaaaaacatt  tattttcatt  gcaatgatgt  atttaaatta  tttctgaata  ttttactaaa
```

FIGURE 15.10 The amino acid sequence derived from the DNA coding sequences in Figure 15.9. The amino acids are represented by one-letter symbols (see Table 9.2). The sequence is 146 amino acids long, and protein databases confirm that this is the sequence of human beta globin.

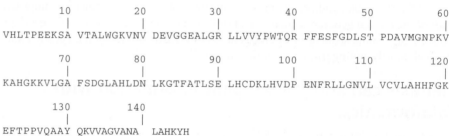

```
        10          20          30          40          50          60
        |           |           |           |           |           |
  VHLTPEEKSA  VTALWGKVNV  DEVGGEALGR  LLVVYPWTQR  FFESFGDLST  PDAVMGNPKV

        70          80          90         100         110         120
        |           |           |           |           |           |
  KAHGKKVLGA  FSDGLAHLDN  LKGTFATLSE  LHCDKLHVDP  ENFRLLGNVL  VCVLAHHFGK

       130         140
        |           |
  EFTPPVQAAY  QKVVAGVANA  LAHKYH
```

Annotation of the human genome is an ongoing process; when it is completed, we will have a final count of how many genes are in our genome—now estimated to be 20,000 to 25,000.

As genes are discovered, the function of their encoded proteins are studied.

After annotation has identified a gene, its amino acid sequence is derived (Figure 15.10) and compared with sequences in protein databases. In our example, the amino acid sequence matches that of the human beta globin gene, confirming the identity of the DNA sequence. In these cases, the match is often made to a protein from another species (see Spotlight on Our Genetic Relative). Functions have been assigned to about 60% of the genes identified in the human genome (Figure 15.11).

15.5 What Have We Learned So Far About the Human Genome?

Several general features of the human genome have been identified in analyzing the sequence information:

- Although the genome is large and contains over 3 billion nucleotides of DNA, only approximately 5% of this DNA encodes genetic information, and only about 1.1% of the genome is composed of exons, the nucleotides that specify amino acids in a protein.

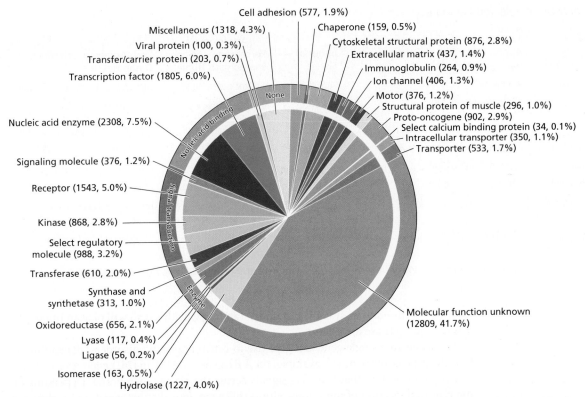

Cell adhesion (577, 1.9%)
Miscellaneous (1318, 4.3%)
Chaperone (159, 0.5%)
Viral protein (100, 0.3%)
Cytoskeletal structural protein (876, 2.8%)
Transfer/carrier protein (203, 0.7%)
Extracellular matrix (437, 1.4%)
Transcription factor (1805, 6.0%)
Immunoglobulin (264, 0.9%)
Ion channel (406, 1.3%)
None
Motor (376, 1.2%)
Structural protein of muscle (296, 1.0%)
Proto-oncogene (902, 2.9%)
Nucleic acid enzyme (2308, 7.5%)
Select calcium binding protein (34, 0.1%)
Intracellular transporter (350, 1.1%)
Transporter (533, 1.7%)
Signaling molecule (376, 1.2%)
Receptor (1543, 5.0%)
Kinase (868, 2.8%)
Select regulatory molecule (988, 3.2%)
Transferase (610, 2.0%)
Synthase and synthetase (313, 1.0%)
Oxidoreductase (656, 2.1%)
Molecular function unknown (12809, 41.7%)
Lyase (117, 0.4%)
Ligase (56, 0.2%)
Isomerase (163, 0.5%)
Hydrolase (1227, 4.0%)

FIGURE 15.11 A preliminary functional assignment for over 26,000 putative genes in the human genome. Just over 12,000 of these genes (41%) have no known function and represent the largest class of genes identified to date, emphasizing the work that needs to be completed before we fully understand our genome. The other classes of genes were assigned because they were known from other experiments or are similar to proteins of known function from other organisms. Among the most common genes are those involved in nucleic acid (DNA and RNA) metabolism (7.5% of all the identified genes).

The rest of the genome (98.9%) does not encode proteins, and aside from introns, most is spacer DNA. About half of our non-coding DNA is composed of sequences that are repeated thousands of times.

- Gene-rich clusters on all 24 chromosomes are separated from each other by long stretches of gene-poor regions. The gene-poor regions correspond to the banded regions of stained chromosomes (see Chapter 6 for a discussion of banding and karyotypes).

- Not all the genes in the human genome have been identified, but humans have 20,000 to 25,000 genes—far fewer than the predicted number of 80,000 to 100,000.

- There are many more different proteins in the human body than there are genes. This discrepancy is explained by the fact that the mRNAs from over 90% of our genes are processed in several different ways so that a genome with 20,000 to 25,000 genes can produce up to 300,000 or more different proteins.

- The human genome is very similar to those of other eukaryotes. We share about half our genes with the fruit fly *Drosophila* (Table 15.4), and more than 98% of our genes are shared with the chimpanzee. We will explore the similarities and differences between the genomes of chimps and humans in Chapter 19.

KEEP IN MIND

The human genome has a surprisingly small number of genes and produces a surprisingly large number of proteins, using a number of different mechanisms.

New disease-related types of mutations have been discovered.

We have already discussed trinucleotide expansion (Chapter 11), a mutation mechanism identified through genomics. Repeat expansion has an important role in heritable

Table 15.4 Comparison of Selected Genomes

Organism	Approximate Size of Genome (Date Completed)	Number of Genes	Approximate Percentage of Genes Shared with Humans	Web Access to Genome Databases
Bacterium (*Escherichia coli*)	4.6 million bp (1997)	4,300	Not determined	http://www.genome.wisc.edu/
Fruit fly (*Drosophila melanogaster*)	165 million bp (2000)	~13,600	50%	http://www.fruitfly.org/ sequence/index.html http://Flybase.bio.indiana.edu/
Humans (*Homo sapiens*)	3,200 million bp (February 2003)	20,000–25,000	100%	http://www.ornl.gov/hgmis/
Mouse (*Mus musculus*)	3,000 million bp (2002)	~30,000	~90%	http://www.informatics.jax.org/
Plant (*Arabidopsis thaliana*)	120 million bp (2000)	~25,000	Not determined	http://www.arabidopsis.org/
Roundworm (*Caenorhabditis elegans*)	97 million bp (1998)	19,000	40%	http://www.genome.wustl.edu/ projects/celegens
Yeast (*Saccharomyces cerevisiae*)	12 million bp (1996)	~6,000	31%	http://www.yeastgenomes.org

Source: Howard Hughes Medical Institute (2001), *The Genes We Share with Yeast, Flies, Worms, and Mice: New Clues to Human Health and Disease.*

disorders of the nervous system, and understanding why this is so will play an important role in treating those disorders.

Mutations in a single gene can give rise to different genetic disorders, depending on how the gene is affected. For example, the *RET* gene (OMIM 164761) encodes a cell-surface receptor protein that transfers signals across the plasma membrane. Depending on the type and location of mutations within this gene, four distinct genetic disorders can result: two types of cancers called multiple endocrine neoplasias (OMIM 171400 and OMIM 162300); another cancer called familial medullary thyroid carcinoma (OMIM 155240); and Hirschsprung disease (OMIM 142623), a disorder in which parts of the large intestine are not connected to the nervous system.

Another significant discovery is that some mutations in DNA repair genes can destabilize distant regions of the genome and make them susceptible to more mutations, often resulting in cancer (these mutations were discussed in Chapter 12). These and similar discoveries have already had an impact on the diagnosis and treatment of and genetic counseling for several groups of genetic disorders.

Nucleotide variation in genomes is common.

The human genome contains several million locations where single nucleotide differences occur in humans. These differences, called single nucleotide polymorphisms (SNPs), were introduced in Chapter 1. Most SNPs in a population are descended from an ancestral mutation that occurred on a specific chromosome and spread through the population. A new SNP becomes associated with other SNPs already on the chromosome in which the mutation occurred. SNPs located close together on a chromosome tend to be inherited together, and this set of SNPs is called a **haplotype**. Although there may be thousands of SNPs in a haplotype, it can be identified by looking for only a few unique combinations of these. SNPs used to identify a haplotype are called tag SNPs. Although there may be between 10 and 12 million SNPs in the human genome, as few as 100,000 tag SNPs can be used to identify most haplotypes.

Because SNP haplotypes are widely distributed in the genome, they can be used as markers in linkage studies to identify disease-causing alleles. If an SNP haplotype is linked to a mutant disease allele, people with the genetic disorder will tend to have a different haplotype than unaffected individuals. Comparison of haplotype differences between affected and unaffected individuals can be used to locate and identify the mutant allele associated with the disorder.

An international effort called the HapMap Project is cataloging and mapping SNPs and SNP haplotypes for use in family and population studies; to date, the project has mapped over 3 million SNPs. Most common diseases are complex (see Chapter 5) and are caused by several genes and environmental interactions, making the genes for these

Haplotype A set of genetic markers located close together on a single chromosome or chromosome region.

disorders very difficult to identify. Even though the HapMap Project is not completed, it has already led to the identification of over 100 genes associated with almost 40 common diseases, including diabetes, cancer, cardiovascular disease, and mental illness. The continued development of this and similar genomic databases will have a significant impact on the diagnosis and treatment of many diseases.

SNPs are variations in single nucleotides, but our genome also contains much larger variations, called **copy number variations (CNVs)**. These variations include duplications or deletions that extend for thousands of base pairs. CNVs can affect gene expression in several ways: they may contain genes and increase or decrease the dosage of such genes, CNVs may insert into coding sequences, creating mutations, or their presence may interfere with long-range gene regulation. CNVs are involved with several complex diseases, including osteoporosis, mental retardation, and prostate cancer. For example, the copy number of a gene called *UGT2B17* (OMIM 601903) varies from zero to two in the human population. The protein encoded by this gene is an enzyme involved in male hormone metabolism, and the deletion allele has been shown to be a risk factor for prostate cancer. Although recently discovered, it is apparent that CNVs may be associated with a significant number of complex genetic disorders.

Copy number variation (CNV) A DNA segment at least 1,000 base pairs long with a variable copy number in the genome.

15.6 Using Genomics to Study a Human Genetic Disorder

In studying any genetic disorder, several important questions must be answered:
- Where is the gene located?
- What is the normal function of the protein encoded by this gene?
- How does the mutant gene or protein produce the disease phenotype?

For some genes, such as the CF (cystic fibrosis) gene, these questions were easy to answer. The CF gene was isolated by positional cloning and mapped to chromosome 7 (Figure 15.12). Its amino acid sequence closely resembled the sequence of membrane proteins that regulated ion transport. Because CF patients have problems with chloride ion flow, the protein function in normal individuals and the problem with chloride ions in CF could be pinpointed.

In other cases, these questions are not as easy to answer. If the amino acid sequence does not match that of any known protein, how can scientists understand what the normal protein does and how a mutant form of the protein causes disease? This is not a hypothetical question; more than half the genes identified by the Human Genome Project have no known function. Let's look at how this problem was solved in the case of one genetic disorder.

Friedreich ataxia (*FRDA*; OMIM 229300) is a progressive and fatal brain disease inherited as an autosomal recessive trait. It occurs with a frequency of 1 in 50,000 births and is a disease of young people, with symptoms appearing between puberty and the age of 25. Affected individuals develop shaky, uncoordinated leg movements and have trouble walking. Later, they may develop loss of sensation in their arms and legs and can have heart disease and diabetes, resulting in early death.

Positional cloning mapped the *FRDA* gene to chromosome 9, and it was isolated, cloned, and sequenced. The mutant allele has an expanded GAA trinucleotide repeat (see Chapter 11 for a discussion of these disorders) that results in abnormally low levels of frataxin, the encoded gene product. Protein databases were searched for similar proteins with a known function in other species. Here is where things quickly reached a dead end; there were no matching proteins in any database. Only matches to proteins of unknown function in a fungus and a nematode were found.

Researchers then decided to rescan protein databases, using only portions of the frataxin amino acid sequence instead of the complete sequence. They hoped to find matches to key portions of the frataxin protein. Using a partial amino acid sequence of frataxin, a match was found with a protein present in several bacterial species.

Friedreich ataxia A progressive and fatal neurodegenerative disorder inherited as an autosomal recessive trait with symptoms appearing between puberty and the age of 25.

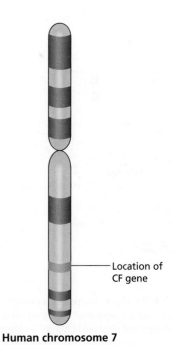

Location of CF gene

Human chromosome 7

FIGURE 15.12 The gene for cystic fibrosis was mapped to the long arm of chromosome 7 by its position relative to known DNA markers.

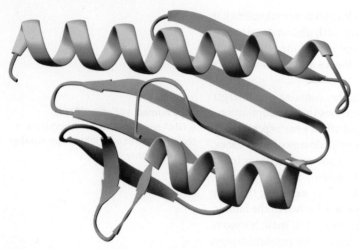

FIGURE 15.13 A three-dimensional model showing the molecular structure of the human frataxin protein, necessary for normal mitochondrial function. A decreased production or function of this protein causes the autosomal recessive disorder Friedreich ataxia. This disorder is one of the first whose molecular basis was uncovered by using analysis of genomic information.

Proteome The set of proteins present in a cell at a specific time under a specific set of conditions.

Proteomics The study of the proteome, the set of expressed proteins present in a cell.

These species were all purple bacteria, the closest living relatives to ancient bacterial species that evolved into mitochondria. From this result, it seemed possible that frataxin might be a mitochondrial protein that plays a role in energy production.

Although this seemed like a long shot, other researchers investigating the cellular location of the frataxin protein discovered that the protein *is* located in mitochondria, and they worked out its three-dimensional structure (Figure 15.13). However, the exact role of frataxin in mitochondrial energy production is still unknown. The protein may protect the cell from free radicals, which are toxic byproducts of energy production. Accumulation of free radicals can lead to cellular malfunction and death, causing the long-term effects of this disorder. The low levels of frataxin also appear to cause deficiencies in iron-containing mitochondrial proteins that are important in energy production.

Experimental treatment with idebenone—a drug that is structurally similar to a molecule that plays a key role in energy production—has shown some promise for treating the symptoms of Friedreich ataxia, but at this time, there is no cure for this genetic disorder.

Genomics played a key role in mapping the gene, identifying the nature of the mutation, establishing its location within the cell, and identifying the structure and some possible functions of the protein. With this information in hand, efforts are now focused on developing new therapies to treat FRDA. This example illustrates the power of genomics and bioinformatics to help unravel the mysteries of human genetic disorders.

15.7 Proteomics Is an Extension of Genomics

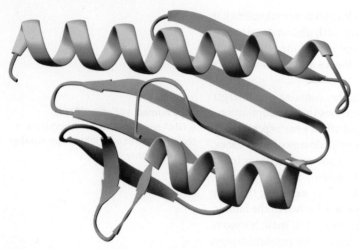

FIGURE 15.14 The proteins expressed in a cell are separated by size and electric charge and displayed on a gel. The spots on this gel represent the proteins in a gene expression profile.

Now that genomics can provide information about the number, location, and organization of genes, the next step is to understand the functions of their encoded proteins. Proteins carry out most of the tasks in the cell and are directly or indirectly responsible for the phenotype of an organism. As we saw in Chapter 9, alternative splicing and protein modifications amplify the number of different proteins that can be produced, and there are many more proteins than genes. As genes are identified and classified into functional groups (see the listing in Figure 15.11), understanding gene function involves more than just identifying the protein that a gene encodes. Proteins work in a cellular environment with many other proteins, and protein–protein interactions and modifications and the chemical linkage of proteins to other molecules such as lipids or carbohydrates are important parts of protein function. Geneticists and molecular biologists are now turning their attention to the **proteome**, the set of all proteins that can be encoded by a genome; this has given rise to a new field called **proteomics**.

To study cellular proteins, proteome profiles are constructed by separating all the proteins in a cell type by their mass and electric charge—some amino acids carry an electrical charge (Figure 15.14). The next step is to identify the individual proteins and gather information about their function. Several methods are available to identify proteins isolated from cells and to analyze functions, including protein–protein interactions.

Proteomics and gene expression profiles provide information for the diagnosis and treatment of disease and are becoming a routine part of medical care. For example, some forms of breast cancer have an expression profile that indicates that they will spread rapidly and have a high fatality rate, whereas other forms have profiles associated with slow growth and high survival rates. With this molecular information in hand, specific therapies can be designed for each form of this cancer.

Proteomics is a rapidly developing field that has several crucial roles:

- Understanding gene function and its changing role in development and aging.
- Identifying proteins that are markers (called biomarkers) for diseases. These proteins can be used in developing diagnostic tests.
- Finding proteins that are targets for the development of drugs to treat diseases and genetic disorders.

Several recent breakthroughs in proteomic techniques, coupled with new developments in information science, are making this new field an essential partner with genomics in medicine.

15.8 Ethical Concerns About Human Genomics

The original planners of the HGP realized that the information generated by a genome project would have many uses, some of which might pose problems (see Exploring Genetics: Who Owns Your Genome?). To study these issues, a program called ELSI (Ethical, Legal, and Social Implications) was established.

ELSI focused on several issues, including:

- Privacy and confidentiality of genetic information.
- Fairness in the use of genetic information by insurers, employers, and others.
- Discrimination caused by someone's genetic status.
- Use of genetic information in reproductive decisions.

The attention given to these potential problems by ELSI and other forums helped formulate and pass legislation about the use of genetic information. The Genetic Information Nondiscrimination Act (GINA) was signed into law in 2008. This federal law prevents health insurance companies and employers from discriminating on the basis of genetic information and prohibits them from asking for or requiring a genetic test.

Although GINA represents a significant step forward, other issues remain unresolved. Should genetic testing be done even if there is no treatment available? Should we test children for diseases that will not develop until adulthood? Will patenting DNA sequences deter the open development of products and make the costs of some genetic tests prohibitive? Even as these issues are being addressed, new areas of concern are surfacing. Advertising campaigns on the internet, on television, and in newspapers are marketing genetic tests directly to consumers. Tests for breast cancer, cystic fibrosis, hemochromatosis, and other genetic disorders are offered. Although this raises public awareness and has increased requests for testing, research has shown that the ads do not convey the tests' ability to accurately detect or predict a genetic disorder. The ads often do not recommend that tests be done in consultation with a health care provider or genetic counselor and may offer tests that ignore ethics statements by professional societies. These developments indicate the need for further discussion about how to inform and educate the public about the way genetic products and services are marketed.

KEEP IN MIND

Genomics is affecting basic research in biology and generating new methods of diagnosis and treatment of disease.

Who Owns Your Genome?

John Moore, an engineer working on the Alaska oil pipeline, was diagnosed in the mid-1970s with a rare and fatal form of cancer known as hairy cell leukemia. This disease causes overproduction of one type of white blood cell known as a T lymphocyte. These cells secrete a growth factor that activates other blood cells that kill cancer cells. Moore went to the UCLA Medical Center for treatment and was examined by Dr. Golde, who recommended that Moore's spleen be removed in an attempt to slow down or stop the cancer. For the next 8 years, John Moore returned to UCLA for checkups. Unknown to Moore, Dr. Golde and his research assistant applied for and received a patent on a cell line and products of that cell line derived from Moore's spleen. The cell line, named Mo, produced a protein that stimulates the growth of two types of blood cells that are important in identifying and killing cancer cells. Arrangements were made with Genetics Institute, a small start-up company, and then Sandoz Pharmaceuticals, to develop the cell line and produce the growth-stimulating protein. Moore filed suit to claim ownership of his cells and asked for a share of the profits derived from the sale of the cells or products from the cells. Eventually, the case went through three courts, and in July 1990—11 years after the case began—the California Supreme Court ruled that patients such as John Moore do not have property rights over any cells or tissues removed from their bodies that are used later to develop drugs or other commercial products.

This case was the first in the nation to establish a legal precedent for the commercial development and use of human tissue. The National Organ Transplant Act of 1984 prevents the sale of human organs. Current laws allow the sale of human tissues and cells but do not define ownership interests of donors. Questions originally raised in the Moore case remain largely unresolved in laws and public policy. These questions are being raised in many other cases as well. Who owns fetal and adult stem-cell lines established from donors, and who has ownership of and a commercial interest in diagnostic tests developed through cell and tissue donations by affected individuals? Who benefits from new genetic technologies based on molecules, cells, or tissues contributed by patients? Are these financial, medical, and ethical benefits being distributed fairly? What can be done to ensure that risks and benefits are distributed in an equitable manner?

Gaps between technology, laws, and public policy developed with the advent of recombinant DNA technology in the 1970s, and in the intervening decades, those gaps have not been closed. These controversies are likely to continue as new developments in technology continue to outpace social consensus about their use.

Genetics in Practice

Genetics in Practice case studies are critical-thinking exercises that allow you to apply your new knowledge of human genetics to real-life problems. You can find these case studies and links to relevant websites at *www.cengage.com/biology/cummings*

CASE 1

James sees an advertisement in a magazine for an at-home genetic test that promises to deliver personalized nutritional advice based on an individual's genetic profile. The company can test for genetic variations, the advertisement states, that predispose individuals to developing health conditions such as heart disease and bone loss or that affect how they metabolize certain foods. If such variations are detected, the company can provide specific nutritional advice that will help counteract their effects. Always keen to take any steps available to ensure the best possible health for their family, James and his wife (Sally) decide that they both should be tested, as should their 11-year-old daughter (Patty). They order three kits and wait for them to arrive.

Once the kits arrive, the family members use cotton swabs to take cell samples from each of their cheeks and place the swabs in three individually labeled envelopes. They mail the envelopes back to the company, along with completed questionnaires regarding their diets. Four weeks later, they receive three individual reports detailing the results of the tests and that provide extensive guidelines about what foods they should eat. Among the results is the finding that James has a particular allele in a gene that may make him vulnerable to the presence of free radicals in his cells. The report suggests that he increase his intake of antioxidants, such as vitamins C and E, and highlights a number of foods that are rich in those vitamins. The tests also show that Sally has several genetic variations that indicate that she may be at risk for elevated bone loss. The report recommends that she try to minimize this possibility by increasing her intake of calcium and vitamin D and lists a number of foods she could emphasize in her diet. Finally, the report shows that Patty has a genetic variation that may mean that she has a lowered ability to metabolize saturated fats, putting her at risk for developing heart disease. The report points to ways in which she can lower her intake of saturated fats and lists various types of foods that would be beneficial for her.

A number of companies now offer genetic-testing services, promising to deliver personalized nutritional or other advice based on people's genetic profiles. Generally, these tests fall into two different categories, with individual companies offering unique combinations of the two. The first type of test detects alleles of known genes that encode proteins that play an established role in, for example, counteracting free radicals in cells or in building up bone. In such cases, it is easy to see why individuals carrying alleles that may encode proteins with lower levels of activity may be more vulnerable to free radicals or more susceptible to bone loss.

A second type of test examines genetic variations that may have no clear biological significance (i.e., they may not occur within a gene or may not have a detectable effect on gene activity) but have been shown to have a statistically significant correlation with a disease or a particular physiological condition. For example, a variation may frequently be detected in individuals with heart disease even though the reason for the correlation between the variation and the disease may be entirely mysterious.

1. Do James and Sally have any guarantees that the tests and recommendations are scientifically valid?

2. Do you think that companies should be allowed to market such tests directly to the public, or do you believe that only a physician should be able to order them?

3. What kinds of regulations, if any, should be in place to ensure that the results of these tests are not abused?

4. Do you think parents should be able to order such a test for their children? What if the test indicates that a child is at risk for a disease for which there is no known cure?

CASE 2

Both the Human Genome Project (HGP) and Celera's genome sequencing project were faced with an interesting dilemma: Whose DNA should be sequenced? Whose genome should serve as a "representative" human genome? The HGP and Celera answered the question in different ways. The HGP started with a collection of DNA samples from a large number of individuals and then randomly selected a small number of samples that were used for sequencing. Celera, in contrast, used a mixture of DNA samples from several individuals of diverse ethnicity (including, as was later revealed, the DNA of Celera's founder, Craig Venter). In both cases, because only a handful of genomes were used in the sequencing, the final sequences represented only a small fraction of the total variation present in humanity.

In 1991, as the genome project got under way, a group of anthropologists and geneticists proposed carrying out an ambitious project to address this issue—namely, to study the genetic variation of all human populations. The Human Genome Diversity Project, as it came to be called, proposed to obtain blood samples from a wide variety of peoples throughout the world and to sequence their genomes in conjunction with the official human genome project. According to the organizers of the project, the information obtained in the project would shed light on the history of different human populations, illuminate the biological relationships between populations, and probably be useful for understanding the causes and genetic features of various diseases.

Although the proposal won some initial support, it soon ran into several major—and ultimately insurmountable—obstacles. Certain scientists questioned its scientific rationale, casting doubt on whether the project would yield important information about human history and disease. But the greatest difficulties came from many of the indigenous populations that the organizers hoped would be participating in the project. The subjects viewed the project suspiciously, raising questions about its true purpose and value: Who would "own" the genetic information that was generated during the project? Would it be patented? Who would benefit from drugs or other products developed by using the information? What other purposes could the genetic information be used for (either good or bad)? Wouldn't the millions of dollars that would be spent on the project be better used trying to help indigenous populations directly?

Despite the attempts of project organizers to address those questions, the critics never relented, and the project was essentially abandoned in the late 1990s. Since then, other alternative approaches have been initiated to try to address issues related to human genetic diversity, such as the HapMap project that officially got under way in 2002.

1. Is it ethical for scientists from developed countries to involve indigenous populations from less developed parts of the world in their research studies, or should they limit their studies to populations living in their own countries?

2. Do you think such a project would be likely to help indigenous populations? Do you think the objections to the project were reasonable?

3. If a scientist makes a medically important discovery using DNA obtained from an indigenous group, should the discovery be patentable? How should any benefits arising from such a discovery be shared?

Summary

15.1 Genome Sequencing Is an Extension of Genetic Mapping

- Because mutant genes are the basis of genetic disorders, mapping helps us identify genes that cause disease and is the first step in developing diagnostic tests and treatments for these disorders.

15.2 Genome Projects Are an Outgrowth of Recombinant DNA Technology

- Instead of finding and mapping markers and disease genes one by one, scientists organized the Human Genome Project (HGP) to sequence all the DNA in the human genome, identify and map the thousands of genes to the 24 chromosomes we carry, and assign a function to all the genes in our genome.

15.3 Genome Projects Have Created New Scientific Fields

- The sheer size of the human genome required the development of new technologies, including automated methods of DNA sequencing and advances in software to collect, analyze, and store the information derived from genome sequencing. The study of genomes by these methods is called genomics. Bioinformatics is the use of software, computational tools, and databases to store, organize, analyze, and visualize genomic information.

15.4 Genomics: Sequencing, Identifying, and Mapping Genes

- Geneticists developed two strategies for genome sequencing. One method, called the map-based method, uses clones from a genomic library that have been arranged to cover an entire chromosome. After the order of the clones is known, they are sequenced. The second method, called whole genome sequencing, randomly selects clones from a genomic library and sequences them. Once the sequence of the clones is known, assembly software organizes them into the genomic sequence.

15.5 What Have We Learned So Far About the Human Genome?

- The Human Genome Project has provided several unexpected findings about our genome. There are about 20,000 to 25,000 genes—far fewer than the 80,000 to 100,000 genes that were predicted.

- Only about 5% of the genome actually encodes proteins, and about half the genome consists of non-gene-containing DNA sequences that are present in hundreds of thousands of copies. What this DNA does is still a mystery.

- Genes are organized in clusters, separated by long stretches of DNA that does not code for any genes.

15.6 Using Genomics to Study a Human Genetic Disorder

- Analysis of genome sequence information has led to the identification of disease-causing genes, including one for Friedreich ataxia, a progressive and fatal neurodegenerative disorder.

15.7 Proteomics Is an Extension of Genomics

- Proteomics is the study of the structure and function of proteins. This new field is playing an important role in analyzing proteins to develop new diagnostic tests for genetic disorders and to identify targets for drug development.

15.8 Ethical Concerns About Human Genomics

- To deal with the impact that genomic information would have on society, the HGP set up the ELSI (Ethical, Legal, and Social Implications) program to ensure that genetic information would be safeguarded and not used in discriminatory ways. ELSI works to develop policy guidelines for the use of genomic information.

Questions and Problems

Genome Sequencing Is an Extension of Genetic Mapping

1. The gene controlling ABO blood type and the gene underlying nail-patella syndrome are said to show *linkage*. What does that mean in terms of their relative locations in the genome? What does it mean in terms of how the two traits are inherited with respect to each other?

2. Hemophilia and color blindness are both recessive conditions caused by genes on the X chromosome. To calculate the recombination frequency between the two genes, you draw a large number of pedigrees that include grandfathers with both hemophilia and color blindness, their daughters (who presumably have one chromosome with two normal alleles and one chromosome with two mutant alleles), and the daughters' sons. Analyzing all the pedigrees together shows that 25 grandsons have both color blindness and hemophilia, 24 have neither of the traits, 1 has color blindness only, and 1 has hemophilia only. How many centimorgans (map units) separate the hemophilia locus from the locus for color blindness?

3. Before the advent of recombinant DNA technology, why was it so difficult for geneticists to map human genes by using pedigrees? How did recombinant DNA technology help move things forward?

Genome Projects Are an Outgrowth of Recombinant DNA Technology

4. In what years did the publicly funded Human Genome Project begin and end? What were the scientific goals of the Human Genome Project?

Genome Projects Have Created New Scientific Fields

5. How many nucleotides does the human genome contain?
6. Which of the following best describes the process of DNA sequencing?
   a. DNA is separated on a gel, and the different bands are labeled with fluorescent nucleotides and scanned with a laser.
   b. A laser is used to fluorescently label the nucleotides present within the DNA, the DNA is run on a gel, and then the DNA is broken into fragments.
   c. Nucleotides are scanned with a laser and incorporated into the DNA that has been separated on a gel, and then the DNA is amplified with PCR.
   d. Fragments of DNA are produced in a reaction that labels them with any of four different fluorescent dyes, and the fragments then are run on a gel and scanned with a laser.
   e. DNA is broken down into its constituent nucleotides, and the nucleotides are then run on a gel and purified with a laser.

7. Which of the following is NOT an activity carried out in the field of bioinformatics?
   a. collecting and storing DNA sequence information produced by various genome sequencing projects
   b. analyzing genomic sequences to determine the location of genes
   c. determining the three-dimensional structure of proteins
   d. comparing genomes of different species
   e. none of the above

Genomics: Sequencing, Identifying, and Mapping Genes

8. How did the sequencing strategy used by the Human Genome Project differ from that used by Celera Corporation?
9. Once an organism's genome has been sequenced, how do geneticists usually go about trying to pinpoint the locations of the genes?

What Have We Learned So Far About the Human Genome?

10. What percentage of the DNA in the genome actually corresponds to genes? How much is actually protein-coding exons? What makes up the rest?
11. When the human genome sequence was finally completed, scientists were surprised to discover that the genome contains far fewer genes than expected. How many genes are present in the human genome? Scientists have also found that there are many more different kinds of proteins in human cells than there are different genes in the genome. How can this be explained?
12. One unexpected result of the sequencing of the human genome was the finding that mutations in a single gene can be responsible for multiple distinct disorders. For example, mutations in the *RET* gene can cause two different types of multiple endocrine neoplasias, familial medullary thyroid carcinoma, and Hirschsprung disease. How do you think mutations in a single gene can have such diverse effects?

Using Genomics to Study a Human Genetic Disorder

13. Sequence comparison studies revealed that the product of the CF (cystic fibrosis) gene has a strong similarity to proteins known to be involved in
   a. transcription.
   b. translation.
   c. transport of ions across the cell membrane.
   d. mRNA splicing.
   e. movement of proteins across the Golgi membrane.
14. You join a laboratory that studies a rare genetic disorder that causes affected individuals to have unusually fast-growing, bright green hair. You are joining the lab at a fortuitous

moment, as the gene causing the disorder has just been cloned. Despite this breakthrough, however, it is still unclear what the function of the gene is, and the lab director asks you for suggestions about how to go about trying to determine this. What do you recommend?

Proteomics Is an Extension of Genomics

15. How does proteomics differ from genomics? What kinds of information can proteomics provide that is not available from genomics studies?

Ethical Concerns About Human Genomics

16. How did the Human Genome Project attempt to deal with the social and ethical issues that were bound to arise from the sequencing of the human genome?

Internet Activities

Internet Activities are critical-thinking exercises using the resources of the World Wide Web to enhance the principles and issues covered in this chapter. For a full set of links and questions investigating the topics described below, visit *www.cengage.com/biology/cummings*

1. *Comparative Genomics.* Access the Department of Energy's *Genome* site. Click on the link titled "Beyond HGP." Scroll down and click on the "Functional and Comparative Genomics Fact Sheet." Read and answer the questions about functional genomics and comparative genomics. Pay close attention to the questions about model organisms and the comparison between humans and mice.

2. *Genomic Analysis.* There are many sites covering the analysis of genomic sequence data. One of the best introductions to

this topic is the *2Can Bioinformatics Educational Resources* page. Access this site and select the tutorial on "Nucleotide Analysis." Follow the directions for checking two unknown sequences to see if they contain vector sequences mixed in with the genomic sequence. On subsequent pages, you can see how well vector sequences detected in your search match known vector sequences already in databases, and you can read about the sequence information stored in genomic databases.

HOW WOULD
YOU VOTE NOW?

As a gift for your twenty-first birthday, your parents announce that they will pay to have your exome sequenced. This gift will allow you to have firsthand information about any genetic disorders you carry in the heterozygous condition and may pass on to your children, as well as define any genetic risk factors for adult-onset disorders. You can use this information to make reproductive decisions and to adjust your lifestyle early enough to reduce your risk of developing adult-onset disorders. The sequence will not be stored in any database and will be given to you on compact discs for you to have read and interpreted in the health care setting of your choice. Now that you know more about the impact of genomics, would you accept your parents' gift and have your genome sequenced? Visit the *Human Heredity* companion website at *www.cengage.com/biology/cummings* to find out more on the issue; then cast your vote online.

16 Reproductive Technology, Genetic Testing, and Gene Therapy

At 11:47 p.m. on July 25, 1978, a 5-pound, 12-ounce baby girl named Louise Brown was born in Oldham, England. Unlike the millions of other children born that year, her birth was an international media event, and she was called "the baby of the century" and a "miracle baby." Although some praised her birth as a medical miracle, others spoke of it as the beginning of a darker era of genetically engineered children and even human-animal hybrids raised in artificial wombs. Why all the fuss? Louise was the first human born after *in vitro* fertilization (IVF), a procedure in which an egg is fertilized outside the body (*in vitro* literally means "in glass") and the embryo is implanted into the uterus to complete its growth.

The development of IVF by Patrick Steptoe and Robert Edwards was a long, slow process. After 9 years and more than 80 failed attempts, Steptoe repeated his procedure one more time in November 1977 by recovering an egg from Lesley Brown. To do this, he inserted a laparoscope through a small abdominal incision to examine the ovary and then made another small incision to remove an egg. He gave it to Edwards, who put it in a sterile glass dish and mixed it with semen from Lesley's husband. The dish containing the fertilized egg in a solution of nutrients was placed in an incubator for 21 days, after which the developing embryo was implanted into Lesley Brown's uterus through a tube inserted into the vagina. For the first time, the procedure was a success, and Louise was born on July 25th of the following year. Louise is now married and the mother of a boy, conceived naturally, born in 2006. In the United States, IVF is now a routine procedure available at over 350 clinics, and to date more than 45,000 babies have started life in a glass dish.

IVF was an important step in the development of methods, collectively called assisted reproductive technologies (ART), used to help infertile couples. Some of these methods are outlined in this chapter. The use of ART coupled with recombinant DNA techniques has created a powerful new technology for reproduction, sex selection, the diagnosis of genetic disorders, and gene therapy.

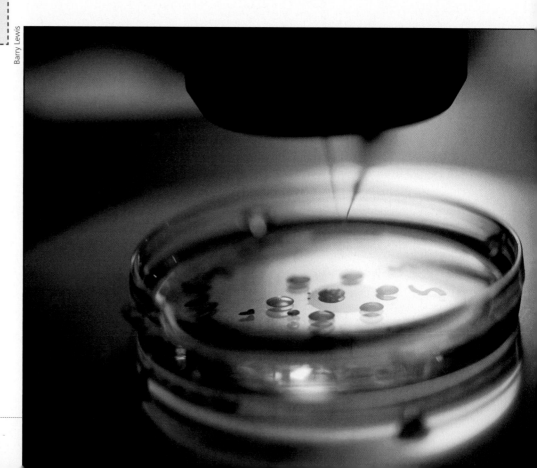

Barry Lewis

Each drop of fluid in this Petri dish contains a single human egg ready for fertilization.

16.1 Infertility Is a Common Problem

Successful reproduction depends on a small number of important factors: healthy gametes (egg and sperm), a place for fertilization to occur (oviduct), and a place for the embryo to grow and develop (the uterus). If any of these components are missing or malfunctioning, infertility may result. In the United States, the number of infertile couples has increased over the past 10 years, and at present, about one in six couples are infertile. Physical or physiological problems can prevent the production of normal sperm or eggs, inhibit fertilization, or hinder implantation of an embryo in the uterus. In about 40% of all cases, the woman is infertile; in another 40% of cases, the male is infertile; and in the remaining 20% of cases, the cause is unexplained or both partners are infertile. Infertility in women becomes more common with increasing age (Figure 16.1); up to one-third of couples in their late 30s are infertile.

HOW WOULD YOU VOTE?

Surplus embryos from IVF are routinely stored in liquid nitrogen. Some are used in subsequent attempts at pregnancy if implantation fails; others are stored in case the couple wants another child at a future date. These embryos can have several fates: They can be stored for years and then discarded if not needed, donated for use in stem-cell research, or donated to other couples. Sweden and Great Britain limit the time embryos can be stored before destruction, but little is known about the extent of unrepaired DNA damage in old embryos. If you were having IVF, what would you do with extra embryos? Visit the *Human Heredity* companion website at *www.cengage.com/biology/cummings* to find out more on the issue; then cast your vote online.

KEEP IN MIND AS YOU READ

- Infertility affects 13% of all couples. Most can be helped by assisted reproductive technologies (ART).

- Gene therapy has not fulfilled its promise of treating genetic disorders.

- Gene therapy has also experienced setbacks and restarts.

- Genetic counseling educates individuals and families about genetic disorders and helps them make decisions about reproductive choices.

Infertility is a complex problem.

Infertility problems fall into two classes: primary and secondary. Infertile couples that have not yet had a child have primary infertility. If a couple has had one child and has trouble conceiving a second, they have secondary infertility; some 3.3 million couples have this condition. The causes are unknown, although age may be a factor in some cases. Our discussion will focus on primary infertility.

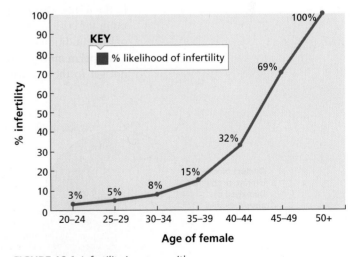

FIGURE 16.1 Infertility increases with age.

355

Infertility in women has many causes.

Earlier, we listed some components of successful reproduction. Three of these are associated with the female reproductive system: the egg, the oviduct, and the uterus. Problems with any of these components can lead to infertility. Figure 16.2 lists causes of infertility in women. One of the most common problems is hormone imbalance that causes ovulation problems. Release of an egg from the ovary (ovulation) depends on the levels of female hormones and their interactions. Several conditions can prevent or interfere with ovulation:

- **Hormone levels.** If estrogen levels are too low, or no estrogen is present, ovulation will not occur. Normally, during a monthly cycle, estrogen levels rise to a peak, trigger ovulation, and stimulate production of another hormone, called luteinizing hormone (LH). Hormonal problems occur in about 50% of all cases in which ovulation does not occur.
- **Ovarian problems.** Damage to the ovaries caused by surgery, inflammation, autoimmune disease, infection, or the presence of ovarian cysts can prevent egg maturation or release. If ovaries have been removed or are not present because of a developmental accident or a genetic disorder, infertility results.
- **Oviduct and uterine problems.** Oviduct blockage occurs in about 15% of all cases of female infertility. Sexually transmitted diseases (STDs), viral infections, and several other agents can cause inflammation, scarring, and blockage of the oviduct. Even seemingly unrelated conditions such as appendicitis or a bowel problem called colitis can cause abdominal inflammation that blocks oviducts. A condition called endometriosis, a problem with the inner lining of the uterus, can also cause infertility. If the endometrium is not formed properly during each menstrual cycle, an embryo may not be able to implant, resulting in a miscarriage.

Infertility in men involves sperm defects.

In males, problems with sperm formation or the structure and function of mature sperm are the main causes of infertility (Figure 16.3). These include:

- **Low sperm count.** Too few sperm (less that 20 million per milliliter of ejaculate) is a condition called oligospermia. Low sperm count can be caused by many factors, including environmental exposure to chemicals or radiation, use of recreational drugs, alcohol consumption, or even underwear that is too tight. Other factors including obesity or hormonal imbalances also contribute to low sperm count. The most common cause of low sperm count is a condition called varicocele, or enlarged veins in the testes. These enlarged veins raise the temperature in the testes above the optimum for critical events in sperm formation, leading to infertility. Surgical correction of this condition often leads to fertility.
- **Low sperm motility.** If sperm move too slowly through the cervical mucus, they cannot get to the egg while it is in the oviduct for fertilization. Some sperm have tail

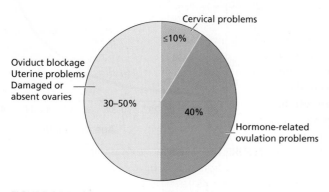

FIGURE 16.2 The major causes of female infertility.

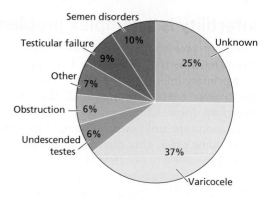

FIGURE 16.3 The major causes of male infertility.

malformations or extra tails that affect motility; others just move too slowly. Infertility tests can diagnose this condition.

- **Genetics plays a role in male infertility.** About 7% of infertile men carry microdeletions of the Y chromosome, but the location and size of these deletions are not always correlated with defects in sperm production.
- **No sperm in semen.** This condition is known as azoospermia. About 40% of all cases involve blockage of the epididymis or vas deferens, which prevent sperm from moving out of the testes. This can be caused by a urinary tract infection, STDs such as chlamydia or gonorrhea, or a vasectomy. Another condition, called nonobstructive azoospermia, is caused by a lack of sperm production. Often, problems with sperm formation are caused by hormonal imbalances, including use of anabolic steroids, developmental problems such as undescended testes, or injury to the testes.

Other causes of infertility.

Personal behavior can have a significant impact on fertility. Smoking and other uses of tobacco lower sperm count in men and increase the risk of low-weight babies and miscarriages in women. Overall, smoking by men or women reduces the chances of conceiving by one-third. Women who are significantly underweight or overweight often have trouble becoming pregnant.

Many couples in the United States postpone pregnancy until their late 30s, or even later. For women, increasing age is associated with decreased fertility. In addition, older women have a higher risk of producing eggs with abnormal chromosome number, leading to miscarriages and an increased risk of having a child with a chromosome abnormality such as Down syndrome (see Chapter 6 for a discussion of these risks).

16.2 Assisted Reproductive Technologies (ART) Expand Childbearing Options

Several methods, grouped under the name **assisted reproductive technologies (ART)**, are available to help infertile or subfertile individuals and couples have children. ART focuses on three areas: retrieval or donation of gametes, fertilization, and implantation of an embryo (Figure 16.4). Some of these methods are discussed below.

Intrauterine insemination uses donor sperm.

Insemination with donor sperm, called IUI (intrauterine insemination) was one of the first methods of ART and was developed to overcome problems of male infertility (see Spotlight on Fatherless Mice). In its simplest form, if a man is infertile, his female partner has sperm from a donor placed in her uterus (Figure 16.5) while she is ovulating, resulting in pregnancy.

Donor sperm can be obtained from sperm banks. Donor profiles, including physical characteristics, blood type, health information, and in some cases educational level and profession are provided to help select a donor. Some banks make it possible for women to use the same donor for subsequent pregnancies.

Although sperm banks in the United States are regulated by the U. S. Food and Drug Administration and most banks screen potential donors for some common genetic disorders, it is impossible to screen for all disorders. In one case, a donor to a Philadelphia sperm bank had a rare genetic disorder called severe congenital neutropenia (OMIM 202700), which affects only about 1 in 5 million individuals. Unfortunately, five children with this disorder were born to four couples who used sperm from this donor.

Assisted reproductive technologies (ART) The collection of techniques used to help infertile couples have children.

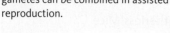

FIGURE 16.4 The many ways gametes can be combined in assisted reproduction.

Artificial Insemination and Embryo Transfer

1. Father is infertile. Mother is inseminated by donor and carries child.

2. Mother is infertile but able to carry child. Donor egg is inseminated by father via IVF. Embryo is transferred and mother carries child.

3. Mother is infertile and unable to carry child. Donor of egg is inseminated by father and carries child.

4. Both parents are infertile, but mother is able to carry child. Donor egg is inseminated by sperm donor via IVF. Embryo is transferred and mother carries child.

KEY

- Sperm from father
- Egg from mother
- Baby born of mother
- Sperm from donor
- Egg from donor
- Baby born of donor (Surrogate)

In Vitro Fertilization (IVF)

1. Mother is fertile but unable to conceive. Egg from mother and sperm from father are combined in laboratory. Embryo is placed in mother's uterus.

2. Mother is infertile but able to carry child. Egg from donor is combined with sperm from father and implanted in mother.

3. Father is infertile and mother is fertile but unable to conceive. Egg from mother is combined with sperm from donor.

4. Both parents are infertile, but mother is able to carry child. Egg and sperm from donors are combined in laboratory (also see number 4, column at left).

5. Mother is infertile and unable to carry child. Egg of donor is combined with sperm from father. Embryo is transferred to donor (also see number 3, column at left).

6. Both parents are fertile, but mother is unable to carry child. Egg from mother and sperm from father are combined. Embryo is transferred to donor.

7. Father is infertile. Mother is fertile but unable to carry child. Egg from mother is combined with sperm from donor. Embryo is transferred to surrogate mother.

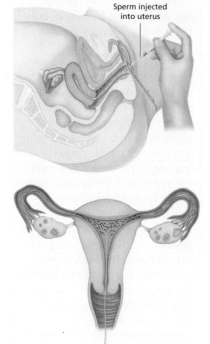

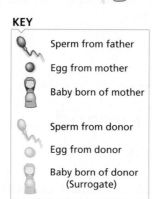

FIGURE 16.5 In intrauterine insemination (IUI), donor sperm is placed into the uterus of an ovulating woman. The sperm swim up the oviduct and fertilize the egg.

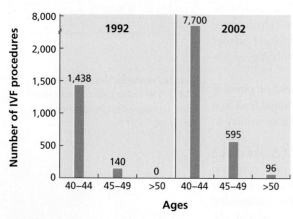

FIGURE 16.6 The use of *in vitro* fertilization (IVF) by older women in Great Britain in 1992 (*left*) and in 2002 (*right*). By 2002, almost 100 women over 50 years of age used IVF to have children.

Egg retrieval or donation is an option.

For infertile women who can ovulate but have blocked oviducts or related problems, several eggs at a time can be collected after hormone treatment and sorted using a microscope to remove those that are too young or too old for use in fertilization. After IVF, one or more embryos are implanted. Extra eggs that will not be fertilized and implanted immediately can be stored in a freezer for later use or donation to other women.

The discovery that the age of the egg, not the age of the reproductive system is responsible for age-related infertility makes it possible for women in their late 50s or even into their 60s to become pregnant using IVF and eggs donated by younger women. Figure 16.6 shows the dramatic increase in the use of ART by older women in Great Britain over a 10-year period. In 1992, no women over 50 had had IVF treatments, but by 2002, nearly 100 women had used the procedure, and 24 children were born to those mothers.

In May 2007, a 60-year-old mother gave birth to twins, becoming the oldest woman to have twins in the United States. The oldest

Alastair Grant/AP Photo

FIGURE 16.7 Children born by IVF and their parents. Louise Brown, the first human born by IVF, is the redheaded woman at the bottom center of the photograph.

woman known to have given birth is a 70-year-old Indian woman, who gave birth to twins in 2008.

Pregnancy in older women poses higher risks of diabetes, stroke, high blood pressure, and heart attacks, and these women have a threefold higher risk of having low-birth-weight and premature infants. A decision by a postmenopausal woman to have a child is usually made after her health has been carefully evaluated.

This discovery also means that younger women can safely postpone childbearing until they are older without having to rely on donated eggs. Women can have their eggs collected and fertilized while they are young and frozen for later use. This form of ART allows younger women to produce embryos at a time when their risks for chromosome abnormalities in the offspring are low. The embryos can be thawed and implanted over a period of years—including after menopause—allowing women to extend their childbearing years.

In vitro fertilization (IVF) is a widely used form of ART.

After the birth of Louise Brown in 1978, IVF quickly became the method of choice for helping many infertile couples become parents. IVF has resulted in the birth of millions of children worldwide (Figure 16.7). For IVF, an egg is collected and placed in a dish. Sperm are added, and if fertilization occurs, the resulting embryo is grown in an incubator before implantation in the uterus of a female partner or a surrogate for development.

The gametes used in IVF can come from a couple, from donors, or from a combination of a couple and donors. The embryo can be transferred to the uterus of the female partner or to a surrogate. In one famous case, a child ended up with five parents. This situation began with an infertile couple that wanted to be parents. They used an egg donor and a sperm donor who contributed the gametes, which were combined using IVF. To carry the child, the couple entered into a contract with a surrogate mother, who gave the child to the infertile couple.

For some cases of male infertility, many couples now use a variation of IVF called **intracytoplasmic sperm injection (ICSI)**. In this procedure, an egg is collected and injected with a carefully selected single sperm from the male partner (Figure 16.8). The embryo develops in an incubator before transfer to the uterus of the female partner. ICSI is used mostly in cases where low sperm count or motility problems are present.

GIFT and ZIFT are based on IVF.

For some couples, IVF is not the only option. In a method known as **GIFT**, or gamete **intrafallopian transfer** (oviducts are also known as fallopian tubes), gametes are collected and placed into the woman's oviduct through a small incision (Figure 16.9a).

Intracytoplasmic sperm injection (ICSI) A treatment to overcome defects in sperm count or motility; an egg is fertilized by microinjection of a single sperm.

Gamete intrafallopian transfer (GIFT) An ART procedure in which gametes are collected and placed into a woman's oviduct for fertilization.

FIGURE 16.8 Injection of a single sperm into an egg. This ART procedure is known as intracytoplasmic sperm injection (ICSI).

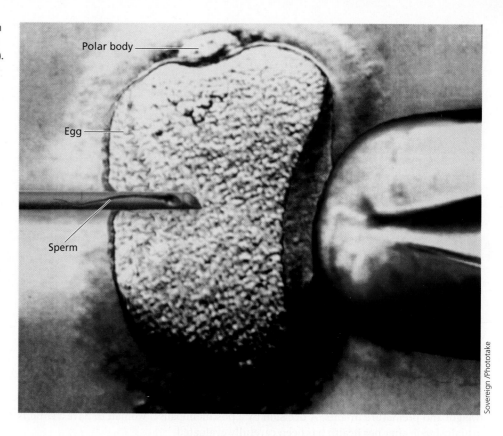

Polar body

Egg

Sperm

Sovereign /Phototake

Fertilization takes place in the oviduct, and the woman carries the child to term. This method can be used where infertility is not the result of oviduct blockage.

Another option is **ZIFT**, or **zygote intrafallopian transfer** (Figure 16.9b). In ZIFT, eggs are collected and fertilized by IVF. The resulting fertilized egg, or zygote, is implanted into the oviduct through a small abdominal incision. ZIFT is used in cases where the woman has ovulation problems, or the man has a low sperm count, and the oviducts are not blocked.

Zygote intrafallopian transfer (ZIFT)
An ART procedure in which gametes are collected, fertilization takes place *in vitro*, and the resulting zygote (fertilized egg) is transferred to a woman's oviduct.

Surrogacy is a controversial form of ART.

ART has altered traditional and accepted patterns of reproduction and redefined the meaning of parenthood. In the United States, surrogate motherhood is a reproductive option for infertile couples. In one form of surrogacy, a woman is artificially inseminated and carries the child to term. After the child is born, she surrenders the child to the

FIGURE 16.9 (a) Gamete intrafallopian transfer (GIFT). Eggs and sperm are collected and implanted into the oviduct (also called the fallopian tube). Fertilization occurs here, and the embryo moves down the oviduct and implants in the uterus. (b) Zygote intrafallopian transfer (ZIFT). Gametes are collected and fertilized by *in vitro* fertilization before transfer to the oviduct.

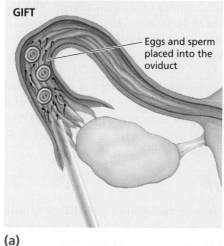

GIFT

Eggs and sperm placed into the oviduct

(a)

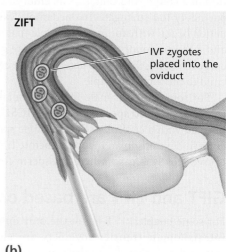

ZIFT

IVF zygotes placed into the oviduct

(b)

FIGURE 16.10 Teresa Anderson, a surrogate mother carrying quintuplets for parents Mr. and Mrs. Gonzales.

father. In this case, the surrogate is both the genetic and the gestational mother of the child. In another form of surrogacy, called gestational surrogacy, a couple provides both the egg and the sperm for IVF. The surrogate mother is implanted with the developing embryo and serves as the gestational mother but is genetically unrelated to the child she bears. After the birth of the child, she surrenders the infant to the couple who contracted for her services (Figure 16.10).

Laws regarding surrogacy vary from state to state. In some states, all forms of surrogacy are legal; in others, only gestational surrogacy is legal; and in still other states, all forms of surrogacy are illegal.

16.3 Ethical Issues in Reproductive Technology

The development and use of ART have developed more rapidly than the social conventions and laws governing their use (see Spotlight on Reproductive Technologies from the Past). In the process, controversy about the moral, ethical, and legal grounds for using these techniques has arisen but has not been resolved. ART has been responsible for more than 3 million conceptions worldwide. Although the benefits of ART have been significant, there are risks associated with the use of these alternative methods of reproduction. Some risks have been well documented, whereas others are still matters for debate and more study. In other cases, the use of ART raises ethical questions (see Exploring Genetics: The Business of Making Babies on page 363). We'll discuss some of these risks and questions in the following sections.

The use of ART carries risks to parents and children.

About 1% of babies born in the United States are the result of IVF. In its 2005 report, the Centers for Disease Control (CDC) pointed out that 49% of these were multiple births. The CDC estimates that these births generate health costs of $1 billion because twins and higher multiples are at high risk of premature birth (Figure 16.11). IVF babies have more than a three-fold higher risk of premature birth, and 42% of IVF births in 2005 (the last year for which figures are available) were premature, compared to about 13% of births in the general population. To avoid problems associated with multiple births, recent guidelines recommend that only one embryo should be transferred after IVF.

IVF risks also include a threefold increase in ectopic pregnancies (a situation in which the fertilized egg implants outside the uterus, and the placenta and embryo begin to develop there).

Spotlight on...

Reproductive Technologies from the Past

Although the most rapid advances in assisted reproductive technologies (ART) began after the 1978 birth of Louise Brown by IVF, one of the first recorded uses of ART occurred in the early 1770s. An Englishman with a malformation of the urethra collected his semen in a syringe and injected it into his wife's vagina; she became pregnant and gave birth to a child. In 1866, Dr. J. Sims reported the first intrauterine insemination in the United States; he later used that procedure over 50 times to aid infertile couples. Dr. William Panacost performed the first artificial insemination using donor sperm in 1884. In 1967, Oklahoma became the first state to legalize artificial insemination.

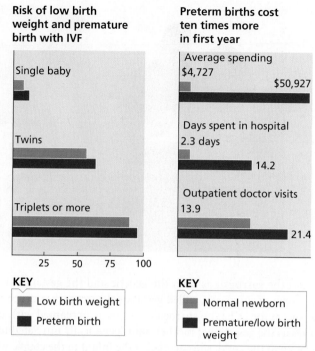

Risk of low birth weight and premature birth with IVF

Single baby

Twins

Triplets or more

25 50 75 100

KEY
- Low birth weight
- Preterm birth

Preterm births cost ten times more in first year

Average spending
$4,727
$50,927

Days spent in hospital
2.3 days
14.2

Outpatient doctor visits
13.9
21.4

KEY
- Normal newborn
- Premature/low birth weight

FIGURE 16.11 (*left*) Risk of premature birth and low birth weight following IVF. (*right*) Increased costs of premature babies in the first year after birth.

Infants born by means of ART have an increased risk of low birth weight and often require prolonged hospital care. When ICSI is used in ART, there is an increased risk of transmitting genetic defects to male children. About 13% of infertile males with a low sperm count carry a small deletion on the Y chromosome. With ICSI, this form of infertility is passed on to their sons. The same is true for some chromosomal abnormalities, such as Klinefelter syndrome. Questions arise as to whether it is ethical to use ICSI to produce sons who will be infertile.

There has been a long-standing debate about whether children conceived using ART have increased risks for birth defects. Although the issue has not been resolved, it is important that couples considering ART be informed of this and other potential risks.

16.4 Genetic Testing and Screening

Genetic testing is done to identify individuals who have or may carry a genetic disease, those at risk of producing a genetically defective child, and those who may have a genetic susceptibility to environmental agents. Genetic screening is done on populations in which there is a risk for a particular genetic disorder. Genetic testing is most often a matter of choice, whereas genetic screening is often a matter of law.

There are several types of testing and screening:

- Newborn screening for infants within 48 to 72 hours after birth for a variety of genetically controlled metabolic disorders
- Carrier testing done on members of families or ethnic groups with a history of a genetic disorder such as sickle cell anemia or cystic fibrosis
- Prenatal testing on a fetus for a genetic disorder such as cystic fibrosis
- Presymptomatic testing, also called predictive testing, for those who will develop adult-onset genetic disorders such as Huntington disease and polycystic kidney disease (PCKD)

The Business of Making Babies

New technology has made the business of human fertilization a part of private enterprise. One in eight couples in the United States is classified as infertile, and most of these couples want to have children. The first successful *in vitro* **fertilization (IVF)** in the United States was done in 1981 at the Medical College of Virginia at Norfolk. Since then, more than 40 hospitals and clinics using IVF and ART have opened. Many of these clinics are associated with university medical centers, but others are operated as freestanding businesses. Most operate only at a single location, but national chains are becoming part of the business. One of the largest, the Sher Institute for Reproductive Medicine, has eight locations nationwide, with plans for more. Some clinics are public companies that have sold stock to raise start-up money or to cover operating costs.

Charges for services in the baby industry include sperm samples ($275), eggs ($10,000 to $50,000), and IVF ($5,000 to

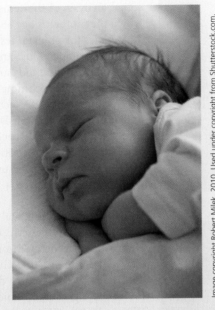

Image copyright Robert Milek, 2010. Used under copyright from Shutterstock.com.

$15,000). Because success rates are less than 50% for each IVF, several attempts are usually required. Because the costs are generally not covered by insurance, IVF is a major expense for couples that want children. If a couple wants a surrogate mother, costs range from $15,000 to $30,000.

Deborah Spar, author of the 2006 book *The Baby Business*, estimates that the ART business is a $3-billion-per-year industry. Because of the high start-up costs and expertise required, it is likely that the field will undergo significant consolidation and eventually be dominated by a small number of companies through franchising agreements. Some investment analysts predict that IVF alone will grow into a $6 billion annual business.

Remarkably, this business has little or no oversight from government agencies or industry groups. There is little consistency from state to state in laws governing the fertility business or insurance coverage for some or all of its procedures and safeguards for the property rights of donors or clients.

Newborn screening is universal in the United States.

All states and the District of Columbia require newborns to be screened for a range of genetic disorders. These programs began in the 1960s with screening for phenylketonuria (PKU) and gradually expanded. Most states screen for 3 to 8 disorders, but states that use newer technology can screen for 30 to 50 heritable metabolic disorders.

In vitro *fertilization (IVF)* A procedure in which gametes are collected and fertilized in a dish in the laboratory; the resulting zygote is implanted in the uterus for development.

Both carrier and prenatal testing are done to screen for genetic disorders.

Prenatal testing can detect genetic disorders and birth defects in the fetus. More than 200 single-gene disorders can be diagnosed by prenatal testing (Table 16.1). In most cases, testing is done only when there is a family history or another indication for testing. If there

Table 16.1 Some Metabolic Diseases and Birth Defects That Can Be Diagnosed by Prenatal Testing

Acatalasemia	Gaucher disease	Niemann-Pick disease
Adrenogenital syndrome	G6PD deficiency	Oroticaciduria
Chediak-Higashi syndrome	Homocystinuria	Progeria
Citrullinemia	I-Cell disease	Sandhoff disease
Cystathioninuria	Lesch-Nyhan syndrome	Spina bifida
Cystic fibrosis	Mannosidosis	Tay-Sachs disease
Fabry disease	Maple syrup urine disease	Thalassemia
Fucosidosis	Marfan syndrome	Werner syndrome
Galactosemia	Muscular dystrophy, X-linked	Xeroderma pigmentosum

Preimplantation genetic diagnosis (PGD) Removal and genetic analysis of a single cell from a 3- to 5-day old embryo. Used to select embryos free of genetic disorders for implantation and development.

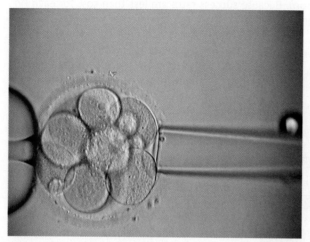

(a)

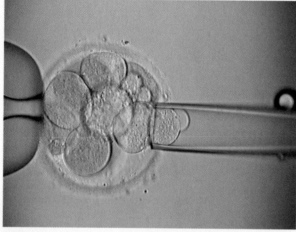

(b)

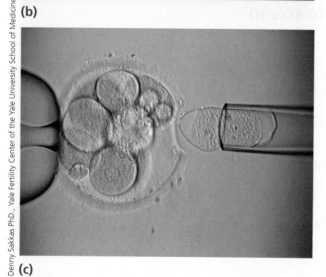

(c)

Denny Sakkas PhD., Yale Fertility Center of the Yale University School of Medicine

FIGURE 16.12 Removal of a single cell from a day-3 embryo for genetic analysis by preimplantation genetic diagnosis (PGD).

is a family history for autosomal recessive disorders such as Tay-Sachs disease or sickle cell anemia, the parents are usually tested to determine if they are heterozygous carriers. If both parents are heterozygotes, the fetus has a 25% chance of being affected. In such cases, prenatal testing can determine the genotype of the fetus.

For other conditions, such as Down syndrome (trisomy 21), chromosome analysis is the most direct way to detect an affected fetus. Testing for Down syndrome is usually done because of maternal age, not because there is a family history of genetic disease. Because the risk of Down syndrome increases dramatically with the age of the mother (see Chapter 6), chromosomal analysis of the fetus is recommended for all pregnancies in which the mother is age 35 or older.

Samples for prenatal testing can be obtained through amniocentesis or chorionic villus sampling (CVS), both of which are described in Chapter 6. The fluids and cells obtained for testing can be analyzed by several techniques, including karyotyping, biochemistry, and recombinant DNA techniques.

Because recombinant DNA technology can analyze the genome directly, it is the most specific and sensitive method currently available. The accuracy, sensitivity, and ease with which recombinant DNA technology can be used to identify genetic diseases and susceptibilities have raised a number of legal and ethical issues that remain unresolved.

In addition to prenatal genetic testing, another method called **preimplantation genetic diagnosis (PGD)** can be used to test embryos for genetic disorders in the earliest stages of embryonic development. In PGD, eggs are fertilized in the laboratory and develop in a culture dish for several days. For testing, one of the six-to-eight embryonic cells (called blastomeres) is removed (Figure 16.12). DNA is extracted from this single cell, amplified by PCR, and tested to determine whether the embryo is homozygous or hemizygous for a genetic disorder. PGD is useful when parents are carriers of autosomal and X-linked disorders that would be fatal to any children born with the disorder (Lesch-Nyhan syndrome or Tay-Sachs disease). PGD can also be used to select the sex of an embryo before implantation.

A related method called polar body biopsy is used to test for genetic disorders even before fertilization takes place. If a woman is heterozygous for an X-linked recessive disorder (such as muscular dystrophy), the X chromosome bearing the mutant allele can segregate during the first meiotic division into the cell destined to be the egg or into the much smaller polar body (see Chapter 7 to review gamete formation). A polar body can be removed by micromanipulation (Figure 16.13) and tested using recombinant DNA technology. If results show that the polar body carries the mutant allele, then the egg must carry the normal allele. Eggs that pass this test can be used for IVF, ensuring that the woman's sons will not be affected with the disorder.

The use of PGD raises ethical issues.

In the early 1990s, Jack and Linda Nash used PGD to screen embryos after their daughter, Molly, was born with Fanconi anemia (OMIM 227650), a fatal bone marrow disorder. In this case, they used PGD so they could have a healthy child, but also one that would be a suitable stem-cell donor for Molly. Umbilical cord blood from their PGD-selected son, Adam, was transfused into Molly, who is now free of Fanconi anemia (Figure 16.14). (See The

Genetic Revolution: Should I Save Cord Blood? on page 367) At the time, bioethicists debated whether it was ethical to have a child who was destined to be a donor for a sibling, a practice called having "savior babies." This case was complicated by the fact that the parents planned to have other children and used PGD to screen out embryos with Fanconi anemia.

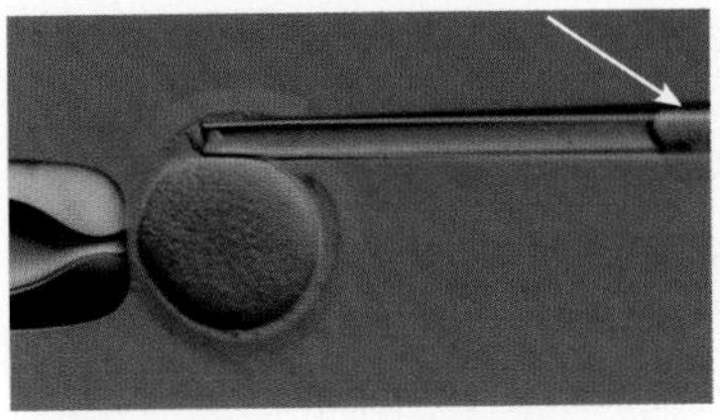

Since then, other couples used IVF and PGD to have babies that were tissue-matched to siblings with leukemia and other diseases. In these cases, the embryos were not screened for genetic disorders—only for alleles that would allow the children produced from the embryos to serve as transplant donors for their siblings. These cases have reignited the debate on whether it is ethical to select for genotypes that have nothing to do with a genetic disorder and whether screening to benefit someone else is acceptable.

Advocates of embryo screening to match transplant donors and recipients say that there are no associated ethical issues, but critics wonder if embryo screening for transplant compatibility will eventually lead to screening for the sex of the embryo or designer baby traits. A survey by the Genetics and Public Policy Center at Johns Hopkins University shows that 61% of Americans surveyed approved of using PGD to select an embryo for the benefit of a sibling, but it also revealed that 80% of those surveyed were concerned that reproductive genetic technologies could get out of control.

Some countries, including Great Britain, now permit PGD screening for breast and ovarian cancer, two genetic diseases with less than a 100% chance of occurrence. Other nations have laws against using PGD for sex selection or for screening embryos to be donors unless they are also screened to avoid a genetic disorder, but the United States has no such restrictions.

Perhaps the most controversial use of PGD is the selection of embryos with conditions that most people would consider disabilities. A 2006 survey reported that a small percentage of clinics (Figure 16.15) used PGD to select embryos that would result in deaf children or children with dwarfism. This procedure allows parents to have children who have the same physical attributes they have, but the ethics of this practice are still being debated.

FIGURE 16.13 Removal of a polar body (*arrow*) for genetic analysis.

Sergei Evsikov, Ph.D., ViaGene Fertility, LLC

FIGURE 16.14 Molly Nash (*right*) and her brother, Adam. Molly's parents used *in vitro* fertilization and prenatal genetic diagnosis to avoid having another child with Fanconi anemia and to select a compatible stem-cell donor for Molly.

Reuters

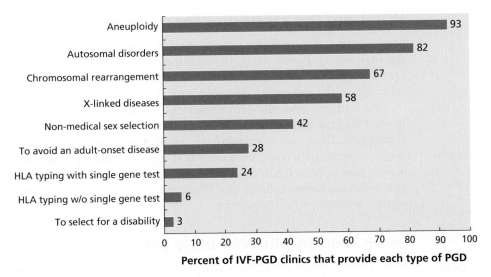

FIGURE 16.15 Reasons for PGD at 137 fertility clinics in the United States. Three clinics reported selecting embryos with specific disabilities, including dwarfism and deafness.

Table 16.2 Cystic Fibrosis Testing for 25 Mutations: Diagnostic Success in Different Ethnic Groups

88% among Ashkenazi Jewish couples
78% among non-Hispanic Caucasian couples
52% among Hispanic Caucasian couples
42% among African American couples
24% among Asian American couples

Gene therapy The transfer of cloned genes into somatic cells as a means of treating a genetic disorder.

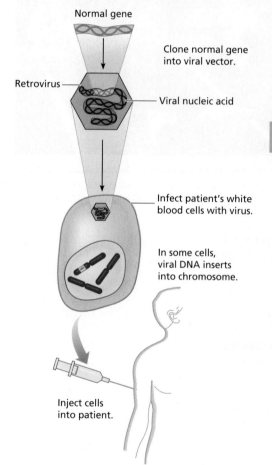

Normal gene

Clone normal gene into viral vector.

Retrovirus

Viral nucleic acid

Infect patient's white blood cells with virus.

In some cells, viral DNA inserts into chromosome.

Inject cells into patient.

FIGURE 16.16 The most widely used method of gene therapy uses a virus as a vector to insert a normal copy of a gene into the white blood cells of a patient who has a genetic disorder. The normal gene becomes active, and the cells are reinserted into the affected individual, curing the genetic disorder. Because white blood cells die after a few months, the procedure has to be repeated regularly. In the future, it is hoped that transferring a normal gene into the mitotically active cells of the bone marrow will make gene therapy a one-time procedure.

Prenatal testing is associated with risks.

Although many genetic disorders and birth defects can be detected with prenatal testing, the technique has some limitations. Prenatal testing poses measurable risks to the mother and the fetus, including infection, hemorrhage, fetal injury, and spontaneous abortion.

Prenatal testing also has limitations. Conventional testing strategies will not always detect the majority of certain defects. Amniocentesis is recommended for all mothers 35 years of age and older to test for Down syndrome, because maternal age is the biggest risk factor for having an affected child. Older women have only about 5% to 7% of all children, but they have 20% of all Down syndrome children. This means that about 80% of all Down syndrome births are to mothers who are not candidates for amniocentesis. It is now recommended that all pregnant women have non-invasive screening for Down syndrome.

Genetic testing on a large scale is not always possible. For disorders such as sickle cell anemia, a single mutation is present in all cases, so testing is efficient and uncovers all cases. In cystic fibrosis (CF), however, over 1,600 different mutations have been identified (see Chapter 11), and testing for all these mutations is impractical. Many of the mutations are found only in one family, and others are found primarily in one ethnic group or another. Using a panel of 25 of the most common mutations to test for CF produces an accurate diagnosis for some mutations but poor results for others (see Table 16.2). Thus, at the moment, CF testing is not widely performed.

16.5 Gene Therapy Promises to Correct Many Disorders

Although PGD and other methods of genetic testing allow couples to have children who are free of genetic disorders, about 5% of all newborns have a genetic or chromosomal disorder. A recombinant DNA–based method called **gene therapy** may be able to treat disorders caused by mutations in single genes. Gene therapy inserts copies of normal genes into cells that carry defective copies. These normal genes make functional proteins that result in a normal phenotype.

What are the strategies for gene transfer?

There are several methods for transferring genes into human cells, including viral vectors (Figure 16.16), chemical methods of DNA transfer across the cell membrane, and physical methods such as microinjection or fusion of cells with vesicles that carry copies of normal genes.

Viral vectors, especially retroviruses, are the most commonly used method for gene therapy. Retroviruses are used because they readily infect human cells. These vectors are genetically modified by removing some viral genes, preventing the virus from causing disease and making room for a human gene to be inserted. Once the recombinant virus carrying a human gene is inside a human cell, the viral DNA inserts itself into a chromosome, where it becomes part of the genome.

Gene therapy showed early promise.

Gene therapy began in 1990, when the human gene for the enzyme adenosine deaminase (ADA) was inserted into a retrovirus and transferred

THE GENETIC REVOLUTION

Should I Save Cord Blood?

Two California boys, Blayke and Garrett LaRue, are alive today thanks to umbilical cord blood donated to cord blood banks by two anonymous mothers—one in New York and the other in Germany. Both boys were diagnosed with a rare and fatal genetic disorder called X-linked lymphoproliferative disorder (XLP; OMIM 308240) and matched with donors through the National Marrow Donor Programs cord blood bank.

Cord blood has advantages over bone marrow for treating some disorders. It is less likely to carry antibodies that can cause incompatibility between the transplant and the recipient. Cord blood is available in cord blood banks, and harvesting blood from the umbilical cord is an easy, noninvasive, and painless procedure.

The question is, if you have a baby, is it worth saving the cord blood? The process is simple: Remove blood from the umbilical cord and store it in a private cord blood bank. Why store cord blood? Stem cells present in cord blood are multipotent and can be used to treat a variety of blood-related disorders including sickle cell anemia and leukemia, as can bone marrow cells.

If you save cord blood, what are the chances you will need it? The answer depends on whom you ask. An ad from one private bank puts the odds at 1 in 27; an editorial in the journal *Obstetrics and Gynecology* put the odds at 1 in 2,500; the American Academy of Pediatrics estimates the odds at 1 in 200,000. There is no recognized database for cord blood transplants, so the odds are difficult to calculate. In addition, most diseases that can be treated with cord stem cells are rare. If your child has a genetic disorder, the cord blood will also carry the disorder, the stem cells will be useless, and cells from other donors will be needed.

What does it cost to extract and store cord blood? There are over 30 private cord blood banks in the United States. The initial charge may be up to $2,000, with monthly fees of about $100. The Institute of Medicine has recommended that the U.S. Congress pass legislation establishing a National Cord Blood Stem Cell Bank Program, but no action has been taken, and costs for this are unknown.

The decision to store or not store cord blood is a personal one. Some people feel that it is like an insurance policy, while others think that there is little chance it will be needed and that other resources are available for transplants. If you are considering storing cord blood, take the time to learn the facts, details, and costs, so you will be able to make an informed decision without feeling pressured by time or the opinions of others.

KEEP IN MIND

- Gene therapy has not fulfilled its promise of treating genetic disorders.
- Gene therapy has also experienced setbacks and restarts.

into the white blood cells of a young girl born with a form of severe combined immunodeficiency disease (SCID; OMIM 102700). She had no functional immune system and was prone to infections, many of which could be fatal. The normal ADA gene, which was inserted into her white blood cells, encodes an enzyme that allows cells of the immune system to mature properly. As a result, she now has a functional immune system and is leading a normal life. Unfortunately, gene therapy for most other children with ADA-related SCID has been unsuccessful.

In the early-to-middle 1990s, gene therapy treatments were started for several genetic disorders, including cystic fibrosis and familial hypercholesterolemia. Over a 10-year period, more than 4,000 people underwent gene therapy. Unfortunately, those trials were largely failures, leading to a loss of confidence in gene therapy. Hopes for gene therapy plummeted even further in September 1999, when an 18-year-old patient died during gene therapy to treat ornithine transcarbamylase deficiency (OMIM 300461). His death was triggered by a massive immune response to the vector, a modified adenovirus (adenoviruses cause colds and respiratory infections).

In 2000, two French children who underwent successful gene therapy for an X-linked form of SCID (OMIM 300400) developed leukemia. In those children, the recombinant virus inserted itself into a gene that controls cell division, activating the gene and causing uncontrolled production of white blood cells and the symptoms of leukemia.

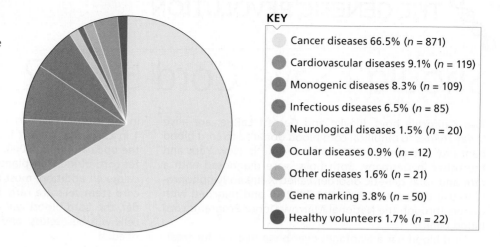

FIGURE 16.17 Target disorders for gene therapy. Most gene therapy trials are for cancer (66%), not for single-gene (monogenic) disorders (8.3%).

KEY

- Cancer diseases 66.5% (*n* = 871)
- Cardiovascular diseases 9.1% (*n* = 119)
- Monogenic diseases 8.3% (*n* = 109)
- Infectious diseases 6.5% (*n* = 85)
- Neurological diseases 1.5% (*n* = 20)
- Ocular diseases 0.9% (*n* = 12)
- Other diseases 1.6% (*n* = 21)
- Gene marking 3.8% (*n* = 50)
- Healthy volunteers 1.7% (*n* = 22)

In 2007, a woman receiving gene therapy for inflammation associated with arthritis died after receiving a second round of therapy. As in the 1999 case, the vector was a modified adenovirus. In the wake of her death, the U.S. Food and Drug Administration (FDA) stopped all gene therapy trials using those vectors until the cause of death is determined.

In 2009, the pendulum began to swing back in favor of gene therapy when it was used successfully to treat blindness associated with a genetic disorder called Leber congenital amaurosis (OMIM 204000). In these cases, the gene (*RPE65*) was carried by a modified adenovirus vector, and patients showed improvements in visual function with no adverse effects.

In spite of its spotty record in treating genetic disorders, gene therapy is used successfully to treat cancer, cardiovascular disease, and HIV infection (Figure 16.17). In fact, gene therapy is used to treat cancer more often than any other condition. At this writing, gene therapy is still an experimental procedure performed only on a few carefully selected patients, under strict regulation by government agencies.

There are ethical issues associated with gene therapy.

Gene therapy is done using an established set of ethical and medical guidelines. All patients are volunteers, gene transfer is started only after the case has undergone several reviews, and the trials are monitored to protect the patients' interests. Newer guidelines instituted after gene therapy deaths have strengthened these protections and coordinated the role of government agencies that regulate gene therapy. Other ethical concerns have not been resolved, as described next.

At present, gene therapy uses somatic cells as targets for transferred genes. This form of gene therapy is called **somatic gene therapy.** Genes are transferred into somatic cells of the body; the procedure involves only a single target tissue, and only one person is treated (only after obtaining informed consent and permission for the treatment).

Two other forms of gene therapy are not yet in use, mainly because the ethical issues surrounding them have not been resolved. One of these is **germ-line gene therapy,** in which cells that produce eggs and sperm are targets for gene transfer. In germ-line therapy, the transferred gene would be present in all the cells of the individual produced from the genetically altered gamete, including his or her germ cells. As a result, members of future generations will be affected by this gene transfer, without their consent. Do we have the right to genetically modify others without their consent? Can we make this decision for members of future generations? These and other ethical concerns have not been resolved, and germ-line therapy is currently prohibited.

The other is **enhancement gene therapy,** which raises even more ethical concerns. If we discover genes that control a desirable trait such as intelligence or athletic ability,

Somatic gene therapy Gene transfer to somatic target cells to correct a genetic disorder.

Germ-line gene therapy Gene transfer to gametes or the cells that produce them. Transfers a gene to all cells in the next generation, including germ cells.

Enhancement gene therapy Gene transfer to enhance traits such as intelligence and athletic ability rather than to treat a genetic disorder.

should we use them to enhance someone's intellectual ability or athletic skills? For now, the consensus is that we should not use gene transfer for such purposes. However, the U.S. Food and Drug Administration allows the use of growth hormone produced by recombinant DNA technology to enhance the growth of children who have no genetic disorder or disease but are likely to be shorter than average adults. Critics point out that approving transfer of a gene for enhancement is only a short step from the current practice of approving a gene product for enhancement.

Gene doping is a controversial form of gene therapy.

The use of performance-enhancing drugs has confounded athletic events in recent years, including cycling's Tour de France and the pursuit of the home-run record in U.S. professional baseball. In the Tour de France, cyclists have been suspended for using erythropoietin (EPO), a hormone that increases the production of red blood cells, which increases the oxygen-carrying capacity of the blood (Figure 16.18). EPO and other drugs can be detected by blood tests.

FIGURE 16.18 The use of EPO (erythropoietin), a hormone that increases red blood cell production to enhance athletic performance, is banned. There is controversy over using gene therapy to transfer the gene for EPO, which would be undetectable.

Concern over the use of genes instead of gene products to enhance athletic performance began in 2001, when the International Olympic Committee (IOC) Medical Commission met to discuss how gene therapy might affect sports competition. Other agencies, including the World Anti-Doping Agency (WADA), have prohibited gene doping as a means of enhancing athletic performance.

An example of gene doping is the use of Repoxygen, a product in which the human *EPO* gene has been inserted into a viral vector adjacent to a control element that regulates expression of the gene. Once in the body, the control element senses low oxygen levels in the blood during strenuous activity and turns on the adjacent EPO gene, increasing the synthesis and release of the hormone, erythropoietin. Repoxygen use may be difficult or impossible to detect, although several athletes at the Turin 2006 Olympic Games were suspected of using this form of gene doping.

Although agencies such as the IOC and WADA prohibit the use of Repoxygen, others are calling for legalization of gene enhancement, arguing that regulating the use of this gene therapy is more effective than attempting to prevent its use. They also argue that gene doping is only an extension of technology such as artificial nutrition and hydration by intravenous fluids, which is already permitted. More than 20 genes have been associated with athletic performance, so many choices are available for gene doping.

16.6 Genetic Counseling Assesses Reproductive Risks

Genetic counseling is a process of communication about the occurrence of or risk for a genetic disorder in a family (Figure 16.19). Counseling involves one or more trained professionals, who help an individual or family understand each of the following:
- The medical facts, including the diagnosis, progression, management, and any available treatment for a genetic disorder
- The way heredity contributes to the disorder and the risk of having children with the disorder

Genetic counseling A process of communication that deals with the occurrence or risk of a genetic disorder in a family.

FIGURE 16.19 In a genetic counseling session, the counselor uses the information from pedigree construction, medical records, and genetic testing to educate and inform a couple about their risks for genetic disorders.

- The alternatives for dealing with the risk of recurrence
- Ways to adjust to the disorder in an affected family member or to the risk of recurrence

Genetic counselors achieve these goals in a nondirective way. They provide the information necessary for individuals and families to make the decisions best suited to them on the basis of their own cultural, religious, and moral beliefs.

> **KEEP IN MIND**
>
> Genetic counseling educates individuals and families about genetic disorders and helps them make decisions about reproductive choices.

Why do people seek genetic counseling?

Typically, individuals or families with a history of a genetic disorder, cancer, birth defect, or developmental disability seek genetic counseling. Women older than 35 years of age and individuals from ethnic groups in which particular genetic conditions occur more frequently are counseled, to teach them about their increased risk for genetic or chromosomal disorders and the availability of diagnostic testing. Counseling is especially recommended for the following individuals or families:

- Women who are pregnant, or are planning to become pregnant, after age 35
- Couples who already have a child with mental retardation, an inherited disorder, or a birth defect
- Couples who would like testing or more information about genetic defects that occur more frequently in their ethnic group
- Couples who are first cousins or other close blood relatives
- Individuals who are concerned that their jobs, lifestyle, or medical history may pose a risk to a pregnancy, including exposure to radiation, medications, chemicals, infection, or drugs
- Women who have had two or more miscarriages or babies who died in infancy
- Couples whose infant has a genetic disease diagnosed by routine newborn screening
- Those who have, or are concerned that they might have, an inherited disorder or birth defect
- Women who have been told that their pregnancies may be at increased risk for complications or birth defects, based on medical tests

How does genetic counseling work?

The counselor usually begins by taking a detailed family and medical history and constructing a pedigree. Prenatal screening and cytogenetic or biochemical tests can be used along with pedigree analysis to help determine what, if any, risks are present. The counselor uses as much information as possible to establish whether the trait in question is genetically determined, and who is at risk.

For genetic traits, the counselor constructs a risk-assessment profile for the couple. In this process, the counselor uses all the information available to explain the risk of having a child affected with the condition or the risk that the individual who is being counseled will be affected with the condition. Often, conditions are difficult to assess because they involve polygenic traits or disorders that have high mutation rates (such as neurofibromatosis).

Genetics in Practice

Genetics in Practice case studies are critical-thinking exercises that allow you to apply your new knowledge of human genetics to real-life problems. You can find these case studies and links to relevant websites at *www.cengage.com/biology/cummings*

CASE 1

Jan, a 32-year-old woman, and her husband, Darryl, have been married for 7 years. They have attempted to have a baby on several occasions. Five years ago, they had a first-trimester miscarriage, followed by an ectopic pregnancy later the same year. Jan continued to see her OB/GYN physician for infertility problems but was very unsatisfied with the response. After four miscarriages, she went to see a fertility specialist, who diagnosed her with severe endometriosis and polycystic ovarian disease (detected by hormone studies). The infertility physician explained that these two conditions were hampering her ability to become pregnant and thus making her infertile. She referred Jan to a genetic counselor.

At the appointment, the counselor explained to Jan that one form of endometriosis (OMIM 131200) can be a genetic disorder and that polycystic ovarian disease can also be a genetic disorder (OMIM 184700) and is one of the most common reproductive disorders among women. The counselor recommended that a detailed family history of both Jan and Darryl would help establish whether Jan's problems have a genetic component and whether any of her potential daughters would be at risk for one or both of these disorders. In the meantime, Jan is taking hormones, and she and Darryl are considering alternative modes of reproduction.

Using the information in Figure 16.4, explain the reproductive options that are open to Jan and Darryl.

1. Would ISCI be an option? Why or why not?

2. Jan is concerned about using ART. She wants to be the genetic mother and have Darryl be the genetic father of any children they have. What methods of ART would you recommend to this couple?

CASE 2

Trudy is a 33-year-old woman who went with her husband, Jeremy, for genetic counseling. Trudy has had three miscarriages. The couple has a 2-year-old daughter who is in good health and is developing normally. Chromosomal analysis was done on tissues recovered from the last miscarriage, which were found to be 46,XY. The last miscarriage occurred in January 2005. Peripheral blood samples for both parents were taken at the time and sent to the laboratory. Trudy's chromosomes were 46,XX, and Jeremy's were 46,XY,t(6;18)(q21;q23). Jeremy appears to have a balanced translocation between chromosome numbers 6 and 18. There is no family history of stillbirths, neonatal death, infertility, mental retardation, or birth defects. Jeremy's parents both died in their 70s from heart disease, and he is unaware of any pregnancy losses experienced by his parents or siblings.

The recurrence risks associated with a balanced translocation between chromosomes 6 and 18 were discussed in detail. The counselor used illustrations to demonstrate the approximately 50% risk of unbalanced gametes; the other 50% of the gametes result in either normal or balanced karyotypes. The family was informed that the relative empirical risk for chromosomally unbalanced conceptions is significantly less than 50%. Prenatal diagnostic procedures were described, including amniocentesis and chorionic villus sampling. The benefits, risks, and limitations of each were described.

The couple expressed a desire to have another child and was interested in proceeding with an amniocentesis.

1. Draw each of the possible combinations of chromosomes 6 and 18 that could be present in Jeremy's gametes, showing how there is an approximately 50% chance that they are normal or balanced and a 50% chance that they are unbalanced.

2. Trudy became pregnant again, and an amniocentesis showed that the fetus received the balanced translocation from her father. Is she likely to have any health problems because of this translocation? Will it affect her in any way?

Summary

16.1 Infertility Is a Common Problem

- In the United States, about 13% of all couples are infertile. Infertility has many causes, including problems with gamete formation and hormonal imbalances.

16.2 Assisted Reproductive Technologies (ART) Expand Childbearing Options

- ART is a collection of techniques used to help infertile couples have children. These techniques have developed ahead of legal and social consensus about their use.

16.3 Ethical Issues in Reproductive Technology

- The use of ART raises several unresolved ethical issues. These issues include health risks to both parents and their offspring resulting from ART and the use of preimplantation genetic diagnosis to select siblings who are suitable tissue or organ donors for other members of the family.

16.4 Genetic Testing and Screening

- Genetic testing identifies individuals with a specific genotype; genetic screening tests general populations to identify heterozygotes carrying specific mutant alleles for a genetic disorder.

16.5 Gene Therapy Promises to Correct Many Disorders

- Gene therapy transfers a normal copy of a gene into target cells of individuals carrying a mutant allele. After initial successes, gene therapy suffered several setbacks, including the death of a participant. Ethical issues surrounding the use of germ-line therapy and enhancement therapy are unresolved, and these therapies are not used.

16.6 Genetic Counseling Assesses Reproductive Risks

- Genetic counseling involves developing an accurate assessment of a family history to determine the risk of genetic disease. In many cases, this is done after the birth of a child affected with a genetic disorder to predict the risks in future pregnancies. Decisions about whether to have additional children, to undergo abortion, or even to marry are always left to those being counseled.

Questions and Problems

CENGAGENOW Preparing for an exam? Assess your understanding of this chapter's topics with a pre-test, a personalized learning plan, and a post-test by logging on to *login.cengage.com/sso* and visiting CengageNOW's Study Tools.

Infertility Is a Common Problem

1. List the common infertility problems in women. What is the major infertility problem in men? Is it correctable?
2. Some fertility clinics limit donated-egg recipients to women who are 55 years of age or younger. Do you think this is an intrusion on reproductive rights?

Assisted Reproductive Technologies (ART) Expand Childbearing Options

3. How does IVF differ from artificial fertilization?
4. What is the difference between gamete intrafallopian transfer (GIFT) and intracytoplasmic sperm injection (ICSI)?

5. Why should women consider collecting and freezing oocytes for use later in life when they want to have children? What are the risks associated with older women having children?

Ethical Issues in Reproductive Technology

6. What do you think are the legal and ethical issues surrounding the use of IVF? How can these issues be resolved? What should be done with the extra gametes that are removed from the woman's body but never implanted in her uterus?

7. Researchers are learning how to transfer sperm-making cells from fertile male mice into infertile male mice in the hopes of learning more about reproductive abnormalities. These donor spermatogonia cells have developed into mature spermatozoa in 70% of cases, and some recipients have gone on to father pups (as baby mice are called). This new advance opens the way for a host of experimental genetic manipulations. It also offers enormous potential for correcting human genetic disease. One potentially useful human application of this procedure is treating infertile males who wish to be fathers.

   a. Do you foresee any ethical or legal problems with the implementation of this technique? If so, elaborate on the concerns.

   b. Could this procedure have the potential for misuse? If so, explain how.

Genetic Testing and Screening

8. What is the difference between genetic testing and genetic screening?

9. Cystic fibrosis is an autosomal disease that mainly affects the white population, and 1 in 20 whites are heterozygotes. Genetic testing can diagnose heterozygotes. Should a genetic screening program for cystic fibrosis be instituted? Should the federal government fund it? Should the program be voluntary or mandatory, and why?

10. You are a governmental science policy adviser, and you learn about a new technique being developed that promises to predict IQ accurately on the basis of a particular combination of genetic markers. You also learn that this technique could potentially be applied to preimplantation genetic diagnosis (PGD), so parents would be able to select an embryo for implantation that is free of a genetic disorder and one that is likely to be relatively smart. What policy recommendations would you make concerning this technology? Do you think parents should have the right to choose any characteristic of their children, or should PGD be limited to ensuring that embryos are free of genetic disorders? Should guidelines be imposed to regulate this process, or should it be banned?

Gene Therapy Promises to Correct Many Disorders

11. Gene therapy involves:

    a. the introduction of recombinant proteins into individuals.

    b. cloning human genes into plants.

    c. the introduction of a normal gene into an individual carrying a mutant copy.

    d. DNA fingerprinting.

    e. none of the above.

12. In selecting target cells to receive a transferred gene in gene therapy, what factors do you think would have to be taken into account?

13. The prospect of using gene therapy to alleviate genetic conditions is still a vision of the future. Gene therapy for adenosine deaminase deficiency has proved to be quite promising, but many obstacles remain to be overcome. Currently, the correction of human genetic defects is done using retroviruses as vectors. For this purpose, viral genes are removed from the retroviral genome, creating a vector capable of transferring human structural genes into sites on human chromosomes within target-tissue cells. Do you see any potential problems with inserting pieces of a retroviral genome into humans? If so, are there ways to combat or prevent these problems?

14. Is gene transfer a form of eugenics? Is it advantageous to use gene transfer to eliminate some genetic disorders? Can this and other technology be used to influence the evolution of our species? Should there be guidelines for the use of genetic technology to control its application to human evolution? Who should create and enforce these guidelines?

Genetic Counseling Assesses Reproductive Risks

15. A couple that wishes to have children visits you, a genetic counselor. There is a history of a deleterious recessive trait in males in the woman's family but not in the man's family. The couple is convinced that because his family shows no history of this genetic disease, they are not at risk of having affected children. What steps would you take to assess this situation and educate the couple?

16. A couple has had a child born with neurofibromatosis. They come to your genetic counseling office for help. After taking an extensive family history, you determine that there is no history of this disease on either side of the family. The couple wants to have another child and wants to be advised about the risks of that child having neurofibromatosis. What advice do you give them?

17. You are a genetic counselor, and your patient has asked to be tested to determine if she carries a gene that predisposes her to early-onset cancer. If your patient has this gene, there is a 50/50 chance that all of her siblings inherited the gene as well; there is also a 50/50 chance that it will be passed on to their offspring. Your patient is concerned about confidentiality and does not want anyone in her family to know she is being tested, including her identical twin sister. Your patient is tested and found to carry a mutant allele that gives her an 85% lifetime risk of developing breast cancer and a 60% lifetime risk of developing ovarian cancer. At the result-disclosure session, she once again reiterates that she does not want anyone in her family to know her test results.

    a. Knowing that a familial mutation is occurring in this family, what would be your next course of action in this case?

    b. Is it your duty to contact members of this family despite the request of your patient? Where do your obligations lie: with your patient or with the patient's family? Would it be inappropriate to try to persuade the patient to share her results with her family members?

18. A young woman (the proband) and her partner are referred for prenatal genetic counseling because the woman has a family history of sickle cell anemia. The proband has the sickle cell trait (*Ss*), and her partner is not a carrier and does not have sickle cell anemia (*SS*). Prenatal testing indicates that the fetus is affected with sickle cell anemia (*ss*). The results of this and other tests indicate that the only way the fetus could have sickle cell disease is if the woman's partner is not the father of the fetus. The couple is at the appointment seeking their test results.

    a. How would you handle this scenario? Should you have contacted the proband beforehand to explain the results and the implications of the results?

    b. Is it appropriate to keep this information from the partner because he believes he is the father of the baby? What other problems do you see with this case?

Internet Activities

Internet Activities are critical-thinking exercises using the resources of the World Wide Web to enhance the principles and issues covered in this chapter. For a full set of links and questions investigating the topics described below, visit *www.cengage.com/biology/cummings*

1. *Overview and History of Genetic Counseling.* At the *Access Excellence: Classics Collection* site, click on the link to the article "Genetic Counseling: Coping with the Impact of Human Disease." This article gives an overview of the history of genetic counseling and the ways in which it is used today. How does the use of genetic information by eugenicists early in the twentieth century compare with the use of genetic information by genetic counselors today? (For review, you may want to refer to Chapter 1, where eugenics was discussed.) What kinds of ethical questions and issues may arise as a result of genetic counseling?

2. *Genetic Counseling Resources.* The New York Online Access to Health program has an excellent home page on genetic disorders and genetic counseling. This site is a good place to start if you or someone in your family has any concerns about genetic disorders.

☑ HOW WOULD YOU VOTE NOW?

Surplus embryos from IVF are routinely stored in liquid nitrogen. Some may be used in subsequent attempts at pregnancy, but many remain in storage. These embryos have several possible fates: They can be stored indefinitely, thawed and discarded, donated to researchers for use in stem-cell research, or donated to other couples. Some nations, such as Sweden and Great Britain, limit the time unused embryos can be stored before destruction, but little is known about the extent of unrepaired DNA damage in old embryos. Now that you know more about IVF, ART, and the issues surrounding reproductive technology, what do you think? If you were having IVF, what would you want done with the extra embryos? Visit the *Human Heredity* companion website at *www.cengage.com/biology/cummings* to find out more on the issue; then cast your vote online.

17 Genes and the Immune System

Xenotransplants Cells, tissues, or organs that are transplanted from one species to another.

A bout every 2 hours, someone in the United States dies while waiting for an organ transplant. At any given time, about 50,000 people are waiting for transplants. Although more Americans are signing pledge cards to become organ donors at death, the demand far outstrips the supply. To address the shortage, scientists and biotechnology companies are developing an alternative source of organs: animals. Nonhuman primates such as baboons and chimpanzees are poor candidates as organ donors; these are endangered species, and they harbor viruses that may cause disease in humans (HIV, for instance, originated in nonhuman primates). Most attention is focused on using a strain of mini-pigs developed over 30 years ago as potential organ donors. Those pigs have major organs (hearts, livers, kidneys, etc.) that are similar in size to those of adult humans and have a compatible physiology.

The major stumbling block to **xenotransplants** (transplants across species) using pigs as organ donors—or using any other animal, for that matter—is rejection by the immune system of the recipient. To overcome this problem, researchers have transferred human genes to pigs so that their organs carry molecular markers found on human organs. Other workers have deleted specific pig genes to make their organs look more like human organs to the human immune system. More radical approaches to making pigs and humans compatible for transplants involves altering the immune system of the human recipient so that a transplanted pig organ will be tolerated. To do this, purified bone marrow cells from the donor pig are infused into the human recipient. After modification in this way, the recipient's immune system accepts the donor pig's organ with fewer complications. Transplant trials across species in animal–animal transplants have been successful, making it likely that this method would work in humans.

Proponents of xenotransplantation point to the lives that will be saved if pig organs can be used for organ transplants. Opponents point out that there is no evidence that pig organs will work properly in humans and that pig organs may harbor harmful viruses that will be transferred to the human recipients. Others question the ethics of genetically modifying animals with human genes or modifying humans by transplanting parts of the pig's immune system.

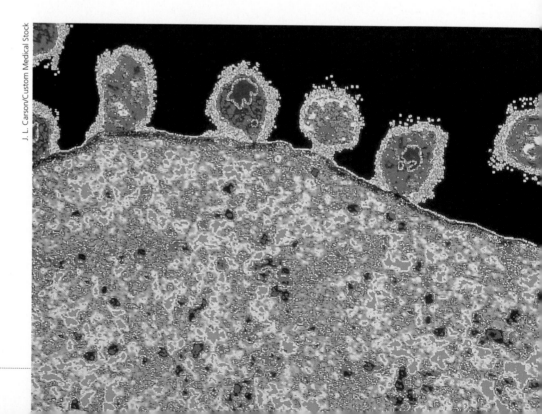

J. L. Carson/Custom Medical Stock

A colorized electron micrograph showing HIV virus particles budding off the surface of an infected T cell.

17.1 The Body Has Three Levels of Defense Against Infection

In the course of an average day, we encounter **pathogens** (disease-causing agents) of many kinds: viruses, bacteria, fungi, and parasites. Fortunately, we possess various levels of defense against infection. Each level brings an increasingly aggressive response to attempts to invade the body and cause damage. Humans have three levels of defense: (1) the skin and the organisms that inhabit it, (2) the innate immune system that uses nonspecific responses such as inflammation, and (3) the adaptive immune system, which mounts specific responses to infection in the form of immune reactions.

HOW WOULD YOU VOTE?

Organ donations are unable to keep up with demand, and thousands of people die each year while waiting for transplants. Using pigs that have been genetically modified to carry human genes that prevent transplant rejection and modifying the immune system of human recipients by injecting pig bone marrow cells are two methods of overcoming the inherent problems of organ transplantation between species. Do you think it is ethical to genetically modify pigs with human genes or to modify humans by giving them a pig immune system to accept transplanted organs? Visit the *Human Heredity* companion website at *www.cengage.com/biology/cummings* to find out more on the issue; then cast your vote online.

KEEP IN MIND AS YOU READ

- Humans have three defenses against infection: the skin and mucous membranes, innate immunity, and the adaptive immune system.

- The adaptive immune response has two components: antibody-mediated immunity and cell-mediated immunity.

- The A and O blood types are the most common, and B and AB are the rarest.

- Disorders of the immune system can be inherited or acquired by infection.

Pathogens Disease-causing agents.

The skin is not part of the immune system but is a physical barrier.

The skin is a physical barrier to infectious agents such as viruses and bacteria and prevents them from entering the body. The skin's outer surface is home to bacteria, fungi, and even mites, but they cannot penetrate the protective layers of dead cells or the tightly interlocked cells in the layers of skin cells. Epithelial cells that line the internal body cavities and ducts (such as the lungs) are coated with mucus that protects against infection. The mucus in some parts of the body contains an enzyme, lysozyme, that breaks down the cell walls of bacteria, adding another layer of protection.

There are two parts to the immune system that protect against infection.

The second line of defense is a series of chemical reactions and cellular responses that respond to pathogens that have entered the body. These reactions, which are part of the innate immune system, are nonspecific, and work against most pathogens.

The nonspecific responses are designed to identify, inactivate, and kill pathogens such as bacteria and viruses. If these defenses do not stop the disease-causing agents, the third

Antibody-mediated immunity Immune reaction that protects primarily against invading viruses and bacteria using antibodies produced by plasma cells.

Cell-mediated immunity Immune reaction mediated by T cells directed against body cells that have been infected by viruses or bacteria.

and most effective defense system is the adaptive immune system. This part of the immune system is specific; it recognizes particular pathogens and responds in a specific way to neutralize or kill the invader. This part of the immune system has two components: **antibody-mediated immunity** and **cell-mediated immunity**. In addition, the immune system plays a major role in the success or failure of blood transfusions and organ transplants.

In this chapter, we examine the components of the innate and adaptive immune systems and explore how they are mobilized to respond to an infection. We also consider how the immune system determines blood groups and affects mother–fetus compatibility. The parts of the immune system that play roles in organ transplants and in risk factors for a wide range of diseases will be discussed. Finally, we describe a number of disorders of the immune system, including how HIV/AIDS acts to cripple the immune response of infected individuals.

> **KEEP IN MIND**
>
> Humans have three defenses against infection: the skin and mucous membranes, innate immunity, and the adaptive immune system.

17.2 The Inflammatory Response Is a General Reaction

If microorganisms penetrate the skin or the epithelial cells lining the respiratory, digestive, and urinary systems, a nonspecific response called the inflammatory response develops (Active Figure 17.1). In the area around a wound, white blood cells called macrophages detect and bind to molecules on the surface of the invading bacteria. Binding activates the macrophage, which then engulfs and destroys the bacteria. Activated macrophages also secrete chemical signal molecules called cytokines. The cytokines, along with **histamine** secreted by mast cells in the area of infection, cause nearby capillaries to dilate, increasing blood flow to the area (that's why the area around a cut or scrape gets red and warm). The heat creates an unfavorable environment for microorganism growth, mobilizes additional white blood cells, and raises the metabolic rate in nearby cells. These reactions promote healing. Additional white blood cells migrate out of the capillaries and flood into the area in response to the chemical signals, engulfing and destroying the invading microorganisms.

Histamine A chemical signal produced by mast cells that triggers dilation of blood vessels.

If infection persists, clotting factors in the plasma trigger a cascade of small blood clots that seal off the injured area, preventing the escape of invading organisms, recruiting more white blood cells (including macrophages) to destroy the invading bacteria. Finally, the area is targeted by white blood cells that clean up dead viruses, bacteria, or fungi and dispose of dead cells and debris. This chain of events, beginning with the release of chemical signals and ending with cleanup, is the **inflammatory response**.

Inflammatory response The body's reaction to invading microorganisms, a nonspecific active defense mechanism that the body employs to resist infection.

The inflammatory response is usually enough to stop the spread of infection. In some cases, however, mutations in genes encoding proteins involved in the inflammatory response can produce clinical symptoms of an inflammatory disease.

Genetic disorders cause inflammatory diseases.

The inner layer of intestinal cells is a physical barrier that prevents bacteria in the digestive system from crossing into the body. Failure of the immune system to monitor or respond to bacteria that somehow cross this barrier results in inflammatory bowel diseases. Inflammatory bowel diseases are genetically complex and involve the interaction of environmental factors with genetically predisposed individuals. Ulcerative colitis (OMIM 191390) and Crohn disease (OMIM 266600) are two forms of inflammatory bowel disease caused by malfunctions in the immune system. Crohn disease occurs with a frequency of 1 in 1,000 individuals—mostly young adults. The frequency of this disorder has increased greatly over the last 50 years, presumably as a result of unknown environmental factors. A genetic predisposition to Crohn disease maps to chromosome 16. The gene for this predisposition has been identified and cloned; the *NOD2* gene encodes a receptor found on the surface of certain cells of the immune system. Normally, the receptor detects the presence of

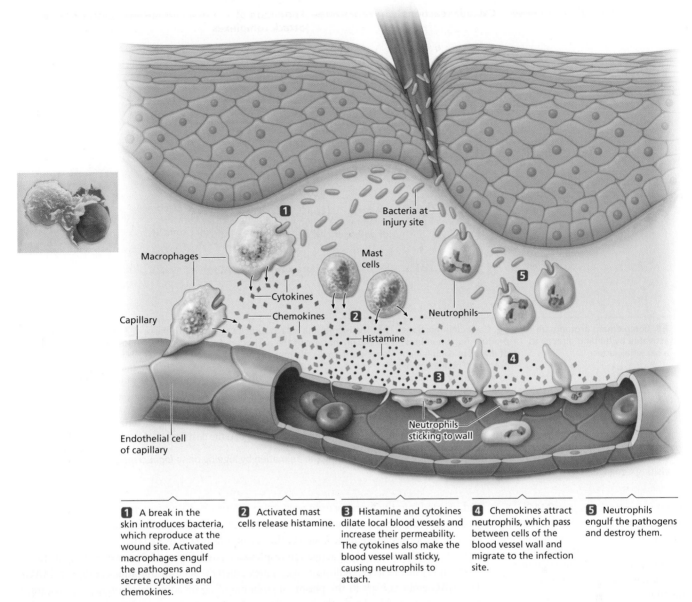

1. A break in the skin introduces bacteria, which reproduce at the wound site. Activated macrophages engulf the pathogens and secrete cytokines and chemokines.

2. Activated mast cells release histamine.

3. Histamine and cytokines dilate local blood vessels and increase their permeability. The cytokines also make the blood vessel wall sticky, causing neutrophils to attach.

4. Chemokines attract neutrophils, which pass between cells of the blood vessel wall and migrate to the infection site.

5. Neutrophils engulf the pathogens and destroy them.

ACTIVE FIGURE 17.1 Stages in the inflammatory response after a bacterial infection.

CENGAGENOW Learn more about the inflammatory response by viewing the animation by logging on to *login.cengage.com/sso* and visiting CengageNOW's Study Tools.

signal molecules on the surface of invading bacteria. Once activated, the receptor signals a protein in the cell nucleus to begin the inflammatory response. In Crohn disease, the protein encoded by the mutant allele is defective and causes an abnormal inflammatory response that damages the intestinal wall. The mutant allele of *NOD2* confers only a predisposition; unknown environmental factors and other genes are probably involved in this disorder.

17.3 The Complement System Kills Microorganisms

The **complement system** is a chemical defense mechanism that works with both the non-specific responses (inflammation) and specific responses (adaptive immune response) to infection. Its name derives from the way it complements the action of the immune system. The complement system consists of some 20 to 30 different proteins synthesized in the liver

Complement system A chemical defense system that kills microorganisms directly, supplements the inflammatory response, and works with (complements) the immune system.

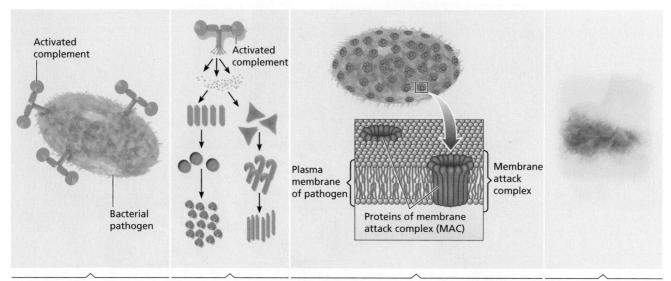

Activation ➡ Cascade reactions ➡ Formation of attack complexes ➡ Lysis of target

1 Complement proteins are activated by binding directly to a bacterial surface.

2 Cascading reactions produce huge numbers of different complement proteins. These assemble to form many membrane attack complexes.

3 The membrane attack complexes insert into the plasma membrane of the pathogen. Each forms a large pore across the membrane.

4 The pores promote lysis of the pathogen, which dies because of the severe disruption of its structure.

ACTIVE FIGURE 17.2 The complement system can be activated by binding directly to the surface of an invading bacterial cell, starting a cascade reaction. This pathway leads to the formation of membrane-attack complexes (MACs) and the destruction of the invading cell.

CENGAGENOW Learn more about the complement system by viewing the animation by logging on to *login.cengage.com/sso* and visiting CengageNOW's Study Tools.

Membrane-attack complex (MAC)
A large, cylindrical multiprotein that embeds itself in the plasma membrane of an invading microorganism and creates a pore through which fluids can flow, eventually bursting the microorganism.

and secreted into the blood plasma as inactive precursors. Complement proteins are activated by contact with certain molecules on the surface of pathogens and respond by mounting one or more responses. Proteins activated at the site of infection activate other nearby complement proteins, starting a cascade of activation responses (Active Figure 17.2). Several components in this pathway form a large, multiprotein complex called the **membrane-attack complex (MAC)**. The MAC embeds itself in the plasma membrane of an invading microorganism, creating a pore (Figure 17.3). Fluid from the blood plasma flows through the pore into the invading cell in response to an osmotic gradient, eventually bursting the cell.

In addition to destroying microorganisms directly, some complement proteins guide white blood cells called phagocytes to the site of infection. The phagocytes engulf and destroy the invading cells. Other parts of the complement system aid the immune response by binding to the surface of microorganisms and marking them for destruction.

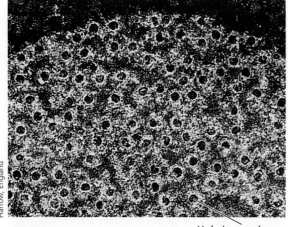

Robert R. Dourmashkin, Courtesy of Clinical Research Centre, Harrow, England

Hole in membrane

FIGURE 17.3 Membrane-attack complexes (MACs) formed by the complement system. The MACs insert themselves into the plasma membrane of the invading cell, forming a pore. This causes water to flow into the cell by osmosis, bursting the cell.

17.4 The Adaptive Immune Response Is a Specific Defense Against Infection

If the nonspecific inflammatory response fails to stop an infection, another, more powerful system—the adaptive immune response—is called into action. The adaptive immune system generates a chemical and cellular response that neutralizes and/or destroys viruses, bacteria, fungi, and cancer cells. The adaptive immune response develops more slowly than the innate response, but it is more effective than the nonspecific defense system and has a memory component that remembers previous encounters with infectious agents (the innate immune system has no memory component). Immunological memory allows a rapid, massive response to a second exposure to a pathogen.

How does the immune response function?

The immune response is mediated by white blood cells called **lymphocytes**. The two main cell types in the immune system are called **B cells** and **T cells**. Both cell types are formed by mitotic division from **stem cells** in bone marrow, and both play important roles in the immune response.

Once formed, B cells mature in the bone marrow. As they develop, each B cell becomes genetically programmed to produce large quantities of a unique protein called an **antibody**. Antibodies are displayed on the surface of the B cell and bind to foreign molecules and microorganisms such as bacterial or fungal cells and toxins in order to inactivate them. Molecules that bind to antibodies are called **antigens** (*anti*body *gen*erators) because they trigger, or generate, an antibody response. Most antigens are proteins or proteins combined with polysaccharides, but *any* molecule, regardless of its source, that can bind to an antibody is an antigen.

T cells are formed in the bone marrow, and while still immature, migrate to the thymus gland where they become programmed to produce unique cell-surface proteins called **T-cell receptors (TCRs)**. These receptors bind to protein markers on the surface of cells infected with viruses, bacteria, or intracellular parasites. Mature T cells circulate in the blood and are also found in lymph nodes and the spleen.

It is important to remember that each B cell makes only one type of antibody and each T cell makes only one type of receptor. Because there are literally billions of possible antigens, there are billions of possible combinations of antibodies and TCRs. When an antigen binds to a TCR or an antibody on the surface of a T cell or B cell, it stimulates that cell to divide, producing a large population of genetically identical descendants, or clones, all with the same TCR or antibody. This process is called clonal selection (Active Figure 17.4).

Specific molecular markers on cell surfaces also play a role in the immune response. Each cell in the body carries recognition molecules that prevent the immune

Lymphocytes White blood cells that originate in bone marrow and mediate the immune response.

B cell A type of lymphocyte that matures in the bone marrow and mediates antibody-directed immunity.

T cell A type of lymphocyte that undergoes maturation in the thymus and mediates cellular immunity.

Stem cells Cells with two properties: the ability to replicate themselves, and the ability to form a variety of cell types in the body.

Antibody A class of proteins produced by B cells that bind to foreign molecules (antigens) and inactivate them.

Antigens Molecules usually carried or produced by viruses, microorganisms, or cells that initiate antibody production.

T-cell receptors (TCRs) Unique proteins on the surface of T cells that bind to specific proteins on the surface of cells infected with viruses, bacteria, or intracellular parasites.

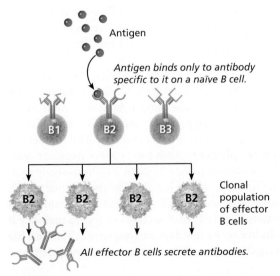

Antigen

Antigen binds only to antibody specific to it on a naïve B cell.

B1   B2   B3

B2   B2.   B2   B2   Clonal population of effector B cells

All effector B cells secrete antibodies.

ACTIVE FIGURE 17.4 Clonal selection. An antigen binds to a specific antibody on a B cell, called a naïve B cell because it has not encountered an antigen before. This encounter triggers mitosis and the buildup of a large population of cells derived from the activated cell. Because all cells in the population are derived from a single ancestor, they are clones.

CENGAGENOW Learn more about clonal selection by viewing the animation by logging on to *login.cengage.com/sso* and visiting CengageNOW's Study Tools.

Major histocompatibility complex (MHC) A set of genes on chromosome 6 that encodes recognition molecules that prevent the immune system from attacking a body's own organs and tissues.

Antibody-mediated immunity Immune reaction that protects primarily against invading viruses and bacteria using antibodies produced by plasma cells.

Cell-mediated immunity Immune reaction mediated by T cells directed against body cells that have been infected by viruses or bacteria.

Helper T cell A lymphocyte that stimulates the production of antibodies by B cells when an antigen is present and stimulates division of B cells and cytotoxic T cells.

Plasma cells Daughter cells of B cells, which synthesize and secrete 2,000 to 20,000 antibody molecules per second into the bloodstream.

Memory B cell A long-lived B cell produced after exposure to an antigen that plays an important role in secondary immunity.

system from attacking our organs and tissues. These markers are encoded by a set of genes on chromosome 6 called the **major histocompatibility complex (MHC)**. The MHC proteins bind to antigens and stimulate the immune response. MHC proteins also play a major role in successful organ transplants, as will be described in a later section.

The immune system has two interconnected parts: **antibody-mediated immunity**, regulated by B-cell antibody production, and **cell-mediated immunity**, controlled by T cells. The two systems are connected by **helper T cells**.

The steps involved in the responses are similar:

1. White blood cells recognize an antigen.
2. The cells become activated and divide to form a clone of identical cells.
3. The clones of activated cells attack and destroy the invading pathogens, clearing the antigens from the body.
4. Some activated cells form memory cells that circulate through the body, ready to mount a rapid and massive response if the same pathogen invades the body again.

Antibody-mediated reactions detect antigens circulating in the blood or body fluids and interact with helper T cells, which signal the B cell with antibodies against that antigen to divide. Helper T cells also activate division of cytotoxic T cells.

Cell-mediated immunity attacks cells of the body infected by viruses or bacteria. T cells also protect against infection by parasites, fungi, and protozoans. One group of T cells also can kill cells of the body if they become cancerous.

Table 17.1 compares the antibody-mediated and cell-mediated immune reactions.

The antibody-mediated immune response involves several stages.

The antibody-mediated immune response has several stages: antigen detection, activation of helper T cells, and division of B cells to form antibody-producing plasma cells (Active Figure 17.5). A specific immune system cell type controls each of these steps. Let's start with a T cell as it encounters an antigen and follow the stages of antibody production and the immune response. In this example, we'll begin with a white blood cell called a dendritic cell, which is a phagocyte—that is, a cell that engulfs and destroys bacteria. Once a dendritic cell engulfs a bacterium, some of the partially digested bacterial proteins bind to dendritic proteins called class II MHC proteins. These protein complexes are displayed on the surface of the dendritic cell, which is now called an antigen-presenting cell (APC). When a T cell with antigen-specific receptors (called T-cell receptors, or TCRs) on its surface encounters a matching antigen on the surface of an APC, the APC cell responds by secreting a cytokine that activates the T cell, which divides to form a large clone of cells called helper T cells. The steps in T-cell activation are summarized in Figure 17.6.

In the next stage of the antibody-mediated immune response, a B cell is activated by the helper T cells. B-cell activation occurs when a B cell with a surface receptor (B cell receptor, or BCR) carrying the antigen is recognized by the helper T cell. This is the same antigen that activated the T cell in the first place.

B-cell activation can begin before an encounter with a helper T cell if the B cell binds to bacterial antigen molecules it encounters in the bloodstream. Once that happens, the receptor and antigen are internalized, and pieces of the antigen bind to class II MHC proteins, which move to the cell surface. When a helper T cell meets a B cell displaying the same antigen, they link together, and the T cell secretes a cytokine called interleukin that activates the B cell. The activated B cell divides to form two types of daughter cells. The first type is **plasma cells** which synthesize and secrete 2,000 to 20,000 antibody molecules *per second* into the bloodstream. The steps in B-cell activation are summarized in Figure 17.7. Plasma cells have cytoplasm filled with rough endoplasmic reticulum—an organelle associated with protein synthesis (Figure 17.8). A second cell type, a **memory B cell**, also forms at this time. Plasma cells live only a few days, but memory cells have a life span of months or even years. Memory cells are part of the immune memory system and are described in a later section.

Table 17.1 Comparison of Antibody-Mediated and Cell-Mediated Immunity

Antibody-Mediated Immunity	Cell-Mediated Immunity
Principal cellular agent is the B cell. B cell responds to bacteria, bacterial toxins, and some viruses.	Principal cellular agent is the T cell; responds to cancer cells, virally infected cells, single-celled fungi, parasites, and foreign cells in an organ transplant.
When activated, B cells form memory cells and plasma cells, which produce antibodies to these antigens.	When activated, T cells differentiate into memory cells, cytotoxic cells, suppressor cells, and helper cells; cytotoxic T cells attack the antigen directly.

Antibody-mediated immune response

T-cell activation

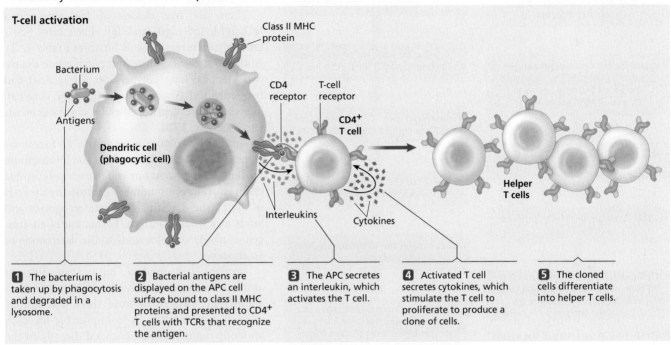

1 The bacterium is taken up by phagocytosis and degraded in a lysosome.

2 Bacterial antigens are displayed on the APC cell surface bound to class II MHC proteins and presented to CD4⁺ T cells with TCRs that recognize the antigen.

3 The APC secretes an interleukin, which activates the T cell.

4 Activated T cell secretes cytokines, which stimulate the T cell to proliferate to produce a clone of cells.

5 The cloned cells differentiate into helper T cells.

B-cell activation and antibody production

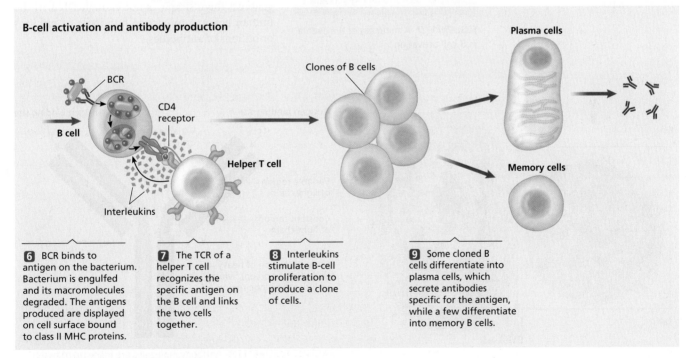

6 BCR binds to antigen on the bacterium. Bacterium is engulfed and its macromolecules degraded. The antigens produced are displayed on cell surface bound to class II MHC proteins.

7 The TCR of a helper T cell recognizes the specific antigen on the B cell and links the two cells together.

8 Interleukins stimulate B-cell proliferation to produce a clone of cells.

9 Some cloned B cells differentiate into plasma cells, which secrete antibodies specific for the antigen, while a few differentiate into memory B cells.

ACTIVE FIGURE 17.5 Overview of the cell–cell interactions in the antibody-mediated immune response.

CENGAGENOW Learn more about the antibody-mediated immune response by viewing the animation by logging on to *login.cengage.com/sso* and visiting CengageNOW's Study Tools.

Antibody-mediated immune response: T-cell activation

Dendritic cell (a phagocyte) is activated by engulfing a pathogen such as a bacterium.

↓

Pathogen macromolecules are degraded in dendritic cell, producing antigens.

↓

Dendritic cell becomes an antigen-presenting cell (APC) by displaying antigens on surface bound to class II MHC proteins.

↓

APC presents antigen to CD4+ T cell and activates the T cell.

↓

CD4+ T cell proliferates to produce a clone of cells.

↓

Clonal cells differentiate into helper T cells, which aid in effecting the specific immune response to the antigen.

FIGURE 17.6 A summary of the events in T-cell activation.

Antibody-mediated immune response: B-cell activation

A BCR on a B cell recognizes antigens on the same bacterial type and engulfs the bacterium.

↓

Pathogen macromolecules are degraded in the B cell, producing antigens.

↓

B cell displays antigens on its surface bound to class II MHC proteins.

↓

Helper T cell with TCR that recognizes the same antigen links to the B cell.

↓

Helper T cell secretes interleukins that activate the B cell.

↓

B cell proliferates to produce a clone of cells.

↓

Some B-cell clones differentiate into plasma cells, which secrete antibodies specific to the antigen, and others differentiate into memory B cells.

FIGURE 17.7 A summary of the events in B-cell activation.

Antibodies are molecular weapons against antigens.

Antibodies secreted by plasma cells are Y-shaped protein molecules that bind to specific antigens in a lock-and-key manner to form an antigen–antibody complex (Active Figure 17.9). Antibodies belong to a class of proteins known as **immunoglobulins (Ig)**.

There are five classes of Igs—abbreviated IgD, IgM, IgG, IgA, and IgE. Each class has a unique structure, size, and function (Table 17.2). Antibody molecules have four polypeptide chains: two identical long polypeptides (H chains) and two identical short polypeptides (L chains). The chains are held together by chemical bonds (Active Figure 17.9).

Antibody structure is related to its functions: (1) recognize and bind an antigen and (2) inactivate the bound antigen. At one end of each polypeptide chain is an antigen-binding site formed by the ends of the H and L chains. This site recognizes and binds to a specific antigen. Formation of an antigen–antibody complex leads to the destruction of an antigen.

Humans can produce billions of different antibody molecules, each of which can bind to a different antigen. Because there are billions of such combinations, it is impossible for each antibody molecule to be encoded directly in the genome; there simply is not enough DNA in the human genome to encode hundreds of millions or billions of antibodies.

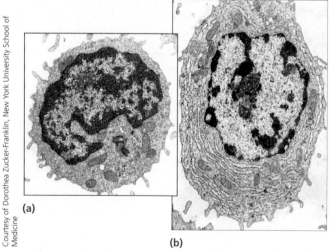

Courtesy of Dorothea Zucker-Franklin, New York University School of Medicine

(a)

(b)

FIGURE 17.8 Electron micrographs of B cells. (a) A mature, unactivated B cell that is not producing antibodies. In this unactivated cell, there is little endoplasmic reticulum. (b) A plasma cell (an activated B cell) that is producing antibodies. The cytoplasm is filled with rough endoplasmic reticulum associated with protein synthesis.

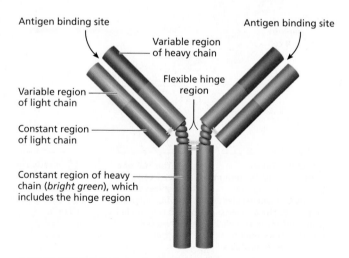

Antigen binding site
Variable region of heavy chain
Antigen binding site
Flexible hinge region
Variable region of light chain
Constant region of light chain
Constant region of heavy chain (*bright green*), which includes the hinge region

ACTIVE FIGURE 17.9 Antibody molecules are made up of two different proteins (an H chain and an L chain). The molecule is Y shaped and forms a specific antigen-binding site at the ends.

CENGAGENOW Learn more about antibodies and antigens by viewing the animation by logging on to *login.cengage.com/sso* and visiting CengageNOW's Study Tools.

Table 17.2 Types and Functions of the Immunoglobulins

Class	Location and Function
IgD	Present on surface of many B cells, but function uncertain; may be a surface receptor for B cells; plays a role in activating B cells.
IgM	Found on surface of B cells and in plasma; acts as a B-cell surface receptor for antigens secreted early in primary response; powerful agglutinating agent.
IgG	Most abundant immunoglobulin in the blood plasma; produced during primary and secondary response; can pass through the placenta, entering fetal bloodstream, thus providing protection to fetus.
IgA	Produced by plasma cells in the digestive, respiratory, and urinary systems, where it protects the surface linings by preventing attachment of bacteria to surfaces of epithelial cells; also present in tears and breast milk; protects lining of digestive, respiratory, and urinary systems.
IgE	Produced by plasma cells in skin, tonsils, and the digestive and respiratory systems; protects against many parasites; overproduction is responsible for allergic reactions, including hay fever and asthma.

Synthesis of a vast number of different antibodies is possible as a result of genetic recombination in three clusters of antibody genes. These are the heavy-chain genes (H genes) on chromosome 14 and two clusters of light-chain genes—the L genes on chromosome 2 and the L genes on chromosome 22. These recombination events take place in B cell nuclei during maturation, producing a unique gene in each B cell that produces one type of antibody. This rearranged gene is stable and is passed on to all daughter B cells. This process of recombination makes it possible to produce billions of possible antibody combinations from only three gene sets.

Immunoglobulins (Ig) The five classes of proteins to which antibodies belong.

T cells mediate the cellular immune response.

The cellular immune response is mediated by cytotoxic, or killer, T cells. Cytotoxic T cells find and destroy cells of the body that are infected with a virus, bacteria, or other infectious agents (Active Figure 17.10).

Cell-mediated immune response

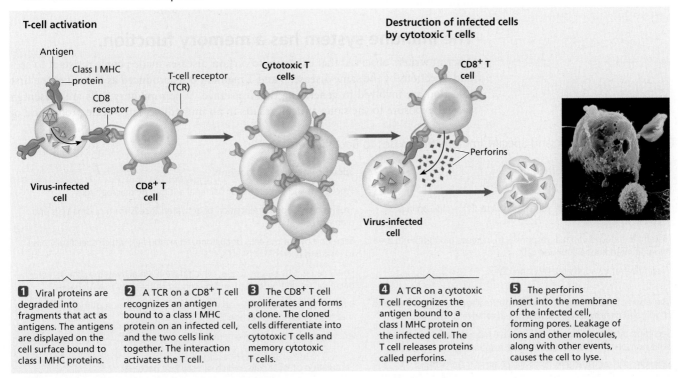

T-cell activation

Antigen
Class I MHC protein
CD8 receptor
T-cell receptor (TCR)
Cytotoxic T cells
Virus-infected cell
CD8⁺ T cell

Destruction of infected cells by cytotoxic T cells

CD8⁺ T cell
Perforins
Virus-infected cell

1 Viral proteins are degraded into fragments that act as antigens. The antigens are displayed on the cell surface bound to class I MHC proteins.

2 A TCR on a CD8⁺ T cell recognizes an antigen bound to a class I MHC protein on an infected cell, and the two cells link together. The interaction activates the T cell.

3 The CD8⁺ T cell proliferates and forms a clone. The cloned cells differentiate into cytotoxic T cells and memory cytotoxic T cells.

4 A TCR on a cytotoxic T cell recognizes the antigen bound to a class I MHC protein on the infected cell. The T cell releases proteins called perforins.

5 The perforins insert into the membrane of the infected cell, forming pores. Leakage of ions and other molecules, along with other events, causes the cell to lyse.

ACTIVE FIGURE 17.10 The cell-mediated immune response.

CENGAGENOW Learn more about the cell-mediated immune response by viewing the animation by logging on to *login.cengage.com/sso* and visiting CengageNOW's Study Tools.

FIGURE 17.11 Killer T cells (*yellow*) attacking a cancer cell (*red*).

Jean Claude Revi/Phototake

When a cell becomes infected with a virus, viral proteins bound to class I MHC proteins appear on the surface, forming an APC cell. Those foreign antigens are recognized by receptors on the surface of a type of T cell called a CD8+ cell. The activated T cell divides to form a clone of cells, some of which form memory T cells. The cytotoxic T cell attaches to the infected APC cell and secretes a protein, perforin, which punches holes in the plasma membrane of the infected cell. The cytoplasmic contents of the infected cell leak out through the holes, and the infected cell dies and is removed by phagocytes.

Cytotoxic T cells also kill cancer cells (Figure 17.11) and transplanted organs if they recognize them as foreign. Table 17.3 summarizes the nonspecific and specific reactions of the immune system.

The immune system has a memory function.

Ancient writers observed that exposure to certain diseases made people resistant to second infections by the same disease. B and T memory cells produced as a result of the first infection are involved in generating this resistance. When memory cells are present, a second exposure to the same antigen results in an immediate, large-scale production of

Table 17.3 Nonspecific and Specific Immune Responses to Bacterial Infection

Nonspecific Immune Mechanisms	Specific Immune Mechanisms
INFLAMMATION	Processing and presenting of bacterial antigen by macrophages
Engulfment of invading bacteria by resident tissue macrophages	Proliferation and differentiation of activated B-cells form plasma cells and memory cells
Histamine-induced vascular responses to increase blood flow to area, bringing in additional immune cells	Secretion by plasma cells of customized antibodies, which specifically bind to invading bacteria
Walling off of invaded area by fibrin clot	Enhancement by helper T cells, which have been activated by the same bacterial antigen processed and presented to them by macrophages
Migration of neutrophils and monocytes/macrophages to the area to engulf and destroy foreign invaders and remove cellular debris	Binding of antibodies to invading bacteria and activation of mechanisms that lead to their destruction
Secretion by phagocytic cells of chemical mediators, which enhance both nonspecific and specific immune responses	Activation of lethal complement system
NONSPECIFIC ACTIVATION OF THE COMPLEMENT SYSTEM	Stimulation of killer cells, which directly lyse bacteria
Formation of hole-punching, membrane-attack complex that lyses bacterial cells	Persistence of memory cells capable of responding more rapidly and more forcefully should the same bacterial strain be encountered again
Enhancement of many steps of inflammation	

antibodies and cytotoxic T cells (Figure 17.12). Because of the presence of the memory cells, the second reaction is faster and more massive and lasts longer than the primary immune response.

The immune response controlled by memory cells is the reason we can be vaccinated against infectious diseases. A **vaccine** stimulates the production of memory cells against a disease-causing agent. A vaccine is really a weakened disease-causing antigen, given orally or by injection, that provokes a primary immune response and the production of memory cells. Often, a second dose is administered to elicit a secondary response that raises, or "boosts," the number of memory cells (that is why such shots are called booster shots).

Vaccines are made from killed or weakened strains (called attenuated strains) of disease-causing agents that stimulate the immune system but do not produce life-threatening symptoms of the disease. Recombinant DNA methods now are used to prepare vaccines against a number of diseases that affect humans and farm animals.

A global vaccination program eliminated smallpox in 1972, and a new effort is attempting to eliminate polio by vaccinating children worldwide. Overall, millions of lives have been saved by vaccination, and it remains one of the foundations of public health.

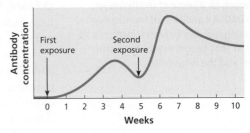

FIGURE 17.12 Antibody levels in the first response to an infection, and the response after a second exposure. The second response is much faster and stronger than the first response, conferring immunity to a second infection.

Vaccine A preparation containing dead or weakened pathogens that elicits an immune response when injected into the body.

17.5 Blood Types Are Determined by Cell-Surface Antigens

Antigens on the surface of blood cells determine compatibility in blood transfusions. There are about 30 known antigens on blood cells, each of which constitutes a blood group, or **blood type**. For successful transfusions, certain critical antigens of the donor and recipient, if present, must be matched. If transfused red blood cells do not have matching surface antigens, the recipient's immune system will produce antibodies against the antigen, clumping the transfused cells. The clumped blood cells block circulation in capillaries and other small blood vessels, with severe and often fatal results. In transfusions, two blood groups are of major significance: the ABO system and the Rh blood group.

Blood type One of the classes into which blood can be separated on the basis of the presence or absence of certain antigens.

ABO blood typing allows for safe blood transfusions.

ABO blood types are determined by a gene I (I for isoagglutinin) encoding an enzyme that alters a cell-surface protein. This gene has three alleles, I^A, I^B, and I^O—often written as A, B, and i. The A and B alleles each produce a slightly different version of the enzyme, and the i allele produces no gene product. Those with type A blood have an A antigen on their red blood cells and do not produce antibodies against this cell-surface marker. However, people with type A blood do have antibodies against the antigen encoded by the B allele (Table 17.4). Those with type B blood carry the B antigen on their red cells and have antibodies against the A antigen. If you have type AB blood, both antigens are present on red blood cells and no antibodies against A and B are made. Those with type

Table 17.4 Summary of ABO Blood Types

Blood Type	Antigens on Plasma Membranes of RBCs	Antibodies in Blood	Safe to Transfuse	
			To	From
A	A	Anti-B	A, AB	A, O
B	B	Anti-A	B, AB	B, O
AB	A + B	None	AB	A, B, AB, O
O	—	Anti-A Anti-B	A, B, AB, O	O

O blood have neither antigen but do have antibodies against both the A antigen and the B antigen.

Because AB individuals carry no antibodies against A or B, they can receive a transfusion of blood of any type. Type O individuals have neither antigen and can donate blood to anyone, even though their plasma contains antibodies against A and B; after transfusion, the concentration of these antibodies is too low to cause problems.

When transfusions are made between people with incompatible blood types, several problems arise. Figure 17.13 shows the cascade of reactions that follows transfusion of someone who has type A with type B blood. Antibodies to the B antigen are in the blood of the recipient. They bind to the transfused red blood cells, causing them to clump. The clumped cells restrict blood flow in capillaries, reducing oxygen delivery. The breakdown of these clumped red blood cells releases large amounts of hemoglobin into the blood. The hemoglobin forms deposits in the kidneys that block the tubules of the kidney and often cause kidney failure.

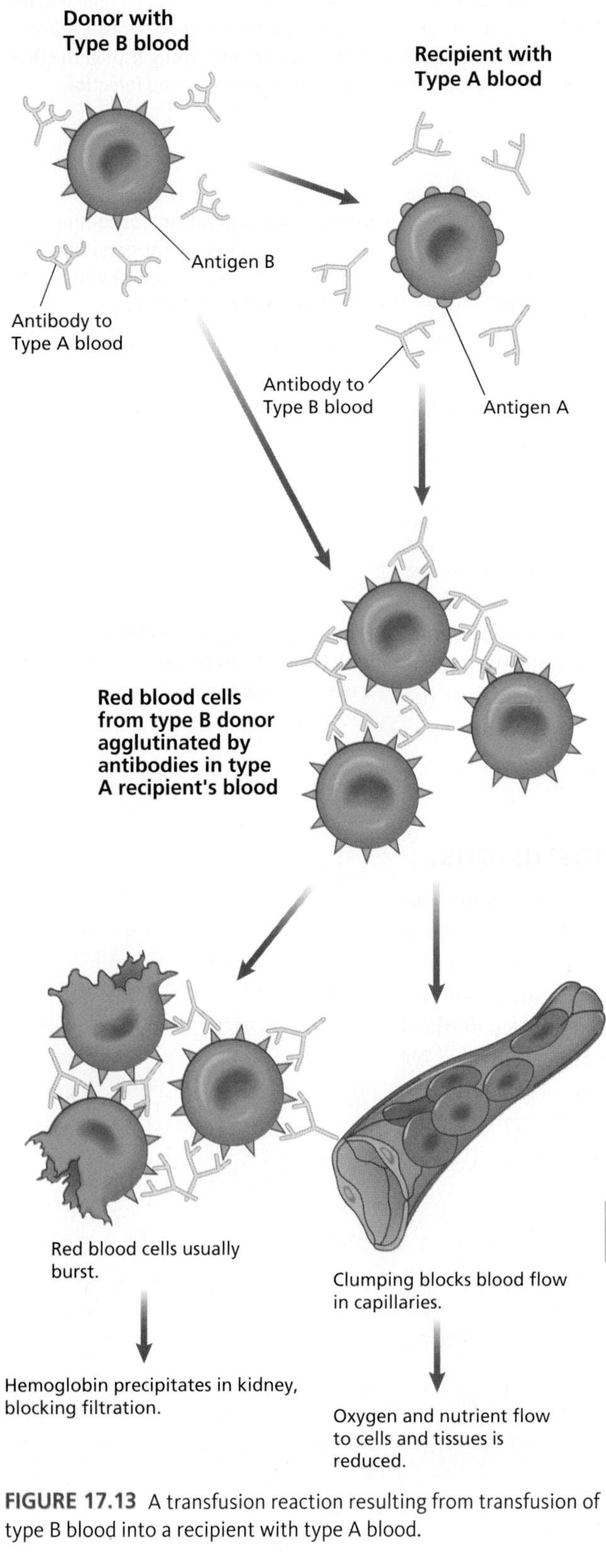

Rh blood types can cause immune reactions between mother and fetus.

The Rh blood group (named for the rhesus monkey, in which it was discovered) includes those who can make the Rh antigen (Rh-positive, Rh⁺) and those who cannot make the antigen (Rh-negative, Rh⁻).

The Rh blood group is a major concern when there is incompatibility between mother and fetus—a condition known as **hemolytic disease of the newborn (HDN).** This occurs most often when the mother is Rh⁻ and the fetus is Rh⁺ (Active Figure 17.14). If Rh⁺ blood from the fetus enters the Rh⁻ maternal circulation, the mother's immune system will produce antibodies against the Rh antigen. If fetal blood mixes with that of the mother during birth, she will make antibodies against the Rh antigen. During a subsequent pregnancy with an Rh⁺ fetus, massive amounts of maternal antibodies cross the placenta in late stages of pregnancy and destroy the fetus's red blood cells, resulting in HDN.

To prevent HDN, Rh⁻ mothers are given an Rh-antibody preparation (RhoGam) during the first pregnancy or after a miscarriage or abortion if the child or fetus is Rh⁺. The injected Rh antibodies destroy any Rh⁺ fetal cells that may have entered the mother's circulation. To be effective, this antibody must be administered before the mother's immune system can make antibodies against the fetal Rh antigen.

Donor with Type B blood

Recipient with Type A blood

Antigen B

Antibody to Type A blood

Antibody to Type B blood

Antigen A

Red blood cells from type B donor agglutinated by antibodies in type A recipient's blood

Red blood cells usually burst.

Clumping blocks blood flow in capillaries.

Hemoglobin precipitates in kidney, blocking filtration.

Oxygen and nutrient flow to cells and tissues is reduced.

FIGURE 17.13 A transfusion reaction resulting from transfusion of type B blood into a recipient with type A blood.

17.6 | Organ Transplants Must Be Immunologically Matched

Successful organ transplants and skin grafts depend on matches between cell-surface antigens of the donor and the recipient. These antigens are proteins found on all cells in the body and serve as identification tags, helping distinguish self from nonself.

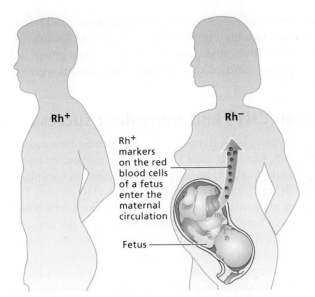

(a) A forthcoming child of an Rh⁻ woman and Rh⁺ man inherits the Rh⁺ allele. During pregnancy or childbirth, some fetal cells bearing the Rh⁺ allele may leak into the maternal bloodstream.

(b) The presence of the Rh⁺ allele stimulates her body to make antibodies. If she gets pregnant again and if this second fetus (or any other) inherits the Rh⁺ allele, the circulating anti-Rh⁺ antibodies will act against it.

ACTIVE FIGURE 17.14 The Rh factor and pregnancy. (a) Rh⁺ cells from the fetus can enter the maternal circulation at birth. The Rh⁻ mother produces antibodies against the Rh factor. (b) In a subsequent pregnancy, if the fetus is Rh⁺, the maternal antibodies cross into the fetal circulation and destroy fetal red blood cells, producing hemolytic disease of the newborn (HDN).

CENGAGENOW Learn more about the Rh factor and pregnancy by viewing the animation by logging on to *login.cengage.com/sso* and visiting CengageNOW's Study Tools.

In humans, a cluster of genes on chromosome 6, known as the *HLA* genes, part of the major histocompatibility complex (MHC), produces these antigens. *HLA* genes play a critical role in the outcome of transplants. The MHC complex contains several *HLA* gene clusters. The class I cluster consists of *HLA-A*, *HLA-B*, and *HLA-C* genes. Adjacent to this is a cluster called class II, which consists of *HLA-DR*, *HLA-DQ*, and *HLA-DP*. A large number of alleles have been identified for each *HLA* gene, making millions of allele combinations possible. The array of *HLA* alleles carried on each copy of chromosome 6 is known as a **haplotype.** Because each of us has two copies of chromosome 6, we each have two *HLA* haplotypes (Figure 17.15).

Haplotype A set of genetic markers located close together on a single chromosome or chromosome region.

Because so many allele combinations are possible, it is rare that any two individuals have a perfect HLA match. The exceptions are identical twins, who will have identical *HLA* allele haplotypes, and siblings, who have a 25% chance of being matched. In the example shown in Figure 17.15, each child receives one haplotype from each parent. As a result, four new haplotype combinations are represented in the children. (Thus, siblings have a one-in-four chance of having the same haplotypes.)

Successful transplants depend on HLA matching.

Successful organ transplants depend largely on matching *HLA* haplotypes between donor and recipient. Because there are so many *HLA* alleles, the best chance for a match is usually between related individuals, with identical twins having a 100% match. The order of preference for organ and tissue donors among relatives is

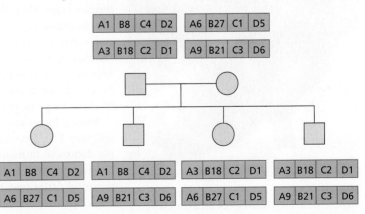

FIGURE 17.15 The transmission of *HLA* haplotypes. In this simplified diagram, each haplotype contains four genes, each of which encodes a different antigen.

identical twin, sibling, parent, and unrelated donor. Among unrelated donors and recipients, the chances for a successful match are only 1 in 100,000 to 1 in 200,000. Because the frequency of *HLA* alleles differs widely across ethnic groups, matches across groups are often more difficult. When HLA types are matched, the survival of transplanted organs is dramatically improved.

Copy number variation (CNV) and transplant success.

One complication of bone marrow transplants is known as graft versus host disease (GVHD). In a bone marrow transplant, stem cells from a donor's bone marrow are transplanted into a recipient who has leukemia or other cancer of the blood. Before the transplant, the recipient's bone marrow stem cells are killed by radiation and chemical treatments. The donor's stem cells migrate to the bone marrow and divide to reconstitute blood cells, blood type, and an immune system. Sometimes, however, after transplantation, the donor immune cells recognize antigens on the cells of the recipient as foreign and attack those cells, causing GVHD. This sometimes occurs even in siblings with closely matched *HLA* alleles.

To investigate the cause of GVHD, researchers scanned the genomes of 1,300 HLA-matched siblings who were bone marrow donor-recipient pairs. The analysis showed that copy number variations (CNVs) for the gene *UGT2B17* (OMIM 601903) was a factor in GVHD when the gene was absent from the donor's genome but present in the recipient's. When this CNV existed, there was a 2.5-fold increase in the incidence of GVHD. It seems likely that other CNVs as well as other genome variations may also be important in transplant compatibility.

Genetic engineering makes animal–human organ transplants possible.

In the United States, about 18,000 organs are transplanted each year, but about 50,000 qualified patients are on waiting lists. Each year, almost 4,000 people on waiting lists die before receiving transplants, and another 100,000 die even *before* they are placed on a waiting list. Although the demand for organ transplants is rising, the number of donated organs is growing very slowly. Experts estimate that more than 50,000 lives would be saved each year if enough organs were available.

One way to increase the supply of organs is to use animal donors for transplants. Animal–human transplants (called **xenotransplants**) have been attempted many times, but with little success. Two important biological problems are related to xenotransplants: (1) complement-mediated rejection and (2) T cell–mediated rejection. In complement rejection, species-specific MHC proteins on the donor organ are detected by the complement system of the recipient. When an animal organ (e.g., from a pig) is transplanted into a human, the pig's MHC proteins are so different that they trigger an immediate and massive immune response known as hyperacute rejection. This reaction, which is mediated by the complement system, usually destroys the transplanted organ within hours.

To overcome this rejection, several research groups isolated and cloned human genes that block the complement reaction. Those genes were injected into fertilized pig eggs, and the resulting transgenic pigs carry human-recognition antigens on all their cells (Figure 17.16). Organs from these transgenic pigs should appear as human organs to the recipient's immune system, preventing a hyperacute rejection. Transplants from genetically engineered pigs to monkey hosts have been successful, but the ultimate step will be an organ transplant from a transgenic pig to a human.

Even if hyperacute rejection can be suppressed, transplanted pig organs will still face T cell–mediated rejection of the transplant. Because transplants from pig donors to humans occur across species, the tendency toward rejection may be stronger and require the lifelong use of immunosuppressive drugs. Those powerful drugs may be toxic when taken over a period of years or will weaken the immune system, paving the way for continuing rounds of infections.

One solution to this problem is to transplant bone marrow from the donor pig to the human recipient. The resulting pig–human immune system (called

Xenotransplants Cells, tissues, or organs that are transplanted from one species to another.

© Dong Min-Jang/EPN/ZUMA Press

FIGURE 17.16 Transgenic pigs are genetically engineered to carry human *HLA* genes, allowing transplantation of pig organs into human recipients.

a chimeric immune system) would recognize the pig organ as "self" and still retain normal human immunity. As farfetched as this may sound, animal experiments using this approach have been successful in preventing rejection for more than 2 years after transplantation without the use of immunosuppressive drugs. This same method has been used in human-to-human heart transplants to increase the chances of successful outcomes (transplants between members of the same species are called *allografts*).

As recently as 10 years ago, the possibility of animal–human organ transplants seemed remote, more suited to science fiction than to medical fact. There are now more than 200 people in the United States who have received xenografts of animal cells or tissues. The advances described here make it likely that xenotransplants of major organs to humans will be attempted in the next few years. Although animal organ donors will probably become common in the near future, guidelines for transgenic donors still need to be developed and problems with immunosuppressive drugs and immune tolerance remain to be solved.

17.7   Disorders of the Immune System

We are able to resist infectious disease because we have an immune system. Unfortunately, failures in the immune system can result in abnormal or even absent immune responses.

KEEP IN MIND

Disorders of the immune system can be inherited or acquired by infection.

The consequences of these failures can range from mild inconvenience to systemic failure and death. In this section, we briefly catalog some ways in which the immune system can fail.

Overreaction in the immune system causes allergies.

Allergies result when the immune system overreacts to weak antigens that do not provoke an immune response in most people (Figure 17.17). These weak antigens, called **allergens**, include a wide range of substances: house dust, pollen, cat dander, certain foods, and even medicines such as penicillin. It is estimated that up to 10% of the U.S. population has at least one allergy (see Exploring Genetics: Peanut Allergies Are Increasing on page 393). Typically, allergic reactions develop after a first exposure to an allergen. The allergen causes B cells to make IgE antibodies instead of IgG antibodies. The IgE antibodies attach to mast cells in tissues, including those of the nose and the respiratory system.

In a second exposure, the allergen binds to IgE antibodies made during the first exposure and the mast cells release histamine, triggering a systemic inflammatory response that causes fluid accumulation, tissue swelling, and mucous secretion. This reaction can be severe in some individuals, and as histamine is released into the circulatory system, it may cause a life-threatening decrease in blood pressure and constriction of airways in the lungs. This reaction, called **anaphylaxis** or anaphylactic shock, most often occurs after exposure to antibiotics, the venom in bee or wasp stings, or certain foods. Prompt treatment of anaphylaxis with antihistamines, epinephrine, and steroids can reverse the reaction. As the name suggests, antihistamines block the action of histamine. Epinephrine opens the airways and constricts blood vessels, raising blood pressure. Steroids, such as prednisone, inhibit the inflammatory response. Some people who have a history of severe reaction to insect stings or foods carry injectable epinephrine with them in a kit.

Allergens Antigens that provoke an inappropriate immune response.

Anaphylaxis A severe allergic response in which histamine is released into the circulatory system.

Autoimmune reactions cause the immune system to attack the body.

One of the most elegant properties of the immune system is its capacity to distinguish self from nonself and destroy what it perceives as nonself. During development, the immune system "learns" not to react against the cells of the body. In some disorders, this immune

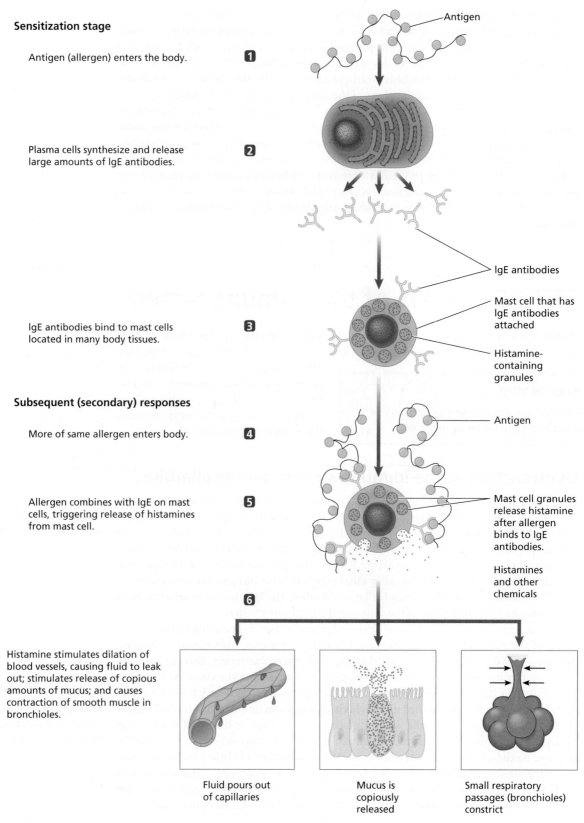

Sensitization stage

Antigen (allergen) enters the body. **1**

Plasma cells synthesize and release large amounts of IgE antibodies. **2**

Antigen

IgE antibodies

IgE antibodies bind to mast cells located in many body tissues. **3**

Mast cell that has IgE antibodies attached

Histamine-containing granules

Subsequent (secondary) responses

More of same allergen enters body. **4**

Antigen

Allergen combines with IgE on mast cells, triggering release of histamines from mast cell. **5**

Mast cell granules release histamine after allergen binds to IgE antibodies.

Histamines and other chemicals

6

Histamine stimulates dilation of blood vessels, causing fluid to leak out; stimulates release of copious amounts of mucus; and causes contraction of smooth muscle in bronchioles.

Fluid pours out of capillaries

Mucus is copiously released

Small respiratory passages (bronchioles) constrict

FIGURE 17.17 The steps in an allergic reaction.

tolerance breaks down, and the immune system attacks and kills cells and tissues in the body. Juvenile diabetes, also known as insulin-dependent diabetes (IDDM; OMIM 222100), is an autoimmune disease. Clusters of cells in the pancreas produce insulin, a hormone that lowers blood sugar levels. In IDDM, the immune system attacks and kills

Peanut Allergies Are Increasing

Allergy to peanuts is one of the most serious food sensitivities and is a growing health concern in the United States. Hypersensitivity to peanuts can provoke a systemic anaphylactic reaction in which the bronchial tubes constrict, closing the airways. Fluids pass from the tissues into the lungs, making breathing difficult. Blood vessels dilate, causing blood pressure to drop, and plasma escapes into the tissues, causing shock. Heart arrhythmias and cardiac shock can develop and cause death within 1 to 2 minutes after the onset of symptoms. About 30,000 cases of food-induced anaphylactic reactions are seen in emergency rooms each year, with 200 fatalities. About 80% of all cases are caused by allergies to peanuts or other nuts. Peanut-sensitive individuals must avoid ingesting peanuts and be trained to recognize the symptoms of anaphylactic reactions. In spite of precautions, accidental exposures caused by cooking pans previously used to cook food with peanuts or the inhalation of peanut dust on airplanes have been reported to cause anaphylactic reactions. Many peanut-sensitive people carry doses of self-injectable epinephrine (EpiPen Autoinjectors and similar products) to stop anaphylaxis in case they are exposed to peanuts.

The number of children and adults allergic to peanuts appears to be increasing. In a 1988–1994 survey of American children, allergic reactions to peanuts were twice as high as they were in a group surveyed from 1980 to 1984. A national survey indicates that about 3 million people in the United States (about 1.1% of the population) are allergic to peanuts, tree nuts, or both. The hypersensitive reaction to one of

three allergenic peanut proteins is mediated by IgE antibodies. Within 1 to 15 minutes of exposure, the IgE antibodies activate mast cells. The stimulated mast cells release large amounts of histamines and chemotactic factors, which attract other white blood cells as part of the inflammatory response. In addition, the mast cells release prostaglandins and other chemicals that trigger an anaphylactic reaction.

What is causing the increase in peanut allergies is unclear. Genetics obviously plays some part, but environmental factors also appear to play a major role. For example, peanut allergies are extremely rare in China, but children of Chinese immigrants have about the same frequency of peanut allergies as children of native-born Americans, pointing to the involvement of environmental factors.

Image copyright SergioZ, 2010. Used under license from Shutterstock.com.

One proposal is that as peanuts have become a major part of the diet in the United States—especially in foods advertised to provide quick energy—exposure of newborns and young children (1 to 2 years of age) to peanuts is now more common. This exposure occurs through breast milk, peanut butter, and other foods. The immune system in newborns is immature and develops over the first few years of life. As a result, food allergies are more likely to develop during the first few years. In the absence of conclusive information, it is recommended that mothers avoid eating peanuts and peanut products during pregnancy and while they are nursing and that children not be exposed to peanuts or other nuts for the first 3 years of life.

Table 17.5 Some Autoimmune Diseases

Addison's disease
Autoimmune hemolytic anemia
Diabetes mellitus, insulin dependent
Graves disease
Membranous glomerulonephritis
Multiple sclerosis
Myasthenia gravis
Polymyositis
Rheumatoid arthritis
Scleroderma
Sjögrens syndrome
Systemic lupus erythematosus

the insulin-producing cells, causing lifelong diabetes and the need for insulin injections to control blood sugar levels.

Other forms of autoimmunity—such as systemic lupus erythematosus (SLE; OMIM 152700)—attack blood cells, organelles such as mitochondria, and DNA-binding proteins in the nucleus. Lupus slowly destroys major organ systems including the kidneys and the heart. Table 17.5 lists some autoimmune diseases.

Genetic disorders can impair the immune system.

The first disease of the immune system was described in 1952 by a physician who examined a young boy who had had at least 20 serious infections in the preceding 5 years. Blood tests showed that the child had no antibodies. Other patients with similar problems were soon discovered. All affected individuals were boys who were highly susceptible to bacterial infections. In all cases, either B cells were completely absent or the B cells were immature and unable to produce antibodies. Without functional B cells, no antibodies can be produced, but there are usually nearly normal levels of T cells. In other words, antibody-mediated immunity is absent or impaired, but cellular immunity is normal. This heritable disorder, called

FIGURE 17.18 David, the "boy in the bubble," had severe combined immunodeficiency and lived in isolation for 12 years. He died of complications after a bone marrow transplant.

X-linked agammaglobulinemia (XLA) A rare, X-linked recessive trait characterized by the total absence of immunoglobulins and B cells.

X-linked agammaglobulinemia (XLA; OMIM 300300), usually appears 5 to 6 months after birth, when maternal antibodies disappear and the infant's B-cell population normally begins to produce antibodies. Patients with XLA are highly susceptible to pneumonia and streptococcal infections and pass from one life-threatening infection to another.

Individuals with XLA lack mature B cells but do have normal populations of immature B cells, indicating that the defective gene controls some stage of development. The *XLA* gene was mapped to Xq21.3–Xq22 and encodes an enzyme that transmits signals from the cell's environment into the cytoplasm. Chemical signals from outside the cell initiate a signal transduction pathway that alters gene expression and helps trigger B-cell maturation. The gene product that is defective in XLA plays a critical role in the signaling process. Understanding the role of the protein in B-cell development may permit the use of gene therapy to treat this disorder.

Severe combined immunodeficiency disease (SCID) A genetic disorder in which affected individuals have no immune response; both the cell-mediated and antibody-mediated responses are missing.

A rare genetic disorder of the immune system causes a complete absence of *both* antibody-mediated and cell-mediated immune responses. This condition is called **severe combined immunodeficiency disease** (SCID; OMIM 102700, 600802, and others). Affected individuals have recurring and severe infections and usually die at an early age from seemingly minor infections. One of the longest known survivors of this condition was David, the "boy in the bubble," who died at 12 years of age after being isolated in a sterile plastic bubble for all but the last 15 days of his life (Figure 17.18).

HIV attacks the immune system.

Acquired immunodeficiency syndrome (AIDS) A collection of disorders that develop as a result of infection with the human immunodeficiency virus (HIV).

The immunodeficiency disorder currently receiving the most attention is **acquired immunodeficiency syndrome (AIDS)**. AIDS is a collection of disorders that develop after infection with the human immunodeficiency virus (HIV) (Active Figure 17.19). Worldwide, about 33 million people are infected with HIV (Table 17.6).

HIV is a retrovirus with three components: (1) a protein coat (which encloses the other two components), (2) RNA molecules (the genetic material), and (3) an enzyme called reverse transcriptase. The virus selectively infects and kills helper T cells, which act as the master "on" switch for the immune system. Once inside a cell, reverse transcriptase transcribes the RNA into a DNA molecule, and the viral DNA is inserted into a human chromosome, where it can remain for months or years until the helper T cell is stimulated by an antigen. Then, the viral genes are activated. Viral RNA and proteins are made, and new viral particles are formed. These particles bud off the surface of the T cell, rupturing and killing the cell and setting off a new round of T-cell infection. Over the

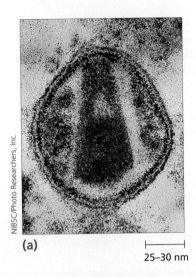

NIBSC/Photo Researchers, Inc.

(a)

|⊢——⊣|
25–30 nm

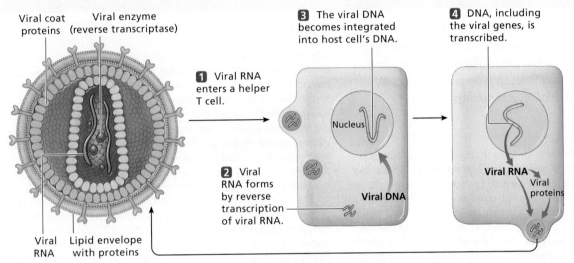

Viral coat proteins

Viral enzyme (reverse transcriptase)

3 The viral DNA becomes integrated into host cell's DNA.

4 DNA, including the viral genes, is transcribed.

1 Viral RNA enters a helper T cell.

Nucleus

2 Viral RNA forms by reverse transcription of viral RNA.

Viral DNA

Viral RNA

Viral proteins

Viral RNA

Lipid envelope with proteins

6 Virus particles that bud from the infected cell may attack a new one.

5 Some transcripts are new viral RNA; others are translated into proteins. Both self-assemble as new virus particles.

(b)

ACTIVE FIGURE 17.19 Steps in HIV replication. (a) Electron micrograph of an HIV particle budding from the surface of an infected T cell. (b) Steps in the HIV life cycle.

CENGAGENOW Learn more about HIV replication by viewing the animation by logging on to *login.cengage.com/sso* and visiting CengageNOW's Study Tools.

Table 17.6 Global HIV and AIDS Cases

Region	AIDS Cases	New HIV Cases
Sub-Saharan Africa	22,500,000	1,700,000
South/Southeast Asia	4,000,000	340,000
Central Asia/Eastern Europe	1,600,000	150,000
Latin America	1,600,000	100,000
North America	1,300,000	46,000
East Asia	800,000	92,000
Western/Central Europe	760,000	31,000
Middle East/North Africa	380,000	35,000
Caribbean Islands	286,000	17,000
Australia/New Zealand	75,000	14,000
Worldwide total	33,200,000	2,500,000

course of HIV infection, the number of helper T4 cells gradually decreases. As the T4-cell population falls, the ability to mount an immune response decreases. The result is increased susceptibility to infection and increased risk of certain forms of cancer. The eventual outcome is premature death brought about by any of a number of diseases that overwhelm the body and its compromised immune system.

HIV is transmitted from infected to uninfected individuals through body fluids, including blood, semen, vaginal secretions, and breast milk. The virus cannot live for more than 1 to 2 hours outside the body and cannot be transmitted by food, water, or casual contact.

Genetics in Practice

Genetics in Practice case studies are critical-thinking exercises that allow you to apply your new knowledge of human genetics to real-life problems. You can find these case studies and links to relevant websites at *www.cengage.com/biology/cummings*

CASE 1

Mary and John Smith went for genetic counseling because John did not believe that their newborn son was his. They both wanted blood tests to help rule out the possibility that someone other than John was the father of this baby. The counselor explained that ABO blood typing could give some preliminary indications about possible paternity. Its use is limited, however, because there are only four possible ABO blood types, and the vast majority of people in any population have only two of those types (A and O). This means that a man may have a blood type consistent with paternity and still not be the father of the tested child. Because the allele for type O can be masked by the genes for A or B, inheritance of blood type can be unclear. More modern (and more expensive) genetic tests, such as DNA typing, would lead to a more reliable conclusion about paternity.

Mary's blood was tested and identified as type O. John's blood was tested and identified as type O. On the basis of these two parental combinations of blood types, the only possible blood type that their son could be is type O. The baby was tested, and his blood type was, indeed, type O.

1. Can any absolute conclusions be drawn on the basis of the results of these blood tests? Why or why not?

2. If not, was it worthwhile doing the test in the first place?

3. Why do you think DNA testing would be more reliable than blood testing for this purpose?

CASE 2

The Joneses were referred to a clinical geneticist because their 6-month-old daughter was failing to grow adequately and was having recurrent infections. The geneticist took a detailed family history (which was uninformative) and a medical history of their daughter. He discovered that their daughter had a history of several ear infections against which antibiotics had no effect, had difficulty gaining weight (failure to thrive), and had an extensive history of yeast infection (thrush) in her mouth. The geneticist did a simple blood test to check their daughter's white blood count and determined that she had severe combined immunodeficiency (SCID).

The geneticist explained that SCID is an immune deficiency that causes a marked susceptibility to infections. The defining characteristic is usually a severe defect in both the T- and B-lymphocyte systems. This results in one or more infections within the first few months of life that are serious and may even be life-threatening. They may include pneumonia, meningitis, and bloodstream infections. Based on the family history, it was possible that their daughter had inherited a mutant allele from each of them and therefore was homozygous for a gene that causes SCID. If so, each time the Joneses had a child, there would be a 25% chance that the child would have SCID. Prenatal testing is available to determine whether the developing fetus has SCID.

1. Genetic testing showed that both parents were heterozygous carriers of a mutant allele of the adenosine deaminase (*ADA*) gene and that the daughter was homozygous for this mutation. Are there any treatment options available for ADA-deficient SCID?

2. If the Joneses want to be certain that their next child will not have SCID, what types of reproductive options do you think they have?

		And the father is			
		A	B	AB	O
If the mother is	A	A or O	A, B, AB, or O	A, B, or AB	A or O
	B	A, B, AB, or O	B or O	A, B, or AB	B or O
	AB	A, B, or AB	A, B, or AB	A, B, or AB	A or B
	O	A or O	B or O	A or B	O
					The child must be

Summary

17.1 The Body Has Three Levels of Defense Against Infection

- The immune system protects the body against infection through a graded series of responses that attack and inactivate foreign molecules and organisms.

17.2 The Inflammatory Response Is a General Reaction

- The lowest level of response to infection involves a nonspecific, local inflammatory response. This response is mediated by cells of the immune system and isolates and kills invading microorganisms. Genetic control of this response is abnormal in inflammatory diseases, including ulcerative colitis and Crohn disease.

17.3 The Complement System Kills Microorganisms

- The complement system participates in both the nonspecific and the specific immune responses in a number of ways, all of which enable it to kill invading cells.

17.4 The Adaptive Immune Response Is a Specific Defense Against Infection

- The immune system has two components: antibody-mediated immunity, which is regulated by B cells and antibody production, and cell-mediated immunity, which is controlled by T cells. The primary function of antibody-mediated reactions is to defend the body against invading viruses and bacteria. Cell-mediated immunity is directed against cells of the body that have been infected by agents such as viruses and bacteria.

17.5 Blood Types Are Determined by Cell-Surface Antigens

- The presence or absence of certain antigens on the surface of blood cells is the basis of blood transfusions and blood types. Two blood groups are of major significance: the ABO system and the Rh blood group. Matching ABO blood types is important in blood transfusions. In some cases, mother–fetus incompatibility in the Rh system can cause maternal antigens to destroy red blood cells of the fetus, resulting in hemolytic disease of the newborn.

17.6 Organ Transplants Must Be Immunologically Matched

- The success of organ transplants and skin grafts depends on matching histocompatibility antigens found on the surface of all cells in the body. In humans, the antigens produced by a group of genes on chromosome 6 (known as the MHC complex) play a critical role in the outcome of transplants.

17.7 Disorders of the Immune System

- Allergies are the result of immunological hypersensitivity to weak antigens that do not provoke an immune response in most people. These weak antigens, known as allergens, include a wide range of substances: house dust, pollen, cat hair, certain foods, and even medicines such as penicillin. Acquired immunodeficiency syndrome (AIDS) is a collection of disorders that develops as a result of infection with a retrovirus known as the human immunodeficiency virus (HIV). The virus selectively infects and kills the T4 helper cells of the immune system.

Questions and Problems

CENGAGENOW Preparing for an exam? Assess your understanding of this chapter's topics with a pre-test, a personalized learning plan, and a post-test by logging on to *login.cengage.com/sso* and visiting CengageNOW's Study Tools.

The Inflammatory Response Is a General Reaction

1. (a) What causes the area around a cut or a scrape to become warm?
   (b) What is the role of this heat in the inflammatory response?

The Complement System Kills Microorganisms

2. The complement system supplements the inflammatory response by directly killing microorganisms. Describe the life cycle of complement proteins, from their synthesis in the liver to their activity at the site of an infection.

The Adaptive Immune Response Is a Specific Defense Against Infection

3. Name the class of molecules that includes antibodies, and name the five groups that make up this class.

4. Discuss the roles of the different types of T cells: helper cells and killer cells.

5. Compare the general inflammatory response, the complement system, and the specific immune response.

6. Distinguish between antibody-mediated immunity and cell-mediated immunity. What components are involved in each?

7. The molecular weight of IgG is 150,000 kd. Assuming that the two heavy chains are equivalent, the two light chains are equivalent, and the molecular weight of the light chains is half that of the heavy chains, what are the molecular weights of each individual subunit?

8. Identify the components of cellular immunity, and define their roles in the immune response.

9. Describe the rationale for vaccines as a form of preventive medicine.

10. Researchers have been having a difficult time developing a vaccine against a certain pathogenic virus as a result of the lack of a weakened strain. They turn to you because of your wide knowledge of recombinant DNA technology and the immune system. How could you vaccinate someone against the virus, using a cloned gene from the virus that encodes a cell-surface protein?

11. It is often helpful to draw a complicated pathway in the form of a flow chart to visualize the multiple steps and the ways in which the steps are connected to each other. Draw the antibody-mediated immune response pathway that acts in response to an invading virus.

12. Describe the genetic basis of antibody diversity.

13. In cystic fibrosis gene therapy, scientists propose the use of viral vectors to deliver normal genes to cells in the lungs. What immunological risks are involved in this procedure?

Blood Types Are Determined by Cell-Surface Antigens

14. A man has the genotype $I^A I^A$, and his wife is $I^B I^B$. If their son needed an emergency blood transfusion, would either parent be able to be a donor? Why or why not?

15. Why can someone with blood type AB receive blood of any type? Why can an individual with blood type O donate blood to anyone?

16. Is it more important that transfused blood have antigens that will not react with the recipient's antibodies, or antibodies that will not react with the recipient's antigens?

17. The following data were presented to a court during a paternity suit: (1) The infant is a universal donor for blood transfusions, (2) the mother bears antibodies against the B antigen only, and (3) the alleged father is a universal recipient in blood transfusions.
    a. Can you identify the ABO genotypes of the three individuals?
    b. Can the court draw any conclusions?

18. A patient of yours has just undergone shoulder surgery and is experiencing kidney failure for no apparent reason. You check his chart and find that his blood is type B, but he has been mistakenly transfused with type A. Explain why he is experiencing kidney failure.

19. Assume that a single gene having alleles that show complete-dominance relationships at the phenotypic level controls the Rh character. An Rh+ father and an Rh⁻ mother have eight boys and eight girls, all Rh+.

a. What are the Rh genotypes of the parents?
b. Should they have been concerned about hemolytic disease of the newborn?

20. How is Rh incompatibility involved in hemolytic disease of the newborn? Is the mother Rh+ or Rh⁻? Is the fetus Rh+ or Rh⁻? Why is a second child that is Rh+ more susceptible to attack from the mother's immune system?

Organ Transplants Must Be Immunologically Matched

21. What mode of inheritance has been observed for the HLA system in humans?

22. A burn victim receives a skin graft from her brother; however, her body rejects the graft a few weeks later. The procedure is attempted again, but this time the graft is rejected in a few days. Explain why the graft was rejected the first time and why it was rejected more rapidly the second time.

23. In the human HLA system there are 23 HLA-A alleles, 47 for HLA-B, 8 for HLA-C, 14 for HLA-DR, 3 for HLA-DQ, and 6 for HLA-DP. How many different human HLA genotypes are possible?

24. In the near future, pig organs may be used for organ transplants. How are researchers attempting to prevent rejection of the pig organs by human recipients?

25. A couple has a young child who needs a bone marrow transplant. They propose that preimplantation screening be done on several embryos fertilized in vitro to find a match for their child.
    a. What do they need to match in this transplant procedure?
    b. The couple proposes that the matching embryo be transplanted to the mother's uterus and serve as a bone marrow donor when old enough. What are the ethical issues involved in this proposal?

Disorders of the Immune System

26. Why are allergens called "weak" antigens?

27. Antihistamines are used as antiallergy drugs. How do these drugs work to relieve allergy symptoms?

28. Autoimmune disorders involve the breakdown of an essential property of the immune system. What is it? How does this breakdown cause juvenile diabetes?

29. A young boy who has had over a dozen viral and bacterial infections in the last 2 years comes to your office for an examination. You determine by testing that he has no circulating antibodies. What syndrome does he have, and what are its characteristics? What component of the two-part immune system is nonfunctional?

30. AIDS is an immunodeficiency syndrome. In the flow chart you drew for Question 11, describe where AIDS sufferers are deficient. Why can't our immune systems fight off this disease?

31. An individual has an immunodeficiency that prevents helper T cells from recognizing the surface antigens presented by macrophages. As a result, the helper T cells are not activated, and they in turn fail to activate the appropriate B cells. At this point, is it certain that the viral infection will continue unchecked?

Internet Activities

Internet Activities are critical-thinking exercises using the resources of the World Wide Web to enhance the principles and issues covered in this chapter. For a full set of links and questions investigating the topics described below, visit *www.cengage.com/biology/cummings*

1. *Immune System Function.* At the *CellsAlive!* website, access the "Antibody" link for a beautiful overview of antibody structure, production, and function.

2. *HIV/AIDS and the Immune System.* The University of Arizona's Biology Project HIV 2001 allows you to run a simulation of the spread of HIV through a population or work through a tutorial on HIV/AIDS and the immune system. The tutorial includes an overview of immune system function.

3. *Autoimmune Diseases.* The National Library of Medicine maintains the *Medline Plus Autoimmune Diseases* Web page. Here you can find links to various resources on autoimmune diseases such as lupus, multiple sclerosis (MS), and rheumatoid arthritis. Scroll down to the "Anatomy/Physiology" link

for a good immune system tutorial from the National Cancer Institute or down to the "Diagnosis/Symptoms" link to find out how the standard antinuclear antibody (ANA) test used in the diagnosis of many autoimmune diseases works.

4. *Which Immune Disorders Have Genetic Bases?* Because the normal functioning of the immune response in humans requires the delicate interplay of B cells, T cells, and phagocytic cells, as well as the actions of several types of immunoglobulins and cytokines, it is easy to see that some genetic disorders are likely to be recognized as being related to absent or abnormal immune function. The National Institute of Allergy and Infectious Disease maintains a fact sheet on "Primary Immune Deficiencies."

 ## HOW WOULD YOU VOTE NOW?

Organ donations are not keeping up with demand, and thousands of people die each year while waiting for transplants. Xenotransplantation techniques may increase the supply of organs by using organs from other animals, such as pigs, but these techniques must address the inherent problem of immune system rejection. Using pigs that have been genetically modified to carry human genes that prevent transplant rejection and modifying the immune system of human recipients by injecting pig bone marrow cells are two methods for overcoming the problems of organ transplantation between species. Now that you know more about the immune system and its role in organ transplants, what do you think? Is it ethical to genetically modify pigs with human genes or to modify humans by giving them a pig immune system to accept transplanted organs? Visit the *Human Heredity* companion website at *www.cengage.com/biology/cummings* to find out more on the issue; then cast your vote online.

18 Genetics of Behavior

Ancient Greece was among the first cultures that observed the link between creativity and madness. The Greek philosopher Socrates wrote:

> If a man comes to the door of poetry untouched by the madness of the Muses, believing that technique alone will make him a good poet, he and his sane compositions never reach perfection, but are utterly eclipsed by the performances of the inspired madman.

By the second century, it was recognized that mania and depression are opposite poles of a cycle, and their familial patterns of inheritance have been known for almost a thousand years. In the last quarter of the twentieth century, advances in genetics established that bipolar disorder is a complex trait with environmental influences. Clearly, not all poets and authors have bipolar disorder—in fact, most do not—and creativity should not be viewed through the filter of genetics. But evidence from authors and poets themselves and advances in medicine and genetics have established that artists, writers, and poets have a much higher rate of depression or bipolar disorder than does the general population.

Vincent Van Gogh's family had an extensive history of psychiatric problems. His brothers, his sister, two of his uncles, and Vincent himself were subject to mental illness—most likely bipolar disorder. All of Vincent's brilliant work as a painter was produced in a 10-year period and conveys to the viewer the intense anguish of his mental illness, which resulted in his suicide at age 37. In writing about his family's illness, he said:

> The root of the evil lies in the constitution itself, in the fatal weakening of families from generation to generation. . . . The root of the evil certainly lies there, and there's no cure for it.

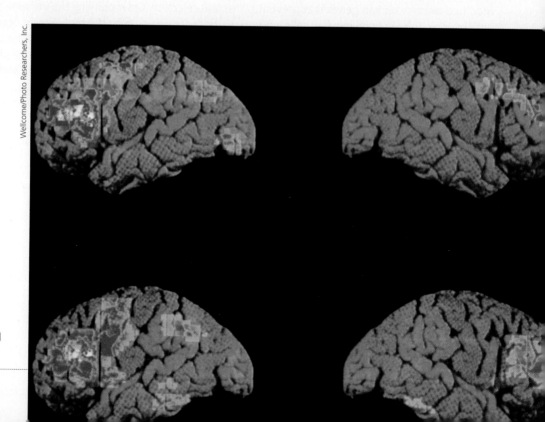

Differences in brain metabolism in the brain of an unaffected individual (*upper*) and a schizophrenic individual (*lower*).

Wellcome/Photo Researchers, Inc.

18.1 Models, Methods, and Phenotypes in Studying Behavior

KEEP IN MIND AS YOU READ

- Most human behaviors are complex traits with environmental influences.

- Transgenic animals carrying human genes are used to develop drugs and treatment strategies for behavioral disorders.

- Evidence from family studies indicates that schizophrenia and bipolar disorder have genetic components, but no major genes controlling these conditions have been identified.

- Human behavior in social settings is complex and often difficult to define.

HOW WOULD YOU VOTE?

Biographical and scientific evidence strongly suggests that in many people, creative abilities in art and literature are linked in a complex way to disorders such as bouts of depression or the onset of manic states. Because of this proposed linkage, it is possible, though not proved, that medicating bipolar disorder may reduce people's creativity. If you were a successful artist, author, or poet who experienced depression or bipolar disorder and a cure for your illness was discovered, would you elect to have the treatment, knowing that your creative abilities might be diminished or even disappear but also knowing that your risk of suicide would be reduced or eliminated? Visit the *Human Heredity* companion website at *www.cengage.com/biology/cummings* to find out more on the issue; then cast your vote online.

Pedigree analysis, family studies, adoption studies, and twin studies suggest that many parts of our behavior are genetically influenced. However, most behaviors with a genetic component are complex traits controlled by several genes, interaction with other genes, and environmental influences. In fact, most behaviors are *not* inherited as single-gene traits, demonstrating the need for genetic models that can explain observed patterns of inheritance. To a large extent, models proposed to explain the inheritance of a trait determine the methods used to analyze its pattern of inheritance and the techniques to be used in mapping and isolating the gene or genes responsible for the trait's characteristic phenotype (discussed later).

The idea that creativity and mental illness are linked is still controversial, illustrating many of the problems geneticists encounter in dissecting the genetic basis of human behavior. This chapter discusses the genetic models and methods used in studying human behavior and the state of our knowledge about the genetic control of behavior.

Table 18.1 Models for Genetic Analysis of Behavior

Model	Description
Single gene	One gene controls a defined behavior
Polygenic trait	Two or more genes contribute equally to the phenotype One or more major genes contribute to the phenotype, with other genes making lesser contributions
Multifactorial trait	Two or more genes interact with each other and/or environmental factors to produce the phenotype

There are several genetic models for inheritance and behavior.

Several models for genetic effects on behavior have been proposed (Table 18.1). The simplest model is a single gene with a dominant or recessive pattern of inheritance that affects a well-defined behavior. Several genetic disorders with behavioral components—Huntington disease, Lesch-Nyhan syndrome (Figure 18.1), fragile-X syndrome, and others—are described by such a model. Multiple-gene models are also possible. The simplest of these is a polygenic additive model in which two or more genes contribute

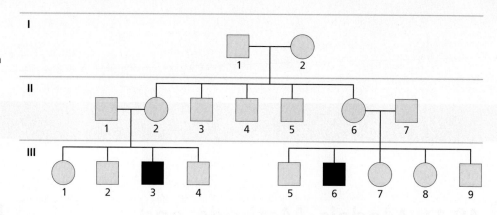

FIGURE 18.1 A pedigree for Lesch-Nyhan syndrome, an X-linked single-gene disorder. Affected males exhibit self-mutilating behavior and bite through their lips and fingers.

Epistasis The interaction of two or more non-allelic genes to control a single phenotype.

equally to the phenotype. In the past, this model has been proposed (along with others) to explain schizophrenia (the inheritance of additive polygenic traits was considered in Chapter 5).

Polygenic models can also include situations in which one or more genes have a major effect and other genes make smaller contributions to the phenotype. In still another polygenic model, two or more gene variants must occur together to produce the behavioral phenotype. This type of interaction is called **epistasis**.

In each of these models, the environment can affect the phenotype significantly, and the study of behavior must take this into account (see Exploring Genetics: Is Going to Medical School a Genetic Trait?). To assess the role of the environment in the phenotype, geneticists use heritability (see Chapter 5) and other methods to measure the genetic and environmental contributions to a trait.

Methods of studying behavior genetics often involve twin studies.

For the most part, methods for studying behavior genetics follow the pattern used for the study of other human traits. If a single gene model is proposed, pedigree analysis and linkage studies, including the use of genomic DNA markers such as SNPs, are the most appropriate methods. However, because many behaviors are complex traits, twin studies play a prominent role in human behavior genetics. Concordance and heritability values based on twin studies have established a genetic link to mental illnesses (schizophrenia and bipolar disorder) and to behavioral traits such as alcoholism. Results of such studies must be interpreted with caution, because there are limitations inherent in interpreting heritability, and these studies often use small sample sizes in which minor variations can have a disproportionately large effect on the outcome.

To overcome these problems, geneticists are adapting twin studies as a genetic tool for studying behavior. One innovation involves studying the children of twins to confirm the existence of genes that predispose a person to a certain behavior. Twin studies also are being coupled with genomic analysis to search for behavior genes, and this combination may prove to be a powerful method for identifying such genes.

Phenotypes: how is behavior defined?

Aside from selecting a model for how a trait is inherited, a second problem is the choice of a consistent phenotype to be used in a study. Phenotypes such as height are easy enough to define, but behavior has many variables that must be considered. The definition of a behavioral phenotype must be precise enough to distinguish it from all other behaviors and from the behavior of the control group. However, the definition cannot be so narrow that it excludes some variations of the behavior.

Is Going to Medical School a Genetic Trait?

Many behavioral traits follow a familial, if not Mendelian, pattern of inheritance. This observation, along with twin and adoption studies, reveals a strong genetic component in complex behavioral disorders. In most cases, these phenotypes are not inherited as simple Mendelian traits. Researchers thus are faced with the task of selecting a model to describe how a behavioral trait is inherited. Using this model, further choices are made to select the methods used in genetic analysis of the trait. A common strategy is to find a family in which the behavior appears to be inherited as a recessive or an incompletely penetrant dominant trait controlled by a single gene. Molecular markers are then used in linkage analysis to identify the chromosome that carries the gene controlling the trait.

If researchers are looking for a single gene when the trait is controlled by more than one gene or genes that strongly interact with the environment, the work may produce negative results, even though preliminary findings can be encouraging. To illustrate some of the pitfalls associated with model selection in behavior genetics, researchers deliberately selected attendance at medical school as a behavioral trait and then determined if the inheritance of this trait in families is consistent with a genetic model. They surveyed 249 first- and second-year medical students. Thirteen percent had first-degree relatives who also had attended medical school, compared with 0.22% of individuals in the general population with such

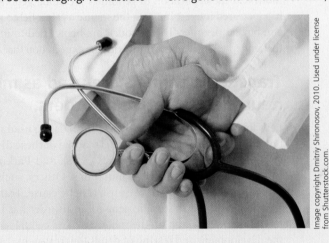

Image copyright Dmitriy Shironosov, 2010. Used under license from Shutterstock.com.

relatives. Thus, the overall risk factor among first-degree relatives for medical-school attendance was 61 times higher for medical students than for the general population, indicating a strong familial pattern. To see if this behavior might behave as an inherited trait, researchers used statistical analysis that supported inheritance and rejected the model of no inheritance. Pedigree analysis supported a single-gene recessive pattern of inheritance, although other models, including polygenic inheritance, were not excluded. Using a further set of statistical tests, the researchers concluded that the recessive inheritance of this trait was just at the border of statistical acceptance.

Similar results of borderline statistical significance are often found in studies of other behavioral traits, and it is usually argued that another, larger study would confirm the results. Although it is true that genetic factors may partly determine whether one will attend medical school, it is highly unlikely that a single recessive gene controls this decision, regardless of the outcome from this family study and segregation analysis of the results.

The authors of this study were not serious in their claims that a decision to attend medical school is a genetic trait, nor did they intend to cast doubts on the methods used in the genetic analysis of behavior. Rather, their work was intended to point out the folly of accepting simple, single-gene explanations for complex behavioral traits.

For some mental illnesses, clinical definitions are provided in the *Diagnostic and Statistical Manual of Mental Disorders* of the American Psychiatric Association. For other behaviors, phenotypes are poorly defined and may not provide clues to the underlying biochemical and molecular basis of the behavior. For example, alcoholism can be defined as deviant behaviors associated with excessive consumption of alcohol. Is this definition explicit enough to be useful as a phenotype in genetic analysis? Is there too much room for interpreting what is deviant behavior or excessive consumption? As we will see, whether the behavioral phenotype is defined narrowly or broadly can affect the outcome of the genetic analysis and even the model of inheritance for the trait.

The nervous system is the focus of behavior genetics.

In Chapter 10, we discussed the role of genes in metabolism. Mutations that disrupt metabolic pathways or interfere with the synthesis of essential gene products can influence the function of cells and thus produce an altered phenotype. If the affected cells are part of the nervous system, the phenotype may include altered behavior. In fact, many genetic disorders affect cells in the nervous system that in turn affect behavior. In the metabolic disorder phenylketonuria (PKU), for example, brain cells are damaged by excess levels of phenylalanine, causing mental retardation and other behavioral deficits.

For much of behavior genetics, the focus is on the structure and function of the nervous system. This emphasis is reinforced by the finding that many disorders with a behavioral phenotype—including Huntington disease, Alzheimer disease, and Charcot-Marie-Tooth disease—alter the structure and/or function of the brain and the nervous system. Other behavior disorders, such as schizophrenia and bipolar disorder, are also disorders of brain structure and function.

18.2 Animal Models: The Search for Behavior Genes

One way to study the genetics of human behavior is to ask whether this behavior can be studied in model systems. If so, then results from experimental organisms can be used to study human behavior. The mouse has been the primary animal used in the study of human psychiatric and other behavioral disorders. Mouse models include strains created by artificial selection for a specific behavior, such as alcohol consumption, and strains created by insertion of a human gene to create a transgenic model of a human behavioral disorder. The genetic similarity between humans and mice (the two species have over 90% of their genes in common) and the application of genomics including genome-wide association studies (GWAS) has increased interest in using the mouse as a model to study illnesses such as schizophrenia and bipolar disorder. These models are being used to investigate the biological mechanisms that underlie specific behaviors, with the expectation that similar mechanisms exist in humans.

Transgenic animals are used as models of human neurodegenerative disorders.

Let's look briefly at how transgenic animals are used in studying members of a group of human neurodegenerative disorders. Some of these disorders, such as Alzheimer disease (AD), amyotrophic lateral sclerosis (ALS), and Parkinson disease (PD), occur sporadically or result from inherited mutations. Others, such as Huntington disease (HD) and spinocerebellar ataxias, have only a genetic cause. Transgenic animal models can be constructed only after a specific gene for a disorder has been identified and isolated. These models allow research on the molecular and cellular mechanisms of the disorder and on the development and testing of drugs for treatment.

ALS (OMIM 105400) is an adult-onset neurodegenerative disorder. About 20% of ALS cases carry a mutation in the *SOD1* (OMIM 14750) gene on chromosome 21 that causes the SOD1 protein to become toxic. Transgenic mice carrying a mutant human *SOD1* allele develop muscle weakness and atrophy similar to that seen in affected humans (Figure 18.2). This mouse model is used to study how the mutant SOD1 protein selectively damages certain nerve cells but leaves others untouched. The transgenic strains are also used to study the effects of drugs to treat ALS. In a later section of this chapter, we will explore the use of mouse models in more detail.

Although mice are the primary model organism used in human behavioral research, human genes transferred to *Drosophila* are also used as models of human neurodegenerative diseases. Flies that carry mutant human alleles for HD and spinocerebellar ataxia 3 are used to study how the mutant proteins kill nerve cells and to identify alleles of other genes that can slow or prevent the loss of cells. Analysis of single-gene behavioral mutants in animal models has provided insight into the structure and function of the human nervous system.

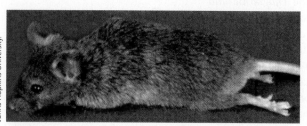

FIGURE 18.2 A transgenic mouse carrying a mutation in the human *SOD1* gene that causes paralysis of the limbs. In humans, the mutation causes amyotrophic lateral sclerosis (ALS), a neurodegenerative disease. The mutant mouse serves as a model for this disease, allowing researchers to explore the mechanism of the disease and to design therapies to treat humans affected with ALS.

18.3 Single Genes Affect the Nervous System and Behavior

In this section we discuss several single-gene disorders with specific effects on the development, structure, and/or function of the nervous system, which affect behavior. After we consider the relationship between these genes and behavior, we will discuss complex behavioral disorders in which the number and functional roles of genes are not well understood and whose effects on the nervous system may be more subtle.

Huntington disease is a model for neurodegenerative disorders.

Huntington disease (HD; OMIM 143100) is an adult-onset neurodegenerative disorder inherited as an autosomal dominant trait which affects about 1 in 10,000 individuals in Europe and the United States. HD was one of the first disorders to be mapped using recombinant DNA techniques (see Chapters 13 and 15). The mutant allele carries an expanded trinucleotide repeat (this topic is covered in Chapter 11), and the disorder shows anticipation (also covered in Chapter 11).

Symptoms of HD usually begin in mid adult life and include personality changes, agitated behavior, dementia, and involuntary muscular movements. Most affected individuals die within 10 to 15 years after the onset of symptoms.

The gene for HD is located on the short arm of chromosome 4. HD is one of several neurodegenerative disorders caused by the expansion of a CAG trinucleotide repeat (see Chapter 11). Mutant alleles carry more CAG triplet repeats than normal alleles. The expansion in the number of CAG repeats causes many more copies of the amino acid glutamine to be inserted into the protein encoded by this gene. This increase—called a polyglutamine expansion—makes the protein toxic, killing cells of the brain and nervous system. People whose *HD* alleles contain fewer than 35 CAG repeats do not develop HD; those who carry 35 to 39 repeats may or may not develop HD. However, those with alleles containing 40 to 60 repeats will develop HD as adults. People with more than 60 repeats will develop HD before age 20.

Anticipation is a pattern of earlier disease onset in successive generations, accompanied by more severe symptoms. In HD, anticipation is associated with expansion of the number of CAG repeats as the *HD* gene is passed from generation to generation. Expansion of paternal repeats is more likely to produce an earlier onset, and juvenile cases of HD are almost always associated with paternal transmission of the mutant allele, but the reasons for this are unclear.

Autopsies of HD victims show damage to the striatum and cerebral cortex regions of the brain. In these regions, cells fill with cytoplasmic and nuclear clusters of the mutant protein and degenerate and die (Figure 18.3). Involuntary movements and progressive personality changes accompany the degeneration and death of these brain cells. The *HD* gene encodes a large protein, huntingtin (Htt). In adult brains, the normal form of Htt enhances production of BDNF, a protein necessary for the survival of cells in the striatum. The mutant protein decreases BDNF production, causing cells of the striatum to degenerate and die.

The Htt protein is expressed in cells throughout the body, and in all regions of the brain. In HD however, although the mutant Htt protein is present in all parts of the brain, its toxic effects are limited to two regions of the brain. Why is mutant Htt toxic to cells in some parts of the brain but not in others? Recent research has uncovered a possible answer to this question. A protein called Rhes (Ras homolog enriched in striatum), is found only in the striatum and to a lesser extent in the cerebral cortex. The Rhes protein binds to the mutant form of Htt (but not to the normal protein), making it cytotoxic. This discovery might be

Huntington disease (HD) An autosomal dominant disorder associated with progressive neural degeneration and dementia. Adult onset is followed by death 10 to 15 years after symptoms appear.

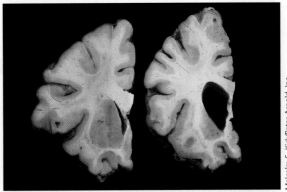

FIGURE 18.3 Section of a normal brain (*left*) and an HD brain (*right*). The HD brain shows extensive damage to the striatum.

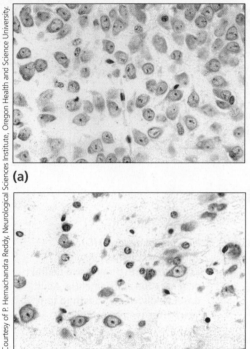

Courtesy of P. Hemachandra Reddy, Neurological Sciences Institute, Oregon Health and Science University.

(a)

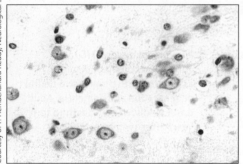

(b)

FIGURE 18.4 Loss of brain cells in a transgenic Huntington-disease mouse. (a) Section of normal mouse-brain striatum showing densely packed neurons. (b) Section of the striatum from an *HD89* mouse showing extensive loss of neurons that accompany this disease.

used to develop drugs that block binding of Rhes to Htt as a way of treating HD. The normal form of Htt is multifunctional and binds to hundreds of other proteins. Whether the mutant Htt binds to other proteins and affects other parts of the disease process in HD remains to be explored.

Transgenic mice that carry and express the mutant human *HD* allele express the human gene in their brains and other organs. These mice show progressive behavioral changes and loss of muscle control. The brains of affected transgenic mice show the same changes that are seen in affected humans: accumulation of Htt clusters leading to degeneration and loss of cells in the striatum and cerebrum (Figure 18.4). These transgenic strains are now used to study the early events in Htt accumulation and to develop treatments that work in presymptomatic stages to prevent cell death.

Experiments show that transplantation of fetal nerve cells into the striatum of transgenic HD mice partially restores nerve connections, muscle control, and behavior. Based on these findings, clinicians are now treating HD patients with transplants of normal human fetal striatal cells on an experimental basis to determine whether the transplanted cells can survive, make connections to other cells, and lead to improvements in muscle control and intellectual functions. Results from the first round of transplants have been encouraging, adding HD to the list of disorders that can be treated with such transplants. This success, however, adds to the debate about fetal stem cells and the direction of stem-cell research in this country.

> **KEEP IN MIND**
>
> Transgenic animals carrying human genes are used to develop drugs and treatment strategies for behavioral disorders.

There is a genetic link between language and brain development.

For over 40 years, linguists, psychologists, and geneticists have argued unsuccessfully over the relationship between language and genetics. About 10 years ago, a large multigenerational family (the KE family), came to the attention of researchers. Members of this family have a very specific speech-and-language disorder, inherited as an autosomal dominant trait (Figure 18.5). Affected members cannot identify language sounds correctly and have difficulty understanding sentences. They also have problems in making language sounds, and it is almost impossible to understand their speech.

With the cooperation of the family, investigators were able to map the disorder to a small region on the long arm of chromosome 7 and named the unknown gene in this region *SPCH1* (SPEECH 1; OMIM 602081). Recently, an unrelated child, CS, was found to

FIGURE 18.5 Pedigree of the KE family in which some members are affected with a severe speech-and-language disorder (darker symbols represent affected members). Asterisks mark individuals who were not analyzed. The pattern of inheritance is consistent with an autosomal dominant trait. The gene for this disorder maps to the long arm of chromosome 7.

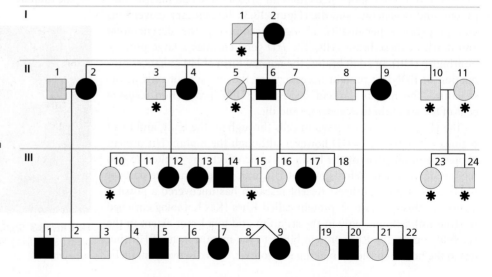

have the same speech deficit as members of the KE family. CS carries a translocation of the long arm of chromosome 7 in the *SPCH1* region. This allowed researchers to identify the gene, now called *FOXP2* (OMIM 605317), because it is a member of a previously identified gene family.

Affected members of the KE family have a single nucleotide change in the *FOXP2* gene that changes one amino acid in the FOXP2 protein, presumably altering the protein's function and resulting in the language deficit.

FOXP2 is a transcription factor, a protein that switches on genes or gene sets, often at specific stages of development. The FOXP2 protein is very active in fetal brains. Affected individuals may have a 50% reduction in the amount of this protein at a critical stage of brain development, leading to an abnormality of language development.

Future work on *FOXP2* may help us learn how the brain understands and processes language and allow the development of therapies to treat language disorders. In addition, comparing the action of *FOXP2* in the developing brains of chimpanzees and other primates (see Chapter 19) may help us understand how language evolved and what separates us from our fellow primates.

18.4 Single Genes Control Aggressive Behavior and Brain Metabolism

In 1993, a new form of X-linked mild mental retardation was identified in a large European family. All affected males showed aggressive, and often violent, behavior (Figure 18.6). Gene mapping and biochemical studies show a direct link between a single-gene defect and a phenotype characterized by aggressive and/or violent behavior. In particular, some males with mild or borderline mental retardation showed a characteristic pattern of aggressive behavior (often violent) triggered by anger, fear, or frustration. Although the levels of violence varied, the behaviors included acts of attempted rape, arson, stabbings, and exhibitionism.

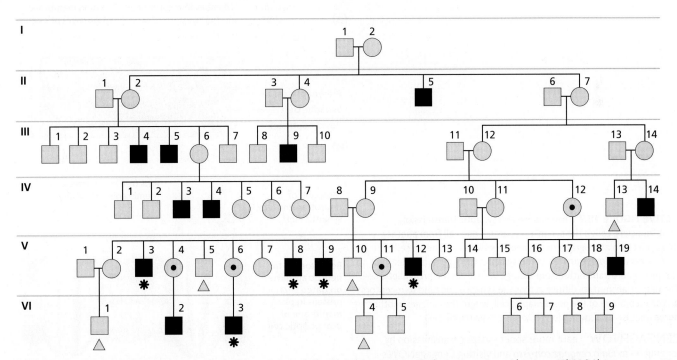

FIGURE 18.6 Cosegregation of mental retardation, aggressive behavior, and a mutation in the *monoamine oxidase type A (MAOA)* gene. Affected males are indicated by the darker symbols. Symbols marked with an asterisk represent males known to carry a mutation of the *MAOA* gene; those marked with a triangle are known to carry the normal allele. Symbols with a dot inside represent females known to be heterozygous carriers.

Table 18.2 Some Common Neurotransmitters

Acetylcholine
Dopamine
Norepinephrine
Epinephrine
Serotonin
Histamine
Glycine
Glutamate
Gamma-aminobutyric acid (GABA)

Geneticists have mapped a gene for aggression.

Using molecular markers, the gene for these behaviors was mapped to the short arm of the X chromosome in region Xp11.23–11.4. A gene in this region encodes an enzyme called monoamine oxidase type A (MAOA) that breaks down a neurotransmitter (Table 18.2). Neurotransmitters are chemical signals that carry nerve impulses across synapses in the brain and nervous system (Active Figure 18.7). Failure to rapidly break down these chemical signals can disrupt the normal function of the nervous system.

The urine of the eight affected individuals contains abnormal levels of compounds indicating that the MAOA enzyme is not functioning properly. The researchers concluded

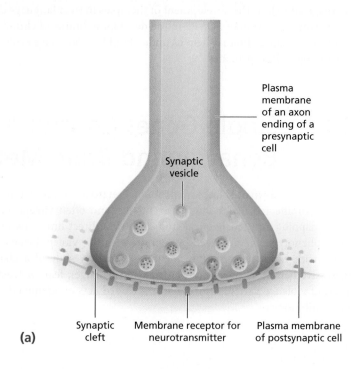

(a)

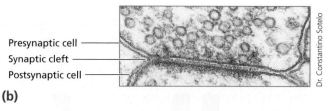

(b)

ACTIVE FIGURE 18.7 The synapse and synaptic transmission.
(a) A thin cleft, called the synapse, separates one cell from another.
(b) An electron micrograph of a synapse between two nerve cells.
(c) As a nerve impulse arrives at the synapse, it triggers the release of a chemical neurotransmitter from storage vesicles in the presynaptic cell. The neurotransmitters diffuse across the synapse and bind to receptors on the membrane of the postsynaptic cell, where they trigger another nerve impulse by allowing ions into the postsynaptic cell.

CENGAGENOW Learn more about synaptic transmission by logging on to *login.cengage.com/sso* and visiting CengageNOW's Study Tools.

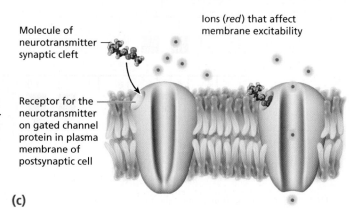

(c)

that the affected males carried a mutation in the *MAOA* gene and that lack of enzymatic MAOA activity is associated with their behavioral pattern.

A follow-up study analyzed the sequence of the *MAOA* gene (OMIM 309850) in five of the eight affected individuals and showed that all five carry a mutation that encodes a nonfunctional gene product. This mutation was also found in two female heterozygotes and is not found in any unaffected males in this pedigree. Loss of MAOA activity in affected males prevents the normal breakdown of certain neurotransmitters, reflected in behavioral problems and elevated levels of toxic compounds in the urine.

Because it is difficult to relate a phenotype such as aggression (e.g., what exactly constitutes aggression?) to a specific genotype, further work is needed to determine whether *MAOA* mutations are associated with altered behavior in other families and in animal models. In addition, the types and amounts of interactions with external factors such as diet, drugs, and environmental stress remain to be established. However, the identification of a specific mutation associated with this behavior pattern is an important discovery and suggests that biochemical or pharmacological treatment for this disorder may be possible.

There are problems with single-gene models for behavioral traits.

Although recombinant DNA markers have been used successfully to identify, isolate, and clone single genes that affect behavior, in other cases this method has produced erroneous results. In 1987, a DNA linkage study mapped a gene for bipolar disorder to a region of chromosome 11. Later, individuals from the study group who did not carry the suspect copy of chromosome 11 developed the disorder, indicating that a related gene was not on that chromosome. Other studies reported linkage between DNA markers on chromosome 5 and schizophrenia; however, the linkage was later found to be coincidental or, at best, could explain the disorder only in a small, isolated population. The recognition that only a very small number of behavioral disorders are caused by single gene mutations led to the development of multi-gene models for complex traits, as described in the next section.

Schizophrenia A behavioral disorder characterized by disordered thought processes and withdrawal from reality. Genetic and environmental factors are involved in this disease.

Bipolar disorder A behavioral disorder characterized by mood swings that vary between manic activity and depression.

18.5 The Genetics of Schizophrenia and Bipolar Disorder

Schizophrenia is a collection of mental disorders characterized by psychotic symptoms, delusions, thought disorders, and antisocial behavior (Figure 18.8). Schizophrenia affects about 1% of the population and usually appears in late adolescence or early adulthood. Family studies and twin studies provide good evidence for a genetic contribution to schizophrenia. Risk factors for relatives of schizophrenics are high (Figure 18.9), confirming the influence of genotype on schizophrenia. Using a narrow definition of schizophrenia, the concordance value for monozygotic (MZ) twins is 46%, versus 14% for dizygotic (DZ) twins. A broader definition has concordance for MZ twins approaching 100% and a 45% risk for siblings, parents, and offspring of schizophrenics. This strongly supports the role of genes in this disorder.

Bipolar disorder is another serious form of mental illness that also affects about 1% of the U.S. population. Onset occurs during adolescence or the second and third decades of life, and males and females are at equal risk for this condition. In bipolar disorder, periods of manic activity alternate with depression. Manic phases are characterized by hyperactivity, acceleration of thought processes, a short attention span, creative impulses, feelings of elation or power, and risk-taking behavior.

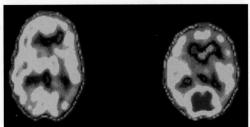

NIH/Science Source/Photo Researchers, Inc.

FIGURE 18.8 Brain metabolism in a schizophrenic individual (*left*) and a normal individual (*right*). These scans of glucose utilization by brain cells are visualized by positron emission tomography (PET scan). The differences lie mainly in regions of the brain where cognitive ability resides.

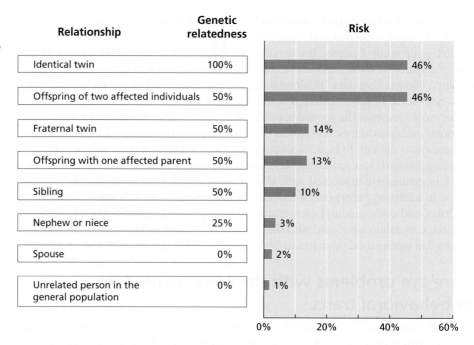

FIGURE 18.9 The lifetime risk for schizophrenia varies with the degree of relationship to an affected individual. The observed risks are more compatible with a multifactorial mode of transmission than with a single-gene or polygenic mode of inheritance.

As with schizophrenia, family, twin, and adoption studies have linked bipolar disorder to genetics. There is 60% concordance for MZ twins and 14% for DZ twins. Adoption studies also indicate that genetic factors are involved in this disorder. Studies have also documented higher risks to first-degree relatives of those with bipolar illness (Figure 18.10). The fact that concordance in MZ twins is not 100% suggests that environmental factors (such as stress) interact with genetic risk.

Genetic models for schizophrenia and bipolar disorders.

Because both schizophrenia and bipolar disorder are diseases of the nervous system, most early models focused on genes that control the transmission of nerve impulses from cell to cell in the nervous system. Some of these candidate genes include neurotransmitters such as monoamine oxidase A (MAOA; OMIM 309850), catechol-O-methyl-transferase (COMT), (OMIM 116790), and 5HT—the serotonin transporter (OMIM 182135). Initially, studies provided some evidence that both schizophrenia and bipolar disorder might be caused by mutations in single genes or a small number of genes. However, more recent work using a number of different techniques clearly showed that both diseases are complex traits caused by a large number of genes, each with a small effect, combined with environmental triggers.

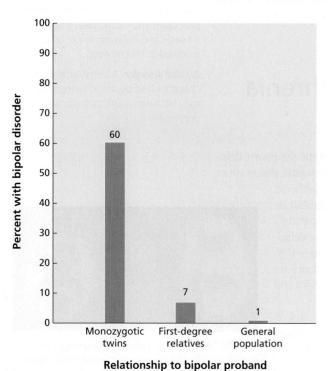

FIGURE 18.10 The frequency of bipolar illness in members of monozygotic twin pairs and in first-degree relatives of affected individuals indicates that genetic factors are involved in this disease.

KEEP IN MIND

Evidence from family studies indicates that schizophrenia and bipolar disorder have genetic components, but no major genes controlling these conditions have been identified.

Genomic approaches to schizophrenia and bipolar disorder.

New strategies using genomics have inspired a concerted effort to screen the human genome for genes that control both schizophrenia and bipolar disorder. Genome-wide association studies (GWAS) are one such approach. If there is an association between certain single nucleotide polymorphism (SNP) markers and chromosome regions that may contain disease genes, those SNPs should occur more frequently in people with schizophrenia and/or bipolar illness than in control populations.

Although still in the early stages of development, whole-genome scans have produced some surprising and provocative findings. GWAS have identified common SNPs that are associated with susceptibility to schizophrenia, bipolar disorder, and autism, and indicated that these disorders may be related to one another. In addition, the studies have revealed rare copy number variations (CNVs) associated with the risk of developing these three disorders. The findings have several implications. They show that common SNPs are linked to all three disorders and that specific CNVs may be important risk factors for autism and schizophrenia, but not bipolar disorder. The results also imply that genes in regions linked to specific SNP markers and genes contained in certain CNVs increase the risk of developing any of these disorders. Genes common to all three disorders encode proteins that control the development and maturation of synapses and the specificity of their connections to other nerve cells, indicating that developmental events in the nervous system may represent common biological pathways that lead to these disorders.

A CNV on chromosome 16 points to schizophrenia and autism as opposite diseases. This CNV spans a 600 kb (1 kb = 1,000 base pairs) region on the short arm of the chromosome. Duplications of this CNV confer a 16.5-fold increase for the risk of schizophrenia, and a deletion of this CNV is a significant risk factor for autism—once again suggesting that these neurodevelopmental disorders share common pathways in the formation and function of the nervous system. Identification and isolation of genes in this region will be important in advancing our understanding of nervous-system development and the nature of both schizophrenia and autism.

18.6 Genetics and Social Behavior

Human geneticists have long been interested in behavior that takes place in a social context; that is, behavior resulting from interactions between and among individuals. Evidence from family, adoption, and twin studies indicates that multifactorial inheritance is an important genetic component of these complex behaviors. Several traits that affect different aspects of social behavior are discussed in the following sections.

KEEP IN MIND

Human behavior in social settings is complex and often difficult to define.

Alzheimer disease is a complex disorder.

Alzheimer disease (AD) is a progressive and fatal neurodegenerative disease that affects almost 2% of the population of the United States. Age is a major risk factor for AD, and as populations in countries with AD age, the world-wide incidence of the disease is expected to increase threefold by 2055. Ten percent of the U.S. population older than 65 years has AD, and the disorder affects 50% of those older than 80 years. The annual cost of treatment and care for AD is close to $100 billion.

Alzheimer disease begins with memory loss, progressive dementia, and disturbances of speech, motor activity, and recognition. Brain lesions accompany the progression of AD

Alzheimer disease (AD) A heterogeneous condition associated with the development of brain lesions, personality changes, and degeneration of intellect. Genetic forms are associated with genes on chromosomes 14, 19, and 21.

FIGURE 18.11 Location of brain lesions in Alzheimer disease. Plaques are concentrated most heavily in the amygdala and hippocampus. These brain regions are part of the limbic system.

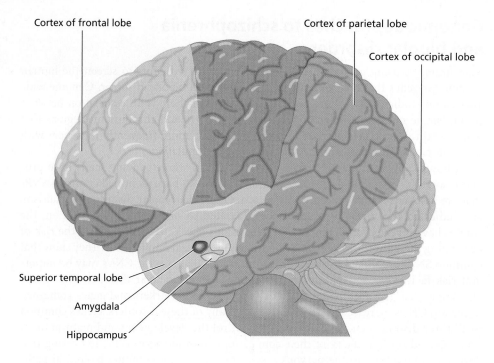

Cortex of frontal lobe

Cortex of parietal lobe

Cortex of occipital lobe

Superior temporal lobe

Amygdala

Hippocampus

(Figure 18.11). The lesions are formed by a protein fragment, amyloid beta-protein, which accumulates outside cells in aggregates known as senile plaques. The plaques cause the degeneration and death of nearby neurons in specific brain regions (Figure 18.12).

There are two forms of AD: one with a strong familial pattern of inheritance, characterized by early onset; a second sporadic form with a late onset. The familial form of AD is caused by mutations in any of three genes associated with the production and processing of the amyloid beta-protein *AD1* (OMIM 104300). However, less than 5% of all cases can be traced to genetic causes. The sporadic, late-onset form of AD may have susceptibility genes, but these have been difficult to identify. Only one such gene that confers an enhanced risk for AD has been identified. This gene, *Apolipo-protein E* (*APOE*; OMIM 107741), encodes a protein involved with cholesterol metabolism, transport, and storage. The *APOE* gene has three alleles (*E*2*, *E*3*, and *E*4*). Those who carry one or two copies of the *APOE*4* allele are at increased risk of AD, but the mechanism is unclear.

Genomic approaches in AD.

Ten GWAS have scanned the genome searching for SNPs linked to chromosome regions and genes associated with AD. Regions associated with AD in some studies were not confirmed in others. However, a number of studies have identified four genes with strong links to AD (Table 18.3). Over two dozen other candidate genes may play a role in susceptibility to AD, but as yet there is no evidence that any of these are directly linked to the disease process. Larger studies and sequencing of these genes in affected and unaffected individuals will be required to confirm the findings.

In summary, we know that AD has several genetic causes and that mutations in any of several genes can produce the AD phenotype. In addition, several unknown factors influence the rate at which the disease

Dr. Dennis Selkoe, Center for Neurologic Diseases, Harvard Medical School

Degenerating neurons

FIGURE 18.12 A lesion called a plaque in the brain of an individual with Alzheimer disease. A ring of degenerating neurons surrounds the deposit of protein.

Table 18.3 Genes and Risk Factors for Alzheimer Disease

Gene	Chromosome	Risk Factor
GAB2	11	?
GALP	14	>10%
PGBD1	6	>20%
TNK1	17	<15%
Unknown	14q31.2	?
Unknown	14q32.13	?
Unknown	6q24.1	?

progresses. Scientists continue to investigate the role of genetic and environmental factors to attempt to define the risk factors and rate of progression for this debilitating condition.

Alcoholism has several components.

As a behavioral disorder, excessive alcohol consumption (OMIM 103780) has two important components. First, drinking in excess over a long period causes damage to the brain, nervous system, and other organs. Over time, the accumulation of damage results in altered behavior, hallucinations, and loss of memory. The second component involves behaviors that lead to excessive alcohol consumption and a loss of the ability to function in social settings, the workplace, and the home.

It is estimated that 75% of the adult U.S. population consumes alcohol, and about 10% of those are classified as alcoholics. From the genetic standpoint, alcoholism is most likely a complex, genetically influenced, multifactorial (genetic and environmental) disorder. The role of genetic factors in alcoholism is indicated by a number of findings:

- In mice, experiments show that alcohol preference can be selected for; some strains of mice choose a solution of 75% alcohol over water, whereas other strains shun all alcohol.
- There is a 25% to 50% risk of alcoholism in the sons and brothers of alcoholic men.
- There is a 55% concordance for alcoholism in MZ twins and a 28% rate in same-sex DZ twins.
- Sons adopted by alcoholic men show a rate of alcoholism closer to that of their biological fathers than that of their adoptive fathers.

Genes that influence alcoholism have not been identified. Early searches focused on genes involved with the metabolism of alcohol (Figure 18.13), including alcohol dehydrogenase (*ADH*; OMIM 103720) and aldehyde dehydrogenase (*ALDH*; OMIM 100640). Alleles of these genes cause a wide variation in the rate of alcohol metabolism. Combinations of these alleles affect alcohol consumption and therefore are risk factors for alcoholism. However, only a fraction of alcoholism is associated with alleles of these genes. In addition to genes involved in alcohol metabolism, the *DRD2* gene (OMIM 126450) has received a lot of attention as a possible gene involved in alcoholism. This gene encodes a receptor for a neurotransmitter, focusing attention on the role of neural pathways in the brain in alcoholism. Results from several studies have been inconsistent, and the role of this gene in alcoholism remains controversial. Over 150 other genes have been identified that may be risk factors for this behavior, indicating that multifactorial inheritance involving several genes and environmental factors may play a larger role in alcohol abuse than individual genes.

The search for genetic factors in alcoholism illustrates the problem of selecting the proper genetic model to analyze behavioral traits. Segregation and linkage studies indicate that there is no single gene responsible for most cases of alcoholism. If a multifactorial model involving a number of genes—each with a small additive effect—is invoked, the problem becomes more complicated. How do you prove or disprove that a specific gene contributes, say, 1% to the behavioral phenotype? At present, the only method to accomplish this would involve studying thousands of individuals to find such effects, and at present, few large-scale GWA studies have been completed.

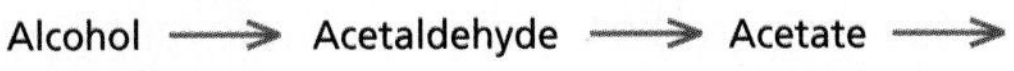

Alcohol $\xrightarrow{\text{ADH}}$ Acetaldehyde $\xrightarrow{\text{ALDH}}$ Acetate $\longrightarrow$

FIGURE 18.13 The pathway of alcohol metabolism. Alcohol is converted to acetaldehyde by the enzyme alcohol dehydrogenase (ADH). In turn, the acetaldehyde is converted to acetate by the enzyme aldehyde dehydrogenase (ALDH). Alleles encoding variations of these enzymes have been the focus of research into alcoholism; these variations explain only a very small percentage of cases.

Summing Up: The Current Status of Human Behavior Genetics

In reviewing the current state of human behavior genetics, several elements are apparent. Almost all studies of complex human behavior have provided only indirect and correlative evidence for the role of single genes.

- Searches for single-gene effects have proved unsuccessful to date, and initial reports of single genes that control bipolar illness, schizophrenia, and alcoholism have been retracted or remain unconfirmed.
- Segregation studies and heritability estimates indicate that most behaviors are complex traits involving many genes, each with a small impact on the phenotype.
- Recent successes using haplotype SNP scans and GWA studies may point the way to identifying genes associated with many behavioral disorders.
- Success in identifying susceptibility genes for behavior should not overshadow the fact that the environment plays a significant role in behavior.

As confirmation of the role of genes in behavior becomes available, investigations into the role of environmental factors cannot be neglected. The history of human-behavior genetics in the eugenics movement of the early part of the last century provides a lesson in the consequences of overemphasizing the role of genetics in behavior.

The identification of genes affecting behavior may lead to improvements in diagnosis and treatment of behavior disorders but also has implications for society at large. As was discussed in Chapter 15, the Human Genome Project has raised questions about the way in which genetic information will be disseminated and used. The same concerns need to be addressed for genes that affect behavior. Many behavioral phenotypes, such as Huntington disease and Alzheimer disease, are clearly regarded as abnormal. Few would argue against the development of treatments for intervention in and prevention of these conditions. The larger question is: When do behavior phenotypes move from being abnormal to being variants? If there is a connection between bipolar disorder or some forms of autism and creativity, to what extent should the condition be treated? If genes that influence sexual orientation are identified, will this behavior be regarded as a variant or as a condition that should be treated and/or prevented?

Although research can provide information about the biological factors that play a role in determining human behavior, it cannot provide answers to questions of social policy. Social policy and laws have to be formulated by using information from research.

Genetics in Practice

Genetics in Practice case studies are critical-thinking exercises that allow you to apply your new knowledge of human genetics to real-life problems. You can find these case studies and links to relevant websites at *www.cengage.com/biology/cummings*

CASE 1

Rachel asked to see a genetic counselor because she was concerned about developing schizophrenia. Her mother and maternal grandmother both had schizophrenia and were institutionalized for most of their adult lives. Rachel's three maternal aunts are all in their 60s and have not shown any signs of this disease. Rachel's father is alive and healthy, and his family history does not suggest any behavioral or genetic conditions. The genetic counselor discussed the multifactorial nature of schizophrenia and explained that many candidate genes have been identified that may be mutated in individuals with the condition. However, a genetic test is not available for presymptomatic testing. The counselor explained that on the basis of Rachel's family history and her relatedness to individuals who have schizophrenia, her risk of developing it is approximately 13%. If an altered gene is in the family and her mother carries the gene, Rachel has a 50% chance of inheriting it.

1. Why do you think it has been so difficult to identify genes underlying schizophrenia?

2. If a test were available that could tell you whether you were likely to develop a disorder such as schizophrenia later in life, would you take the test? Why or why not?

A genetic counselor was called to the pediatric ward of the hospital for a consultation. Her patient, an 8-year-old boy, was having a "temper tantrum" and was biting his own fingers and toes. The nurse called after she noticed that he was a patient of a clinical geneticist at another institution. The counselor reviewed the boy's chart and noted a history of growth retardation and self-mutilation since age 3. His movements were very "jerky," and he was banging his head against the bedpost. The nurses were having a very difficult time controlling him. The counselor immediately recognized these symptoms as part of a genetic disorder known as Lesch-Nyhan syndrome.

Lesch-Nyhan syndrome is an X-linked recessive condition (Xq26) caused by mutation in the *hypoxanthine phosphoribosyltransferase*

gene, and it affects about 1 in 100,000 males. Symptoms usually begin between the ages of 3 and 6 months. Prenatal testing is available, but there is no treatment for Lesch-Nyhan, and most affected individuals die by the second decade of life.

1. If your child had Lesch-Nyhan syndrome and you heard about an experimental gene therapy technique that had shown some promise in treating the disease but also had significant associated risks, would you attempt to enroll your child in a clinical trial of the technique? Explain.

2. Lesch-Nyhan syndrome is quite rare (1 in 100,000 males), but its effects are devastating. Would you support an effort to screen every developing fetus for this disorder? Explain.

Summary

18.1 Models, Methods, and Phenotypes in Studying Behavior

- Many forms of behavior represent complex phenotypes. The methods used to study inheritance of behavior encompass classical methods of linkage and pedigree analysis, newer methods of recombinant DNA analysis, and new combinations of techniques such as twin studies together with molecular methods. Refined definitions of behavior phenotypes are also being used in the genetic analysis of behavior.

18.2 Animal Models: The Search for Behavior Genes

- Results from work on experimental animals indicate that behavior is under genetic control and have provided estimates of heritability. The molecular basis of single-gene effects in some forms of behavior has been identified and provides useful models to study gene action and behavior. Transgenic animals carry mutant copies of human genes and are studied to understand the action of the mutant alleles and develop drugs for the treatment of these behavioral disorders.

18.3 Single Genes Affect the Nervous System and Behavior

- Several single-gene effects on human behavior are known. Most of them affect the development, structure, or function of the nervous system and consequently affect behavior. Huntington disease serves as a model for neurodegenerative disorders. Language and brain development are linked by genes that encode transcription factors.

18.4 Single Genes Control Aggressive Behavior and Brain Metabolism

- Most forms of mental retardation are genetically complex multifactorial disorders. One form of X-linked retardation that is

associated with aggressive behavior is linked to abnormal metabolism of a neurotransmitter (a chemical that transfers nerve impulses from cell to cell).

18.5 The Genetics of Schizophrenia and Bipolar Disorder

- Schizophrenia and bipolar disorder are common mental illnesses, each affecting about 1% of the population. Simple models of single-gene inheritance for these disorders have not been supported by extensive studies of affected families, and polygenic models for these diseases have been developed. Genomic scans have identified genes involved in schizophrenia, opening the way for the development of treatments.

18.6 Genetics and Social Behavior

- Multifactorial traits that affect behavior include Alzheimer disease and alcoholism. Genome scans have identified several genes that may increase risk for nonfamilial forms of Alzheimer disease, but few such studies have been completed for alcoholism.

18.7 Summing Up: The Current Status of Human Behavior Genetics

- The evidence for genetic control of complex behaviors is indirect. Although some progress has been made in showing linkage between certain chromosome regions and disorders such as bipolar disorder and schizophrenia, no specific genes contributing to these or other behaviors such as alcohol abuse have been discovered. Newer methods using larger data sets and more SNP markers may offer higher resolution of candidate chromosome regions and identify specific alleles that cause behavior.

Questions and Problems

Models, Methods, and Phenotypes in Studying Behavior

1. What are the major differences in the methods used to study the behavior genetics of single-gene traits versus polygenic traits?
2. In human behavior genetics, why is it important that the trait under study be defined accurately?
3. One of the models for behavioral traits in humans involves a form of interaction known as epistasis. What is epistasis?

Animal Models: The Search for Behavior Genes

4. What are the advantages of using *Drosophila* in the study of behavior genetics? Can this organism serve as a model for human behavior genetics? Why or why not?
5. You are a researcher studying an autosomal dominant neurodegenerative disorder. You have cloned the gene underlying the disorder and have found that it encodes an enzyme that is overexpressed in the neurons of individuals who have the disorder. To better understand how this enzyme causes neurodegeneration in humans, you make a strain of transgenic *Drosophila* whose nerve cells overexpress the enzyme.
   a. How might you use these transgenic flies to try to gain insight into the disease or identify drugs that might be useful in the treatment of the disease?
   b. Can you think of any potential limitations of this approach?

Single Genes Affect the Nervous System and Behavior

6. What type of mutation causes Huntington disease? How does this mutation result in neurodegeneration?
7. Perfect pitch is the ability to identify a note when it is sounded. In a study of this behavior, perfect pitch was found to predominate in females (24 out of 35 in one group). In one group of seven families, individuals in each family had perfect pitch. In two of those families, the affected individuals included a parent and a child. In another group of three families, three or more members (up to five) of each had perfect pitch, and in all three families, two generations were involved. Given this information, what, if any, conclusions can you draw about whether this behavioral trait might be genetic? How would you test your conclusion? What further evidence would be needed to confirm your conclusion?
8. The opposite of perfect pitch is tone deafness—the inability to identify musical notes. In one study, a bimodal distribution in populations was found, with frequent segregation in families and sibling pairs. The author of the study concluded that the trait might be dominant. In a family study, segregation analysis suggested an autosomal dominant inheritance of tone deafness with imperfect penetrance. One of the pedigrees is presented here. On the basis of the results, do you agree with this conclusion? Could perfect pitch and tone deafness be alleles of a gene for musical ability?

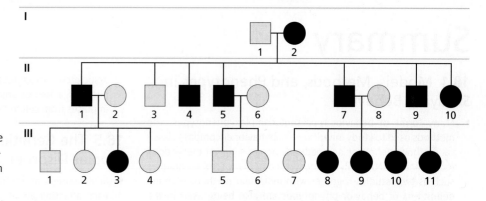

Single Genes Control Aggressive Behavior and Brain Metabolism

9. Name three genes whose mutations lead to an altered behavioral phenotype. Briefly describe the normal function of the mutated gene as well as the altered phenotype.
10. Mutations in the gene encoding monoamine oxidase type A (MAOA) have been linked to aggressive and sometimes violent behavior. On the basis of this finding, it is conceivable that a genetic test could be developed that could identify individuals likely to exhibit such behaviors. Do you think such a test would be a good idea? What would some of the ethical and societal implications of the test be?

The Genetics of Schizophrenia and Bipolar Disorder

11. Two common behavioral disorders are schizophrenia and bipolar disorder. What are the essential differences and similarities between these disorders?
12. A pedigree analysis was performed on the family of a man with schizophrenia. On the basis of the known concordance statistics, would his MZ twin be at high risk for the disease? Would the twin's risk decrease if he were raised in an environment different from that of his schizophrenic brother?
13. You are a researcher studying bipolar disorder. Your RFLP data shows linkage between a marker on chromosome 7 and bipolar illness. Later in the study, you find that a number of individuals lack this RFLP marker but still develop the disease. Does this mean that bipolar disorder is not genetic?

14. A region on chromosome 6 has been linked to schizophrenia, but researchers have not found a specific gene associated with this disease. What steps would be necessary to locate the gene?

Genetics and Social Behavior

15. Of the following findings, which does *not* support the idea that alcoholism is genetic?
   a. Some strains of mice select alcohol over water 75% of the time, whereas others shun alcohol.
   b. The concordance value is 55% for MZ twins and 28% for DZ twins.
   c. Biological sons of alcoholic men who have been adopted have a rate of alcoholism more like that of their adoptive fathers.
   d. There is a 20% to 25% risk of alcoholism in the sons of alcoholic men.
   e. none of the above

16. A woman diagnosed with Alzheimer disease wants to know the probability that her children will inherit the disorder. Explain to her the complications of determining heritability for this disease.

17. In July 1996, *The Independent*, a popular newspaper published in London, England, reported a study conducted by Dr. Aikarakudy Alias, a psychiatrist who had been working on the relationship between body hair and intelligence for 22 years. Dr. Alias told the 8th Congress of the Association of European Psychiatrists that hairy chests are more likely to be found among the most intelligent and highly educated than in the general population. According to this new research, excessive body hair could also mean higher intelligence. Is correlating body hair with intelligence a valid method for studying the genetics of intelligence? Why or why not? What factors are known to contribute to intelligence? Is it logical to assume that individuals with little or no body hair are consistently less intelligent than their hairy counterparts? What type of study could be done to prove or disprove this idea?

Internet Activities

Internet Activities are critical-thinking exercises using the resources of the World Wide Web to enhance the principles and issues covered in this chapter. For a full set of links and questions investigating the topics described below, visit *www.cengage.com/biology/cummings*

1. *The Genetics of Personality*. The *Personality Research Site* presents an overview of scientific research programs in personality psychology. Follow the "Behavior Genetics" link. The study of human behavior and behavioral disorders is complex and must account for both environmental and genetic influences. What types of studies do researchers use to attempt to tease out these differences?

HOW WOULD YOU VOTE NOW?

Biographical and scientific evidence strongly suggests that in many people, creative abilities in art and literature are linked in a complex way to bouts of depression or the onset of manic states. In light of this proposed linkage, it is possible that medicating mood disorders may reduce people's creativity. Now that you know more about the genetic and environmental factors that affect behavior in general and mood disorders in particular, what do you think? If you were a successful artist, author, or poet who had depression or bipolar illness and a cure for your illness was discovered, would you elect to have the treatment, knowing that your creative abilities might be diminished or even disappear but also knowing that your risk of suicide would be reduced or eliminated? Visit the *Human Heredity* companion website at *www.cengage.com/biology/cummings* to find out more on the issue; then cast your vote online.

19 Population Genetics and Human Evolution

I n 1775, a typhoon swept across the Pacific Ocean and devastated the Pingelap atoll, a small island located northeast of Papua New Guinea, killing 90% of the population. One of the 20 survivors was a heterozygous carrier of the mutant allele for complete achromatopsia, a recessive disorder that causes complete color blindness and other vision problems in affected individuals. Achromatopsia (known in the Pingalese language as *maskum*, meaning "not see") did not appear until the fourth generation after the typhoon, when about 3% of the population was affected. As the population grew, the frequency of this disorder increased from generation to generation. Today, the population on Pingelap, descended from the 20 survivors, has the highest frequency of achromatopsia in the world. Between 5% and 10% of the 3,000 people on the island have the disorder, and another 30% are heterozygous carriers (for comparison, the frequency in the United States is 1 in 33,000). All affected members of the population can trace their ancestry to the heterozygous male survivor. The story of this disorder and its impact on the Pingelapese is told in Oliver Sacks' book *The Island of the Colorblind*.

The mutation causing achromatopsia among the Pingelese has been identified and mapped to chromosome 8. Although genetic testing is possible, the test is not available on the island, making it difficult for individuals and couples to receive genetic counseling for the disorder.

In this chapter, we consider how to measure allele frequencies in populations and how these measurements are used to analyze the genetic structure of populations and to explore the factors that lead to different allele frequencies in different populations. We will also discuss how evidence from several disciplines is providing insight into the origin of our species and its dispersal over the earth.

Tim Davis/Stone/Getty Images

Members of one population in our species.

19.1 How Can We Measure Allele Frequencies in Populations?

HOW WOULD YOU VOTE?

Tests for about 900 genetic disorders are available through public and private testing laboratories. There is no testing available for many rare diseases. In the United States, rare diseases are defined by law as those that affect about 1 in 1,500 people or fewer. Thus, achromatopsia, which affects 1 in 33,000, is a rare disease. Many rare diseases develop early in childhood, and as many as 30% of those with rare diseases die before the age of 5. Even though there may be a low number of those affected with a recessively inherited disease in a population, heterozygotes can be quite frequent. For example, if a disease has a frequency of 1 in 1,500, 1 in 20 members of the population are heterozygotes.

If your family history showed the presence of a rare disease that is fatal in early childhood and testing was not available to determine if you carried the mutation, what would you do? There are several options open: You can take a chance that your mate is not a heterozygote; you can go ahead and have children knowing that if your mate is a heterozygote, there is only a 25% chance that a child will be affected; you can decide not to have children; or you can decide to adopt. Visit the *Human Heredity* companion website at *www.cengage.com/biology/cummings* to find out more on the issue; then cast your vote online.

If a genetic disorder such as achromatopsia is caused by a recessive allele, we cannot directly identify those members of the population that carry the mutant allele without DNA testing. Why? Because when we look at phenotypes, we can't tell who has a heterozygous genotype and who has a homozygous dominant genotype. An achromotopsial heterozygote has an *Aa* genotype with normal vision. Someone with the genotype *AA* also has normal vision. As a result, we can't directly count how many heterozygotes there are in the **population** and therefore, we can't measure the frequency of the recessive allele. To solve this problem, early in the twentieth century, Godfrey Hardy and Wilhelm Weinberg independently developed a formula that can be used to determine recessive **allele frequencies** in populations when a number of conditions (described later) are met. This method is known as the **Hardy-Weinberg Law**. This formula is used by geneticists, clinicians, field biologists, population geneticists, and others to measure the frequency of alleles and genotypes in populations, without the need to use DNA testing.

Population A local group of organisms belonging to a single species, sharing a common gene pool

Allele frequency The frequency with which alleles of a particular gene are present in a population.

Hardy-Weinberg law The statement that allele and genotype frequencies remain constant from generation to generation when the population meets certain assumptions.

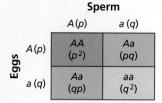

Sperm

		A (p)	a (q)
Eggs	A (p)	AA (p²)	Aa (pq)
	a (q)	Aa (qp)	aa (q²)

FIGURE 19.1 The frequency of the dominant and recessive alleles in the gametes of the parental generation determines the frequency of the alleles and the genotypes of the next generation.

Genetic equilibrium The situation when the allele frequency for a particular gene remains constant from generation to generation.

We can use the Hardy-Weinberg law to calculate allele and genotype frequencies.

Before we begin using the Hardy-Weinberg law, let's look at the assumptions that will be made about the population under study:

- The population is very large.
- No genotype is better than any other; that is, all genotypes have equal ability to survive and reproduce.
- Mating in the population is random (see Spotlight on Selective Breeding Gone Bad).
- Other factors that change allele frequency, such as mutation and migration, are absent or are rare events and can be ignored.

Let's see how the model works in a population carrying an autosomal gene with two alleles—a dominant allele A and a recessive allele a. In Hardy-Weinberg calculations, p represents the frequency of the dominant allele ($p = A$), and q represents the frequency of the recessive allele ($q = a$). Because the sum of p and q represents 100% of the alleles for that gene in the population, $p + q = 1$. A Punnett square can be used to predict the genotypes produced by the random combination of these gametes (Figure 19.1).

A look at the Punnett square shows that the frequency of homozygous genotypes with two copies of the dominant allele (AA) is represented as p^2 ($p \times p = p^2$); the frequency of heterozygous genotypes with one copy of each allele (Aa) is represented as $2pq$ ($pq + qp$); and the frequency of homozygous recessive genotypes is represented as q^2 ($q \times q = q^2$). When added together, the frequencies of these three genotypes represent 100% of the genotypes for this gene in the population, and therefore $p^2 + 2pq + q^2 = 1$.

This equation formulates the Hardy-Weinberg Law and, based on its assumptions, says that both allele and genotype frequencies will remain constant from generation to generation in a large, interbreeding population in which mating is random and there is no selection, migration, or mutation. The Hardy-Weinberg formula can be used to calculate allele frequencies (A and a in our example) and the frequency of the various genotypes (AA, Aa, and aa) in a population.

KEEP IN MIND

Without DNA testing, the frequency of recessive alleles in a population cannot be measured directly.

Populations can be in genetic equilibrium.

If the frequencies of A and a in the new generation are the same as they are in the parents' generation, and remain constant, the population is in **genetic equilibrium** for that gene and its alleles. This doesn't mean that the population is in equilibrium for all other genes. On the contrary, if mutation, selection, or migration is operating, the frequency of other alleles may change from one generation to the next.

Equilibrium helps maintain genetic variability in the population. If at equilibrium the A allele has a frequency of 0.6 (60%) and the a allele has a frequency of 0.4 (40%), we can be assured that these frequencies will be maintained in generation after generation. The presence of genetic variability is important to the process of evolution.

19.2 Using the Hardy-Weinberg Law in Human Genetics

The Hardy-Weinberg law is one of the foundations of population genetics and has many applications in human genetics and human evolution. We will consider only a few uses—primarily those that apply to measuring allele and genotype frequencies.

The Hardy-Weinberg law can be used to calculate the frequency of alleles and genotypes.

If a trait is inherited recessively, we can use the Hardy-Weinberg law to calculate the frequency of the recessive allele in the population. To do this, we begin by counting the number of homozygous recessive individuals in the population. For example, cystic fibrosis is an autosomal recessive trait, and homozygous recessive individuals can be identified by their distinctive phenotype. Suppose that 1 in 2,500 members of a population of European descent is affected with cystic fibrosis. We know that an affected individual has the genotype aa. According to the Hardy-Weinberg equation, the frequency of this genotype in the population is equal to q^2. The frequency of the a allele in this population is therefore equal to the square root of q^2:

$$q^2 = \frac{1}{2,500} = 0.0004$$

$$q = \sqrt{0.0004}$$

$$q = 0.02 = \frac{1}{50}$$

Because $p + q = 1$, once we know that the frequency of the a allele is 0.02 (2%), we can calculate the frequency of the dominant allele A by subtraction:

$$p = 1 - q$$
$$p = 1 - 0.02$$
$$p = 0.98 \quad 98\%$$

In this population, 98% of the alleles for gene A are dominant (A) and 2% are recessive (a). This method can be used to calculate the allele frequencies for any dominant or recessive trait.

For many reasons, it is important to know the population frequency of heterozygotes carrying a deleterious recessive allele. Calculating the frequency of heterozygotes is an important application of the Hardy-Weinberg Law because for rare traits, most disease-causing alleles are carried by heterozygotes. To calculate the frequency of heterozygous carriers for such recessive traits, we must first know the frequency of the alleles in the population. We have already calculated that the frequency of the normal CF allele is 0.98 (98%), and the frequency of the mutant allele is 0.02, or 2% (among Americans of European ancestry—the disease is much rarer among Americans of African and Asian descent). Knowing the allele frequencies, we can use the Hardy-Weinberg equation to calculate genotype frequencies. Recall that in the Hardy-Weinberg equation, $2pq$ is the frequency for the heterozygous genotype. Using the values we have calculated for p and q, we can determine the frequency of heterozygotes as follows:

$$2pq = 2(0.98 \times 0.02)$$
$$2pq = 2(0.0196)$$
$$\text{Heterozygote frequency} = 0.039 = 3.9\%$$

This means that 3.9%, or approximately 1 in 25 Americans of European ancestry, carry the gene for cystic fibrosis.

Sickle cell anemia is an autosomal recessive disorder that affects 1 in 500 Americans of West African descent. Using the Hardy-Weinberg equation, we can calculate that 8.5% of this population, or 1 in 12, are heterozygous carriers for sickle cell anemia. Table 19.1 lists the frequencies of heterozygotes, with a frequency range from 1 in 10 to 1 in 10,000,000. Table 19.2 lists the heterozygote frequencies for some common human autosomal recessive traits.

Heterozygotes for many genetic disorders are common in the population.

Many people are surprised to learn that heterozygotes for recessive traits are so common in the population. If a genetic disorder is relatively rare (say 1 in 10,000 individuals), they generally assume that the number of heterozygotes must also be rather low. In fact, if 1 in 10,000 members of a population are homozygous for a recessively inherited disorder, it

Table 19.1 Heterozygote Frequencies for Recessive Traits

Frequency of Homozygous Recessive Individuals (q^2)	Frequency of Heterozygous Individuals ($2pq$)
1/100	1/5.5
1/500	1/12
1/1,000	1/16
1/2,500	1/25
1/5,000	1/36
1/10,000	1/50
1/20,000	1/71
1/100,000	1/158
1/1,000,000	1/500
1/10,000,000	1/1,582

Table 19.2 Frequency of Heterozygotes for Some Recessive Traits in Several Populations

Trait	Heterozygote Frequency
Cystic fibrosis	1/25 whites; much lower in blacks, Asians
Sickle cell anemia	1/12 blacks; much lower in most whites and in Asians
Tay-Sachs disease	1/30 among descendants of Eastern European Jews; 1/350 among others of European descent
Phenylketonuria	1/55 among whites; much lower in blacks and those of Asian descent
Albinism	1/10,000 in Northern Ireland; 1/67,800 in British Columbia

turns out that 1 in 50 (2%) members of the population is a heterozygote, and there are about 200 times as many heterozygotes as there are homozygous recessive individuals.

Let's take this a step further. What are the chances that two heterozygotes will mate and have an affected child? The chance that two heterozygotes will mate is $1/50 \times 1/50 = 1/2,500$. Because they are heterozygotes, the chance that they will produce an affected child is one in four. Therefore, the chance that they will mate and produce an affected child is $1/2,500 \times 1/4 = 1/10,000$. In other words, if the disorder is present in 1 of every 10,000 individuals, 1 in 50 individuals *must* be a heterozygote.

The relationship between allele frequency and genotype frequency is shown in Figure 19.2. As the frequencies of p and q move away from zero, the percentage of heterozygotes in the population increases rapidly. This again illustrates the point that in many disorders, the majority of the recessive alleles are carried by heterozygotes.

> **KEEP IN MIND**
>
> Estimating the frequency of heterozygotes in a population is an important part of genetic counseling.

Calculating the frequency of X-linked alleles.

Human females carry two X chromosomes and have two copies of all X-linked genes. Males have only one X chromosome and have only one copy of all genes on the X chromosome. Thus, X-linked genes are not distributed equally in the population: Females

FIGURE 19.2 The relationship between allelic frequency and genotypic frequency in a population that is in Hardy-Weinberg equilibrium. As the frequencies of the homozygous genotypes (p^2 and q^2) decline, the frequency of the heterozygote genotype ($2pq$) rises.

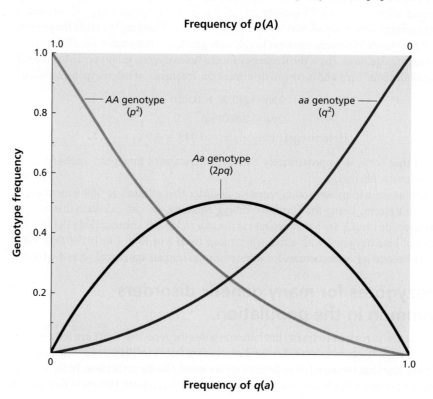

(and their gametes) carry two-thirds of the alleles, and males (and their gametes) carry one-third of the alleles. Can the Hardy-Weinberg formula be used to calculate genotype frequencies for recessive X-linked traits? The answer is yes, but for males it is not necessary to use the formula. Because males carry only one copy of all genes on the X chromosome, the allele frequency for recessive X-linked traits is equal to the number of males with the recessive phenotype. For example, in the United States, about 8% of males are color blind. Therefore, the frequency of the color-blindness allele in the male population is 0.08 ($q = 0.08$).

Because females carry two X chromosomes, genotypic frequencies for X-linked recessive traits in females can be calculated by using the Hardy-Weinberg equation. If color blindness in males has a frequency of 8% ($q = 0.08$), we expect color blindness in females to have a frequency of q^2, or 0.0064 (0.64%). In a population of 10,000 males, 800 would be color blind; but in a population of 10,000 females, only 64 would be color blind. This example reemphasizes the fact that males are at much higher risk for disorders carried on the X chromosome. The relative values for the frequency of X-linked recessive traits in males and females are listed in Table 19.3.

Table 19.3 Frequency of X-Linked Recessive Traits in Males and Females

Males	Females
1/10	1/00
1/100	1/10,000
1/1,000	1/1,000,000
1/10,000	1/100,000,000

19.3 Measuring Genetic Diversity in Human Populations

Understanding our evolutionary history depends on identifying factors that lead to variations in allele frequencies between populations and the way in which those variations are acted on by natural selection. In the following sections, we explore how genetic variation is produced and the role of natural selection and culture as a force in changing allele frequencies.

Mutation generates new alleles but has little impact on allele frequency.

As the **gene pool** is reshuffled each generation to produce the genotypes of the offspring, new combinations of alleles are generated by recombination and Mendelian assortment, but these events do not create any new alleles. Mutation is the ultimate source of all new alleles and is the origin of all genetic variability.

Gene pool The set of genetic information carried by the members of a sexually reproducing population.

If the mutation rate for a gene is known, we can use the Hardy-Weinberg formula to calculate the change in allele frequency resulting from new mutations in that gene in each generation. Let's use the dominant trait achondroplasia as an example. For this calculation, we assume that initially only homozygous recessive individuals with the genotype *dd* (normal stature) were present in the population and that mutation has added new mutant (*D*) alleles to each generation at the rate of 1×10^{-5}. The change in allele frequency over time that results from this rate of mutation is shown in Figure 19.3. To change the frequency of the recessive allele (*d*) from 1.0 (100%) to 0.5 (50%) at this rate of mutation will require 70,000 generations, or 1.4 million years. Our conclusion in this case is that mutation alone has a minimal impact on the genetic variability present in a population.

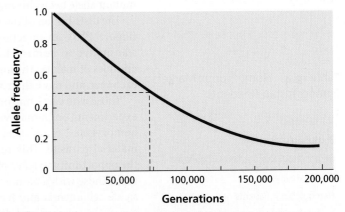

FIGURE 19.3 The rate of replacement of a recessive allele *d* by the dominant allele *D* by mutation alone. Even though the initial rate of replacement is high, it will take about 70,000 generations, or 1.4 million years, to drive the frequency of the allele *d* from 1.0 to 0.5. This curve is asymptotic, and will only reach zero at an infinite number of populations.

KEEP IN MIND

Mutation generates all new alleles, but drift, migration, and selection determine the frequency of alleles in a population.

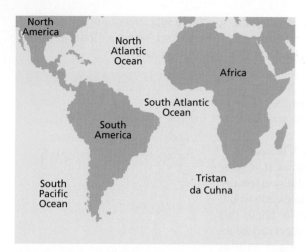

FIGURE 19.4 Location of the island of Tristan da Cuhna, first discovered by a Portuguese admiral in 1506.

Founder effects Allele frequencies established by chance in a population that is started by a small number of individuals (perhaps only a fertilized female).

Genetic drift The random fluctuations of allele frequencies from generation to generation that take place in small populations.

Evolution Changes in allele frequencies in a population over time.

Natural selection The differential reproduction shown by some members of a population that is the result of differences in fitness.

Table **19.4** Homozygous Markers among Tristan Residents

| Transferrins |
| Phosphoglucomutase |
| 6-Phosphogluconate dehydrogenase |
| Adenylate kinase |
| Hemoglobin A variants |
| Carbonic anhydrase (two forms) |
| Isocitrate dehydrogenase |
| Glutathione peroxide |
| Peptidase A, B, C, D |

Genetic drift can change allele frequencies.

As we saw with the aftermath of the typhoon on Pingelap atoll, populations sometimes start with a small number of individuals, or founders. Alleles carried by the founders, whether they are advantageous or detrimental, become established in the new population. These events take place simply by chance and are known as **founder effects**.

Changes in allele frequency that occur by chance from generation to generation in small populations are examples of **genetic drift**. The impact of genetic drift is magnified in small populations. On Pingelap, frequency of the mutant allele increased from 2.5% to 18–20% in just over 200 years.

The island of Tristan da Cuhna (Figure 19.4) is in the southern Atlantic Ocean, about 2,000 miles from the nearest land. In the early 1800s, Corporal William Glass, his wife, and his two daughters emigrated to the island, and others soon followed. The growth of the isolated and highly inbred population that formed there can be traced with great accuracy.

One effect of reproduction in a small gene pool is the random loss of alleles, resulting in an increase in homozygosity for recessive traits. Table 19.4 lists some genetic markers for which all the islanders tested are homozygous.

This brief example illustrates how genetic drift can be responsible for changing allele frequencies in populations that are isolated, inbred, and stable for long periods. Most human populations, however, do not live on remote islands and are not subject to prolonged isolation and inbreeding. Yet there are differences in the distribution and frequency of alleles among populations, indicating that founder effects and drift are not the only factors that can change allele frequencies.

Natural selection acts on variation in populations.

In formulating their theory of **evolution**, Alfred Russel Wallace and Charles Darwin recognized that some members of a population are better adapted to the environment than others. These better-adapted individuals have an increased chance of surviving and leaving more offspring than those with other genotypes. The differential survival and reproduction of better-adapted genotypes is the basis of **natural selection**.

The relationship between the allele for sickle cell anemia and malaria is an example of how natural selection changes allele frequencies. Sickle cell anemia is an autosomal recessive disorder associated with a mutant form of hemoglobin (see Chapters 10 and 11 for a review). Even though many untreated victims of sickle cell anemia die before passing the mutant allele to another generation, it is present in very high frequencies in certain populations. In some West African countries, 20% to 40% of the population is heterozygous. If homozygous individuals die before they reproduce, why hasn't this mutant allele been eliminated from the population? The answer: natural selection.

The distribution of the sickle cell allele matches the geographic distribution of malaria (Figure 19.5), an infectious disease caused by a protozoan parasite (Figure 19.6) and transmitted to humans by infected mosquitoes. Once infected, victims have recurring episodes of malaria throughout their lives. More than 2 million people die from malaria each year, and more than 300 million individuals worldwide are infected with the disease.

Those who carry one or two copies of the sickle cell allele are resistant to malaria, and experiments on human volunteers have confirmed this. In heterozygotes and recessive homozygotes, the plasma membrane of red blood cells is resistant to infection by the malarial parasite. This resistance gives heterozygotes a better chance at survival and the opportunity to leave more offspring than those with a homozygous dominant genotype. Those with a homozygous recessive genotype are resistant to malaria but also have sickle cell anemia and have a much lower survival rate. In this case, natural selection favors the survival and differential reproduction of heterozygotes at the expense of the recessive homozygous and homozygous dominant genotypes.

Because of their resistance to malaria and the ability of heterozygotes to leave more offspring, the sickle cell allele is spread through the population and maintained at a high frequency.

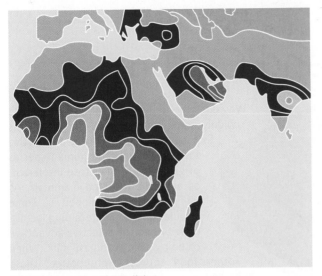

Allele frequencies of HbS allele

KEY

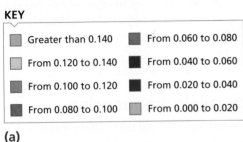

Greater than 0.140	From 0.060 to 0.080
From 0.120 to 0.140	From 0.040 to 0.060
From 0.100 to 0.120	From 0.020 to 0.040
From 0.080 to 0.100	From 0.000 to 0.020

(a)

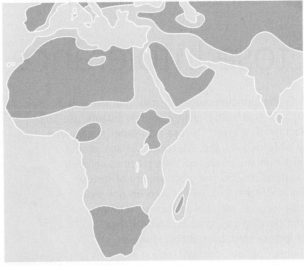

KEY

Regions with malaria

(b)

FIGURE 19.5 (a) The distribution of sickle cell anemia in the Old World. (b) The distribution of malaria overlaps the distribution of sickle cell anemia.

19.4 Natural Selection Affects the Frequency of Genetic Disorders

Many genetic disorders are disabling or fatal, so why are they so common? In other words, what keeps natural selection from eliminating the deleterious alleles responsible for those disorders? In analyzing the frequency and population distribution of human genetic disorders, it is clear that there is no single answer. One conclusion, drawn from the Hardy-Weinberg Law, is that rare lethal or deleterious recessive alleles survive because the vast majority of them are carried in the heterozygous condition and thus are hidden in the gene pool. Other factors, however, can cause the differential distribution of alleles in human populations, and several of those factors are discussed in the following paragraphs (see Exploring Genetics: Lactose Intolerance and Culture).

Natural selection can cause detrimental alleles to have high frequencies in large, well-established populations. Heterozygote advantage in sickle cell anemia is an example. For other genetic disorders with a high frequency of the disease allele, the heterozygote selective advantage may be less obvious or perhaps no longer exists.

Tay-Sachs disease (OMIM 272800) is an autosomal recessive disorder that is fatal in early childhood. Although this is a rare disease in most populations, Ashkenazi Jews (those who live in or have ancestors from eastern Europe) have a tenfold higher incidence of the disease than members of the general population. In these populations, heterozygote frequency can be as high as 11%. There is indirect evidence that Tay-Sachs heterozygotes are more resistant to tuberculosis, a disease endemic to cities and towns where most European Jews lived in the past. As in sickle cell anemia, the death of homozygous Tay-Sachs individuals is the genetic price paid by the population to allow the higher **fitness** and survival of the more numerous heterozygotes.

Fitness A measure of the relative survival and reproductive success of a specific individual or genotype.

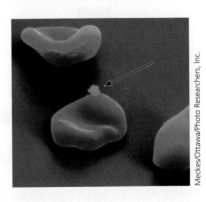

FIGURE 19.6 *Plasmodium* parasites (*yellow*) attacking and infecting red blood cells (*arrow*). Infection by *Plasmodium* causes malaria.

Lactose Intolerance and Culture

Lactose is the principal sugar in milk (human milk is 7% lactose) and is a ready energy source. Infants use the enzyme lactase to convert lactose into other sugars that are easily absorbed by the intestine. Lactase production in most humans slows and then stops as children grow into adults (review the biochemistry of lactose breakdown in Chapter 10). Adults who do not produce lactase cannot metabolize lactose (OMIM 223100) and develop gas, cramps, and diarrhea if they eat lactose-containing foods. However, in some human populations, lactase is produced throughout adulthood. The ability to make lactase as an adult is inherited as an autosomal dominant trait. Across different populations, the frequency of this trait varies from 0.0% to 100%.

Why does the frequency of this allele vary so widely across populations? The answer is that cultural practices are a selective force. Adults of the human species originally were all lactose-intolerant, as are all other land mammals. As dairy herding developed in some populations, adults retaining lactase activity had the selective advantage of being able to digest milk. When other food sources were scarce, these adults had an enhanced chance of survival and success in leaving offspring. In other words, the cultural practice of maintaining dairy herds was the selective factor that provided an advantage for lactase production in adults.

Image copyright Arti_Zav, 2010. Used under license from Shutterstock.com.

Selection can rapidly change allele frequencies.

The effects of selection can be seen in geographically separated subpopulations that originate from a single ancestral population. The emigration of Jews from what is now Israel over the past 2,500 years—forming subpopulations in Europe, North Africa, the Middle East, and Asia—is an example of such a dispersal. The frequency of a glucose-6-phosphate dehydrogenase (*G6PD*) deficiency allele (MIM/OMIM 305900) in these subpopulations is now very different than in the original population (Figure 19.7), ranging from near zero to around 70%.

Two explanations are possible: founder effects and selection. For several reasons, the evidence favors selection as the answer. First, the distribution of the *G6PD* allele and malaria are very similar. Second, the rapid change in allele frequency occurred in only 100 to 125 generations. As with sickle cell anemia, malaria is the selective force here, because G6PD-deficient homozygous females and hemizygous males are resistant to malaria. If the ancient ancestral population in Israel had a low frequency of the *G6PD* deficiency allele, then selection has rapidly changed the allele frequency in areas with malaria, illustrating the power of natural selection.

These examples illustrate that several factors contribute to disease-allele frequency and that each disease must be analyzed individually. In some cases, migration and founder effects change allele frequencies, while other alleles are spread through the population and maintained at a high frequency by natural selection that favors heterozygotes.

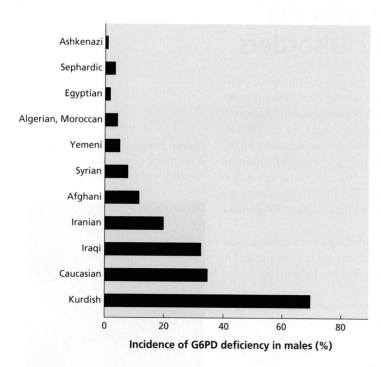

Incidence of G6PD deficiency in males (%)

FIGURE 19.7 Distribution of glucose-6-phosphate dehydrogenase (G6PD) deficiency in various Jewish populations. Because intermarriage with native populations is rare, the differences in frequency are attributed to selection. G6PD deficiency confers resistance to malaria, and the frequency of the allele is highest in regions where malaria is endemic.

KEEP IN MIND

Survival and differential reproduction are the basis of natural selection.

19.5 Genetic Variation in Human Populations

Genomics has helped quantify the amount of genetic variation present in the human genome. Before the development of genomics, geneticists studied this variation by surveying populations and cataloging differences in allele frequencies to study the extent and nature of genetic variation across geographic regions. In many cases, these differences provided clues about the origin and migrations of populations. For example, the *B* allele of the ABO blood group is distributed across Asia and Europe in a gradient (geneticists call these gradients *clines*) from East to West (Figure 19.8). This gradient reflects waves of migration from central Asia into Europe from ancient times up to the Middle Ages. Gradients of alleles of other genes accompanied the spread of agriculture from the Middle East northwestward across Europe. There is also genetic variation among populations. How much variation is present, and what is its significance?

Are there human races?

In the nineteenth century, biologists used the term *race* to describe populations within a species that were phenotypically different from other populations in that species. Clearly, there are phenotypic differences among human populations. Residents of the Kalahari Desert are rarely mistaken for close relatives of Aleutian seal hunters, for example. In the twentieth century, as population genetics emerged as a discipline, the phenotypic

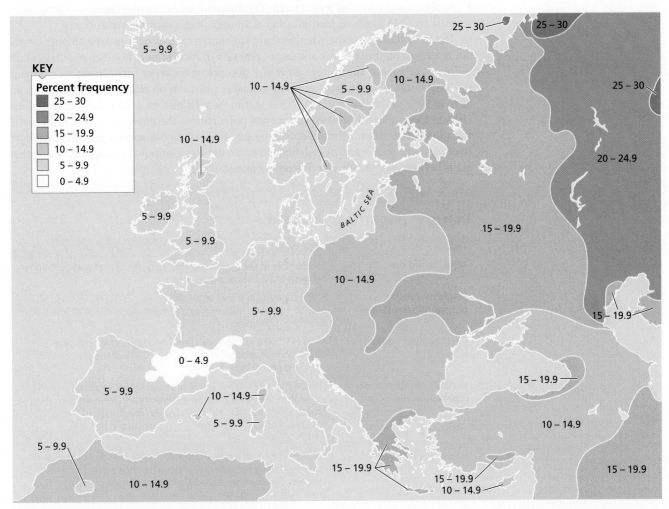

FIGURE 19.8 The gradient of the *B* allele of the ABO blood groups from East to West is due to waves of migration from Asia to Europe.

definition of race was replaced with a genetic definition. To a geneticist, races are populations of the same species between which allele frequencies differ by 25% to 30%.

In one sense, it is easy to assume that the physical differences we see in human populations are evidence of underlying genetic differences. However, any decision to divide our species into races depends on showing that there are significant genetic differences among the races, not simply phenotypic differences.

In the early 1970s, studies of protein variation in different populations showed that 85% of the detected variation is present within populations and that less than 7% of the variation is present between populations classified into racial groups. Beginning in the 1990s, genomic methods and the results of genome sequencing have revolutionized the study of human populations. Genomics has expanded our knowledge of the types and extent of human genome variations—between individuals, in populations, and across continents. Some of these variations are listed in Table 19.5.

Genomic studies that screened thousands of genomes using tens of thousands of SNPs have identified haplotypes that can be used to identify individuals and populations. The results of these studies can be summarized as follows:

- There is very little genetic variation in the human genome. Two randomly selected chimpanzees differ from each other in about 1 in every 500 nucleotides; two randomly selected humans differ in every 1 in 1,000 to 1 in 1,500 nucleotides, indicating that we are a very young species.
- Variation in the human genome is continuously distributed, and there are few sharp boundaries separating populations.
- Most genetic variation is widely shared among populations, but a small amount is geographically clustered, and its continental distribution roughly corresponds to traditional racial classifications.
- Because some variation is geographically clustered, analysis using ancestry-informative-markers (AIMs) allows identification of an individual's continent of ancestry.

The current debate about whether there are human races is not centered on phenotypes but is focused on the type and amount of genetic variation present between populations on different continents. The average difference in genomic sequence between any two humans is 0.1%. Of this, what proportion varies among populations on different continents? About 85% to 90% of this variation is found within populations on a continent, and the rest (10% to 15%) is found between continental populations. The question is, does this genetic variation between continental populations justify dividing our species into racial groups? In other words, can "race" be redefined to mean the small amount of genetic variation that exists between populations on different continents?

Table 19.5   Human Genome Variations

Type	Description
Allelic variation	The number of alleles of a given gene, or the sum of allelic differences between individuals or populations.
Single nucleotide polymorphisms (SNPs)	Single base changes that occur about every 1,000 nucleotides. There may be 10 million SNPs in the genome, only about 1% of which have functional significance.
Copy number variations (CNVs)	Short (>1,000 base pairs) deletions, insertions, inversions, and duplications of genomic regions. About 0.4% of the genomes of unrelated people differ in CNVs.
Variable number tandem repeats (VNTRs)	Short nucleotide sequences organized as tandem repeats, showing variation in repeat length between individuals. There are two groups of VNTRs: minisatellites (15–100 base pairs) and microsatellites (5–15 base pairs). Both are used in personal identification.
Epigenetic markers	Epigenetic markers are modifications in DNA that do not change the nucleotide sequence. These changes alter gene expression. Most epigenetic markers are erased and remarked each generation, but some markers are passed to the next generation.

Supporters of this view emphasize the importance of these genetic differences for the study of the evolution of our species, the dispersal of populations across the globe, and medical care based on genetic variation. Opponents point out that relying on the small amount of variation present between continental groups can lead to misidentification, and that accurate classification of individuals from different populations with continuously varying levels of genetic variation is impossible. For example, in 38% of cases in one study, Europeans were more similar to Asians than to other Europeans based on genetic variation in the alleles studied and for these individuals, their continent of origin would be identified as Asia instead of Europe.

Given the very low levels of genetic variation present in the human genome and its distribution (Figure 19.9), is there any reason to divide our species into races? At present, the vast majority of geneticists would answer no to that question and agree that there is currently no genetic basis for subdividing our species into racial groups.

FIGURE 19.9 The amount of genetic variation within populations is far greater than the variation between populations. Each colored circle represents genetic variation within a population classified as a racial group. The variations overlap greatly, with few or no genetic differences belonging to a single racial group. Because most variation is found within groups, many geneticists believe that there is no basis for classifying humans into racial groups.

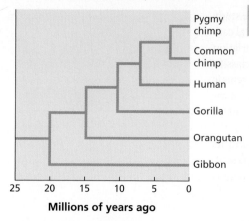

FIGURE 19.10 A phylogenetic tree showing the evolutionary relationships among hominoids. The evidence shows that chimpanzees and humans shared a common ancestor about 7 million years ago and that chimpanzees are our closest hominoid relative.

Hominoids The superfamily of primates that includes apes and humans.

Hominins A classification that includes all bipedal primates from australopithecines to our species.

The Evolutionary History and Spread of Our Species (*Homo sapiens*)

Tracing the origins of our species is a multidisciplinary task, using the tools and methods of genomics, anthropology, paleontology, archaeology, and even satellite mapping from space. These methods are being used to reconstruct the origins and ancestry of populations of *H. sapiens* to determine how and when our species originated, to establish our genetic relationship to other primates and other human species, and to piece together the migrations that dispersed us across the globe.

Our evolutionary heritage begins with hominoids.

Africa is home to many modern species of primates, including baboons, gorillas, and chimpanzees. These species and their ancestors are called **hominoids**. The fossil record of hominoids leads to our species, but there are many gaps in the record. Is there a way to reconstruct the evolutionary relationships and history of these lineages and our species, and to provide a time scale for these events? It turns out that the genomes of living hominoid species contain clues to those evolutionary relationships. These genomic differences can be used to reconstruct these relationships among species, producing a phylogenetic tree that shows the relationships among species with common ancestors. Figure 19.10 outlines the genome-based evolutionary relationships among humans and related primates. It shows that the hominoid lineage began about 25 million years ago and that chimpanzees and humans last shared a common ancestor about 7 million years ago, making chimpanzees our closest relative among other primates. Evolutionary events after the split from the chimpanzee line have been reconstructed from the fossil record.

Early humans emerged almost 5 million years ago.

Over a period of several million years after the split from the line that led to chimpanzees, three different species groups, collectively called **hominins**, appeared in Africa (Figure 19.11). The original group, called australopithecines, were small and had ape-like features with small brain cases but walked upright. About 2.5 million years ago,

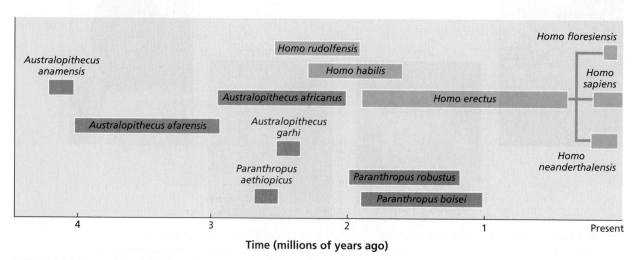

FIGURE 19.11 Estimates of the dates of origin and extinction of the three main groups of hominins (*green, blue,* and *orange*). The australopithecines split into two groups about 2.5 to 2.7 million years ago. One of those groups, the genus *Homo*, contained the ancestors to our species, *Homo sapiens*.

two other groups split from the australopithecines. One of these was a cluster of species that became extinct about 1 million years ago. The second—all members of the genus *Homo*—is our ancestral group.

One early human species known only from fragmentary fossils is called *Homo habilis* ("handy man") because it is credited as the first to use manufactured tools. In response to as-yet-unknown selection pressures and/or advantageous mutations at that time, there was a rapid increase in brain size, and *Homo habilis* had a brain about one-third larger than australopithecines. A successor species, *Homo erectus*, was taller than *H. habilis* and had an even larger brain. *H. erectus* made relatively sophisticated tool kits, used fire, and may have hunted animals. About 2 million years ago, this very successful species migrated out of Africa to the Middle East, Europe, and Asia, reaching China and Indonesia.

Our species, *Homo sapiens*, originated in Africa.

Originally, there were two opposing views on how and where our species originated. One school of thought used fossil evidence to argue that after leaving Africa, populations of *H. erectus* formed a network of interbreeding populations that gradually transformed into our species, *H. sapiens*. The other school of thought used a combination of genetic and fossil evidence to argue that our species originated in Africa from *H. erectus* between 200,000 and 100,000 years ago. The genetic evidence shows that small groups of *H. sapiens* spread from Africa by about 60,000 years ago and replaced earlier populations of other human species, including *H. erectus* and Neanderthals (*H. neanderthalensis*), and perhaps *H. floresiensis*, driving them into extinction. The two opposing ideas can be summarized as follows: One favors evolution and transition within a single species (the multiregional model), and the other favors speciation in Africa, followed by migration and replacement of other species (the out-of-Africa model).

The evidence now overwhelmingly supports the out-of-Africa model, identifying southwestern Africa as the most likely place where modern humans originated and East Africa as the point of migration from Africa. According to this model, modern human populations are all derived from a single speciation event that took place in a restricted region within Africa. As a result, the human populations which remained in Africa have the highest degree of genetic diversity, and all populations outside Africa show a high degree of genetic relatedness because they are derived from the relatively small migrant population (Figure 19.12).

The recent discovery of fossil human skeletons in Indonesia (called "hobbits" in the popular press because they were very short) has added to the number of species in our genus but does not change the strong case for the origin of our species in Africa. This new species, called *Homo floresiensis*, adapted to life on an island and became extinct only about 13,000 years ago. Their relationship to *H. erectus* and our species is still being investigated.

Ancient migrations dispersed humans across the globe.

There is strong evidence that *H. sapiens* originated in Africa and spread from there to other parts of the world. There may have been one primary migration or several from a base in eastern Africa. The emigrants carried a subset of the variation present in the African population, consistent with the finding that present-day non-African populations have a small amount of genetic variation compared to African populations. These non-African populations also carry a small genetic contribution from Neanderthals.

Using genetic markers from mitochondrial DNA and from the Y chromosome, these ancient migrations (Figure 19.13) have been reconstructed in great detail (see The Genetic Revolution: Tracing Ancient Migrations on page 433).

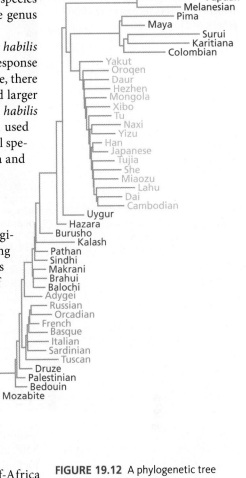

FIGURE 19.12 A phylogenetic tree showing that all populations outside Africa are derived from African populations. The colors correspond to the major continental regions.

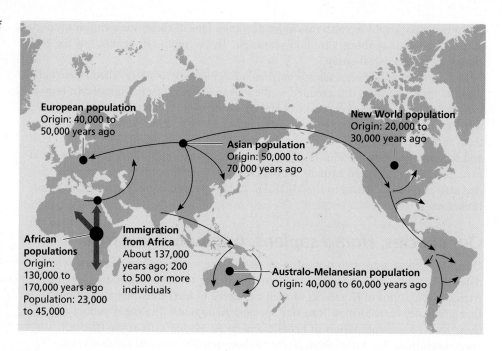

FIGURE 19.13 The origin and spread of modern *H. sapiens*, reconstructed from genetic and fossil evidence.

European population
Origin: 40,000 to 50,000 years ago

Asian population
Origin: 50,000 to 70,000 years ago

New World population
Origin: 20,000 to 30,000 years ago

African populations
Origin: 130,000 to 170,000 years ago
Population: 23,000 to 45,000

Immigration from Africa
About 137,000 years ago; 200 to 500 or more individuals

Australo-Melanesian population
Origin: 40,000 to 60,000 years ago

19.7 Genomics and Human Evolution

Genomic techniques including DNA sequencing, bioinformatics, the development of SNP markers, and microarray technology have revolutionized many areas of genetics, including evolutionary genetics. Once the human genome sequence was completed, efforts were directed at sequencing the genomes of closely related hominoid primates, including the gorilla and the chimpanzee. Newer technology now allows researchers to extract and sequence fossil DNA, including DNA from Neanderthal fossils. These methods are being used to dissect many aspects of human evolution. In the following sections, we will discuss what has been learned about our evolutionary history from comparing the genome sequences of the chimpanzee (our closest primate relative), the genome of our closest human relative (the Neanderthals), and our own genome.

The human and chimpanzee genomes are similar in many ways.

The chimpanzee and human genomes have been separated for about 7 million years. Now that the genome sequences of these two species are available, analysis shows many similarities but also many subtle differences:

- In spite of their long separation from a common ancestor, the human and chimpanzee genome sequences are 98.8% identical.
- Genome variations including insertions, deletions, and duplications differ between the species. Many of these are present in one species but not the other, potentially changing gene dosage.
- There is only about 1% difference in the coding sequences of genes analyzed to date.
- Phenotypic differences between humans and chimps cannot be explained only by differences in coding sequences and probably involve changes in patterns of gene expression and regulation. Significant differences between the two genomes in promoter sequences and transcription factors (see Chapter 9) have been identified.
- Alterations in certain structural genes and the target genes controlled by transcription factors also help explain what separates us from chimpanzees.

THE GENETIC REVOLUTION

Tracing Ancient Migrations

Genomic information is being used to trace the paths followed by ancient migrations out of Africa to all parts of the Earth. A logical question is: How can we map out events that occurred thousands of years ago that left no written records? The answer is written in the genomes of present-day populations. To work out these routes, geneticists use a set of genetic markers. The markers used in this work are Y chromosome sequences, which are passed directly from father to son, and mitochondrial sequences, which are passed from a mother to all her children. These markers allow men to trace their paternal heritage and men and women to trace their maternal heritage. Because these DNA markers do not undergo recombination during meiosis, mutations that arise in these DNA sequences become heritable markers. These new mutations spread through the population; after many generations, a specific marker will be carried by most members of a population living in a particular geographical region. If people leave that region, they carry that marker with them and pass it on to their offspring, making its path traceable. The relative ages of markers can be established by assuming that mutations in the markers are random and occur at a constant rate. This assumption is more reliable for Y chromosome markers than for mitochondrial markers but is still useful for establishing the relative ages of each marker.

Ancient migration routes are traced by cataloging the markers present in existing populations. Knowing the markers characteristic of many indigenous populations provides a starting point from which researchers work back to track the markers through different populations. DNA samples donated by about 10,000 members of indigenous and traditional peoples from around the world form the starting-point database. The sets of markers we carry each represent an ancient point of origin and an end point (where we are now) along a path of migration. By surveying many people in present-day populations, the track of each marker can be reconstructed.

What this means for all of us is that it is now possible to trace our heritage far beyond grandparents and great-grandparents to ancestors who lived thousands of years ago, and to follow the path of their ancient migrations that lead to us and where we live now. The Genographic Project is assembling the largest database for these studies. Part of the database is made up of DNA samples from the 5,000-or-so indigenous populations that have lived in particular regions for many generations and have maintained their languages and cultures. However, the project is also selling kits to those who wish to contribute their DNA, using swabs to collect cheek cells. Online vendors offer similar kits. Others offer autosomal DNA testing using SNPs to provide a large-scale view of someone's heritage, but these tests do not have the specificity of tests using Y chromosome and mitochondrial markers.

Modern forms of *H. sapiens* spread through central Asia some 50,000 to 70,000 years ago and into Southeast Asia and Australia about 40,000 to 60,000 years ago. *H. sapiens* moved into Europe some 40,000 to 50,000 years ago, displacing the Neanderthals who had lived there from about 100,000 years ago to about 30,000 years ago.

Genetic data and recent archaeological findings indicate that North America and South America were populated by three or four waves of migration that occurred 15,000 to 30,000 years ago. Migrations from Asia across the Bering Sea are well supported by archaeological and genetic findings, but Asia may not have been the only source of the first Americans. Some skeletal remains, such as Kennewick Man and the Spirit Cave mummy, have features that more closely resemble Europeans than Asians. Evidence from a mitochondrial DNA variant called haplotype X, found only in Europeans, and a reinterpretation of stone-tool technology make it seem likely that Europeans migrated to North America more than 10,000 years ago. Although a model with migrations from two sources explains most of the data available, there are other issues that remain to be resolved. Nonetheless, genetic analysis of present-day population—coupled with anthropology, archaeology, and linguistics—has proven to be a powerful tool for reconstructing the history of our species.

Analysis of just over 7,500 genes found in both human and chimpanzee genomes shows that more than 1,500 of these have evolved differently. Differences in one of these—the *FOXP2* gene, associated with language—will be explored in a later section of this chapter.

Neanderthals are not closely related to us.

Three known hominin species followed *H. erectus*: *H. neanderthalensis*, *H. sapiens*, and *H. floresiensis*. Fossil evidence indicates that Neanderthals lived in the Middle East, western Asia, and Europe 300,000 to about 30,000 years ago and for some of this time lived alongside *H. sapiens*, raising several questions: (1) Were Neanderthals our ancestors? (2) Was there interbreeding between the two species? (3) Can we compare our genome to the Neanderthal genome?

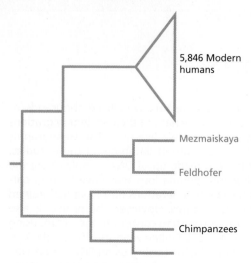

FIGURE 19.14 A phylogenetic tree constructed from DNA analysis of over 5,000 present-day individuals, Neanderthal fossils (*orange*), and chimpanzees. The evidence shows that Neanderthals are distant relatives to modern humans.

Analysis of mitochondrial DNA sequences from Neanderthals, modern humans, and chimpanzees (Figure 19.14) clearly shows that we are not descended from Neanderthals; in fact, they are distant relatives, last sharing a common ancestor with us about 700,000 years ago (Figure 19.15).

Sequencing of most of the Neanderthal genome in 2010 revealed more information about the relationship to our sister species:

- Small amounts of interbreeding did occur, probably in the Middle East, after modern humans left Africa.
- Gene flow between the species took place before modern humans expanded into Europe and Asia.
- As a result, Neanderthals are more closely related to present-day non-Africans than to Africans. About 1–4% of the genes carried by non-Africans are from Neanderthals.
- Several genome regions in ancestral modern humans may have been subject to positive selection, including genes involved in cognition, skeletal development, and metabolism, further separating us from our Neanderthal cousins.

Analysis shows that the genomes of Neanderthals and modern humans are more than 99.5% identical.

Chimpanzees, modern humans, and Neanderthals share a gene important in language development.

With the availability of genomic sequences from chimpanzees and humans and the identification of a gene called *FOXP2*, involved in the development of human speech, it is now possible to define the network of genes controlled by *FOXP2* in both chimps and humans as a way of exploring why humans have complex spoken languages and chimps do not.

FOXP2 evolved rapidly after the separation of the chimp and human lines from a common ancestor about 7 million years ago. The timing of these changes may have occurred around the time of language development in humans. To investigate whether these changes caused functional differences between chimps and humans, researchers

FIGURE 19.15 Genomic and fossil evidence has been used to estimate the times of divergence of human and Neanderthal genomic sequences relative to landmark events in both human and Neanderthal evolution.

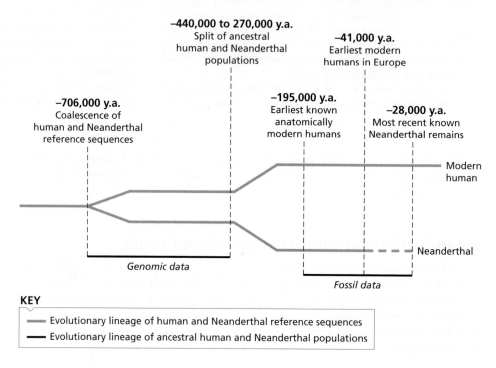

used whole-genome microarrays to identify the changes in gene expression that occur under the control of chimp or human *FOXP2*. The *FOXP2* gene is a transcription factor, a molecular switch that controls the expression of a set of genes known as hub genes. When *FOXP2* is turned on, the hub genes are activated, and they in turn change the pattern of expression of other genes (Figure 19.16). In human brain cells, the downstream targets of *FOXP2* include genes involved in the development of brain structure and function, craniofacial formation, and cartilage and connective tissue formation.

The research team then investigated whether there were changes in the network of expressed genes when the chimp version of *FOXP2* was expressed in human brain cells. The results identified 116 genes that were linked only to the network of the human version of *FOXP2* and were not expressed in the network controlled by the chimp *FOXP2* gene. Because expression of human *FOXP2* is essential for the development of speech, the genes differentially expressed by the human and chimp versions may represent networks and pathways important in the development of speech and language. Using these genes as a starting point, researchers can explore the evolution of our species.

There has been a long-standing debate about whether Neanderthals had a complex spoken language. Analysis of DNA recovered from Neanderthal fossils shows that members of this closely related human species had a version of the *FOXP2* gene identical to that of our species in parts of the gene in which the human and chimpanzee versions differ. However, until a complete Neanderthal genome is available for comparison of the *FOXP2* target genes in Neanderthals and our species, the question of whether or not Neanderthals had the capacity for a spoken language remains open.

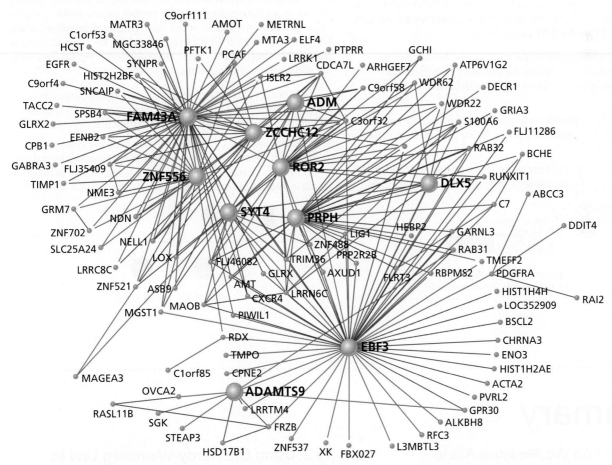

FIGURE 19.16 Differential expression of human genes under the control of the human and chimpanzee *FOXP2* genes. Target genes expressed under the control of both the human and chimpanzee genes are shown with red lines, genes expressed under the control of the human gene but not the chimpanzee gene are shown with blue lines. The large green dots represent 10 major hub genes that control the expression of many other genes.

Genetics in Practice

Genetics in Practice case studies are critical-thinking exercises that allow you to apply your new knowledge of human genetics to real-life problems. You can find these case studies and links to relevant websites at *www.cengage.com/biology/cummings*

CASE 1

Jane, a healthy woman, was referred for genetic counseling because she had two siblings, a brother Matt and a sister Edna, with cystic fibrosis who died at the ages of 32 and 16, respectively. Jane's husband, John, has no family history of cystic fibrosis. Jane wants to know the probability that she and John will have a child with cystic fibrosis. The genetic counselor used the Hardy-Weinberg model to calculate the probability that this couple will have an affected child.

The counselor explained that there is a two-in-three chance that Jane is a carrier for the mutant *CFTR* allele; she used a Punnett square to illustrate this. The probability that John is a carrier is equal to the population carrier frequency ($2pq$). The probability that John and Jane will have a child who has cystic fibrosis equals the probability that Jane is a carrier (2/3), multiplied by the probability that John is a carrier ($2pq$), multiplied by the probability that they will have an affected child if they are both carriers (1/4).

1. Using the heterozygote frequency for cystic fibrosis among white Americans to estimate the probability that John is a carrier, what is the likelihood that their child would have the disease?

2. If you were their genetic counselor, would you recommend that Jane and John be genetically tested before they attempt to have any children?

3. It is now possible to use preimplantation testing, which involves *in vitro* fertilization plus genetic testing of the embryo before implantation, to ensure that a heterozygous couple has a child free of cystic fibrosis. Do you see any ethical problems or potential future dangers associated with this technology?

Case 2

Natural selection alters genotypic frequencies by increasing or decreasing fitness (i.e., differential fertility or mortality). There are several examples of selection associated with human genetic disorders. Sickle cell anemia and other abnormal hemoglobins are the best examples of selection in humans. Carriers of the sickle and other hemoglobin mutations are more resistant to malaria than is either homozygous class. Therefore, in areas where malaria is endemic, carriers are less likely to die of malaria and will have proportionally more offspring than will homozygotes, thus passing on more genes. Balancing selection may also have influenced carrier frequencies for more "common" recessive diseases, such as cystic fibrosis in Europeans and Tay-Sachs in the Ashkenazi Jewish population, but the selective agent is not known for certain.

Selection may favor homozygotes over heterozygotes, resulting in an unstable polymorphism. One example is selection against heterozygous fetuses when an Rh⁻ mother carries an Rh⁺ (heterozygous) fetus. This should result in a gradual elimination of the Rh⁻ allele. However, the high frequency of the Rh⁻ allele in so many populations suggests that other, unknown factors may maintain the Rh⁻ allele in human populations.

1. If you suspected that heterozygous carriers of a particular disease gene had a selective advantage in resisting a type of infection, how would you go about testing that hypothesis?

2. If allele frequencies in the hemoglobin gene are influenced by sickle cell anemia on the one hand and by resistance to malaria on the other hand, what factors may cause a change in these allele frequencies over time?

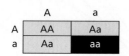

Among healthy offspring of carrier parents, 2/3 are carriers

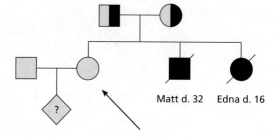

Matt d. 32   Edna d. 16

Summary

19.1 How Can We Measure Allele Frequencies in Populations?

- Aside from DNA testing, the frequency of recessive alleles in populations cannot be determined directly. The Hardy-Weinberg Law provides a means of measuring allele frequencies within populations.

19.2 Using the Hardy-Weinberg Law in Human Genetics

- The Hardy-Weinberg Law also can be used to estimate the frequency of autosomal recessive and X-linked recessive alleles in a population. It can also be used to detect when allele frequencies are shifting in the population. Changing allele frequencies in a population represent evolutionary change. One of the law's most common uses is to measure the frequency of heterozygous carriers of deleterious recessive alleles in a population. This information can be used to calculate the risk of having an affected child.

19.3 Measuring Genetic Diversity in Human Populations

- All genetic variants originate by mutation, but mutation is an insignificant force in bringing about changes in allele frequency. Other forces, including genetic drift, act on the genetic variation in the gene pool and are responsible for changing the frequency of alleles in the population (which is evolution). Drift is a random process that acts in small, isolated populations to change allele frequency from generation to generation. Examples include island populations and those separated from the general population by socioreligious practices. Natural selection acts on genetic diversity in populations to drive the process of evolution by changing allele frequencies.

19.4 Natural Selection Affects the Frequency of Genetic Disorders

- Selection increases the reproductive success of fitter genotypes. As these individuals make a disproportionate contribution to the gene pool of succeeding generations, allele frequencies and genotype frequencies change. The differential reproduction of fitter genotypes is known as natural selection. Wallace and Darwin identified selection as the primary force in evolution that leads to evolutionary divergence and the formation of new species. The high frequency of genetic disorders in some populations is the result of selection that often confers increased fitness on heterozygotes.

19.5 Genetic Variation in Human Populations

- The biological concept of race changed from an emphasis on phenotypic differences to an emphasis on genotypic differences. Information from variations in proteins, microsatellites, and nuclear genes shows that most human genetic variation is present within populations rather than between populations. For this reason, there is no clear genetic basis for dividing our species into races.

19.6 The Evolutionary History and Spread of Our Species (*Homo sapiens*)

- A combination of anthropology, paleontology, archaeology, and genetics is being used to reconstruct the dispersal of human populations around the globe. The evidence available suggests that North America and South America were populated by waves of migration sometime during the last 15,000 to 30,000 years.

19.7 Genomics and Human Evolution

- Genomic methods are being used to compare the similarities and differences in the genomes of humans, chimpanzees, and Neanderthals. These methods are also revealing details about differences in the regulation of genes associated with brain development and language in chimpanzees and humans.

Questions and Problems

CENGAGENOW Preparing for an exam? Assess your understanding of this chapter's topics with a pre-test, a personalized learning plan, and a post-test by logging on to *login.cengage.com/sso* and visiting CengageNOW's Study Tools.

How Can We Measure Allele Frequencies in Populations?

1. Define the following terms:
   a. population
   b. gene pool
   c. allele frequency
   d. genotype frequency
2. The MN blood group is a single-gene, two-allele system in which each allele is codominant. Why are such codominant alleles ideal for studies of allele frequencies in a population?
3. Explain the connection between changes in population allele frequencies and evolution, and relate this to the observations made by Wallace and Darwin concerning natural selection.
4. Can populations evolve without changes in allele frequencies?
5. Design an experiment to determine if a population is evolving.
6. What are four assumptions of the Hardy-Weinberg Law?
7. Drawing on your newly acquired understanding of the Hardy-Weinberg equilibrium law, point out why the following statement is erroneous: "Because most of the people in Sweden have blond hair and blue eyes, the genes for blond hair and blue eyes must be dominant in that population."
8. In a population where the females have the allelic frequencies $A = 0.35$ and $a = 0.65$ and the frequencies for males are $A = 0.1$ and $a = 0.9$, how many generations will it take to reach Hardy-Weinberg equilibrium for both the allelic and the genotypic frequencies? Assume random mating and show the allelic and genotypic frequencies for each generation.

Using the Hardy-Weinberg Law in Human Genetics

9. Suppose you are monitoring the allelic and genotypic frequencies of the MN blood group locus (see question 2 for a description of the MN blood group) in a small human population. You find that for 1-year-old children, the genotypic frequencies are MM = 0.25, MN = 0.5, and NN = 0.25, whereas the genotypic frequencies for adults are MM = 0.3, MN = 0.4, and NN = 0.3.
   a. Compute the M and N allele frequencies for 1-year-olds and adults.
   b. Are the allele frequencies in equilibrium in this population?
   c. Are the genotypic frequencies in equilibrium?

10. Using Table 19.1, determine the frequencies of p and q that result in the greatest proportion of heterozygotes in a population.

11. In a given population, the frequencies of the four phenotypic classes of the ABO blood groups are found to be A = 0.33, B = 0.33, AB = 0.18, and i = 0.16. What is the frequency of the i allele?

12. If a trait determined by an autosomal recessive allele occurs at a frequency of 0.25 in a population, what are the allelic frequencies? Assume Hardy-Weinberg equilibrium and use A and a to symbolize the dominant and recessive alleles, respectively.

Measuring Genetic Diversity in Human Populations

13. Why is it that mutation, acting alone, has little effect on gene frequency?

14. Successful adaptation is defined by:
   a. evolving new traits.
   b. producing many offspring.
   c. leaving more offspring than others.
   d. moving to a new location.

15. What is the relationship between founder effects and genetic drift?

16. How would a drastic reduction in a population's size affect that population's gene pool?

17. The major factor causing deviations from Hardy-Weinberg equilibrium is
   a. selection.
   b. nonrandom mating.
   c. mutation.
   d. migration.
   e. early death.

18. A specific mutation in the BRCA1 gene has been estimated to be present in approximately 1% of Ashkenazi Jewish women of Eastern European descent. This specific alteration, 185delAG, is found about three times more often in this ethnic group than the combined frequency of the other 125 mutations found to date. It is believed that the mutation is the result of a founder effect from many centuries ago. Explain the founder principle.

19. The theory of natural selection has been summarized popularly as "survival of the fittest." Is this an accurate description of natural selection? Why or why not?

Natural Selection Affects the Frequency of Genetic Disorders

20. Will a recessive allele that is lethal in the homozygous condition ever be completely removed from a large population by natural selection?

21. Do you think that our species is still evolving, or are we shielded from natural selection by civilization? Is it possible that misapplications of technology will end up exposing our species to more rather than less natural selection (consider the history of antibiotics)?

Genetic Variation in Human Populations

22. a. Provide a genetic definition of race.
   b. Using this definition, can modern humans be divided into races? Why or why not?

The Evolutionary History and Spread of Our Species (Homo sapiens)

23. a. Briefly describe the two major theories discussed in this chapter about the origin of modern humans.
   b. Which of these two theories would predict a closer relationship for the various modern human populations?
   c. Which of the two theories is best supported by the genetic evidence?

Genomics and Human Evolution

24. The human and chimpanzee genomes are 98.8% identical. If this is so, why are the phenotypes of chimps and humans so different?

25. The development of language in humans depends in part on expression of the transcription factor gene FOXP2. Research indicates that Neanderthals had a version of the FOXP2 gene identical to that of our species in regions where human and chimpanzee genes differ. Is this enough evidence to conclude that Neanderthals had a complex spoken language? Why or why not?

Internet Activities

Internet Activities are critical-thinking exercises using the resources of the World Wide Web to enhance the principles and issues covered in this chapter. For a full set of links and questions investigating the topics described below, visit *www.cengage.com/biology/cummings*

1. *Comparing DNA Sequences.* GenBank is the National Institutes of Health's (NIH) database of all known nucleotide and protein sequences, including supporting bibliographic and biological data. Use GenBank's Entrez system to search for a DNA sequence and BLAST to find similar sequences in GenBank.
2. *Exploring the Hardy-Weinberg Equilibrium Equation.* The *Access Excellence Activities Exchange* site includes several Hardy-Weinberg–related exercises. To see how selection can affect a population's allele frequencies, try the *Fishy Frequencies* activity. This exercise can be done alone or as part of a group—and you get to eat fish crackers as you work!
3. *DNA, Archaeology, and Human History.* Read the article "Scientists Rough Out Humanity's 50,000-Year-Old History" at the *New York Times Learning Network* site.

HOW WOULD YOU VOTE NOW?

Tests for about 900 genetic disorders are available through public and private testing laboratories. There is no testing available for many rare diseases. In the United States, rare genetic disease are defined by law as those that affect 1 in 1,500 people or fewer. Thus, achromatopsia, which affects 1 in 33,000 is a rare disease. Many rare diseases develop early in childhood, and as many as 30% of those with rare diseases die before the age of 5. Even though there may be a low number of those affected with a recessively inherited disease in a population, heterozygotes can be quite frequent. For example, if a disease has a frequency of 1 in 1,500, 1 in 20 members of the population are heterozygotes.

If your family history showed the presence of a rare genetic disease that is fatal in early childhood and testing was not available to determine if you carried the mutation, what would you do? There are several options open: You can take a chance that your mate is not a heterozygote; you can go ahead and have children, knowing that if your mate is a heterozygote, there is only a 25% chance that a child will be affected; you can decide not to have children; or you can decide to adopt. Visit the *Human Heredity* companion website at *www.cengage.com/biology/cummings* to find out more on the issue; then cast your vote online.

Appendix

Answers to Selected Questions and Problems

Chapter 1

2. Population genetics studies genetic variations found in individuals in a population and the forces that alter the frequency of these variations as they are passed from generation to generation.

5. A genome is the haploid set of DNA sequences carried by an individual.

6. Genomics is the study of genomes and their genetic content, organization, function, and evolution.

Chapter 2

3. There are 44 autosomes in a body (somatic) cell and 22 autosomes in a gamete.

5. d

7. Cells undergo a series of events involving growth, DNA replication, and division that are repeated by the daughter cells, forming a cycle, called the cell cycle. During S phase, DNA synthesis occurs. During M phase, mitosis and cytokinesis take place.

8. a, e

10. Meiosis II, the division responsible for the separation of sister chromatids, would no longer be necessary. Meiosis I, wherein homologues segregate, would still be required.

18. Cell-cycle gene products regulate the process of cell division. If a gene normally promotes cell division, mutant alleles can cause too much cell division. If a gene normally turns off cell division, mutant alleles may no longer repress cell division. Each of these errors in cell cycle regulation may lead to the uncontrolled cell divisions characteristic of cancer.

19.

Attribute	Mitosis	Meiosis
Number of daughter cells produced	2	4
Number of chromosomes per daughter cell	$2n$	n
Number of cell divisions	1	2
Do chromosomes pair? (Y/N)	N	Y
Does crossing over occur? (Y/N)	N	Y
Can the daughter cells divide again? (Y/N)	Y	N
Do the chromosomes replicate before division? (Y/N)	Y	Y
Type of cell produced	SOMATIC	GAMETE

20.

24. a. Mitosis   b. Meiosis I   c. Meiosis II

27. Meiotic anaphase I: no centromere division, chromosomes consisting of two sister chromatids are migrating; Meiotic anaphase II: centromere division, the separating sister chromatids are migrating. Meiotic anaphase II more closely resembles mitotic anaphase by these two criteria.

Chapter 3

1. a. A gene is the fundamental unit of heredity. The gene encodes a specific gene product (e.g., a pigment involved in determining eye color). Alleles are alternate forms of a gene that may cause various phenotypic effects. For example, a gene for eye color may have blue, brown, and green eye-color alleles. The locus is the position of a gene on a chromosome. In a normal situation, all alleles of a gene would have the same locus.

b. Genotype refers to the genetic constitution of the individual (*AaBb* or *aabb*). Notice that the genotype always includes at least two letters, each representing one allele of a gene pair in a diploid organism. A gamete would contain only one allele of each gene because of its haploid state (*Ab* or *ab*). Phenotype refers to an observable trait. For example, *Aa* (the genotype) will cause a normal pigmentation (the phenotype) in an individual, whereas *aa* will cause albinism.

c. Dominance and recessiveness are comparative terms applied to alleles. Dominant alleles are expressed in the heterozygous condition. Therefore, only one copy of a dominant allele needs to be present to express the phenotype. Recessiveness refers to a trait that is not expressed in the heterozygous condition. It is masked by the dominant allele. To express a recessive trait, two copies of the recessive allele must be present in the individual.

d. Complete dominance occurs when a dominant allele completely masks the expression of a recessive allele. For example, in pea plants, yellow seed color is dominant to green. In a heterozygous state, the phenotype of the seeds is yellow. This is the same phenotype seen in seeds homozygous for the yellow allele.

Incomplete dominance occurs when the phenotype of the heterozygote is intermediate between the two homozygotes. For example, in *Mirabilis*, a red flower crossed with a white flower will give a pink flower.

In codominant inheritance, there is full expression of both alleles in the heterozygous condition. For example, in the AB blood type, the gene products of the *A* allele and the *B* allele are expressed and present on the surface of blood cells.

2. Phenotypes: b, d; Genotypes: a, c, e

5. a. 1/2 *A*, 1/2 *a*
b. All *A*
c. All *a*

11. a. All F1 plants will be long-stemmed.
b. Let *S* = long stemmed and *s* = short stemmed. The long-stemmed P1 genotype is *SS*, the short-stemmed P1 genotype is *ss*, the long-stemmed F1 genotype is *Ss*.
c. Approximately 225 long stemmed and 75 short stemmed
d. The expected genotypic ratio is 1 *SS*:2 *Ss*:1 *ss*.

13. a. 1/2 *A_B_*, 1/2 *A_bb*
b. 1/4 *A_B_*, 1/4 *A_bb*, 1/4 *aaB_*, 1/4 *aabb*
c. 9/16 *A_B_*, 3/16 *A_bb*, 3/16 *aaB_*, 1/16 *aabb*

15. a. Both are 3:1
b. 9:3:3:1
c. Swollen is dominant to pinched, yellow is dominant to green.
d. Let *P* = swollen and *p* = pinched; *C* = yellow and *c* = green. Then: P1 = *PPcc* × *ppCC*; F1 = *PpCc*

17. a. Let *S* = smooth and *s* = wrinkled; *Y* = yellow and *y* = green. The genotype of the smooth, yellow parent is *SsYy*.
b. The genotypes of the offspring are *SsYy* (smooth, yellow), *Ssyy* (smooth, green), *ssYy* (wrinkled, yellow), and *ssyy* (wrinkled, green).

21. 3/4 for *A* × 1/2 for *b* × 1 for *C* = 3/8 for *A*, *b*, *C*

23. During meiotic prophase I, the replicated chromosomes synapse, or pair, with their homologues. These paired chromosomes align themselves at the equator of the cell during metaphase I. During anaphase I, it is the homologues (each containing two chromatids) that separate from each other. There is no preordained orientation for this process—it is equally likely that a maternal or a paternal homologue will migrate to a given pole. This provides the basis for the law of random segregation. Independent assortment results from the fact that the polarity of one set of homologues has absolutely no influence on the orientation of a second set of homologues. For example, if the maternal homologue of chromosome 1 migrated to a certain pole, it would have no bearing on whether the maternal or paternal homologue of chromosome 2 migrated to that same pole.

28. The P1 generation is *FF* × *ff*. The F1 generation is *Ff*. The mode of inheritance is incomplete dominance.

31. Because neither species produces progeny resembling a parent, simple dominance is ruled out. The species producing pink-flowered progeny from red and white (or very pale yellow)

suggests incomplete dominance as a mode of inheritance. However, in the second species, the production of orange-colored progeny cannot be explained in this fashion—because orange would result from an equal production of red and yellow pigments. Instead, in this case, codominant inheritance is suggested, with one parent producing bright red flowers and the other producing pale yellow flowers.

Chapter 4

2. d

3. a. Female
b. Yes
c. 3 siblings; The proband is the youngest child.

5. Autosomal dominant with incomplete penetrance

7. a.

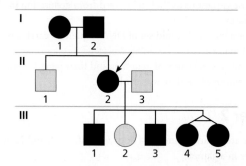

b. The mode of inheritance is consistent with an autosomal dominant trait. Both of the proband's parents are affected. If this trait were recessive, all their children would have to be affected (*aa* × *aa* can only produce *aa* offspring). As we see in this pedigree, the brother of the proband is not affected, indicating that this is a dominant trait. His genotype is *aa*, the proband's genotype is *AA* or *Aa*, and both parents' genotype is most likely *Aa*.
c. Because the proband's husband is unaffected, he is *aa*.

9. a. This pedigree is consistent with autosomal recessive inheritance.
b. If inheritance is autosomal recessive, the individual in question is heterozygous.

10.

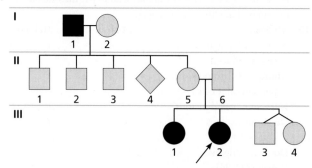

14. Due to the rarity of the disease, we will assume that the paternal grandfather is heterozygous for the allele responsible for Huntington disease. His son, now in his twenties, has a 50% chance of inheriting the mutant allele. In turn, should he carry the *HD* allele, his son would have a 50% chance of inheriting it. Therefore, at present, the child has a 1/2 × 1/2 = 1/4 chance of having inherited the *HD* allele.

16. a. 50% chance for sons, 50% for all children

    b. 50% for daughters, 50% for all children

19. In the autosomal case, the parents are *Aa* and *Aa* and the disorder is inherited as a recessive trait. They can have an affected (*aa*) son or daughter, or unaffected (*AA* or *Aa*) children. If the trait is inherited in an X-linked recessive manner, the parents would be $X^AX^a \times X^AY$. Daughters would be X^AX^a and unaffected; sons would be affected and be X^aY. Because an unaffected daughter and affected son are possible in each case, this limited information is not enough to determine the inheritance pattern.

20. a. autosomal dominant or **c.** X-linked dominant patterns of inheritance are both possible in this case.

21. Strictly speaking, this trait could be inherited in the following ways: **a.** autosomal dominant; **b.** autosomal recessive; **d.** X-linked recessive. Some possibilities are more probable than others, depending on the frequency of the trait in a population.

23. Mitochondria contain DNA carrying genetic information and are maternally inherited.

27. 20% of 90 = 18

Chapter 5

2. a. Height in pea plants is determined by a single pair of genes with dominant and recessive alleles. Height in humans is a complex trait, involving a number of genes and environmental factors.

    b. For traits controlled by several genes, the offspring of matings between extreme phenotypes show a tendency to regress toward the mean phenotype in the population.

5. a. F1 genotype = *A'AB'B*, phenotype = height of 6 ft.

    b. *A'AB'B* × *A'AB'B*

Genotypes	Phenotypes
A'A'B'B'	7 ft.
A'A'B'B	6 ft. 6 in.
A'A'BB	6 ft.
A'AB'B'	6 ft. 6 in.
A'AB'B	6 ft.
A'ABB	5 ft. 6 in.
AAB'B'	6 ft.
AAB'B	5 ft. 6 in.
AABB	5 ft.

6. In the case of traits controlled by several genes, expression of the trait depends on the interactions of many genes, each of which contributes a small amount to the phenotype. Thus, the differences between genotypes are often not clearly distinguishable. In the case of monogenic determination of a trait, the alleles of a single locus have major effects on the expression of the trait, and the differences between genotypes are usually easy to discern.

8. In the multifactorial threshold model, liability is caused by a number of genes acting to produce the defect. If exposed to certain environmental conditions, the person above the threshold will most likely develop the disorder. The person below the threshold is not predisposed to the disorder and will most likely not develop the disorder.

12. Relatives are used because the proportion of genes held in common by relatives is known.

14. No. Dizygotic twins arise from two separate fertilized eggs. Only monozygotic twins can be Siamese, because they originate from the same fertilized egg and are genetically identical.

15. b

19. a. The study included only men who were able to pass a physical exam that eliminated markedly obese individuals, so the conclusions cannot be generalized beyond the group of men inducted into the armed forces.

    b. To design a better study, include MZ and DZ twin men and women, maybe even children, and include a cross section of various populations (ethnic groups, socioeconomic groups, weight classifications, etc.). Control the diet so that it remains a constant. Another approach is to study MZ and DZ twins who were reared apart (and presumably in different environments), or adopted and natural children who were raised in the same household (same environment). There are other possible answers.

24. First, intelligence is difficult to measure. Also, for such a complex trait, many genes and a significant environmental component are likely to be involved.

26. The heritability difference observed between the racial groups for this trait cannot be compared because heritability measures variation within one population at the time of the study. Heritability cannot be used to estimate genetic variation between populations.

Chapter 6

1. a. Chemical treatment of chromosomes resulting in unique banding patterns

    b. Q banding and G banding with Giemsa

    c. Karyotypes provide information about the sex of the individual and the presence of any abnormalities in the number of specific chromosomes, as well as any detectable duplications, deletions, inversions, or translocations. The ability to provide this information for specific chromosomes is possible because the development of chromosome banding allows identification of specific chromosomes.

5. Advanced maternal age, previous aneuploid child, presence of a chromosomal rearrangement, presence of a known genetic disorder in the family history

7. Triploidy

12. The embryo will be tetraploid. Inhibition of centromere division results in nondisjunction of an entire chromosome set. After cytoplasmic division, some cytoplasm is lost in an inviable product lacking genetic material, and the embryo develops from the tetraploid product.

17. Condition 2 is most likely lethal. This condition involves a chromosomal aberration: trisomy. This has the potential for interfering with the action of all genes on the trisomic chromosome. Condition 1 involves an autosomal dominant lesion to a single gene, which is more likely to be tolerated by the organism.

21. Turner syndrome (45,X) is monosomy for the X chromosome. A paternal nondisjunction event could contribute a gamete lacking a sex chromosome to result in Turner syndrome. The complementary gamete would contain both X and Y chromosomes. This gamete would contribute to Kleinfelter syndrome (47,XXY).

22.
a.	Loss of a chromosome segment	deletion
b.	Extra copies of a chromosome segment	duplication
c.	Reversal in the order of a chromosome segment	inversion
d.	Movement of a chromosome segment to another, nonhomologous chromosome	translocation

24. In theory, the chances are 1/2.

25. Several possibilities should be considered. The child could be monosomic for the relevant chromosome. The child has the paternal copy carrying the allele for albinism (father is heterozygous), and a nondisjunction event resulted in failure to receive a chromosomal copy from the homozygous mother. Autosomal monosomy, however, is fatal, and this possibility can be ruled out. The second possibility is that the maternal chromosome carries a small deletion, allowing the albinism to be expressed. The third possibility is that the child represents a new mutation, inheriting one albino allele and having the other by mutation. A fourth possibility is that the phenotype results from uniparental disomy for a paternal chromosome.

Chapter 7

1. Secondary oocytes have completed meiosis I when they are ovulated and contain 23 replicated chromosomes, each consisting of two sister chromatids held together by a single centromere.

3. Meiosis began before the birth of the parent and is completed shortly after fertilization. The time taken is therefore approximate. Shortest time: from January 1, 1980, to July 1, 2004—24.5 years. Longest time: from about June 1, 1979, to July 1, 2004—25.16 years.

5. Significant economic and social consequences are associated with FAS, including the costs of surgery for facial reconstruction, treatment of learning disorders and mental retardation, and caring for institutionalized individuals. Prevention depends on the education of pregnant women and the early treatment of pregnant women with alcohol dependencies. Other answers are possible.

8. d

9. Female. The 1:1 ratio of purple:yellow-eyed offspring indicates that females are heterogametic.

11. A mutation causing the loss of the *SRY* gene, testosterone, or testosterone receptor gene function can each cause an XY individual to be phenotypically female. Also, a defect in the conversion from testosterone to DHT can cause the female external phenotype until puberty.

15. Pattern baldness acts as an autosomal dominant trait in males and an autosomal recessive trait in females. The pattern of expression is affected by hormonal differences in males and females.

18. Random inactivation occurs in females, so the genes from both X chromosomes are active in the body as a whole. In rare cases, inactivation is skewed, resulting in females heterozygous for X-linked recessive disorders having a mutant phenotype.

Chapter 8

2. Chromosomes contain both proteins and DNA, but the organization of DNA, involving only 4 different nucleotides, seemed too simple to carry genetic information. Cells contain hundreds or thousands of different proteins; only 2 main types of nucleic acids.

4. Protease destroyed any small amounts of protein contaminants in the transforming extract. Similarly, treatment with RNAse destroyed any RNA present in the mixture. Most importantly, treatment with DNAse destroyed any DNA in the mixture and was the only enzyme treatment to abolish transforming ability in the extract.

6. The process is transformation, discovered by Frederick Griffith. The P bacteria contain genetic information that is still functional even though the cell has been heat killed. However, it needs a live recipient host cell to accept its genetic information. When heat-killed P and live D bacteria are injected together, genetic information from the dead P bacteria can be transferred to the live D bacteria. As a result, the D bacteria are transformed into P bacteria and can now cause polka dots.

8. Chargaff's rule: A = T and C = G
If A = 27%, then T must equal 27%.
If G = 23%, then C must equal 23%.
Base composition:
A = 27%
T = 27%
C = 23%
G = 23%
  100%

10. b and e

11. b

13. c

18.

		DNA	RNA
a.	Number of chains:	2	1
b.	Bases used:	A, C, G, T	A, C, G, U
c.	Sugar used:	deoxyribose	ribose
d.	Function:	blueprint of genetic information	transfer of genetic information from nucleus to cytoplasm

21. a

Chapter 9

2. Replication is the process of making DNA from a DNA template. transcription makes RNA from a DNA template, and translation makes an amino acid chain (a polypeptide) from an mRNA template. Replication and transcription happen in the nucleus, and translation occurs in the cytoplasm.

3. There would be 4^4, or 256, possible amino acids encoded.

8. b

9. 1. Removal of introns: to generate a contiguous coding sequence that can make an amino acid chain
   2. Addition of the 5' cap: ribosome binding
   3. Addition of the 3' poly-A tail: mRNA stability

12. Codons are triplets of bases on an mRNA molecule. Anticodons are triplets of bases on a tRNA molecule and are complementary in sequence to the nucleotides in codons.

13. Answer: 25%   Total length: 10 kb
                             Coding region: 2.5 kb

16.

tRNA:	UAC	UCU	CGA	GGC
mRNA:	AUG	AGA	GCU	CCG
DNA:	TAC	TCT	CGA	GGC
protein:	met	arg	ala	pro

Hydrogen bonds present in the DNA: 31
7 GC pairs × 3 = 21
5 AT pairs × 2 = 10

24. **a.** No
    b. Yes

Chapter 10

3. c

4. **a.** Buildup of substance A, no substance B or C
    b. Buildup of substance B, no substance C
    c. Buildup of substance B, as long as A is not limiting factor
    d. 1/2 the amount of C

5. **a.** Yes. Each would carry the normal gene for the other enzyme. (Individual 1 would be mutant for enzyme 1 but normal for enzyme 2. This is because two different genes encode enzymes 1 and 2.)
    b. Let D = dominant mutation in enzyme 1; let normal allele = d
    Let A = dominant mutation in enzyme 2; let normal allele = a

$Ddaa$	×	$ddAa$
Offspring:	$DdAa$	mutation in enzyme 1 and 2, A buildup, no C
	$Ddaa$	mutation in enzyme 1, A buildup, no C
	$ddAa$	mutation in enzyme 2, B buildup, no C
	$ddaa$	no mutation, normal

Ratio would be 1:2:1 for substance B buildup, no C : substance A buildup, no C : normal

6. Alleles for enzyme 1: A (dominant, 50% activity); a (recessive, 0% activity). Alleles for enzyme 2: B (dominant, 50% activity); b (recessive, 0% activity).

	Enzyme 1	Enzyme 2	Compound A	Compound B	Compound C
1 *AABB*	100	100	N	N	N
2 *AaBB*	50	100	N	N	N
4 *AaBb*	50	50	N	N	N
2 *AABb*	100	50	N	N	N
1 *AAbb*	100	0	N	B	L
2 *Aabb*	50	0	N	B	L
1 *aaBB*	0	100	B	L	L
2 *aaBb*	0	50	B	L	L
1 *aabb*	0	0	B	L	L

N, normal; B, buildup; L, less

12. b

17. No, because individuals who are G^D/G^D show 50% activity. The g allele abolishes enzyme activity, so G^+/g heterozygotes have 50% activity and are normal. It is not until the level of activity falls below 50% (G^D/g or g/g) that the mutant phenotype is observed.

20. It would cause a frameshift mutation very early in the protein. Most likely, the protein would lose all of its functional capacity.

22. Drugs usually act on proteins. Different people have different forms of proteins. Different proteins are inherited as different alleles of a gene.

23. People have different abilities to smell and taste chemical compounds such as phenylthiocarbamide (PTC); some people are unable to smell skunk odors; there are different reactions to succinylcholine, a muscle relaxant. Others are sensitive to the pesticide parathion.

Chapter 11

1. Mutation rate measures the occurrence of mutations per gene per generation.

2. 245,000 births represent 490,000 copies of the achondroplasia gene, because each child carries two copies of the gene. The mutation rate is therefore 10/490,000, or 2×10^{-5}, per generation.

7. Muscular dystrophy is an X-linked disorder. A son receives an X chromosome from his mother and a Y chromosome from his father. In this case, the mother was a heterozygous carrier of muscular dystrophy and passed the mutant gene to her son. The father's exposure to chemicals in the workplace is unrelated to his son's condition.

10. Missense, same; nonsense, shorter; sense, longer

16. **a.**

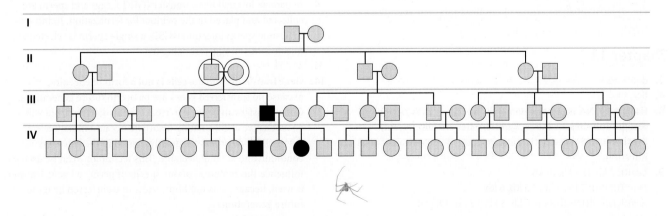

b. No.

c. Mutation in the *RB1* gene causes a dominantly inherited predisposition to retinoblastoma, but expression of the tumor requires mutation of the normal allele in a retinal cell, This second mutational event occurs in about 85–90% of those carrying an inherited mutant allele of the *RB1* gene.

19. DNA polymerase has a proofreading function that repairs many errors introduced during replication. In addition, thymine dimers, induced by exposure to ultraviolet light, are corrected by several other DNA repair systems.

Chapter 12

3. a

6. d

9. A proto-oncogene is a normal gene whose products promote cell division when conditions are right. When mutated to an oncogene, cell division is promoted continuously, resulting in cancer. A tumor suppressor gene is a normal gene that stops cell division when it is not needed. If mutated, a tumor-suppressor gene can lose its function and no longer be able to control cell division. The result is uncontrolled cell division that can lead to cancer.

10. The inheritance of the predisposition to retinoblastoma is a dominant trait because the presence of one mutant allele causes a predisposition to retinoblastoma. However, the second allele must also be mutated in at least one eye cell to produce the disease. Therefore, the expression of retinoblastoma is recessive.

15. Conditions **a** and **d** would produce cancer. The loss of function of a tumor-suppressor gene would allow cell growth to go unchecked. The overexpression of a proto-oncogene would promote more cell division than normal.

19. Mutations in *APC* form hundreds or thousands of benign tumors. If a specific set of mutations occur in a cell in one of these tumors, the benign growths can progress to cancer. The large number of benign tumors makes the chance of acquiring other mutations likely.

21. c-*myc* lies at the breakpoint of a translocation involving chromosome 8 and chromosome 14, 22, or 2. The translocation places the *myc* gene in an altered chromosomal environment and alters its normal expression. Altered expression of c-*myc* is thought to be necessary for the production of Burkitt's lymphoma.

27. Diet is suspected as the cause in both cases. When Japanese move to the United States and adopt an American diet, the rate of breast cancer goes up, but the rate of colon cancer goes down. The reverse is also the case. Conclusion: The Japanese diet in Japan predisposes to colon cancer but not to breast cancer.

Chapter 13

2. d

4. b

6. Bacterial DNA is either chemically modified at restriction recognition sites, or the recognition sites are not present. Bacteria contain restriction enzymes as a defense against viral infection.

9. *Eco*RI: 2 kb, 11.5 kb, 10 kb

*Hin*dIII/*Pst*I: 7 kb, 3 kb, 7.5 kb, 6 kb

*Eco*RI/*Hin*dIII/*Pst*I: 2 kb, 5 kb, 3 kb, 3.5 kb, 4 kb, 6kb

13. b

15. a

20. DNA derived from individuals with sickle cell anemia will lack one fragment contained in the DNA from normal individuals. In addition, there will be a large (uncleaved) fragment not seen in normal DNA.

Chapter 14

6. c

10. a. $1/100 \times 1/500 = 1/50,000$ individuals with this combination of alleles are present in the population.

b. This is not very convincing, because in a large city, say, with a population of 3 million, there will be approximately 60 individuals with this profile.

c. The lab should test two or more additional loci to reduce the probability of another individual having this profile to a much lower number, such as 1 in 50 million or more.

d. The answer is an opinion and a point of discussion.

13. The answers to **a**, **b**, and **c** are matters of opinion but can be supported by scientific evidence.

Chapter 15

1. Genes that are said to show linkage are located near each other on the same chromosome. Linked genes tend to be inherited together.

5. The human genome contains about 3.2 billion nucleotides.

10. Genes make up about 5% of the total DNA sequence of the human genome. Only about 1.1% of the genome is composed of protein-coding exons. It is not clear what, if any, function the remaining DNA has. About half the total DNA in the genome is made up of various kinds of repeated sequences.

13. c

16. In addition to the scientific elements of the project, the organizers of the HGP set up a program called ELSI to address the ethical, legal, and social implications of genomics research. This project has used meetings, grants, workshops, and other forums to discuss various issues related to genomics research and to help bring about legislation to protect against the abuse of genetic information.

Chapter 16

2. Answer is a matter of opinion.

4. In gamete intrafallopian transfer (GIFT), eggs and sperm are collected and placed in the oviduct for fertilization. In intracytoplasmic sperm injection (ICSI), a single sperm is selected and injected into an egg, fertilizing it.

11. c

14. Gene transfer to somatic cells is not a form of eugenics, because no transferred genes are passed to future generations. Gene transfer can be used to treat genetic disorders, but will be ineffective at eliminating disorders because mutant alleles for recessive disorders are mostly carried in the heterozygous condition and are undetectable. This technology can be used to influence the evolution of our species if germ cell gene transfer is used, because this method results in transferred genes to future generations.

If germ cell gene transfer becomes an accepted practice, guidelines will hopefully be part of the process of approving this procedure. Most likely, these guidelines would be created and enforced by the federal government, as are the current guidelines for somatic cell gene transfer.

18. a. The story of nonpaternity during family genetic counseling is familiar to genetic counselors. When deciding how to approach this type of unexpected findings, counselors need to weigh the benefits and harms of nondisclosure against those of disclosure. The first considerations include the relevance of the information to the patient's situation and the consequences of the findings. The 1983 President's Commission recommends that patients be advised before testing that unexpected information may be revealed.

    b. It is reasonable for the counselor to call the woman beforehand and explain the results and the implications of the findings. Given the sensitivity of this information, the long-term effect on the couple's relationship may be dramatic, but the couple must be told that the child will have sickle cell anemia and should be treated for this condition.

Chapter 17

1. a. Microorganisms that penetrate the skin infect cells, which then release chemical signals such as histamine. This causes an increased blood flow into the area, resulting in an increase in temperature.

    b. The heat serves to inhibit microorganism growth, mobilize white blood cells, and raise the metabolic rate in nearby cells, thereby promoting healing.

3. Immunoglobulins: IgD, IgM, IgG, IgA, IgE

4. Helper T cells activate B cells to produce antibodies. Cytotoxic (killer) T cells target and destroy infected cells.

10. Express the cloned gene to make the protein product, isolate the protein, and inject it into humans as a vaccine. The immune system should make antibodies to that protein. When the actual live virus is encountered, the immune system will have circulating antibodies and T cells that will recognize the protein (antigen) on the surface of the virus, initiating a strong secondary response.

16. The antigens of the donor/recipient are more important. The antigen of the donor will be rejected if the recipient does not have the same antigen. The antigen of the recipient determines which antibodies can be produced. For example, a blood type A individual will make B antibodies if exposed to the B antigen.

20. If the mother is Rh$^-$, and her fetus is Rh$^+$, she will produce antibodies against the Rh antigen. This happens when blood from the fetus enters the maternal circulation. The mother already has circulating antibodies against the Rh antigen from her first Rh$^+$ child. She can mount a greater immune response against the second Rh$^+$ child by generating a large number of antibodies.

24. One approach is to clone the human gene that suppresses hyperacute rejection and inject it into pig embryos. The hope is that the pig's cells will express this human protein on the cell surface. The human recipient may then recognize the transplanted organ as "self." In addition, transplants of bone marrow from donor pigs into human recipients may help in preventing rejection mediated by T cells. This dual bone-marrow system will recognize the pig's organ as "self" but will retain the normal human immunity.

25. a. They need to test one cell of each eight-cell embryo for an ABO and Rh blood type match and also an HLA complex match. If a match exists, they will implant the embryo(s) into the mother and hope that pregnancy occurs. When the baby is born, bone marrow will be extracted and transplanted into the existing child.

    b. Ethically, it is difficult to imagine having a child for the primary purpose of being a bone marrow donor. The new child may come to feel demeaned or less valued. Also, what happens to the embryos that are not a match to the couple's existing child? These embryos are completely healthy; they simply have the wrong blood type and histocompatibility complex. However, if the couple will love and provide for this new child, it may be a wonderful experience that the new child has the opportunity to save the life of his or her sibling.

27. Allergens cause a release of IgE antibodies, which bind to mast cells. These cells release histamine, which causes fluid accumulation, tissue swelling (such as swollen airways or eyes), and mucous secretion (such as a runny nose). Antihistamines are chemicals that block the production or action of histamine, treating the symptoms of allergies.

30. The HIV virus infects and kills helper T4 cells, the very cells that normally trigger the antibody-mediated immune response. Therefore, as the infection progresses, the immune system gets weaker and weaker as more T cells are killed. The AIDS sufferer is then susceptible to various infections and certain forms of cancer.

Chapter 18

2. The definition must be precise enough to distinguish the behavior from other similar behaviors and from the behavior of the control group. The definition of the behavior can significantly affect the results of the genetic analysis and even the proposed pattern of inheritance of the trait.

4. *Drosophila* has many advantages for the study of behavior. Mutagenesis and screening for behavior mutants allow the recovery of mutations that affect many forms of behavior. The ability to perform genetic crosses and recover large numbers of progeny over a short period also enhances the genetic analysis of behavior. This organism can serve as a model for human behavior, because cells of the nervous systems in *Drosophila* and humans use similar mechanisms to transmit impulses and store information.

6. Huntington disease is caused by expansion of a CAG trinucleotide repeat within the *HD* gene. Expansion of the repeat causes an increase in the number of glutamines in the encoded protein, causing the protein to become toxic to neurons. In regions of the nervous system expressing the mutant protein, cells fill with clusters of the protein, degenerate, and die.

10. It may be argued that using such a test to identify potentially violent individuals would allow them to be given appropriate therapy (such as drugs to increase MAOA activity) before they harmed anyone. However, the use of such a test would result in individuals being labeled as aggressive or violent based not on their behavior but on their genotype. This could have significant consequences in their work and personal lives. On the other hand, attorneys are now using the presence of mutations in the MAOA gene as a defense in criminal trials, claiming the violent behavior is genetically determined.

13. No, it means that there are probably other genes involved or the environment plays a significant role. The linkage to chromosome 7 is still valid, and the next goal would be to find the gene on chromosome 7 that is linked to bipolar disorder. Also, finding other genes involved is important. A researcher may find that a subset of those with bipolar disorder have a defect in the gene on chromosome 7 and another subset have a defect in a gene or genes on other chromosomes. Mutations in several genes can contribute to the same disease.

16. The heritability of Alzheimer disease, a multifactorial disorder, cannot be established because of interactions between genetic and environmental factors. Less than 50% of Alzheimer cases can be attributed to genetic causes, indicating that the environment plays a large role in the development of this disease.

17. This is probably not a valid method, because correlations must eventually be related to a causal relationship—in this case, between body hair and intelligence. Many factors contribute to intelligence, including environmental factors. Lacking an explanation for the relationship between body hair and intelligence, it is not logical to assume that there is a relationship. Testing would depend on the definition of intelligence used in the study and may involve IQ testing by individuals who know nothing about the person's body hair. Alternatively, testing could be done for *g*, a measure of cognitive ability in blind testing, where the presence of the subject's body hair is unknown to the test administrator.

Chapter 19

1. **a.** Population: local groups of individuals occupying a given space at a given time
   b. Gene pool: the set of genetic information carried by a population
   c. Allele frequency: the frequency of occurrence of particular alleles in the gene pool of a population
   d. Genotype frequency: the frequency of occurrence of particular genotypes among the individuals of a population

7. In a population at equilibrium, the frequency of an allele has no relationship to its mode of inheritance. Selection can change allele frequency, but not whether the allele and its phenotype are inherited as a dominant or a recessive trait.

9. **a.** Children: $M = 0.5$, $N = 0.5$; Adults: $M = 0.5$, $N = 0.5$
   b. Yes. Allelic frequencies are unchanged.
   c. No. The genotypic frequencies are changing within each generation.

19. No. It is not a particularly accurate description. Natural selection depends not just on an ability to survive but also to leave more offspring than others. It is the differential reproduction of some individuals that is the essence of natural selection.

21. This is an open question. Culture, in the form of society and technology, has shielded humans from many forms of selective forces in the environment but has also created new forms of selection for humans and for infectious agents. The abuse of antibiotics may increase the effect of selection on human populations, as will the expansion of the human population into new geographic areas, increasing exposure to endemic agents of disease.

22. **a.** Genetically speaking, races are populations with significant differences in allele frequencies compared with other populations.
    b. No. No systematic differences have been identified in allele frequencies within modern human populations that are large enough to justify the use of the term *race*. Studies into the level of genetic variation within and between populations have consistently found that there is much more variety within each population than between them.

23. **a.** The out-of-Africa hypothesis holds that modern humans first appeared in Africa and then left the continent to replace all of the then-existing hominid populations in the world. The multiregional hypothesis proposes that, through a network of interbreeding populations, *Homo erectus* gradually evolved into modern *Homo sapiens* in different regions of the world.
    b. The out-of-Africa hypothesis
    c. Both mitochondrial and nuclear DNA evidence favor the out-of-Africa hypothesis.

Glossary

Acquired immunodeficiency syndrome (AIDS) A collection of disorders that develop as a result of infection with the human immunodeficiency virus (HIV).

Acrocentric Describes a chromosome whose centromere is placed very close to, but not at, one end.

Adenine One of two nitrogen-containing purine bases found in nucleic acids, along with guanine.

Adult stem cells Stem cells recovered from bone marrow and other organs of adults. These cells can differentiate to form a limited number of adult cells, and are called multipotent cells.

Alkaptonuria An autosomal recessive trait with altered metabolism of homogentisic acid. Affected individuals do not produce the enzyme needed to metabolize this acid, and their urine turns black.

Allele One of the possible alternative forms of a gene, usually distinguished from other alleles by its phenotypic effects.

Allele frequency The frequency with which alleles of a particular gene are present in a population.

Allelic expansion Increase in gene size caused by an increase in the number of trinucleotide repeat sequences.

Allergens Antigens that provoke an inappropriate immune response.

Alpha thalassemia Genetic disorder associated with an imbalance in the ratio of alpha and beta globin caused by reduced or absent synthesis of alpha globin.

Alzheimer disease (AD) A heterogeneous condition associated with the development of brain lesions, personality changes, and degeneration of intellect. Genetic forms are associated with genes on chromosomes 14, 19, and 21.

Amino acid One of the 20 subunits of proteins. Each contains an amino group, a carboxyl group, and an R group.

Amino group A chemical group (NH_2) found in amino acids and at one end of a polypeptide chain.

Amniocentesis A method of sampling the fluid surrounding the developing fetus by inserting a hollow needle and withdrawing suspended fetal cells and fluid; used in diagnosing fetal genetic and developmental disorders; usually performed in the sixteenth week of pregnancy.

Anaphase A stage in mitosis during which the centromeres split and the daughter chromosomes begin to separate.

Anaphylaxis A severe allergic response in which histamine is released into the circulatory system.

Aneuploidy A chromosomal number that is not an exact multiple of the haploid set.

Annotation The analysis of genomic nucleotide sequence data to identify the protein-coding genes, the nonprotein-coding genes, their regulatory sequences, and their function(s).

Antibody A class of proteins produced by B cells that bind to foreign molecules (antigens) and inactivate them.

Antibody-mediated immunity Immune reaction that protects primarily against invading viruses and bacteria using antibodies produced by plasma cells.

Anti-Müllerian hormone (AMH) A hormone produced by the developing testis that causes the breakdown of the Müllerian ducts in the embryo.

Anticipation Onset of a genetic disorder at earlier ages and with increasing severity in successive generations.

Anticodon A group of three nucleotides in a tRNA molecule that pairs with a complementary sequence (known as a codon) in an mRNA molecule.

Antigens Molecules usually carried or produced by viruses, microorganisms, or cells that initiate antibody production.

Assisted reproductive technologies (ART) The collection of techniques used to help infertile couples have children.

Assortment The result of meiosis I that puts random combinations of maternal and paternal chromosomes into gametes.

Autosomes Chromosomes other than the sex chromosomes. In humans, chromosomes 1 to 22 are autosomes.

B cell A type of lymphocyte that matures in the bone marrow and mediates antibody-directed immunity.

Background radiation Radiation in the environment that contributes to radiation exposure.

Barr body A densely staining mass in the somatic nuclei of mammalian females; an inactivated X chromosome.

Base analog A purine or pyrimidine that differs in chemical structure from those normally found in DNA or RNA.

Beta thalassemia Genetic disorder associated with an imbalance in the ratio of alpha and beta globin caused by reduced or absent synthesis of beta globin.

Bioinformatics The use of computers and software to acquire, store, analyze, and visualize the information from genomics.

Biotechnology The use of recombinant DNA technology to produce commercial goods and services.

Bipolar disorder A behavioral disorder characterized by mood swings that vary between manic activity and depression.

Blastocyst The developmental stage at which the embryo implants into the uterine wall.

Blastomere A cell produced in the early stages of embryonic development.

Blood type One of the classes into which blood can be separated on the basis of the presence or absence of certain antigens.

Bulbourethral glands Glands in the male that secrete a mucus-like substance that provides lubrication for intercourse.

Camptodactyly A dominant human genetic trait that is expressed as immobile, bent, little fingers.

Cap A modified base (guanine nucleotide) attached to the 5′ end of eukaryotic mRNA molecules.

Carbohydrates Macromolecules including sugars, glycogen, and starches composed of sugar monomers linked and cross-linked together.

Carboxyl group A chemical group (COOH) found in amino acids and at one end of a polypeptide chain.

Cell cycle The sequence of events that takes place between successive mitotic divisions.

Cell-mediated immunity Immune reaction mediated by T cells directed against body cells that have been infected by viruses or bacteria.

Centimorgan (cM) A unit of distance between genes on chromosomes. One centimorgan equals a frequency of 1% crossing over between two genes.

Centromere A region of a chromosome to which spindle fibers attach during cell division. The location of a centromere gives a chromosome its characteristic shape.

Cervix The lower neck of the uterus, opening into the vagina.

Chorion A two-layered structure formed during embryonic development from the trophoblast.

Chorionic villus sampling (CVS) A method of sampling fetal chorionic cells by inserting a catheter through the vagina or abdominal wall into the uterus. Used in diagnosing biochemical and cytogenetic defects in the embryo. Usually performed in the eighth or ninth week of pregnancy.

Chromatid One of the strands of a duplicated chromosome, joined by a single centromere to its sister chromatid.

Chromatin The DNA and protein components of chromosomes, visible as clumps or threads in nuclei.

Chromatin remodeling The set of chemical changes to the DNA and histones that activate and inactivate gene expression.

Chromosomes The threadlike structures in the nucleus that carry genetic information.

Clinodactyly An autosomal dominant trait that produces a bent finger.

Clone-by-clone method A method of genome sequencing that begins with genetic and physical maps and sequences overlapping clones after they have been placed in a linear order. More commonly known as *map-based sequencing*.

Clones Genetically identical molecules, cells, or organisms, all derived from a single ancestor.

Codominance Full phenotypic expression of both members of a gene pair in the heterozygous condition.

Codon Triplets of nucleotides in mRNA that encode the information for a specific amino acid in a protein.

Color blindness Defective color vision caused by reduction or absence of visual pigments. There are three forms: red, green, and blue color blindness.

Comparative genomics Compares the genomes of different species to look for clues to the evolutionary history of genes or a species.

Complement system A chemical defense system that kills microorganisms directly, supplements the inflammatory response, and works with (complements) the immune system.

Complete androgen insensitivity (CAIS) An X-linked genetic trait that causes XY individuals to develop into phenotypic females.

Complex traits Traits controlled by multiple genes, the interaction of genes with each other, and with environmental factors where the contributions of genes and environment are undefined.

Concordance Agreement between traits exhibited by both twins.

Continuous variation A distribution of phenotypic characters that is distributed from one extreme to another in an overlapping, or continuous, fashion.

Copy number variation (CNV) A DNA segment at least 1,000 base pairs long with a variable copy number in the genome.

Correlation coefficients Measures the degree of interdependence of two or more variables.

Covalent bonds Chemical bonds that result from electron sharing between atoms. Covalent bonds are formed and broken during chemical reactions.

Cri du chat syndrome A deletion of the short arm of chromosome 5 associated with an array of congenital malformations, the most characteristic of which is an infant cry that resembles a meowing cat.

Crossing over A process in which chromosomes physically exchange parts.

C-terminus The end of a polypeptide or protein that has a free carboxyl group.

Cystic fibrosis An often fatal recessive genetic disorder associated with abnormal secretions of the exocrine glands.

Cytogenetics The branch of genetics that studies the organization and arrangement of genes and chromosomes by using the techniques of microscopy.

Cytokinesis The process of cytoplasmic division that accompanies cell division.

Cytosine One of three nitrogen-containing pyrimidine bases found in nucleic acids along with thymine and uracil.

Deoxyribonucleic acid (DNA) A molecule consisting of antiparallel strands of polynucleotides that is the primary carrier of genetic information.

Deoxyribose One of two pentose sugars found in nucleic acids. Deoxyribose is found in DNA, ribose in RNA.

Dermatoglyphics The study of the skin ridges on fingers, palms, toes, and soles.

Diploid (2n) The condition in which each chromosome is represented twice as a member of a homologous pair.

Discontinuous variation Phenotypes that fall into two or more distinct, nonoverlapping classes.

Dizygotic (DZ) Twins derived from two separate and nearly simultaneous fertilizations, each involving one egg and one sperm. Such twins share, on average, 50% of their genes.

DNA A helical molecule consisting of two strands of nucleotides that is the primary carrier of genetic information.

DNA fingerprint Detection of variations in minisatellites used to identify individuals.

DNA microarray A series of short nucleotide sequences placed on a solid support (such as glass) that have several different uses, such as detection of mutant genes or differences in the pattern of gene expression in normal and cancerous cells.

DNA polymerase An enzyme that catalyzes the synthesis of DNA using a template DNA strand and nucleotides.

DNA profile The pattern of STR allele frequencies used to identify individuals.

DNA sequencing A technique for determining the nucleotide sequence of a DNA molecule.

Dominant trait The trait expressed in the F1 (or heterozygous) condition.

Dosage compensation A mechanism that regulates the expression of sex-linked genes.

Ecogenetics A branch of genetics that studies genetic traits related to the response to environmental substances.

Ejaculatory duct In males, a short connector from the vas deferens to the urethra.

Embryonic stem cells (ESC) Cells in the inner cell mass of early embryos that will form all the cells, tissues, and organs of the adult. Because of their ability to form so many different cell types, these are called pluripotent cells.

Endometrium The inner lining of the uterus that is shed at menstruation if fertilization has not occurred.

Endoplasmic reticulum (ER) A system of cytoplasmic membranes arranged into sheets and channels whose function is to synthesize and transport gene products.

Enhancement gene therapy Gene transfer to enhance traits such as intelligence or athletic ability rather than to treat a genetic disorder.

Environmental variance The phenotypic variance of a trait in a population that is attributed to differences in the environment.

Enzyme replacement therapy Treatment of a genetic disorder by providing a missing enzyme encoded by the mutant allele responsible for the disorder.

Epididymis A part of the male reproductive system where sperm are stored.

Epigenetics Reversible chemical modifications of chromosomal DNA (such as methylation of bases) and/or associated histone proteins that change the pattern of gene expression without affecting the nucleotide sequence of the DNA.

Epistasis The interaction of two or more non-allelic genes to control a single phenotype.

Essential amino acids Amino acids that cannot be synthesized in the body and must be supplied in the diet.

Eugenics The attempt to improve the human species by selective breeding.

Evolution Changes in allele frequencies in a population over time.

Exons DNA sequences that are transcribed, joined to other exons during mRNA processing, and translated into the amino acid sequence of a protein.

Expressivity The range of phenotypes resulting from a given genotype.

Familial adenomatous polyposis (FAP) An autosomal dominant trait resulting in the development of polyps and benign growths in the colon. Polyps often develop into malignant growths and cause cancer of the colon and/or rectum.

Familial hypercholesterolemia Autosomal dominant disorder with defective or absent LDL receptors. Affected individuals are at increased risk for cardiovascular disease.

Fertilization The fusion of two gametes to produce a zygote.

Fetal alcohol syndrome (FAS) A constellation of birth defects caused by maternal alcohol consumption during pregnancy.

Fitness A measure of the relative survival and reproductive success of a specific individual or genotype.

Follicle A developing egg surrounded by an outer layer of follicle cells, contained in the ovary.

Founder effects Allele frequencies established by chance in a population that is started by a small number of individuals (perhaps only a fertilized female).

Fragile X An X chromosome that carries a gap, or break, at band q27; associated with mental retardation in males.

Frameshift mutations Mutational events in which a number of bases (other than multiples of three) are added to or removed from DNA, causing a shift in the codon reading frame.

Friedreich ataxia A progressive and fatal neurodegenerative disorder inherited as an autosomal recessive trait with symptoms appearing between puberty and the age of 25.

Galactosemia A heritable trait associated with the inability to metabolize the sugar galactose. If it is left untreated, high levels of galactose-1-phosphate accumulate, causing cataracts and mental retardation.

Gamete intrafallopian transfer (GIFT) A procedure in which gametes are collected and placed into a woman's oviduct for fertilization.

Gametes Unfertilized germ cells.

Gene The fundamental unit of heredity and the basic structural and functional unit of genetics.

Gene pool The set of genetic information carried by the members of a sexually reproducing population.

Gene therapy The transfer of cloned genes into somatic cells as a means of treating a genetic disorder.

General cognitive ability Characteristics that include verbal and spatial abilities, memory, speed of perception, and reasoning.

Genetic code The sequence of nucleotides that encodes the information for amino acids in a polypeptide chain.

Genetic counseling A process of communication that deals with the occurrence or risk of a genetic disorder in a family.

Genetic drift The random fluctuations of allele frequencies from generation to generation that take place in small populations.

Genetic equilibrium The situation when the allele frequency for a particular gene remains constant from generation to generation.

Genetic library In recombinant DNA terminology, a collection of clones that contains all the DNA in an individual.

Genetic map A diagram of a chromosome showing the order of genes and the distance between them based on recombination frequencies (centimorgans).

Genetic screening The systematic search for individuals in a population who have certain genotypes.

Genetic testing The use of methods to determine if an individual has a genetic disorder, will develop one, or is a carrier.

Genetic variance The phenotypic variance of a trait in a population that is attributed to genotypic differences.

Genetically modified organisms (GMOs) A general term used to refer to transgenic plants or animals created by recombinant DNA techniques.

Genetics The scientific study of heredity.

Genome The set of DNA sequences carried by an individual.

Genome-wide association study (GWAS) Analysis of genetic variation across an entire genome, searching for associations (linkages) between variations in DNA sequence and a genome region encoding a specific phenotype.

Genomic imprinting Phenomenon in which the expression of a gene depends on whether it is inherited from the mother or the father; also known as genetic or parental imprinting.

Genomic library In recombinant DNA terminology, a collection of clones that contains all the genetic information in an individual.

Genomics The study of the organization, function, and evolution of genomes.

Genotype The specific genetic constitution of an organism.

Germ-line gene therapy Gene transfer to gametes or the cells that produce them. Transfers a gene to all cells in the next generation, including germ cells.

Golgi complex Membranous cellular organelles composed of a series of flattened sacs. They sort, modify, and package proteins synthesized in the ER.

Gonads Organs where gametes are produced.

Guanine One of two nitrogen-containing purine bases found in nucleic acids, along with adenine.

Haploid (n) The condition in which each chromosome is represented once in an unpaired condition.

Haplotype A set of genetic markers located close together on a single chromosome or chromosome region.

Hardy-Weinberg law The statement that allele and genotype frequencies remain constant from generation to generation when the population meets certain assumptions.

Helper T cell A lymphocyte that stimulates the production of antibodies by B cells when an antigen is present and stimulates division of B cells and cytotoxic T cells.

Hemizygous A gene present on the X chromosome that is expressed in males in both the recessive and the dominant conditions.

Hemoglobin variants Alpha and beta globins with variant amino acid sequences.

Hemolytic disease of the newborn (HDN) A condition of immunological incompatibility between mother and fetus that occurs when the mother is Rh⁻ and the fetus is Rh⁺.

Hereditarianism The mistaken idea that human traits are determined solely by genetic inheritance, ignoring the contribution of the environment.

Hereditary nonpolyposis colon cancer (HNPCC) An autosomal dominant trait associated with genomic instability of microsatellite DNA sequences and a form of colon cancer.

Heritability An expression of how much of the observed variation in a phenotype is due to differences in genotype.

Heterozygous Carrying two different alleles for one or more genes.

Histamine A chemical signal produced by mast cells that triggers dilation of blood vessels.

Histones DNA-binding proteins that help compact and fold DNA into chromosomes.

Hominins A classification that includes all bipedal primates from australopithecines to our species.

Hominoids The superfamily of primates that includes apes and humans.

Homologous chromosomes Chromosomes that physically associate (pair) during meiosis. Homologous chromosomes have identical gene loci.

Homozygous Having identical alleles for one or more genes.

Huntington disease (HD) An autosomal dominant disorder associated with progressive neural degeneration and dementia. Adult onset is followed by death 10 to 15 years after symptoms appear.

Hydrogen bond A weak chemical bonding force between hydrogen and another atom.

Immunoglobulins (Ig) The five classes of proteins to which antibodies belong.

Imprinting A phenomenon in which expression of a gene depends on whether it is inherited from the mother or the father.

In vitro fertilization (IVF) A procedure in which gametes are collected and fertilized in a dish in the laboratory; the resulting zygote is implanted in the uterus for development.

Inborn error of metabolism The concept advanced by Archibald Garrod that many genetic traits result from alterations in biochemical pathways.

Incomplete dominance Expression of a phenotype that is intermediate to those of the parents.

Independent assortment The random distribution of genes into gametes during meiosis.

Induced pluripotent stem cells (iPS) Adult cells that can be reprogrammed (induced) by gene transfer to form cells with most of the developmental potential of embryonic stem cells. Because of this developmental potential, such cells are pluripotent.

Inflammatory response The body's reaction to invading microorganisms, a nonspecific active defense mechanism that the body employs to resist infection.

Initiation complex Formed by the combination of mRNA, tRNA, and the small ribosome subunit. The first step in translation.

Inner cell mass A cluster of cells in the blastocyst that gives rise to the embryonic body. The inner cell mass contains the embryonic stem cells.

Intelligence quotient (IQ) A score derived from standardized tests that is calculated by dividing the individual's mental age (determined by the test) by his or her chronological age and multiplying the quotient by 100.

Interphase The period of time in the cell cycle between mitotic divisions.

Intracytoplasmic sperm injection (ICSI) A treatment to overcome defects in sperm count or motility; an egg is fertilized by microinjection of a single sperm.

Introns DNA sequences present in some genes that are transcribed but are removed during processing and therefore are not present in mature mRNA.

Ionizing radiation Radiation that produces ions during interaction with other matter, including molecules in cells.

Karyotype A complete set of chromosomes from a cell that has been photographed during cell division and arranged in a standard sequence.

Killer T cells T cells that destroy body cells infected by viruses or bacteria. These cells also can attack viruses, bacteria, cancer cells, and cells of transplanted organs directly.

Klinefelter syndrome Aneuploidy of the sex chromosomes involving an XXY chromosomal constitution.

Leptin A hormone produced by fat cells that signals the brain and ovary. As fat levels become depleted, secretion of leptin slows and eventually stops.

Linkage A condition in which two or more genes do not show independent assortment. Rather, they tend to be inherited together. Such genes are located on the same chromosome. When the degree of recombination between linked genes is measured, the distance between them can be determined.

Lipids A class of cellular macromolecules including fats and oils that are insoluble in water.

Lipoproteins Particles that have protein and phospholipid coats that transport cholesterol and other lipids in the bloodstream.

Locus The position occupied by a gene on a chromosome.

Lod method A probability technique used to determine whether two genes are linked.

Lod score The ratio of probabilities that two genes are linked to the probability that they are not linked, expressed as a $\log_{10}$. Scores of 3.0 or higher are taken as establishing linkage.

Loss of heterozygosity (LOH) In a cell, the loss of normal function in one allele of a gene where the other allele is already inactivated by mutation.

Lymphocytes White blood cells that originate in bone marrow and mediate the immune response.

Lyon hypothesis The proposal that dosage compensation in mammalian females is accomplished by partially and randomly inactivating one of the two X chromosomes.

Lysosomes Membrane-enclosed organelles in eukaryotic cells that contain digestive enzymes.

Macromolecules Large cellular polymers assembled by chemically linking monomers together.

Mad-cow disease A prion disease of cattle, also known as bovine spongiform encephalopathy, or BSE.

Major histocompatibility complex (MHC) A set of genes on chromosome 6 that encode recognition molecules that prevent the immune system from attacking a body's own organs and tissues.

Map-based sequencing A method of genome sequencing that begins with genetic and physical maps; clones are sequenced after they have been placed in order.

Marfan syndrome An autosomal dominant genetic disorder that affects the skeletal system, the cardiovascular system, and the eyes.

Meiosis The process of cell division during which one cycle of chromosomal replication is followed by two successive cell divisions to produce four haploid cells.

Membrane-attack complex (MAC) A large, cylindrical multiprotein that embeds itself in the plasma membrane of an invading microorganism and creates a pore through which fluids can flow, eventually bursting the microorganism.

Memory B cell A long-lived B cell produced after exposure to an antigen that plays an important role in secondary immunity.

messenger RNA (mRNA) A single-stranded complementary copy of the amino acid-coding nucleotide sequence of a gene.

Metabolism The sum of all biochemical reactions by which cells convert and utilize energy.

Metacentric Describes a chromosome that has a centrally placed centromere.

Metaphase A stage in mitosis during which the chromosomes become arranged near the middle of the cell.

Metastasis A process by which cells detach from the primary tumor and move to other sites, forming new malignant tumors in the body.

Millirem A rem is a measure of radiation dose equal to 1,000 millirems.

Minisatellite Nucleotide sequences 14 to 100 base pairs long organized into clusters of varying lengths, on many different chromosomes; used in the construction of DNA fingerprints.

Missense mutations Mutations that cause the substitution of one amino acid for another in a protein.

Mitochondria (singular: mitochondrion) Membrane-bound organelles, present in the cytoplasm of eukaryotic cells, that are sites of energy production.

Mitosis Form of cell division that produces two cells, each of which has the same complement of chromosomes as the parent cell.

Molecular genetics The study of genetic events at the biochemical level.

Molecules Structures composed of two or more atoms held together by chemical bonds.

Monosomy A condition in which one member of a chromosomal pair is missing; having one less than the diploid number $(2n - 1)$.

Monozygotic (MZ) Twins derived from a single fertilization involving one egg and one sperm; such twins are genetically identical.

Multifactorial traits Traits that result from the interaction of one or more environmental factors and two or more genes.

Multiple alleles Genes that have more than two alleles.

Multipotent The restricted ability of a stem cell to form only one or a few different cell types.

Muscular dystrophy A group of genetic diseases associated with progressive degeneration of muscles. Two of these, Duchenne and Becker muscular dystrophy, are inherited as X-linked allelic recessive traits.

Mutation rate The number of events that produce mutated alleles per locus per generation.

Natural selection The differential reproduction shown by some members of a population that is the result of differences in fitness.

Nitrogen-containing base A purine or pyrimidine that is a component of nucleotides.

Nondisjunction The failure of homologous chromosomes to separate properly during meiosis or mitosis.

Nonsense mutations Mutations that change an amino acid specifying a codon to one of the three termination codons.

N-terminus The end of a polypeptide or protein that has a free amino group.

Nucleic acids A class of cellular macromolecules composed of nucleotide monomers linked together. There are two types of nucleic acids, deoxyribonucleic acid (DNA) and ribonucleic acid (RNA), which differ in the structure of the monomers.

Nucleolus (plural: nucleoli) A nuclear region that functions in the synthesis of ribosomes.

Nucleosome A bead-like structure composed of histone proteins wrapped with DNA.

Nucleotide The basic building block of DNA and RNA. Each nucleotide consists of a base, a phosphate, and a sugar.

Nucleotide substitutions Mutations that involve replacement of one or more nucleotides in a DNA molecule with other nucleotides.

Nucleus The membrane-bound organelle in eukaryotic cells that contains the chromosomes.

Oncogenes Genes that induce or continue uncontrolled cell proliferation.

Oocyte A cell from which an ovum develops by meiosis.

Oogenesis The process of oocyte production.

Oogonia Cells that produce primary oocytes by mitotic division.

Open reading frame (ORF) The codons in a gene that encode the amino acids of the gene product.

Organelles Cytoplasmic structures that have a specialized function.

Ovaries Female gonads that produce oocytes and female sex hormones.

Oviduct A duct with fingerlike projections partially surrounding the ovary and connecting to the uterus. Also called the fallopian or uterine tube.

Ovulation The release of a secondary oocyte from the follicle; usually occurs monthly during a female's reproductive lifetime.

Ovum The haploid cell produced by meiosis that becomes the functional gamete.

Pathogens Disease-causing agents.

Pattern baldness A sex-influenced trait that acts like an autosomal dominant trait in males and an autosomal recessive trait in females.

Pedigree A diagram listing the members and ancestral relationships in a family; used in the study of human heredity.

Pedigree analysis The construction of family trees and their use to follow the transmission of genetic traits in families. It is the basic method of studying the inheritance of traits in humans.

Pedigree construction Use of family history to determine how a trait is inherited and to estimate risk factors for family members.

Penetrance The probability that a disease phenotype will appear when a disease-related genotype is present.

Pentose sugar A five-carbon sugar molecule found in nucleic acids.

Peptide bond A covalent chemical link between the carboxyl group of one amino acid and the amino group of another amino acid.

Pharmacogenetics A branch of genetics concerned with the identification of protein variants that underlie differences in the response to drugs.

Pharmacogenomics A branch of genetics that analyzes genes and proteins to identify targets for therapeutic drugs.

Phenotype The observable properties of an organism.

Phenylketonuria (PKU) An autosomal recessive disorder of amino acid metabolism that results in mental retardation if untreated.

Philadelphia chromosome An abnormal chromosome produced by translocation of parts of the long arms of chromosomes 9 and 22.

Phosphate group A compound containing phosphorus chemically bonded to four oxygen molecules.

Physical map A diagram of a chromosome showing the order of genes and the distance between them measured in base pairs.

Plasma cells Daughter cells of B cells, which synthesize and secrete 2,000 to 20,000 antibody molecules per second into the bloodstream.

Pluripotent The ability of a stem cell to form any fetal or adult cell type.

Polar bodies Cells produced in the first and second meiotic division in female meiosis that contain little cytoplasm and will not function as gametes.

Poly-A tail A series of A nucleotides added to the 3′ end of mRNA molecules.

Polygenic traits Traits controlled by two or more genes.

Polymerase chain reaction (PCR) A method for amplifying DNA segments using cycles of denaturation, annealing to primers, and DNA polymerase-directed DNA synthesis.

Polypeptide A molecule made of amino acids joined together by peptide bonds.

Polyploidy A chromosomal number that is a multiple of the normal haploid chromosomal set.

Polyps A fleshy growth in the lining of the nose, colon, uterus, and other organs.

Polysomes A messenger RNA (mRNA) molecule with several ribosomes attached.

Population A local group of organisms belonging to a single species, sharing a common gene pool.

Population genetics The branch of genetics that studies inherited variation in populations of individuals and the forces that alter gene frequency.

Positional cloning A recombinant DNA–based method of mapping and cloning genes with no prior information about the gene product or its function.

Preimplantation genetic diagnosis (PGD) Removal and genetic analysis of a single cell from a 3- to 5-day-old embryo. Used to select embryos free of genetic disorders for implantation and development.

pre-messenger RNA (pre-mRNA) The transcript made from the DNA template that is processed and modified to form messenger RNA.

Primary structure The amino acid sequence in a polypeptide chain.

Prion A protein folded into an infectious conformation that is the cause of several disorders, including Creutzfeldt-Jakob disease and mad-cow disease.

Proband First affected family member who seeks medical attention for a genetic disorder.

Probe A labeled nucleic acid used to identify a complementary region in a clone or genome.

Product The specific chemical compound that is the result of enzymatic action. In biochemical pathways, a compound can serve as the product of one reaction and the substrate for the next reaction.

Promoter region A region of a DNA molecule to which RNA polymerase binds and initiates transcription.

Prophase A stage in mitosis during which the chromosomes become visible and contain sister chromatids joined at the centromere.

Prostaglandins Locally acting chemical messengers that stimulate contraction of the female reproductive system to assist in sperm movement.

Prostate gland A gland that secretes a milky, alkaline fluid that neutralizes acidic vaginal secretions and enhances sperm viability.

Proteins A class of cellular macromolecules composed of amino acid monomers linked together and folded into a three-dimensional shape.

Proteome The set of proteins present in a cell at a specific time under a specific set of conditions.

Proteomics The study of the proteome, the set of expressed proteins present in a cell.

Proto-oncogenes Normal genes that initiate or maintain cell division and that may become cancer genes (oncogenes) by mutation.

Pseudogenes Nonfunctional genes that are closely related (by DNA sequence) to functional genes present elsewhere in the genome.

Pseudohermaphroditism An autosomal genetic condition that causes XY individuals to develop the phenotypic sex of females.

Purine A class of double-ringed organic bases found in nucleic acids.

Pyrimidine A class of single-ringed organic bases found in nucleic acids.

Quaternary structure The structure formed by the interaction of two or more polypeptide chains in a protein.

R group Each amino acid has a different side chain, called an R group. An R group can be positively or negatively charged or neutral.

Radiation The process by which electromagnetic energy travels through space or a medium such as air.

Recessive trait The trait unexpressed in the F1 but re-expressed in some members of the F2 generation.

Recombinant DNA technology A series of techniques in which DNA fragments from an organism are linked to self-replicating vectors to create recombinant DNA molecules, which are replicated or cloned in a host cell.

Regression to the mean In a polygenic system, the tendency of offspring of parents with extreme differences in phenotype to exhibit a phenotype that is the average of the two parental phenotypes.

Rem The unit of radiation exposure used to measure radiation damage in humans. It is the amount of ionizing radiation that has the same effect as a standard amount of X-rays.

Restriction enzyme A bacterial enzyme that cuts DNA at specific sites.

Retinoblastoma A malignant tumor of the eye arising in retinoblasts (embryonic retinal cells that disappear at about 2 years of age). Because mature retinal cells do not transform into tumors, this is a tumor that usually occurs only in children.

Ribonucleic acid (RNA) A nucleic acid molecule that contains the pyrimidine uracil and the sugar ribose. The several forms of RNA function in gene expression.

Ribose One of two pentose sugars found in nucleic acids. Deoxyribose is found in DNA, ribose in RNA.

ribosomal RNA (rRNA) RNA molecules that form part of the ribosome.

Ribosomes Cytoplasmic particles that aid in the production of proteins.

RNA interference (RNAi) A mechanism of gene regulation that controls the amounts of mRNA available for translation.

Schizophrenia A behavioral disorder characterized by disordered thought processes and withdrawal from reality. Genetic and environmental factors are involved in this disease.

Scrotum A pouch of skin outside the male body that contains the testes.

Secondary oocyte The large cell produced by the first meiotic division.

Secondary structure The pleated or helical structure in a protein molecule generated by the formation of bonds between amino acids.

Segregation The separation of members of a gene pair from each other during gamete formation.

Semen A mixture of sperm and various glandular secretions containing 5% spermatozoa.

Semiconservative replication A model of DNA replication that provides each daughter molecule with one old strand and one newly synthesized strand. DNA replicates in this fashion.

Seminal vesicles Glands in males that secrete fructose and prostaglandins into the sperm.

Seminiferous tubules Small, tightly coiled tubes inside the testes where sperm are produced.

Sense mutations Mutations that change a termination codon into one that codes for an amino acid. Such mutations produce elongated proteins.

Severe combined immunodeficiency disease (SCID) A collection of genetic disorders in which affected individuals have no immune response; both the cell-mediated and antibody-mediated responses are missing.

Sex chromosomes In humans, the X and Y chromosomes that are involved in sex determination.

Sex-influenced traits Traits controlled by autosomal genes that are usually dominant in one sex but recessive in the other sex.

Sex-limited genes Loci that produce a phenotype in only one sex.

Sex ratio The proportion of males to females, which changes throughout the life cycle. The ratio is close to 1:1 at fertilization, but the ratio of females to males increases as a population ages.

Short tandem repeat (STR) Short nucleotide sequences 2 to 9 base pairs long found throughout the genome that are organized into clusters of varying lengths; used in the construction of DNA profiles.

Shotgun sequencing A method of genome sequencing that selects clones at random from a genomic library and, after sequencing them, assembles the genome sequence by using software analysis.

Sickle cell anemia A recessive genetic disorder associated with an abnormal type of hemoglobin, a blood transport protein.

Signal transduction A cellular molecular pathway by which an external signal is converted into a functional response.

Single nucleotide polymorphism (SNP) Single nucleotide differences between and among individuals in a population or species.

Sister chromatids Two chromatids joined by a common centromere. Each chromatid carries identical genetic information.

Somatic cell nuclear transfer A cloning technique that transfers a somatic cell nucleus to an enucleated egg, which is stimulated to develop into an embryo. Inner cell mass cells are collected from the embryo and grown to form a population of stem cells. Also called therapeutic cloning.

Somatic gene therapy Gene transfer to somatic target cells to correct a genetic disorder.

Southern blot A method for transferring DNA fragments from a gel to a membrane filter, developed by Edwin Southern for use in hybridization experiments.

Sperm Male gamete.

Spermatids The four haploid cells produced by meiotic division of a primary spermatocyte.

Spermatocytes Diploid cells that undergo meiosis to form haploid spermatids.

Spermatogenesis The process of sperm production.

Spermatogonia Mitotically active cells in the gonads of males that give rise to primary spermatocytes.

SRY A gene, called the sex-determining region of the Y, located near the end of the short arm of the Y chromosome that plays a major role in causing the undifferentiated gonad to develop into a testis.

Start codon A codon present in mRNA that signals the location for translation to begin. The codon AUG functions as a start codon and codes for the amino acid methionine.

Stem cells Cells with two properties: the ability to replicate themselves, and the ability to form a variety of cell types in the body.

Stop codon A codon in mRNA that signals the end of translation. UAA, UAG, and UGA are stop codons.

Structural genomics A branch of genomics that generates three dimensional structure of proteins from their amino acid sequences.

Submetacentric Describes a chromosome whose centromere is placed closer to one end than the other.

Substrate The specific chemical compound that is acted on by an enzyme.

Sugar In nucleic acids, either ribose, found in RNA, or deoxyribose, found in DNA. The difference between the two sugars is an OH group present in ribose and absent in deoxyribose.

Suppressor T cells T cells that slow or stop the immune response of B cells and other T cells.

T cell A type of lymphocyte that undergoes maturation in the thymus and mediates cellular immunity.

T-cell receptors (TCRs) Unique proteins on the surface of T cells that bind to specific proteins on the surface of cells infected with viruses, bacteria, or intracellular parasites.

Telomerase An enzyme that adds telomere repeats to the ends of chromosomes, keeping them the same length after each cell division.

Telomere Short repeated DNA sequences located at each end of chromosomes.

Telophase The last stage of mitosis, during which the chromosomes of the daughter cells decondense and the nucleus re-forms.

Template The single-stranded DNA that serves to specify the nucleotide sequence of a newly synthesized polynucleotide strand.

Teratogen Any physical or chemical agent that brings about an increase in congenital malformations.

Termination sequence The nucleotide sequence at the end of a gene that signals the end of transcription.

Tertiary structure The three-dimensional structure of a protein molecule brought about by folding on itself.

Testes Male gonads that produce spermatozoa and sex hormones.

Testosterone A steroid hormone produced by the testis; the male sex hormone.

Tetraploidy A chromosomal number that is four times the haploid number, having four copies of all autosomes and four sex chromosomes.

Thalassemias Disorders associated with an imbalance in the production of alpha or beta globin.

Thymine One of three nitrogen-containing pyrimidine bases found in nucleic acids, along with uracil and cytosine.

Thymine dimer A molecular lesion in which chemical bonds form between a pair of adjacent thymine bases in a DNA molecule.

Trait Any observable property of an organism.

Transcription Transfer of genetic information from the base sequence of DNA to the base sequence of RNA, mediated by RNA synthesis.

transfer RNA (tRNA) A small RNA molecule that contains a binding site for a specific type of amino acid and has a three-base segment known as an anticodon that recognizes a specific base sequence in messenger RNA.

Transformation The process of transferring genetic information between cells by DNA molecules.

Transforming factor The molecular agent of transformation; DNA.

Transgenic Refers to the transfer of genes between species by recombinant DNA technology; transgenic organisms have received such a gene.

Translation Conversion of information encoded in the nucleotide sequence of an mRNA molecule into the linear sequence of amino acids in a protein.

Transmission genetics The branch of genetics concerned with the mechanisms by which genes are transferred from parent to offspring.

Trinucleotide repeats A form of mutation associated with the expansion in copy number of a nucleotide triplet in or near a gene.

Triploidy A chromosomal number that is three times the haploid number, having three copies of all autosomes and three sex chromosomes.

Trisomy A condition in which one chromosome is present in three copies, whereas all others are diploid; having one more than the diploid number ($2n + 1$).

Trisomy 21 Aneuploidy involving the presence of an extra copy of chromosome 21, resulting in Down syndrome.

Trophoblast The outer layer of cells in the blastocyst that gives rise to the membranes surrounding the embryo.

Tumor-suppressor genes Genes encoding proteins that suppress cell division.

Turner syndrome A monosomy of the X chromosome (45,X) that results in female sterility.

Uniparental disomy (UPD) A condition in which both copies of a chromosome are inherited from one parent.

Uracil One of three nitrogen-containing pyrimidine bases found in nucleic acids, along with thymine and cytosine.

Urethra A tube that passes from the bladder and opens to the outside. It functions in urine transport and, in males, also carries sperm.

Uterus A hollow, pear-shaped muscular organ where an early embryo will implant and develop throughout pregnancy.

Vaccine A preparation containing dead or weakened pathogens that elicits an immune response when injected into the body.

Vagina The opening that receives the penis during intercourse and also serves as the birth canal.

Vas deferens A duct connected to the epididymis, which sperm travels through.

Vasectomy A contraceptive procedure for men in which each vas deferens is cut and sealed to prevent the transport of sperm.

Vectors Self-replicating DNA molecules that are used to transfer foreign DNA segments between host cells.

Whole genome sequencing A method of genome sequencing that selects clones at random from a genomic library and, after sequencing them, assembles the genome sequence by using software analysis.

X inactivation center (Xic) A region on the X chromosome where inactivation begins.

Xenotransplants Cells, tissues, or organs that are transplanted from one species to another.

X-linked The pattern of inheritance that results from genes located on the X chromosome.

X-linked agammaglobulinemia (XLA) A rare, X-linked recessive trait characterized by the total absence of immunoglobulins and B cells.

XYY karyotype Aneuploidy of the sex chromosomes involving XYY chromosomal constitution.

Yeast artificial chromosome (YAC) A cloning vector that has telomeres and a centromere that can accommodate large DNA inserts and uses the eukaryote yeast as a host cell.

Y-linked The pattern of inheritance that results from genes located only on the Y chromosome.

Zygote The fertilized egg that develops into a new individual.

Zygote intrafallopian transfer (ZIFT) An ART procedure in which gametes are collected, fertilization takes place *in vitro*, and the resulting zygote (fertilized egg) is transferred to a woman's oviduct.

Index

in Human Genome Project, *338, 344*
intelligence studies, 114
transgenic research and, 322
drug sensitivities, 235
Duchenne muscular dystrophy, 82, *83,* 172
mutation rates for, *249*
dwarfism, 219
dystrophin, 82
DZ (dizygotic), 104

E

ear, development of, *158*
Eastern European Jews, and Tay-Sachs disease, 64
ecogenetics, 232, 236–238
Edwards, Robert, 354
Edwards syndrome (Trisomy 18), *133,* 133–134
Edward VII, 85
Egeland, Borgny, 218
Egeland, Liv and Dag, 218, 222
egg nucleus, *155*
Ehlers-Danlos syndrome, *74*
ejaculatory duct, 150, *150*
electrophoresis in Southern blot procedure, 304, *304,* 305
Elizabeth II, England, 325
elongation phase
of transcription, 201–202
of translation, 204, 207
embryonic stem cells, 316, *316,* 316–318
embryos
preimplantation genetic diagnosis and, 364–365
sex differentiation in, 165, *165*
embryo splitting, 106
endometrium, 152, 153, *156*
endoplasmic reticulum (ER), *25*
defined, 24
structure and function, 24
endoxifen, 236
enhancement gene therapy, 368
environment
cancer and, 286–287
factors, and multifactorial traits, 100–103
mutation rates and, 249–253
and peanut allergies, 391–393
phenylketonuria (PKU) and, 223
environmental variance, 103
enzyme replacement therapy, 315
enzymes, functions of, 220–221
epidermal growth factor, *314*
epididymis, 150, *150,* 152
epigenetics, 263
epinephrine, *408*
epistasis, 63, 402
Epstein-Barr virus (EBV), 287
ER (endoplasmic reticulum), 24
erythropoietin, *314,* 369
Escherichia coli, 179
and cloning, 296
genome of, 344
in Human Genome Project, 338
and irradiated foods, 245
essential amino acids, 221
ethics
of assisted reproductive technologies, 361–362
and behavior genetics, 414

and biotechnology, 325, 326
of gene therapy, 368–369
of genetic screening and testing, 45, 354–365
and Human Genome Project, 347
of preimplantation genetic diagnosis (PGD), 362
of reproductive technology, 361–362
ethylene glycol poisoning, 44
eugenics, 10
eugenic sterilization, 13
immigration laws and, 11–12
Nazi Germany and, 13
reproductive rights and, 12–13
eukaryotes, and cancer, *273*
evolution, 424, 432–435
exons, 202
expressed sequence tags (ESTs), 192
expressivity, 87–88, *88*
extracellular fluid, *23*
eye
color, *99*
color blindness, *81*

F

Fabry disease, 83
Fairchild, Greg and Tierney, 6
familial adenomatous polyposis (FAP), 278–279
defined, 278
heritable predispositions to, *271*
familial hypercholesterolemia, 64, *74,* 367
family planning, and genetic testing, 14
family studies, multifactorial traits and, 106–108
family trees. *See* pedigree analysis
Fanconi anemia, *74, 280*
FAP (familial adenmatous polyposis), 278
FAS (fetal alcohol syndrome), 161, *161*
fatal familial insomnia, 210
FBI (Federal Bureau of Investigation) DNA profiles and, 326
female reproductive system
anatomy of, *152*
components and functions of, *153*
fertilization, 155, *155, 156*
fetal alcohol syndrome (FAS), 161, *161*
fetus, *158, 159*
teratogens as risk to development of, 160–161, *161*
F2 generation, 52–54
fitness, 425
flame retardants, 253
flowers, smell of, 235, *235*
FMRI gene expansion, 256, *256*
focal dermal hypoplasia, 172
folate, 101
follicle, 152, *155*
foods
allergies, 391–393
genetic modification of, 15
irradiated foods, 244–245
peanut allergies, 391, 393
foot blistering, 246, *246*
foot plate, *158*
Forbes disease, *227*
forebrain, *158*
forensic use of DNA profiles, 323–324, 326
forked line method, 54

founder effects, 424
FOXP2 gene, 407, 434–435
fragile X syndrome, *143,* 256
defined, 143
FMR1 gene expansions, 256, *256*
and simple gene model, 401
and trinucleotide repeats, 256
frameshift mutations, 253–254, *256*
Franklin, Rosalind, 184, 185–186
frataxin, 346, *346*
frataxin protein, 346
FRDA gene (Friedreich ataxia), 345
Frederick the Great, Prussia, 94
Frederick William I, Prussia, 94, 99
Friedreich ataxia, 345–346
fructose, 226, *226*
fruit fly. *See Drosophila melanogaster* (fruit fly)
F2 generation
pea plant studies, 47, *48*
predicting genotypes of, 50

G

galactose, 226, *226*
galactosemia, *74,* 225–227
galactose-1-phosphate, *227*
galactose-1-phosphate uridyl transferase, *227*
Galileo, 50
Galton, Francis, 10–11
gamete intrafallopian transfer (GIFT), 359
gametes, 33
defined, 149
formation of, 38–39, 154
oogenesis, 38–39, *39*
spermatogenesis, 38, *39*
and translocations in chromosomes, 139–140, *140*
Garrod, Archibald, 197, 220, 232, 238
Gaucher disease, 20, 21, 26, 317
G-banding, *126*
gene chip, *14*
gene doping, 369
gene expression
androgen insensitivity, 167–168
camptodactyly, 87–88, *88*
mosaic pattern, 170
penetrance, 87–88, *88*
profile, *346*
regulating mechanisms, 210–213
gene pool, 423–425
general cognitive ability, 114
genes
defined, 4, 49
description of, 4–6, *5, 6*
disease-causing, 192
and enzymes, 197
hybrid genes and leukemia, 280–282
immune system and, 276–399
imprinted, 172
patents, 186, 192
and protein, 198
splicing defects, 201
study of, 6–8
transmission of, 7
gene silencing, 211
gene therapy, 9, 366
enhancement gene therapy, 368–369
ethics of, 368–369
future of, 368–369
germ-line gene therapy, 368

HLA genes, 388–390
   haplotypes, transmission of, 389, *389*
HMGA2 gene, 110
HNPCC (hereditary nonpolyposis colon cancer), 278
Holmes, Oliver Wendell, 12
hominin, 430
hominoid, 430
Homo erectus, 431
homologous chromosomes, 33
Homo sapiens, 430–432
   in Human Genome Project, *338, 344*
homozygous, 51
hormones, 165–167
Howard University, 3
human development, 155–159
   from fertilization through implantation, *156*
   organ formation, 157
   organ maturation, 157
   rapid growth, 158
   risks to fetus, 160–161
   trimesters, 157–158
human embryo, *15*
Human Genome Project, 14, 337–338
   behavior genetics and, 414
   controversy, 192
   DNA sequencing and, 303–305
   ethics and, 347
   intelligence studies, 114
   origins of, 336–338, *338*
   race/ethnicity and, 427–429
   recombinant DNA technology in, 9
   scientific fields created by, 339–340
   size of genome, 340
   timeline for, *338*
   yeast artificial chromosomes (YACs), 298
human growth hormone, *314*
human insulin, *314*
Huntington disease, *74,* 87, *87,* 317, 405
   brain cells, loss of, *405,* 405–406
   and simple gene model, 401, 405–407
   transgenic animals as models for, *323,* 406
huntington (Htt) protein, 405–406
hybrid genes and leukemia, 280–282
hydrogen bond, 181
hydroxyurea, 231
Hyman, Flo, *77*

I

Icelandic Health Sector Database (HSD), 2–3
ichthyosis, 83
ICSI (intracytoplasmic sperm injection), 359
idebenone, 346
Ig (immunoglobulins), 385
Immigration and Nationality Act of 1965, 12
Immigration Restriction Act of 1924, 11
immune response
   nonspecific mechanisms, *386*
   as specific defense against infection, *386*
   specific mechanisms, *386*
immune system, 376. *See also* immune response
   allergies and, *392,* 393
   antibody-mediated immunity, 382
   autoimmune reactions, 391–393
   cell-mediated immunity, 382, *383*
   disorders of, 391–396
   genetic disorders and, 391–393
   HIV/AIDS and, 394, 396
   inflammatory diseases, 378–379

memory function of, 386–387
   severe combined immunodeficiency disease, 367
   and xenotransplants, 376
immunoglobulins (Ig), 384, 385, *385*
implantation, *156*
imprinting, 172
inborn error of metabolism, 220
incomplete dominance, 61–62, *62*
incontinentia pigmenti, 172
independent assortment, 51–54, 58
induced pluripotent stem cells (iPS), 15
induced pluripotent stem (iPS) cells, 15, 316
infertility, 355–357
inflammation and immune system, 378
inflammatory diseases, 378–379
inflammatory response, 378, *379,* 380
informed consent, genetic research and, 3
Ingram, Vernon, 233
initiation complex, 204
initiation phase
   of transcription, 201–202
   of translation, 204, 206
inner cell mass, 156
insulin, human, 314
intelligence, 114
intelligence quotient (IQ)
   defined, 112
   race/ethnicity and, 112–114
intercalating agents, 253
interferons, *314*
interleukins, *314*
interphase, 27, 28, *28,* 33
intracellular fluid, *23*
intracytoplasmic sperm injection (ICSI), 359
intrauterine insemination, 357
introns, 202
in vitro fertilization (IVF), 14, *15,* 354, *358, 359,* 363
   as business, 361–362
   for older women, 357
   and PKU females, 224
ionizing radiation, 250
iPS (induced pluripotent stem cells), 316
IQ. *See* intelligence quotient (IQ)
irradiation of food, 244
Itano, Harvey, 233
IVF. *See in vitro* fertilization (IVF)

J

Jeffreys, Alec, 312, 323
Joan of Arc, 168
Johns Hopkins University, 86

K

karyotypes, *9*
   and amniocentesis, 127–128
   banding methods, 125, *126*
   with banding pattern, *123*
   cells obtained for, 126, *280*
   chorionic villus sampling (CVS), 128–129
   chromosome abnormality identification and, 121
   construction and analysis, *124,* 124–129
   defined, 8, 122
   of XYY syndrome, *138*
Kearns-Sayre syndrome, 85, 89

kinetochore, 191, *191*
King, Mary-Claire, 276
Klinefelter syndrome, 136, *137*
Kunze, J., 5

L

labium major, 152, *152*
labium minor, 152, *152*
lactase, 226
lactose, 226, *226,* 227
lactose intolerance, 227, *426*
language and brain development, 406–407, 434–435
Las Meninas (Velasquez), 248
laws. *See* public policy and laws
Leber optic atrophy (LHON), 85
Leigh syndrome, 85
Lejeune, Jerome, 134
leptin
   defined, 108
   ovulation and, 108
   production of, *109*
Lesch-Nyhan syndrome, 83, 401, *402*
leukemia, and Down syndrome, 40
Li-Fraumeni syndrome, *271*
limbs, development of, *158*
Lincoln, Abraham, 70, 77–78
linkages, 333–334
   lod scores measuring, 335–336
   nail-patella syndrome and ABO blood type and, *334, 335*
lipids, *23*
   defined, 21
   function, 22
   subclasses and functions, 22
locus, 57
lod method, 335–336
lod score, 335–336
LOH (loss of heterozygosity), 271
loss of heterozygosity (LOH), 271
LP gene, 108
lung cancer
   mutations associated with, *279*
   smoking and, 287
   transgenic animal studies, *323*
lupus erythematosus, 393, *393*
lymphocytes, 381
Lynch syndrome, 280
Lyon, Mary, 169–170
Lyon hypothesis, 170
lysosomes, *23*
   defined, 25
   structure and function, 24–26, *25*

M

MacLeod, Colin, 178
MAC (membrane-attack complex), 380
macromolecules, 21
macrophages, *379*
mad-cow disease, 196, 210, 213
Maddox, Brenda, 186
major histocompatability complex (MHC), 382, 389
male-lethal X-linked dominant traits, 172
male reproductive system, 149–152
   anatomy of, *150*

soybeans, genetically modified, 15
Spar, Deborah, 363
SPCH1 gene, 406
sperm, 149, *151*, 154, 356
spermatids, 38
spermatocytes, 38, *38*, 150
spermatogenesis, 38, *39*, 150, 154
spermatogonia, 38
S phase, 191
spina bifida, 101, 173
spinal cord injuries, 32
splicing of mRNA (messenger RNA), 200–201, *202*
squamous cell carcinoma, *272*
SRY gene, 164–166
S (synthesis) phase, 28
Stallings, Patricia, 44, 45
start codon, 199
stem-cell research, 15
stem cells, 381
  adult stem cells, 316
  B cells from, 381
  cancer stem cells, 285
  cell division in, 32
  cord blood transplants, 317
  gene therapy and, 317
  T cells from, 381
Steptoe, Patrick, 354
sterilization, 12–13, 17
Stern, Wilhelm, 112
sterols, 22
stop codon, 199
STR (short tandem repeat), 323
structural genomics, 340
submetacentric, 122
substrate, 220
succinycholine sensitivity, 235
sucrose, 226, *226*, 227
sugar, 182
sugars. *See also* specific types, 226, *226*
superoxide dismutase, *314*
surrogate parenthood, 224, 360–361
Sutton, Walter, 57
synapses, *408*
synaptic transmission, *408*, 408–409
systemic lupus erythmatosus, 393, *393*

T

Tackett, John, *33*
tamoxifen, 236
taste buds, 234
taste difference, 232–235
Tatum, Edward, 197
Tay-Sachs disease, 64, *74*, 364
T-cell receptors (TCRs), 381
T cells, 381, 385–386. *See also* cell-mediated immunity
(TCGA) Cancer Genome Atlas, 286
telomerase, 191
telomere, 123
telophase
  of meiosis, 33, *34*, *35*
  of mitosis, 31, *31*
template, 187
Tepper, Beverly, 234
teratogen, 160, *160*
termination phase
  of transcription, 201–202
  of translation, 204, 207

tertiary structure, 208
tertiary structure of protein, 208, *209*
test anxiety, 46
testis/testes, *150*, *152*
  defined, 149, 150
  development of, 165, *166*
testosterone, 165
tetraploidy, 131
thalassemias, *74*, 230, 255
thymine dimers, 259, *259*
thymine (T), 4, *5*, 181, *252*
tissue plasminogen activator, *314*
traits
  defined, 4
  pea plant studies of, 7
transcription, *196*, 200, 201–202
  elongation phase of, 201
  genetic messages and, 201–202
  initiation phase, 201
  termination phase of, 201
transfer RNA (tRNA), 204
transformation, 178
transforming factor, 178
The Transforming Principle: Discovering That Genes Are Made of DNA (McCarty), 178
transgenic, 322
transgenic animals, 322–323
  and human neurodegenerative disorders, 328, 404
  list of diseases studied in, *323*
  process for making, 322–323
  xenotransplants, 390–391
transgenic organisms, 9, *10*
transgenic plants, 315–316, 321. *See also* genetically modified organisms (GMOs)
translation, 200, 203–204
  phases of, 203–204
  proteins, production of, 205–208
  steps in process, 204
translocations
  and cancer, 280, *281*
  and leukemia, 282
transmission genetics, 8, 9
Trial of Rehabilitation, 168
trinucleotide repeats, 249
tripeptides, 203
triploidy, *131*
  defined, 131
  karyotype of, 131
  miscarriages and, *140*, 141
trisomy, 132
  defined, 129
  maternal age and, 134–136, *135*
trisomy 21. *See* Down syndrome
Trisomy 18 (Edwards syndrome), *133*, 133–134
Trisomy 13 (Patau syndrome), 132–133, *133*
tRNA (transfer RNA), 203–205
  and antibiotics, 205
trophoblast, 156
Tsui, Lap-Chee, 39–40
tumors, *283*. *See also* cancer
tumor-suppressor genes, 273
Turner syndrome, 136, *136*, *137*
twin studies
  behavior research and, 402
  of bipolar disorder, 409, *410*

concordance rates, *105*, 105–106
  and multifactorial traits, 106–108
  obesity, heritability estimates for, *106*, 106–107
  of schizophrenia, 409–411
tyrosenemia, 225
tyrosine and phenylketonuria (PKU), 224

U

ulcerative colitis, 378
umbilical cord, *158*
uniparental disomy (UPD), 141–142, *142*
  and Prader-Willi syndrome, 142
UPD (uniparental disomy). *See* uniparental disomy (UPD)
uracil, 181, 251, *252*
urethra, *152*
  defined, 150
  female, *152*
  male, *150*
urinary bladder, *150*, *152*
urine, alkaptonuria, 197, *198*
uterus, 152, 153, *155*
UV light, 287

V

vaccines, 387
vacuole, *23*
vagina, 152, *152*, 153, *155*
valproic acid, 173
VANGL1, 101
Van Gogh, Vincent, 400
variable expressivity, 88
vas deferens, 150, *150*, *152*
vectors, 297
  cloning and, 297
  yeast artificial chromosomes (YACs), 298
Velasquez, Diego, 249
verbena flowers, 235, *235*
Victoria, queen of England, 246, *247*
viruses
  bacterial, genetic information in, 179–180
  as teratogens, 160–161
Von Gierke disease, 227
Von Hippel-andau disease
  heritable predispositions to, *271*
  mutation rates for, *249*

W

Wallace, Alfred Russel, 424
Wambaugh, Joseph, 324
Watson, James, 181, 183–186, 233
Watson-Crick model of DNA structure, 181, 183–186, *185*
Weinberg, Wilhelm, 419–420
Werner syndrome, 33
whole genome sequencing, 340
Wilkins, Maurice, 184, 185–186
William of Ockham, 50
Wilms tumor, *139*
  heritable predispositions to, *271*
  number of mutations associated with, *279*
Wolffian duct, 165, *166*, 168
World Anti-Doping Agency (WADA), 369